The age of the molecule

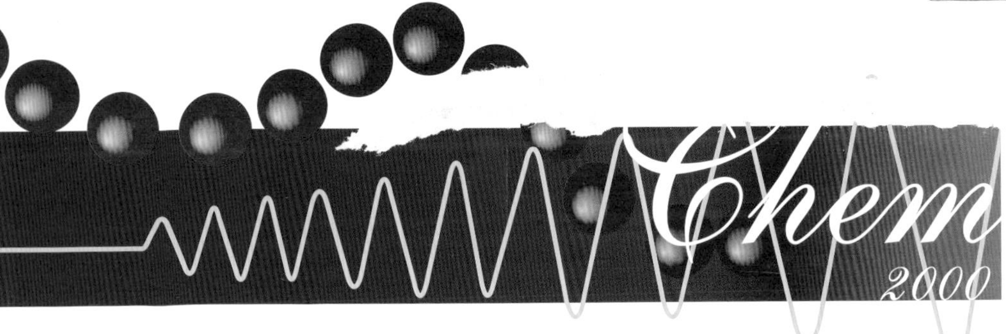

RS•C

ROYAL SOCIETY OF CHEMISTRY

RS•C
Royal Society of Chemistry

The age of the molecule
Edited by Nina Hall

Designed by GBW Marketing Design Ltd
Printed by Butler & Tanner Ltd
Published by the Royal Society of Chemistry

For further information on other educational activities undertaken by the Royal Society of Chemistry write to:
Education Department
Royal Society of Chemistry
Burlington House, Piccadilly
London W1V 0BN
United Kingdom

ISBN 0-85404-945-2
British Library Cataloguing in Publication Data.
A catalogue for this book is available from the British Library.

Cover pictures courtesy of PhotoDisc and DuPont.

10 DOWNING STREET, LONDON SW1A 2AA

Education *is* the Government's top priority. For Britain to compete and succeed in the modern world, and for all our people to fulfil their potential, education is the key to our futures.

But the new competitive world and the new opportunities opening for people are increasingly driven by fast-changing science and technology. So education in science is crucial. To sustain and develop the UK's competitive edge in global markets, Britain's science base is vital – the bedrock of our economic performance, generating the knowledge, technology and skills we need.

The Government is taking action to help: the extra funding we have announced as a result of our comprehensive spending review will, I believe, be a big enhancement to science and engineering in Britain.

As we stand on the threshold of the 21st century, scientific progress has made possible real advances for all of us – better health, increased life expectancy and a higher standard of living, which have improved the quality of our lives. It will do much more. Chemistry is one of the core scientific disciplines. Advances in chemistry contribute directly to our everyday lives – to the medicines we take, the food we eat, the clothes we wear, the environment we live in. In combination with other sciences, chemistry will help underpin much of the progress we can make in the future – and the prosperity which will flow from it.

That's why I welcome this new publication from the Royal Society of Chemistry. *The age of the molecule* contains wonderful examples of the exciting achievements of chemists worldwide, as well as alerting us to the challenges which lie ahead.

I hope it will help us inspire a genuine interest in chemistry, and in science generally, among young people. I want to see them reach out and take up the challenges and the opportunities of the future which are before them. I believe that they will. For their own benefit – and for the benefit of Britain.

Acknowledgements

This book arises from many discussions within the Royal Society of Chemistry and represents the dedication of numerous people. The Steering Committee consulted widely on the topics to be included and, with much advice from many interested parties, selected a series of topics. Its role has been crucial to the success of this publication. Particular thanks are due to Professor Jim Turner who chaired the Steering Committee, and whose enthusiasm and dedication to the project has ensured its success.

Usually a specialist in each topic was chosen to oversee the writing of each chapter. Their task was to identify suitable authors and to ensure the chapter included appropriate coverage of the topic. With such a disparate group of authors it was necessary to have an editor and we were able to persuade Nina Hall to undertake this.

The authors were given the task of writing about the developments in their area at a level which can be understood by a wide-ranging scientific audience. They have done a splendid job. Nina Hall has drawn the book together and edited it in a style to be enjoyed by those with an interest in the molecular sciences.

Special thanks are due to Dr Denise Rafferty who managed this project throughout. The book is fully illustrated and thanks are due to all those who contributed photographs and diagrams.

Steering Committee

Professor J J Turner (Chairman)
Dr A D Ashmore
Professor D Gani
Professor J H Holloway
Professor Sir Harry Kroto
Dr B T Pierce
Dr B J Price
Dr D Rafferty
Professor D J Waddington
Professor K Wade

Chapter "specialists"

Professor P D Bartlett
Professor D W Bruce
Professor P P Edwards
Professor W J Feast
Professor J W Goodby
Professor G W Gray
Professor R K Harris
Professor P M Maitlis
Professor J Mann
Professor M J Pilling
Professor G W Richards
Professor M W Roberts
Professor P J Sadler
Professor J P Simons
Professor N S Simpkins

Contributors

Dr J P Attfield
Dr P R Birkin
Dr P R Davies
Professor C Dobson
Dr M J Duer
Dr P J Dyson
Professor W J Feast
Dr H H Fielding
Professor C D Garner
Professor S E Gibson
Professor M L H Green
Professor K J Hale
Dr A Haynes
Dr C T Imrie
Dr R L Johnston
Professor Sir Harry Kroto
Dr A R Leach
Professor D A Leigh
Professor J Mann
Professor K Prassides
Professor S M Roberts
Dr J Sloan
Dr K Stott
Dr N K Terrett
Dr N R Thomas
Professor J M J Williams

Anthony Ledwith

Professor Anthony Ledwith
President, Royal Society of Chemistry.

RS•C

Royal Society of Chemistry

The Royal Society of Chemistry (RSC) is the learned society for chemistry and the professional body for chemists in the United Kingdom, with over 45 000 members worldwide. It was established as the world's first chemical society, in London in 1841. The Society is charged by Royal Charter, with the advancement of chemical science and its application, and to serve the public interest by acting as an adviser on matters relating to the science and practice of chemistry. The Society has a comprehensive educational programme at all levels, organising a wide range of scientific meetings and is a major publisher of scientific research.

GlaxoWellcome

The Royal Society of Chemistry gratefully acknowledges sponsorship from GlaxoWellcome, as part of GlaxoWellcome's on-going support of science education, towards the cost of design and distribution of this publication.

Contents

6 Introduction: What is chemistry?

13 Make me a molecule

41 Analysis and structure of molecules

73 Chemical marriage-brokers

97 Following chemical reactions

117 The power of electrochemistry

137 The age of plastics

161 The world of liquid crystals

181 New science from new materials

209 Computational chemistry and the virtual laboratory

229 The chemistry of life

261 Epilogue: Chemistry – architecture of the microcosmos

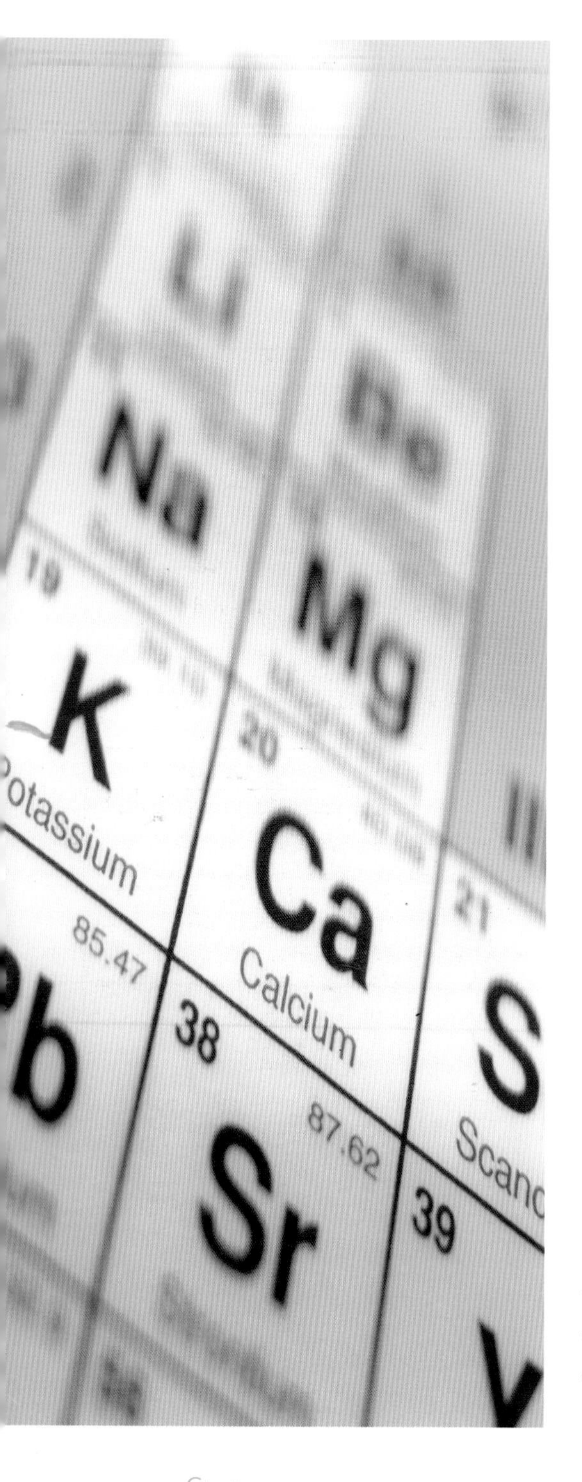

Introduction

What is chemistry?

In the words of the eminent chemist Sir Jack Baldwin: "Chemistry is about making forms of matter that have never existed before". These new forms of matter, from plastics and detergents to contraceptives and anticancer drugs, have had an extraordinary impact on the lives of everyone. We now take so many of these modern products for granted that we forget that they would not exist without the chemical knowledge used to make them.

Although designing and making new molecules is at the heart of chemistry, another important aspect is analysing substances and working out how and why chemical reactions happen. In this way chemistry has also contributed to our quality of life by providing the means for quality control in manufacturing, for monitoring the environment and assessing our health needs, and even for crime detection. On a more philosophical level, chemistry is also about seeking a deeper understanding of our place in the grand scheme of things – by revealing the molecular complexity of the world around us and its subtle relationships with ourselves and other living organisms.

Now that we are starting a new millennium, it is worth looking back at the rise of chemistry and how it has affected our social and cultural development. In fact, we have to go back a long way to the earliest times when humans first started to manipulate their environment – through fire, a chemical reaction, to cook, and later, to extract metals. Many of the first chemical reactions used were probably accidental discoveries. Primitive forms of both soap and man-made glass were probably first made by accident thousands of years ago – soap from a mixture of goat tallow and wood ash, and glass from sand, seaweed and salt in the ashes of camp fires on the beach.

At that time, of course, people had no idea how one material appeared to change into another. However, the Ancient Greeks were very interested in understanding the material nature of the world but they took the wrong course. Democritus suggested that matter was made of indivisible atoms, but, unfortunately his ideas were swamped by those of other philosophers such as Aristotle who thought that matter consisted of the so-called four elements Earth, Air, Fire and Water. Later, in the Middle Ages, the alchemists (the predecessors of the modern chemist),

The Alchemist.

supported Aristotle and spent much of their time trying to convert base metals such as lead into gold. Not surprisingly, they were unsuccessful. Nevertheless, we should not scorn their efforts since they did develop methods that were to form the basis of real chemistry such as filtration, crystallisation, distillation, and they studied a wide range of chemical reactions.

The end of the 18th century

Chemistry as we know it started towards the end of the 18th century. Largely because of the work of the Frenchman Antoine Lavoisier, chemists accepted the fundamental nature of combustion – as a reaction with oxygen. Lavoisier's efforts to put chemical research on a quantitative basis led to the modern concept of an element as a substance that cannot be further divided into parts. Lavoisier also produced the first textbook to interpret chemistry using these new ideas.

The new chemical thinking coincided with the beginning of the Industrial Revolution. The textile mills, which covered the landscape in parts of Northern England, created the need for new chemistry to convert fibre into finished cloth. Soap-making and glass-making also expanded. Thus a need developed for chemicals on a scale not satisfied by traditional methods. Sulfuric acid, required for textile finishing and for making soda and alkali, was produced by the 'lead chamber' process – introduced by John Roebuck of Birmingham – in which the basic reaction of sulfur and nitre was carried out in huge wooden boxes lined with lead. Bleaching was an important aspect of textile finishing and Charles Tennant together with Charles Mackintosh, developed bleaching powder by absorbing chlorine gas in lime. The alkali industry also started in earnest at this time with a new process, introduced in 1790 by Nicolas Leblanc, for producing sodium carbonate (for soap and glass-making), previously obtained by burning the barilla plant. His method involved first converting common salt into sulfate, followed by conversion of the sulfate to crude soda with charcoal and chalk. These reactions, together with one or two others, were really the limit of the chemical industry; and almost all these techniques were later superseded.

The end of the 19th century

Chemistry really took off in the 19th century. Rather than give a potted history it is instructive to summarise a few things that were known and understood at the end of the century, and, equally revealing, what was not known.

The Periodic Table, the foundation of our understanding of chemical behaviour, had been introduced by Mendeleef and had shown its power by predicting the existence of then unknown elements which were later isolated. The figure below shows the Periodic Table in 1900 (and overleaf for comparison the Periodic Table at the end of the 20th century). The atomic theory was, after some skirmishes, firmly rooted and atomic weights were accurately known for most elements.

The notion of a mysterious life force, supposedly associated with organic material, had been dispatched as chemists realised that organic compounds could be made from inorganic precursors. By studying chemical behaviour, researchers were beginning to understand the structure of organic compounds. They realised that carbon was the major constituent of organic molecules and was very often surrounded by four atoms arranged tetrahedrally – which controlled the structure of the molecule. One of the most important chemical structures in organic chemistry – the structure of benzene – had been discovered

TABLE II.

THE ATOMIC WEIGHTS OF THE ELEMENTS

Distribution of the Elements in Periods

Groups	Higher Salt-forming Oxides	Typical or first small period	Large Periods				
			1st	2nd	3rd	4th	5th
I.	R_2O	Li =7	K 39	Rb 85	Cs 133	—	—
II.	RO	Be =9	Ca 40	S 87	Ba 137	—	—
III.	R_2O_3	B =11	Sc 44	Y 89	La 138	Yb 173	—
IV.	RO_2	C =12	Ti 48	Zr 90	Ce 140	—	Th 232
V.	R_2O_5	N =14	V 51	Nb 94	—	Ta 182	—
VI.	RO_3	O =16	Cr 52	Mo 96	—	W 184	Ur240
VII.	R_2O_7	F =19	Mn 55	—	—	—	—
VIII.			Fe 56	Ru 103	—	Os 191	—
			Co 58.5	Rh 104	—	Ir 193	—
			Ni 59	Pd 106	—	Pt 196	—
I.	R_2O	H=1. Na =23	Cu 63	Ag 108	—	Au 198	—
II.	RO	Mg=24	Zn 65	Cd 112	—	Hg 200	—
III.	R_2O_3	Al =27	Ga 70	In 113	—	Tl 204	—
IV.	RO_2	Si =28	Ge 72	Sn 118	—	Pb 206	—
V.	R_2O_5	P =31	As 75	Sb 120	—	Bi 208	—
VI.	RO_3	S =32	Se 79	Te 125	—	—	—
VII.	R_2O_7	Cl =35.5	Br 80	I 127	—	—	—
		2nd small Period	1st	2nd	3rd	4th	5th
			Large Periods				

1 H hydrogen																		2 He helium
3 Li lithium	4 Be beryllium												5 B boron	6 C carbon	7 N nitrogen	8 O oxygen	9 F fluorine	10 Ne neon
11 Na sodium	12 Mg magnesium												13 Al aluminium	14 Si silicon	15 P phosphorus	16 S sulfur	17 Cl chlorine	18 Ar argon
19 K potassium	20 Ca calcium		21 Sc scandium	22 Ti titanium	23 V vanadium	24 Cr chromium	25 Mn manganese	26 Fe iron	27 Co cobalt	28 Ni nickel	29 Cu copper	30 Zn zinc	31 Ga gallium	32 Ge germanium	33 As arsenic	34 Se selenium	35 Br bromine	36 Kr krypton
37 Rb rubidium	38 Sr strontium		39 Y yttrium	40 Zr zirconium	41 Nb niobium	42 Mo molybdenum	43 Tc technetium	44 Ru ruthenium	45 Rh rhodium	46 Pd palladium	47 Ag silver	48 Cd cadmium	49 In indium	50 Sn tin	51 Sb antimony	52 Te tellurium	53 I iodine	54 Xe xenon
55 Cs caesium	56 Ba barium	57-70 *	71 Lu lutetium	72 Hf hafnium	73 Ta tantalum	74 W tungsten	75 Re rhenium	76 Os osmium	77 Ir iridium	78 Pt platinum	79 Au gold	80 Hg mercury	81 Tl thallium	82 Pb lead	83 Bi bismuth	84 Po polonium	85 At astatine	86 Rn radon
87 Fr francium	88 Ra radium	89-102 **	103 Lr lawrencium	104 Rf rutherfordium	105 Db dubnium	106 Sg seaborgium	107 Bh bohrium	108 Hs hassium	109 Mt meitnerium	110 Uun ununnilium	111 Uuu unununium	112 Uub ununbium						

* lanthanides	57 La lanthanum	58 Ce cerium	59 Pr praseodymium	60 Nd neodymium	61 Pm promethium	62 Sm samarium	63 Eu europium	64 Gd gadolinium	65 Tb terbium	66 Dy dysprosium	67 Ho holmium	68 Er erbium	69 Tm thulium	70 Yb ytterbium
** actinides	89 Ac actinium	90 Th thorium	91 Pa protactinium	92 U uranium	93 Np neptunium	94 Pu plutonium	95 Am americium	96 Cm curium	97 Bk berkelium	98 Cf californium	99 Es einsteinium	100 Fm fermium	101 Md mendelevium	102 No nobelium

to be an unusual, flat, six-sided ring. With these insights, synthetic organic chemistry produced some wonderful new molecules, including aspirin, marketed in 1899.

All areas that we now consider modern chemistry had rapidly expanded. The Swiss chemist Alfred Werner had begun his pioneering work on showing how metals could bind with groups of atoms to form complex coordination compounds, now so important in catalysis. A quantitative understanding of chemical reactions had developed in terms of rates of reactions (kinetics) and the energetic processes that determine whether a reaction happens (thermo-dynamics). The relationships between electricity and chemistry (electrochemistry) had been thoroughly investigated, and analytical chemistry, mostly based on 'wet' methods and the flame test, had become sophisticated.

Industrial chemistry had expanded beyond recognition. One of the greatest developments was the synthetic dye industry, the origin of which tells a revealing story. William Henry Perkin, a young man of 18, working under the directorship of August Hofmann, was trying to make quinine by oxidising derivatives of anthracene from coal tar pitch – in modern terminology:

$$2C_{10}H_{13}N + 3[O] \longrightarrow C_{20}H_{24}N_2O_2 + H_2O$$

Modern chemists know the detailed structure of the starting material (allyl toluidine) and of quinine; they bear no relation to each other, so Perkin's efforts were doomed to failure, and he obtained a brown sludge. However, Perkin thought it might be interesting to try a simpler starting material and chose aniline, which had been isolated from coal tar by Hofmann. Working in his own laboratory at home, Perkin obtained a mauve compound which he sent off to a firm of dyers. Their enthusiasm persuaded Perkin to set up a business exploiting this new dye, now called mauveine. This was the foundation of the dye industry based on synthetic chemistry rather than natural materials.

This story is revealing in two ways. First, it shows the importance of accident (or serendipity). Secondly, it shows that the accident must be appreciated by someone – as Louis Pasteur put it: "Where observation is concerned, chance favours the prepared mind". Other major industrial developments included the replacement, in 1870, of the Leblanc process for producing sodium carbonate by the Solvay process (from the reaction of carbon dioxide with ammonium salts); the replacement of the lead chamber process for sulfuric acid by the contact process (from the reaction of sulfur dioxide and oxygen over a catalyst); the discovery of the vulcanisation of rubber by Charles Goodyear (in 1841); and the Castner-Kellner electrochemical method for the manufacture of both chlorine and sodium hydroxide.

What techniques did chemists then have available? These were very limited. They included conventional laboratory methods such as distillation and crystallisation, and analytical techniques based on gravimetric, titrimetric, combustion or flame methods, and also some ultraviolet/visible spectroscopy and crude infrared spectroscopy.

What was not known? There was no understanding of atomic structure in terms of a nucleus and electrons, and thus no understanding of the basis of the Periodic Table. There was no idea of what constitutes a chemical bond. There was no understanding of the mechanism of chemical reactions. On the practical front, there was no polymer industry (except for cellulose and bakelite); Paul Ehrlich had not made his 'magic bullet' Salvarsan (arsphenamin, for syphilis), and the sulfonamides had not been made; there were no synthetic fertilisers, which had to wait for the Haber process. There were none of the modern physical analytical techniques we have today such as X-ray crystallography, and nuclear magnetic resonance (NMR). As Sir Harry Kroto points out in the epilogue to this book, it is a constant source of amazement that the chemists of the 19th century could make so many advances in the absence of so much.

And now, the end of the 20th century

The 20th century has seen extraordinary strides in chemistry. This book aims to explore some of these advances. It cannot be in any way comprehensive – chemistry is now too large a subject. Instead we have selected some areas of chemical endeavour in which there have been significant developments, particularly in the past 50 years, and which are likely to lead to more scientific breakthroughs in the 21st century. We have also chosen areas that illustrate the overall framework of chemical knowledge and the interdependent relationships between different areas of chemistry, and between academic research and industry. While there is this general theme to the book, each chapter stands on its own. It covers a particular area and is written by chemists researching at the forefront of that field. The content of the chapters represents, often in a historical context, what they think is exciting and important, and what they think the next developments may be in the future.

We planned the book so that it will give the reader some insights into how chemists think and what they do. The Nobel Prize-winning chemist Roald Hoffmann put it thus in his book *The Same but not the Same*: "Chemists make molecules ... they study the properties of these molecules; they analyse, they form theories as to why molecules are stable, why they have the shapes and colours they do; they study mechanisms, trying to find out how molecules react." But, as he says "at the heart of their science is the molecule that is made". In chemical research, there is a great deal of intuition, often a lot of luck, work in one area may have unexpected consequences for another, and it is all driven by insatiable curiosity.

Ronald Breslow, recently President of the American Chemical Society, calls chemistry "The central, useful and creative science". It is central because it underpins many other scientific disciplines such as biology, geology and materials science (the interests of scientists who choose to describe themselves as chemists, physicists or biologists increasingly integrate and overlap). It is useful in providing many of the materials of everyday life, the knowledge to offer better healthcare and better food, and to solve environmental problems. It is creative in that it designs original structures often with new and unique properties. We hope that this book will reveal all these aspects.

The book starts with a chapter on how chemists set about making molecules. *Make me a molecule* gives a flavour of the ingenious strategies that modern synthetic chemists have developed. It emphasises the great contributions made in the early part of this century by an extraordinary group of individuals who were able to synthesise large naturally-occurring molecules using only the limited techniques of the time. Perhaps the most fascinating aspects of modern synthesis are the chemist's ability to tailor-make highly complex molecules with the exact required three-dimensional geometry, and the automation of laboratory processes which, in some areas, is rapidly changing the way synthesis is being done.

Synthetic chemistry is totally dependent on knowing the structure of the end-product. In addition to the 'classical' methods of the 19th century, a modern synthetic chemist possesses a wide battery of techniques, most of them derived from some new physical principle. The chapter on *Analysis and structure of molecules* gives an historical introduction in the first half and outlines what can be learned from modern techniques. The second half of the chapter concentrates on nuclear magnetic resonance (NMR) which has become an indispensable analytical technique for the synthetic chemist. What is interesting is the way this technique is continually being developed in quite new directions with applications ranging from solid-state chemistry and physics to probing the human body via the method known as magnetic resonance imaging (MRI).

An important aspect of making molecules is the use of catalysts – compounds that can accelerate the rates at which chemical reactions occur. Most of the chemical industry depends on catalysts, and life could not exist without the biological catalysts called enzymes. We have, therefore, included a chapter on catalysis – *Chemical marriage-brokers*. Chemists are continually looking for new catalysts which are both selective (make only the desired product), and efficient. They also need to be robust, environmentally-friendly and to work under mild conditions. Enormous effort is expended on studying just how catalysts work at the molecular level in order to improve their efficiency.

To find better catalysts, and to understand chemical reactivity in general, it is extremely important to know just how a reaction proceeds and what the intermediate stages are. Understanding the details of chemical reactions is now a very important area of chemistry. Many reactions can be extremely fast yet chemists have managed to study them using ingenious techniques. The chapter, *Following chemical reactions,* describes these techniques which have become available over the past 50 years. Most of the recent work is dominated by the ubiquitous laser, which has allowed chemists to probe the exact details of how chemical bonds break and reform during a reaction – including biological reactions such as photosynthesis in plants.

Photosynthesis is one of several crucially important biological reactions which involve the transfer of energy via the movement of electrons. This is the area of electrochemistry which makes use of the intimate relationship between electricity and chemistry. The chapter on *The power of electrochemistry* describes how electricity is produced in batteries from chemical reactions. We all use batteries, whether in the car or the personal stereo, and chemists are trying hard to improve them by finding new materials that store energy more efficiently. One of the aims is to develop a lightweight battery with a high energy density for the electric car. A competitor to the battery is the fuel cell – a related electrochemical device which depends on a continuous flow of reacting materials.

Perhaps the most visible results of chemical advances in the 20th century have been the synthetic materials, in particular plastics, now used by everyone. The chapter, *The age of plastics,* plots the progress of polymer research from the time when the early pioneers of this area of chemistry were working on a new idea – that of large polymeric molecules – which the chemical 'establishment' regarded as ludicrous, to the latest developments in light-emitting polymers for electronic devices. As is often the case chance observations have played a part. Polythene, which we now see everywhere, was discovered because some chemists at ICI were studying the reactions of ethene under high pressure and temperature; they noticed a completely unanticipated waxy solid in the reaction vessel. Today, there are dozens of different types of plastic, each developed for particular uses.

Another type of material that is also common are liquid crystals. As described in *The world of liquid crystals,* these substances consist of molecules that show some kind of orientational order as in a crystal but can move around freely as in a liquid. They were not discovered until the 19th century, although Nature long ago employed liquid crystals for making cell membranes and other materials such as beetle cuticles and spiders' webs. Because certain liquid crystals respond to electric fields, they are used in visual displays for watches and miniature television sets. This modern application of liquid crystals had to wait for synthetic chemistry to produce molecules with the correct tailor-made properties. Among the more important recent developments are liquid crystal polymers such as Kevlar which are extraordinarily strong.

The chapter on *New science from new materials* selects some examples of novel molecules – most made only recently – which possess surprising and often unexpected electronic and magnetic properties. These novel species have tremendous potential in leading to new types of electronic devices. They include metal oxides whose electrical resistance changes hugely in an applied magnetic field (colossal magnetoresistance) and the celebrated high-temperature superconductors which lose all electrical resistance below liquid nitrogen temperatures. Another group of materials that currently fascinate scientists are minute clusters of atoms – nanoclusters – whose properties are neither like isolated atoms nor bulk material. The most famous clusters, however, are the carbon-based fullerenes, such as C_{60} – a molecule shaped like a soccer ball. Fullerenes and the related 'nanotubes' can incorporate other chemical species and have potentially exciting uses in electronics and nanotechnology.

Our understanding of the way that newly discovered materials behave depends upon theories of chemical bonding and the behaviour of electrons in atoms. The cornerstone of these theories is quantum mechanics. However, it is a formidable task to apply quantum ideas to complex atoms and molecules because the equations are difficult to solve. Fortunately, the computer has come to the rescue as shown in the chapter *Computational chemistry and the virtual laboratory*. Today's powerful computers have revolutionised this area of theoretical chemistry. Chemists can also use computers to display molecular structures as colourful graphics, and also simulate the dynamical behaviour of groups of atoms or molecules. These studies have had enormous influence on molecular biology, pharmaceutical and materials design.

Perhaps the most important part of modern chemistry is its intimate relationship with biology. Living organisms consist of complex chemicals which behave in a self-organising way. The science of molecular biology has grown out of chemistry and still depends on fundamental chemical ideas and techniques. The final chapter, *The chemistry of life,* traces this relationship by revealing how chemical studies have led to an understanding of living processes and therefore to new medical treatments. It describes the development of antibiotics, contraceptives and anticancer drugs. The chapter also documents how the structure of DNA was discovered and gave rise to the new science of genetics.

This book also has an epilogue. The British chemist Sir Harry Kroto who shared the 1996 Nobel Prize for Chemistry for the discovery of C_{60} gives a personal view of what he thinks is exciting about science in general and chemistry in particular. We think it makes a fitting end to the book.

The end of the 21st century

What will chemistry produce in the decades ahead? Predicting the scientific future is a hazardous business. Several things, however, are clear. There will be discoveries that no one could have foreseen; it may be that some of these will be so revolutionary that the lifestyle of our descendants will be unimaginably different from ours. In the absence of such foreknowledge, we can only speculate about the obvious. There will be materials with almost magical properties; advances in computing and laser power will have permitted even more understanding of chemical structure and interactions; extraordinary medicines and medical therapies will have been developed; miniature devices based on the behaviour of individual molecules, with fantastically sophisticated properties, will have been made; the chemical industry will be totally environmentally friendly; and sunlight, via its interaction with chemicals, will be routinely used for many energy purposes. It will certainly be an interesting time!

It is worth adding a final comment: what this book is NOT about. It does not have any political agenda and it is not a defence of chemistry. To the average citizen the word 'chemical' can conjure up all kinds of environmental and 'unnatural' horrors. We, therefore, offer a gentle reminder that everything around us is made of chemicals including ourselves. Chemistry is the study and manipulation of all matter at the molecular level including the natural world. That is not to say that there are not serious and legitimate issues concerning the production of chemicals and the environment. There is no doubt that the impact of many chemicals, starting with gaseous hydrochloric acid in the Leblanc process, has been deleterious. The fact that this book does not address these issues directly is not because they are not important – they are – but because we believe that the benefits of chemistry far far outweigh the deficits. ***The age of the molecule*** *is simply about why chemistry is exciting and important and while it will be even more so in the future.*

James J Turner

Further reading

I am indebted to the following book,

The History of Chemistry, John Hudson, Macmillan, London: 1992.

The Chemical Industry, ed A. Heaton (ed), 2nd edn, Chapman and Hall, London: 1994.

The Same but not the Same, Roald Hoffmann, Columbia University Press, New York: 1994.

Chemistry, Today and Tomorrow, Ronald Breslow, American Chemical Society, Washington: 1997.

Make me a molecule

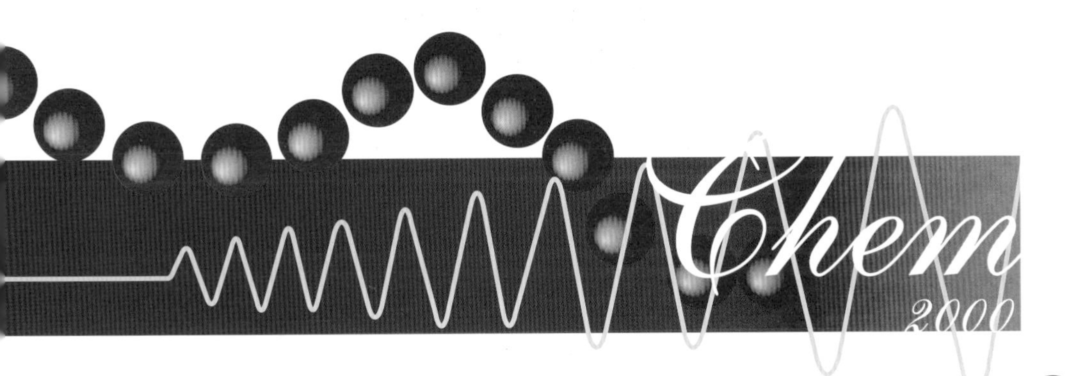

Chemists continually strive to meet the challenges of preparing new and exciting molecules. Virtual libraries and 'molecular train sets'

ensure that research in synthesis will lead to even more exciting discoveries.

Professor Sue E Gibson
King's College London
Professor Karl J Hale
University College London
Professor David A Leigh
University of Warwick
Dr Nick K Terrett
Pfizer
Professor Jonathan Williams
University of Bath

Constructing complex molecules is both an art and a science which stretches the chemist's knowledge and insight, practical skills and imagination. Chemists are constantly developing novel synthetic methods of breathtaking sophistication to make new forms of matter which not only bring huge social and economic benefits but widen our understanding of Nature.

Johnson Matthey

GlaxoWellcome

Everyone is familiar with the Hollywood cliche of the 'mad scientist' crouching over convoluted glassware in which fuming green liquids bubble away. This is inevitably a chemist who is synthesising an 'elixir of life' or some similar potion. For the average person, synthesis is what chemistry is all about and, of course, it is the bedrock on which the chemical and pharmaceutical industries are built. Many of the materials we use today are made by chemical synthesis. Although chemists haven't exactly come up with an elixir of life yet, they do create biologically active molecules that are life-saving drugs. Similarly important chemicals such as pesticides (see *The chemistry of life*) and plastics (see *The age of plastics*), as well as specialist materials like liquid crystals (see *The world of liquid crystals*), are the result of new synthetic discoveries.

Most of the new compounds made today are organic, that is, they are composed largely of carbon. This element has a unique ability in being able to form strong chemical bonds between not only carbon atoms but also many other elements such as hydrogen, oxygen, nitrogen, chlorine and sulfur. The electronic structure of a carbon atom is such

GlaxoWellcome

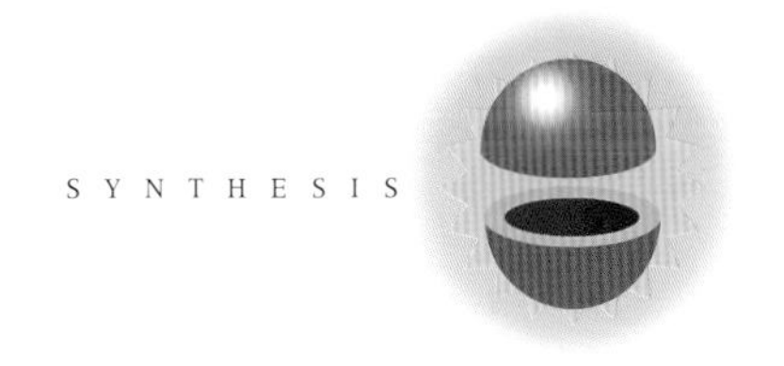

that it can bond with up to four other atoms, forming single, double and triple bonds, in many permutations to form an infinite variety of structures – linear and branched chains, rings of all sizes and more complex bridged structures. Indeed, the chemistry of life is largely a result of the incredible versatility of carbon and is, of course, why carbon chemistry is called *organic chemistry.* Friedrich Wöhler, who in the early 19th century first showed that organic compounds could be synthesised without the need for a 'vital force' (see *The chemistry of life*), wrote to the Swedish chemist Jon Berzelius: "Organic chemistry just now is enough to drive one mad. It gives me an impression of a primaeval tropical forest, full of the most remarkable things, a monstrous boundless thicket, with no way to escape, into which one may well dread to enter".

Today, there are thousands of new organic molecules recorded in the scientific literature every year. People have tried to estimate the number of different chemical compounds that could be formed, and the estimates of 10^{200} possible structures postulated by some chemists are simply impossibly large numbers to comprehend. The entire Universe contains too little material to make even one molecule of all these possible structures!

Chemists, therefore, have an almost infinite molecular landscape before them to explore, so how do they go about making new organic molecules? Synthesis involves breaking and making new chemical bonds to create new chemical structures. Much of organic chemistry is concerned with understanding these processes and then using that knowledge to design new kinds of reactions. In this way chemists build up a vast repertoire of reactions. Often they have a synthetic target in mind such as a possible drug molecule or a natural product, in which case they then have to develop a synthetic strategy – a series of reaction steps – to reach their final compound. In the case of

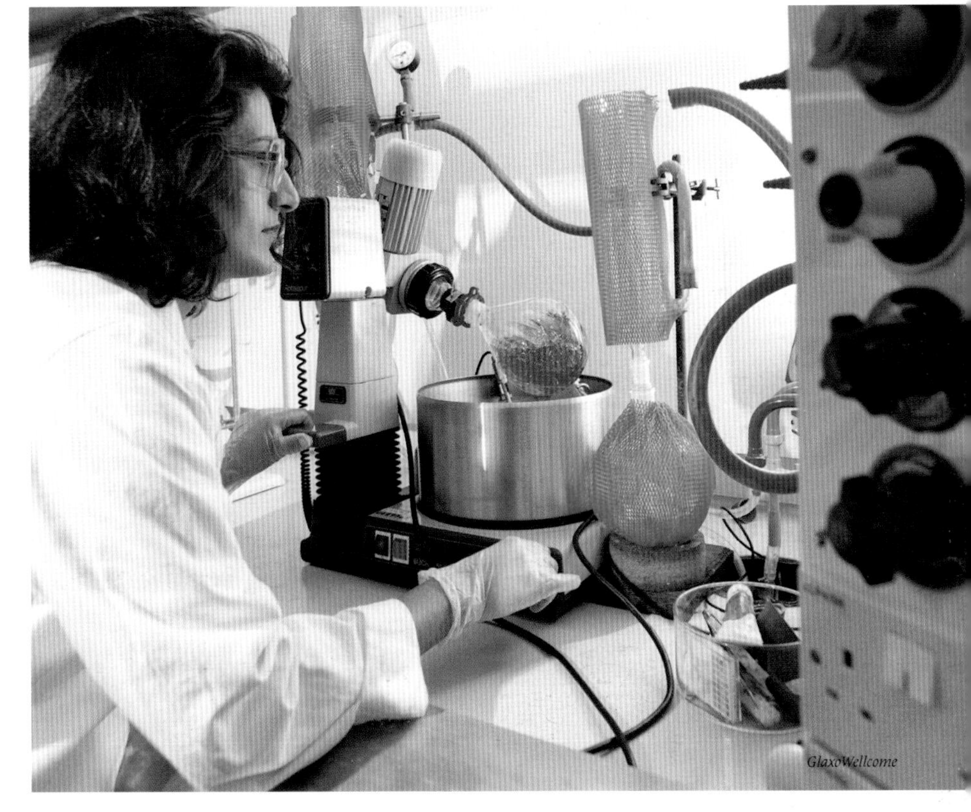

GlaxoWellcome

large complex molecules, this may involve dozens of steps and a great deal of ingenuity, skill and perseverance.

Organic synthetic chemistry is a truly creative process, and organic chemists pride themselves in designing synthetic routes that are both elegant and efficient. This means carrying out the synthesis in the least number of steps using simple reagents and obtaining the maximum yield of product at each stage. To achieve the latter, *selectivity* becomes the key word. Many reactions involve several competing pathways leading to different products. These products are often *isomers*, in other words, compounds with the same chemical formula but different structural arrangements. The chemist tries to tune the conditions of the reaction to obtain as much of the required product as possible.

The basic concepts of synthesis

There are two concepts in organic chemistry that synthetic chemists work from. One is the notion of a *functional group*. This is a group of atoms in an organic molecule that tends to behave as a single chemical entity in a chemical reaction. Examples are the carbonyl group which is a carbon atom doubly bonded to an oxygen atom (–C=O) or a primary amino group which consists of nitrogen linked to two hydrogens ($-NH_2$). Functional groups have specific electronic attributes that lead them to undergo a particular set of chemical transformations.

The other vitally important characteristic that has to be taken into account is that the carbon atom forms bonds in a definite three-dimensional array. The single bonds formed by carbon are arranged tetrahedrally. This means that complex structures can have equally complex geometries in three dimensions *(stereochemistry)*, which synthetic chemists must take on board. For instance, a bulky functional group can affect chemical behaviour involving nearby bonds by blocking a particular reaction pathway.

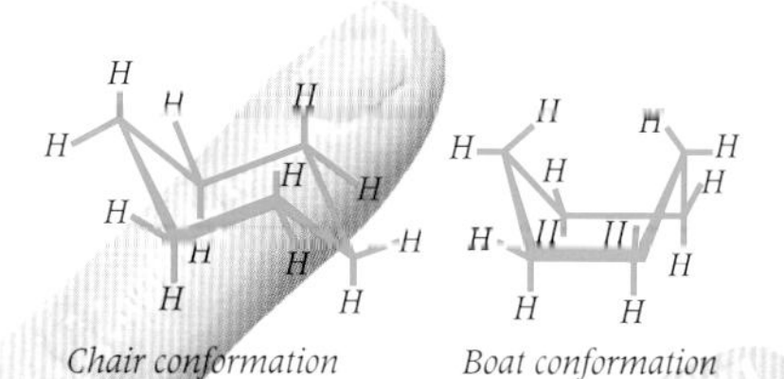

Figure 1. The rotation of carbon bonds lead different conformations as shown for cyclohexane which can adopt a so-called boat or a chair conformation.

Also of importance, is the fact that single carbon bonds can also rotate, leading to different shapes or *conformations* for the same molecule. An example is the molecule cyclohexane (C_6H_6) – a six-membered carbon ring that can adopt a so-called boat or a chair conformation (Figure 1). These conformations become significant when considering large natural molecules which are made up of such rings (for example, steroids). Studying the geometry of the constituent bonds and their affect on reactivity is called *conformational analysis*.

Another consequence of the three-dimensional geometry of carbon compounds is that compounds with the same basic structure can be arranged differently in space. In particular, a carbon atom with four different functional groups attached to it can exist in mirror-image isomers which are non-superimposable, like right and left hands (Figure 2). This property of handedness is called *chirality*, and is of great significance in synthesis as we shall see later.

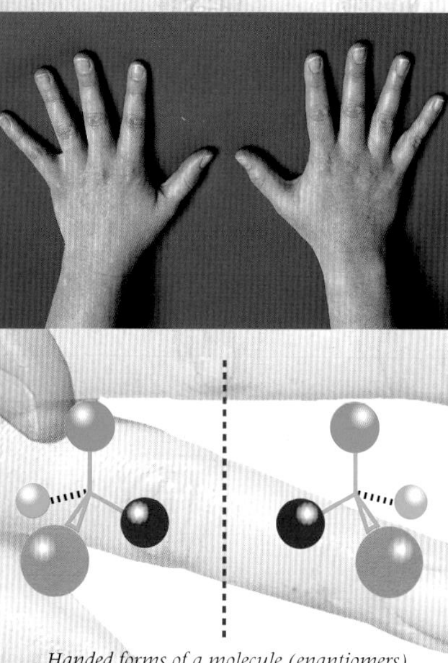

Figure 2. Molecules often have a handedness. Like our hands, the molecules can be mirror images, but not identical.

Chiral forms of molecules, or *enantiomers* were discovered in the earliest days of organic chemistry.

Figure 3. Louis Pasteur was able to separate crystals of tartaric acid into two mirror images. Note that a wedge shape implies a bond coming out of the page and a dashed line going into the page.

Reproduced courtesy of the Library and Information Centre, Royal Society of Chemistry.

In 1848, Louis Pasteur showed that salts of tartaric acid crystallised into mirror-image forms which could be separated by using a pair of tweezers (Figure 3). When Pasteur re-dissolved the separate types of crystals he noticed a remarkable thing. One solution made polarised light twist to the right, while the other rotated it to the left. For this reason, simple chiral

isomers like the tartaric acid salts are often called *optical isomers* and are said to be *optically active*. Twenty years later, Jacobus Henrikus van't Hoff and Josephe-Achille Le Bel independently explained the origins of the right and left-handed forms of the crystals in terms of mirror-image molecules based on the tetrahedral nature of carbon bonding.

Making complex molecules

It was during this period in the 1800s that organic chemistry began to be set on a sound structural framework, allowing chemists for the first time to plan organic syntheses. Using the new stereochemical ideas, the German chemist Emil Fischer set about trying to define the absolute configuration of a set of organic molecules that were of great interest in the late 19th century, the optically active sugars. In the process Fischer achieved the first synthesis of the first truly complicated organic molecule, the sugar molecule D-glucose in 1890 (Figure 4).

O=C–H
H–C–OH
HO–C–H
H–C–OH
H–C–OH
CH_2OH

Figure 4. D-Glucose.

Fischer was a great chemical genius and superb experimentalist, and his synthesis of D-glucose was visionary for a number of reasons. First, it yielded D-glucose as a single optical isomer. Secondly, it introduced three new reactions into organic chemistry which enabled the synthesis to be completed. Thirdly, it employed multiple sequences of reactions that assembled the sugar rapidly and efficiently all in one step, and fourthly, it proved the structure of D-glucose beyond all doubt. With this great achievement, Fischer showed his fellow organic chemists the types of complex synthesis problem they should be addressing, and he delivered the message that seemingly insoluble chemical issues can indeed be overcome by coupling chemical ingenuity with a spirit of chemical invention and adventure. Fischer's work on the *total synthesis* of D-glucose is regarded as the catalyst for the development of synthetic organic chemistry in the 20th century.

Emil Fischer
Reproduced courtesy of the Library and Information Centre, Royal Society of Chemistry.

Following Fischer's seminal work, synthetic organic chemists made huge progress in creating ever more complex structures. In the early-to-middle part of this century a British chemist, Robert Robinson at the University of Oxford, synthesised a range of complicated natural products. His most brilliant synthetic exploit was to make the alkaloid tropinone. This is a rather tricky molecule, because it is a seven-sided carbon ring with a carbon-oxygen double bond on one side (a carbonyl group) and a

Robert Robinson

Methylamine
NMe =O ⟹ CHO CHO H_2NMe CH_3 CH_3 =O
Succindialdehyde
Acetone

Figure 5. Robert Robinson's retrosynthetic analysis of tropinone (Me stands for methyl, CH_3).

'nitrogen-atom bridge' across the ring. Based on his own view of how Nature prepares this molecule, Robinson designed a synthesis that involves just one step. When Robinson's route to tropinone was published in 1917, it was instantly hailed as revolutionary (Figure 5). It not only made the previous synthesis of tropinone, by a German chemist Richard Willstatter, seem

obsolete and cumbersome (Willstatter's synthesis took 19 steps), but also more important, it alerted chemists to a better method of planning a synthetic route. This was to look at the target molecule and try to imagine how the molecule could be constructed from simpler chemical units. You can then design a strategy starting from those basic building blocks. This approach is now called *retrosynthetic analysis,* a term coined by another great organic chemist, Elias Corey at Harvard University, in the 1960s. Robinson made full use of retrosynthetic analytical principles during his planning of the tropinone synthesis, a fact that we can ascertain from his own words in his original research paper describing the synthesis of tropinone: "By imaginary hydrolysis at the points indicated by the dotted lines, the substance (tropinone) may be resolved into succindialdehyde, methylamine and acetone".

Prior to Robinson's brilliant achievement on tropinone, organic chemists had almost universally planned their synthetic routes only in a forward direction. Typically, they selected a chemical starting material that bore a structural resemblance to the target molecule in question, and then they attempted to think of reactions that could convert that starting compound into the target. In other words, they allowed the starting material to dictate their synthetic plan rather than the target molecule. Often this classical approach would fail or lead to very lengthy syntheses, as exemplified by Willstatter's 19-step synthesis of tropinone. By introducing this reverse way of thinking into organic chemistry, Robinson captured everyone's imagination and shaped the way total synthesis would be done for the remainder of the 20th century.

Christopher Ingold

In parallel with his inspirational synthetic work, Robinson also made fundamental contributions to the theory of how organic reactions happen. Robinson, with Christopher Ingold of University College London, formulated an approach which describes the mechanism by which bonds are broken or made in organic molecules in terms of their electronic structures (or more specifically in terms of electron re-distribution between the constituent atoms and groups of atoms). This produced sets of rules that helped chemists to predict the outcomes of new reactions they wanted to use. This approach continues to be used today. It was also Ingold, and Saul Winstein of the University of California at Los Angeles, who put this mechanistic organic chemistry on a firm theoretical footing through their meticulous experimental studies of the kinetics and rates of organic reactions, and they are now regarded by many chemists as the founding fathers of *physical organic chemistry*.

As a result of a better understanding of the mechanistic principles of organic reactions, by the 1940s and 1950s the field of total synthesis really took off. The superstar of this period and probably the greatest American organic chemist of this century was Robert Burns Woodward of Harvard University. Woodward brought a new dimension to organic synthesis. Not only did he take on immensely complex molecules with a very limited range of reactions and analytical techniques at his disposal, he usually finished these synthetic ventures in record time.

Robert Burns Woodward

Chlorophyll-a

First steroid to be synthesised

Strychnine

(-)-Vitamin B_{12}

Figure 6. Some of the complex molecules made by Woodward.

Woodward brought philosophy and art to organic synthesis. Indeed he received the Nobel Prize for Chemistry in 1965 for "Achievements in the art of organic synthesis". Like Fischer, Woodward developed much new chemistry in order to complete his syntheses. He had a prodigious, almost encyclopaedic, memory for what would seem to many as esoteric and unimportant facts of organic chemistry. However, the quality that set Woodward apart from all his peers was his phenomenal ability to tie together the threads of these seemingly unrelated chemistry facts. This would allow him to create, for example, a new chemical reagent for performing a specific chemical reaction, or to propose a powerful new theory for organic chemistry. This is what singled out Woodward as one of the greatest synthetic organic chemists of all time.

Woodward's landmark syntheses are too numerous to mention here, but some of the feats he is most remembered for include his first synthesis of a steroid, his first synthesis of the hemlock poison strychnine, his glorious route to the plant pigment chlorophyll, and his total synthesis of vitamin B_{12}, which can only be described as a breathtaking masterpiece. The latter was developed with the aid of Albert Eschenmoser at the Federal Institute of Technology (ETH) in Switzerland, and to this day, remains the crowning achievement of Woodward's entire career. Vitamin B_{12} is an extremely complex molecule as can be seen in Figure 6, with a cobalt atom at the centre of a flat, so-called porphyrin ring (a similar structure to that found in the chlorophyll molecule).

The importance of stereochemistry in synthesis

It is fair to say that Woodward was the first organic chemist to make a serious effort to control the relative spatial relationship between the functional groups in the molecules he was building. Woodward was aided in his planning by his great understanding of the Ingold-Robinson electronic theories of organic chemistry, and by his acute awareness of the molecular shape and topography of organic molecules. The latter aspect of Woodward's understanding stemmed from his great friendship with Derek Barton at Imperial College London, the primary architect of the concept of conformational analysis mentioned earlier.

By carefully planning all his synthetic routes with great precision, Woodward would often obtain only one product isomer at each stage. The majority of Woodward's syntheses produced the desired target compound as a single enantiomer (provided the target molecule existed in this form). Woodward's usual approach for obtaining single enantiomers generally entailed synthesising the compound as a 50:50 mixture of opposite enantiomers (a *racemic mixture*), and then separating that mixture by *optical resolution*. This involves reacting the racemic mixture with an enantiomerically pure resolving agent. Optical resolution is not an easy technique to apply but Woodward was a master of it. Woodward was not alone in using

this approach, for the majority of organic chemists up until the early 1970s regularly performed such resolutions in their syntheses to make them specific for one enantiomer. Indeed, Woodward's research team was able to make a synthetic version of the naturally occurring compound reserpine, used to treat high blood pressure, controlling the chirality at six carbon positions in this molecule – of which there are 64 possible isomers (Figure 7).

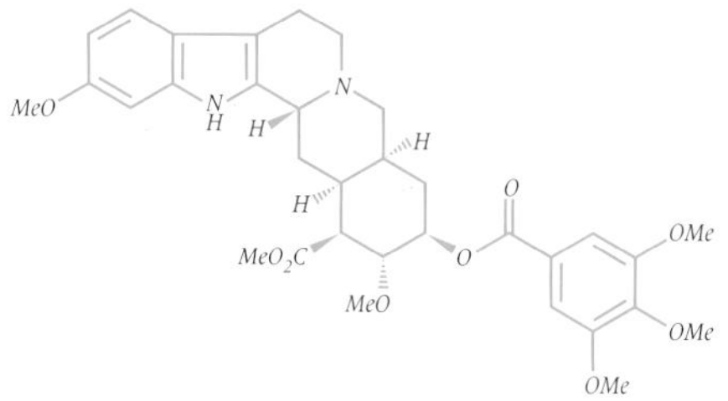

Figure 7. Reserpine.

By the mid-1970s, most organic chemists preferred to employ a *'chiral pool'* of starting materials to carry out their *enantiospecific syntheses*. The chiral pool is a metaphorical term used to describe the pool of readily available chemical starting materials that exist in Nature as

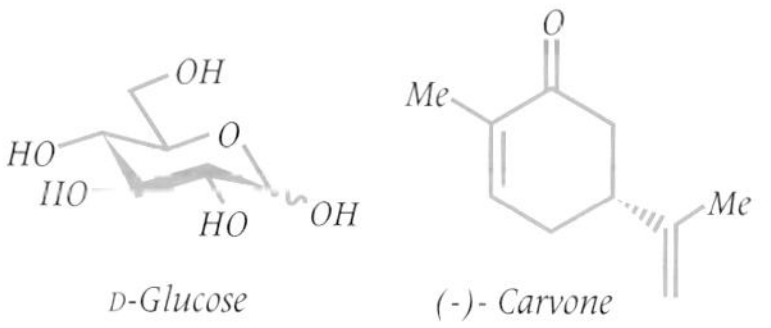

Figure 8. Two examples of chiral materials used in synthesis.

individual pure enantiomers. For example, D-glucose (mentioned earlier) and (-)-carvone (Figure 8) are two examples of chiral materials that belong to the so-called chiral pool.

In the 1970s and 1980s, eminent organic chemists such as Corey, Kishi, Ireland, Clark-Still, and at the Scripps Research Institute in La Jolla, California, Nicolaou all made extensive use of the chiral pool to do their complex molecule synthesis work, and by treading this path they developed much new chemistry. Figure 9 shows some of the molecules that were synthesised in this era by these five great chemists.

Making molecules with mirror images

While today, the chiral pool continues to be a major source of chiral starting materials for organic chemists, many are developing single enantiomer reagents which can react with molecules without chiral centres to create new molecules that exist in single enantiomeric form. Such a process is called *asymmetric*

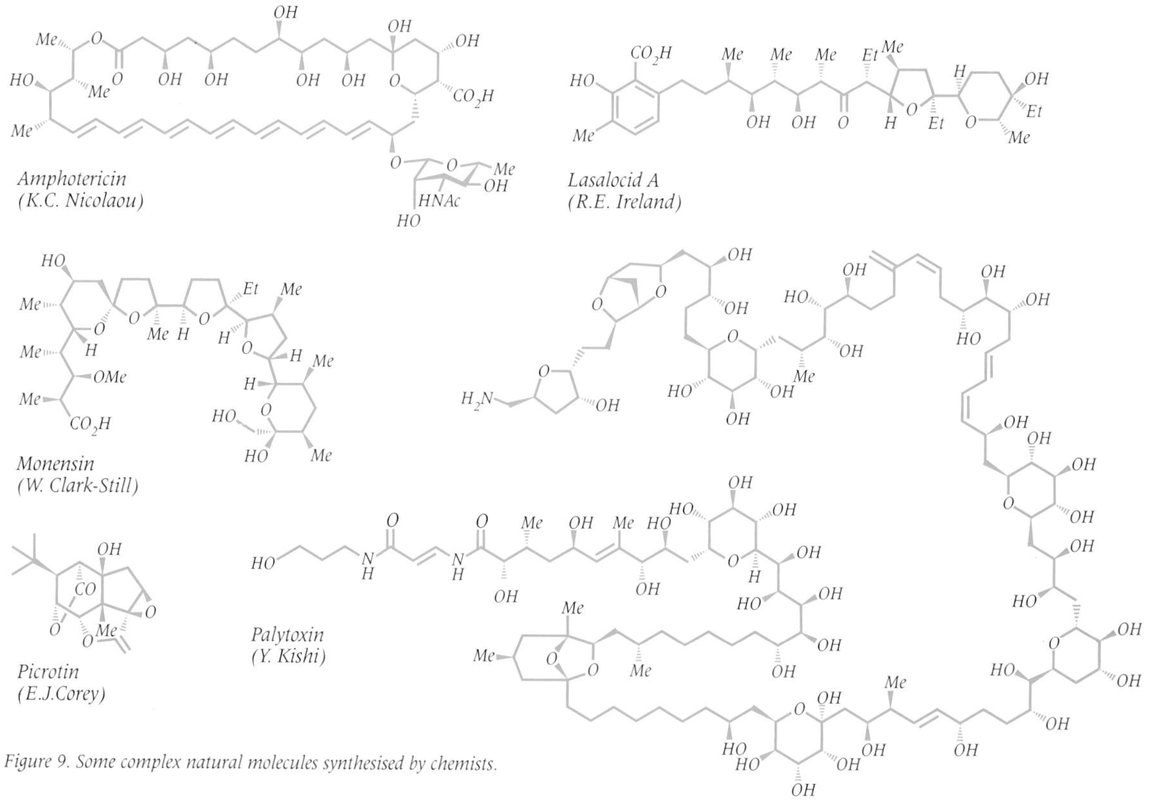

Figure 9. Some complex natural molecules synthesised by chemists.

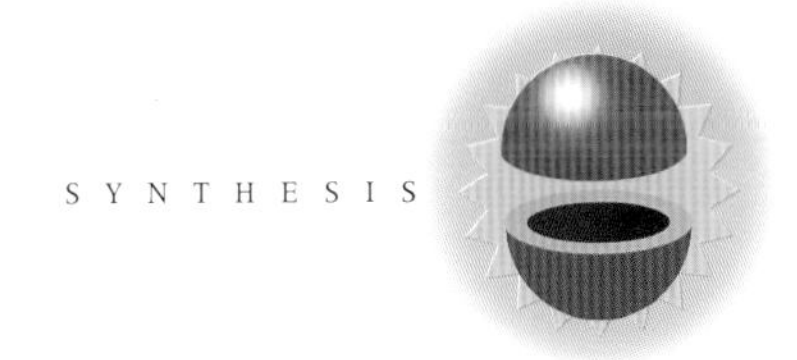

synthesis, and this approach is now being used successfully to construct not only important complex natural products but also drug molecules for the pharmaceutical industry.

Finding new methods of asymmetric synthesis has in the past 20 or 30 years become a key activity for organic chemists. The reason is that different enantiomeric forms of a natural product or a drug may have very different biological or pharmacological properties. It was the tragic case of the thalidomide drug which was a racemic mixture that brought this home to the pharmaceutical industry in particular (Box 1). Today, it is usually thought essential to isolate the desired enantiomeric form of a compound. Although enantiomers can be separated, half the product is then wasted. Industrial companies, who are concerned about disposing of the unwanted compound and also the inefficiency and cost involved, prefer, therefore, to find efficient methods of asymmetric synthesis.

Box 1 Mirror molecules and life

Many of the compounds associated with living organisms are handed – including vital molecular components such as DNA, enzymes, antibodies and hormones. Each enantiomer may have distinctly different characteristics. This is despite the fact that they may have the same physical-chemical properties such as boiling point or solubility in solvents, and when analysed even give the same spectra. This is true of limonene, a compound which is formed naturally in both chiral forms, but one of the enantiomers (S)-(-)-limonene smells of lemons, while the mirror-image compound (R)-(+)-limonene smells of oranges (*S* stands for *sinister* and *R* for *rectus*). Similarly, the compound (+)-nootkatone, which is responsible for the smell of grapefruit, smells 750 times more strongly than the other enantiomer.

R-(+)-limonene Smells of oranges

S-(−)-limonene Smells of lemons

We are able to smell the difference between mirror-image molecules.

The reason we can distinguish between these enantiomers is that our nasal receptors are also made up from chiral molecules that recognise the difference. In fact, insects, which have a phenomenal sense of smell, sometimes use chiral chemical messengers (pheromones) as sex attractants. Just recently, chemists discovered that one form of the insect pheromone, olean, attracts male fruit flies, while its mirror image works on the female of the species.

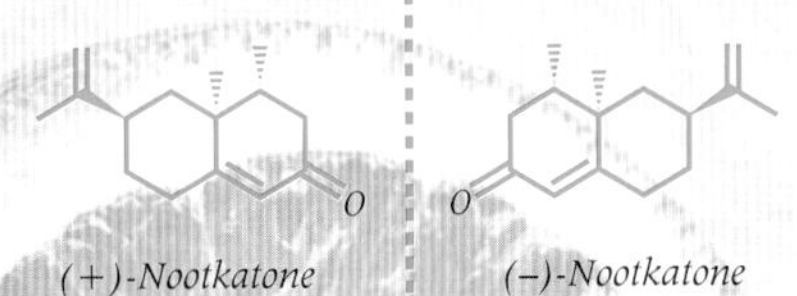

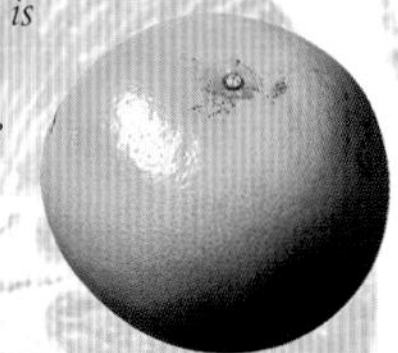

The smell of grapefruit is much stronger in one mirror image molecule than the other.

Not surprisingly, because biology is so sensitive to chirality, the activity of drugs also depends on which chiral form is used. However, the significance was not always fully taken on board. In the early 1960s, the drug Thalidomide was prescribed to alleviate morning sickness in pregnant women. Tragically, the drug also caused deformities in the limbs of children born to these women. It seems that one enantiomer of Thalidomide was beneficial while the other caused the birth defects. There is considerable controversy about the details of this argument, partly because the two mirror image forms of Thalidomide can interconvert easily in the body. Even so, pharmaceutical companies now make sure that both handed forms of a drug are tested for their biological activity and toxicity before they are marketed.

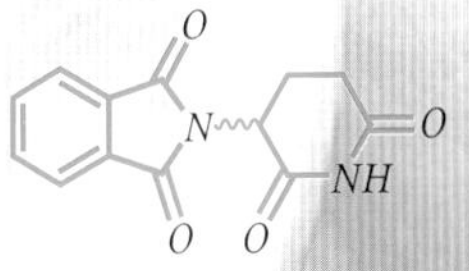

Thalidomide – a drug to treat morning sickness was sold as a mixture of handed forms.

One current chiral drug recently tested is levobupivacaine, a long-acting anaesthetic (of duration 6 to 10 hours) which would be ideal for use in dental surgery and for treating deep post-operative pain. At the moment, the only long-lasting local anaesthetic available is the racemic version bupivacaine. This has limited use because it is toxic to the cardiovascular and central nervous systems. The left-handed version of bupivacaine, however, shows considerably reduced heart and neural toxicity and can be used at relatively low dose rates. It has been developed by the chemical company Chiroscience in Cambridge.

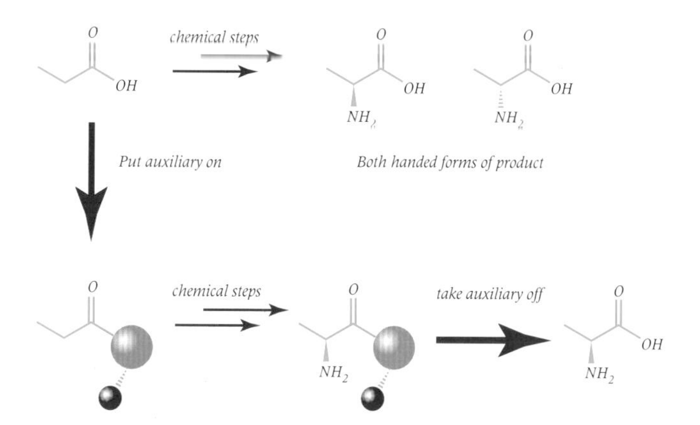

Figure 10. Adding a chiral auxiliary to flat molecules controls which handed form is made.

A non-chiral molecule can be converted into a chiral molecule by simple chemical steps with the aid of a chemical unit called a *chiral auxiliary*. For example, Figure 10 shows a reaction starting with propionic acid which doesn't have right- and left-handed forms. By attaching an auxiliary to propionic acid, it creates the stereochemical conditions that force the chemical steps to follow a certain geometrical path. Once the handedness of the new molecule has been set, the auxiliary can be taken off (or better, recycled) leaving behind the product molecule in a chiral form.

Of the many auxiliaries that are available, the one prepared by David Evans and colleagues at Harvard University has found the most widespread use, and is prepared from simple, naturally occurring molecules. The Evans' team has used this approach in the synthesis of fascinating and complex molecules, such as the antibiotic compound shown in Figure 11. Researchers in the UK have also developed auxiliaries, and Steve Davies at the University of Oxford is one of the world leaders in this area. In the 1980s his research team developed the iron-containing auxiliary shown in Figure 12, which is especially good for use on a small scale.

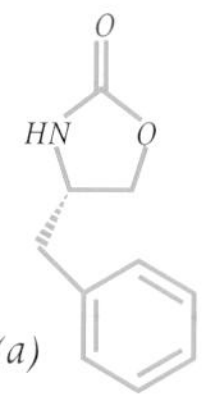

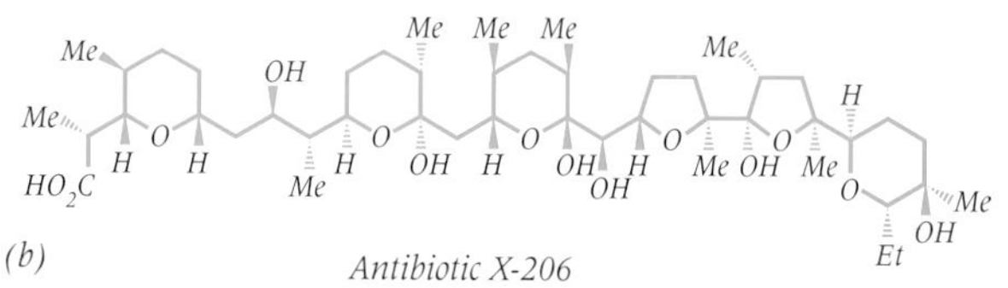

Figure 11. David Evans has used auxiliaries such as (a) to synthesise complicated molecules such as (b).

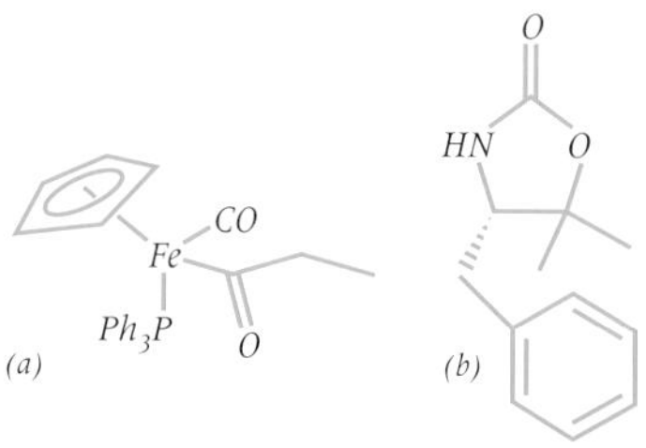

Figure 12. Steve Davies' chiral auxiliaries (a) an iron-containing complex and (b) a Superquat.

More recently, the Davies group has developed a version of the Evans auxiliaries which are easy to use. These are known as superquats. Davies set up a company, now known as Oxford Asymmetry International, which sells these auxiliaries, and prepares chiral molecules on a large scale.

One drawback of attaching auxiliaries to flat organic molecules, is that an extra step is needed to add the auxiliary, and then another extra step to remove it. An alternative approach uses carefully designed reagents to control the handedness of the reaction. Herbert Brown at Purdue University in the US, was awarded a Nobel prize in 1979 for his work involving organic reactions using boron-containing reagents. His research group has developed the reagent Ipc_2BCl which can be used to convert carbonyl groups selectively to alcohols – in other words, C=O is converted into CHOH (Figure 13).

Herbert Brown.

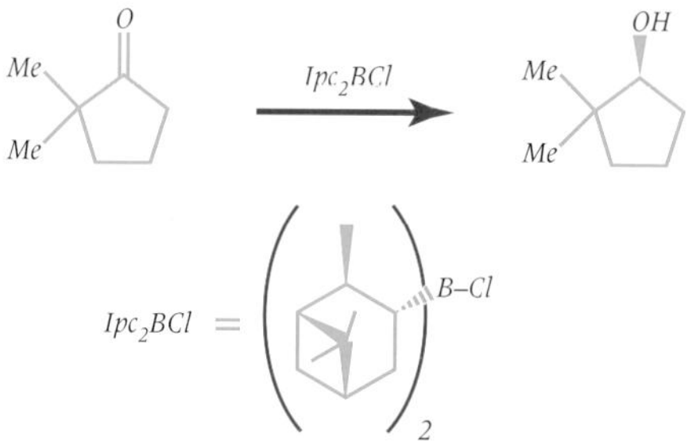

Figure 13. Herbert Brown's boron-containing reagent, Ipc_2BCl is used to convert carbonyl groups to alcohols in a spatially selective way.

Ideally, a chiral agent should behave as catalyst. (This means it is needed in only small amounts and is regenerated at the end of the reaction.) Nature controls the handedness of molecules by using enzymes to catalyse reactions, changing geometrically 'flat' molecules into three-dimensional chiral molecules; the enzyme faithfully produces the same chiral form of product all the time. Although Nature has been exploiting this idea for a long time, it is only recently that it has been possible to design three-dimensional catalysts (affectionately called 'molecular robots') which can provide enzyme-like levels of selectivity. The advantage of being able to use a catalyst is that a very small amount of handed information can generate a large amount of product. In the best cases much less than 1 per cent of catalyst can be used.

One of the most important catalytic reactions is the addition of hydrogen to an alkene (a molecule with a carbon-carbon double bond – see *Chemical marriage-brokers*). This catalytic reaction has been used by the famous Japanese chemist Ryoji Noyori in Nagoya, Japan. One example is the catalysed addition of hydrogen to the alkene shown, where with superb selectivity one chiral form of the anti-inflammatory drug naproxen is synthesised. The handedness comes from a ruthenium catalyst which is attached to a chiral molecule called 'BINAP' (Figure 14). In fact, BINAP is useful not only for catalysed reactions with hydrogen but also with other metals in other catalysed reactions.

Ru
MeO
CO_2H
'Flat' precursor
Hydrogen
Me
H
MeO
CO_2H
Naproxen anti-inflammatory drug

Figure 14. Ruthenium catalysts have been used to add hydrogen selectively to give pharmaceutical drugs. The catalyst used is a ruthenium dihalide attached to a phosphine group.

Ryoji Noyori.
Courtesy of Chemistry Department, Nagoya University.

Alkenes can also be converted into oxygen-containing compounds (epoxides and diols). Barry Sharpless, now at the Scripps Research Institute in La Jolla, California, is the world leader in achieving these reactions catalytically and with control of the handedness of the products. Some of the oxidation reactions carried out by the Sharpless research group have reached outstanding levels of selectivity and use only tiny amounts of catalyst. Using a titanium-based catalyst in the presence of derivatives of tartaric acid, the Sharpless group has shown that it is possible to convert flat alkenes selectively into the single-handed form of the product epoxide (Figure 15).

The development of the stereoselective dihydroxylation of alkenes (two hydroxyl groups are added to a double bond) by the Sharpless group is internationally recognised as one of the major achievements in chemistry in the 1990s. The catalyst is based on an osmium compound, but the reactivity of this catalyst is dramatically improved in the presence of nitrogen-containing

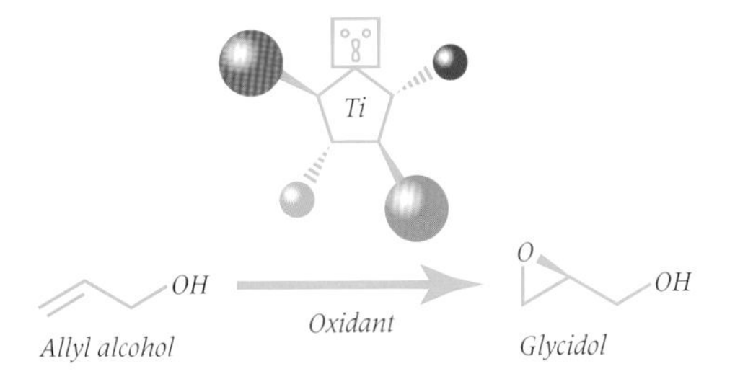

Figure 15. The Sharpless epoxidation reaction was an innovation in asymmetric catalysis. A titanium catalyst brings together the flat starting material and oxidant, generating the three-dimensional product with control over handedness.

Barry Sharpless (left) with former student Tedros Amanios.
Courtesy of The Scripps Reasearch Institute.

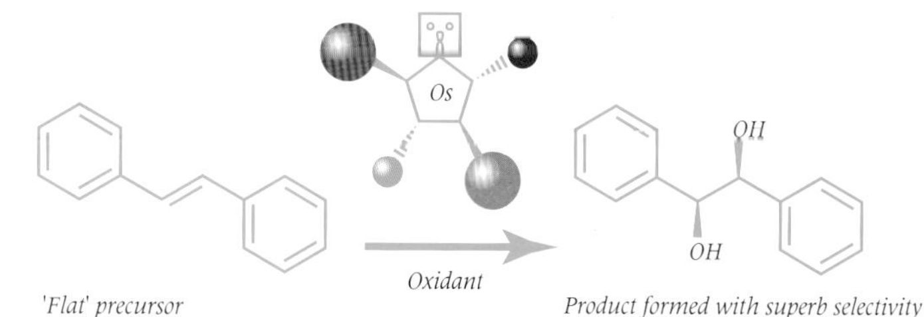

Figure 16. Osmium catalysts have been used to add three-dimensional shape to flat molecules with superb selectivity.

molecules which interact with the osmium. These molecules are derived from the naturally occurring alkaloids quinine and the closely related quinidine. In the example shown, the product formed can itself be used to control the three-dimensional geometry during other metal-catalysed reactions (Figure 16).

Chemists are now applying these new asymmetric synthetic strategies to many complicated molecules. For example, the anticancer drug taxol has a particularly complex molecular structure (Figure 17). However, the total synthesis of taxol using the principles of asymmetric synthesis has been achieved independently by the research groups of K C Nicolaou at the Scripps Research Institute in California and Robert Holton at Florida State University, and subsequently by other research groups. The preparation of such complicated structures is a testament not only to the skills of the research teams who have developed the synthesis, but also to the many research groups around the world who have developed useful synthetic methods for the construction of molecules in a single-handed form. Taxol can be isolated from yew trees, but in such small quantities that there is not enough compound available to treat many cancer patients effectively. The need for useful synthetic routes capable of delivering commercial quantities then becomes very important.

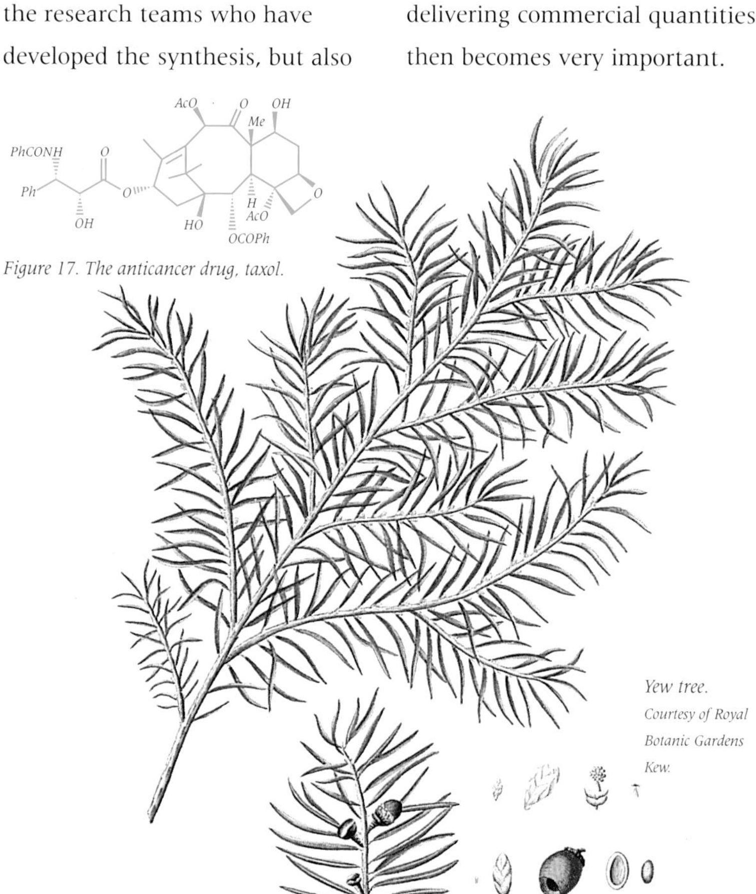

Figure 17. The anticancer drug, taxol.

Yew tree. Courtesy of Royal Botanic Gardens Kew.

Organometallic compounds in synthesis

As mentioned earlier, searching for new approaches to synthesising organic molecules, is driven largely by a desire for greater efficiency (higher yields of products) and selectivity (the elimination of unwanted side products). Many different approaches have been investigated in the second half of the 20th century and been incorporated into modern synthetic methodology. One area that has had a significant impact on our current way of thinking about the construction of complex organic molecules is *organometallic chemistry,* an area of chemistry concerned with compounds containing both metallic and organic fragments. You may have noticed that many of the asymmetric syntheses just described are based on compounds containing a transition metal atom. The development in synthesis of organometallic reagents containing transition metals has been closely intertwined with this area.

The early work on organometallic compounds, however, was fundamental in nature, and was carried out well away from the gaze of organic chemists. In the 1970s, ideas emerging from this work began to be incorporated into mainstream organic synthesis (not without some resistance it must be said!). Eventually, the development in the 1980s and 1990s of many new types of organometallic reagents and catalysts led to a general acceptance of organometallic compounds by organic chemists, and now it is almost impossible to identify an innovative synthesis of an organic molecule which doesn't rely heavily on at least one organometallic reagent or catalyst.

Geoffrey Wilkinson

The use of transition metal-based organometallic compounds in organic synthesis can trace its origins back to the birth of organometallic chemistry. The insight of British chemist, Geoffrey Wilkinson into the organometallic 'sandwich' compound ferrocene (see *Chemical marriage-brokers*) led to a period of unparalleled activity among synthetic chemists. Hundreds, if not thousands, of new compounds were made within a very short period of time – they contained a wide variety of organic components and an impressive range of metals at their core. The excitement was contagious and many chemists joined the race to map out the uncharted areas of the rapidly unfolding subject. Quite naturally the pioneers of organometallic chemistry began to explore the properties of their new compounds, probing, for example, how they reacted with organic compounds. These experiments led in some cases to unexpected changes in the organic molecules and thus the potential of organometallic compounds in organic chemistry was recognised.

How are organometallic compounds used in organic chemistry? Well, in a number of ways. They are employed as user-friendly sources of highly unstable organic molecules. For example, cyclobutadiene, a ring formed from four carbon atoms and four hydrogen atoms, is so strained and thus unstable that it does not exist at normal temperatures and pressures. In fact it reacts with itself to form a larger ring. Under the same conditions, however, a 'complex' formed with iron tricarbonyl, is a well-behaved and easily-manipulated compound (Figure 18). (Carbonyls are molecules of the gas carbon monoxide bound to a metal through their carbon atoms.) Because the iron and the cyclobutadiene are readily separated, the iron complex provides organic chemists with a convenient source of cyclobutadiene.

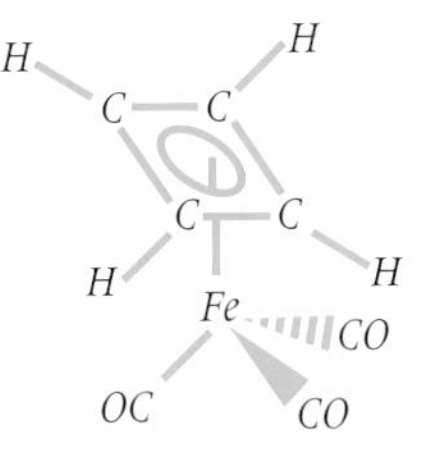

Figure 18. Unstable cyclobutadiene forms a stable complex with iron tricarbonyl, making it easier to manipulate.

Organic chemists also use organometallic compounds as templates to construct sophisticated 'designer' molecules. They exploit the fact that attaching a metal to an organic molecule often induces a dramatic character change in the organic molecule. For example, benzene, a ring made up of six carbon atoms and six hydrogen atoms, is surrounded by a cloud of negative electrons and so, as opposites attract, it prefers to interact with positively-charged molecules.

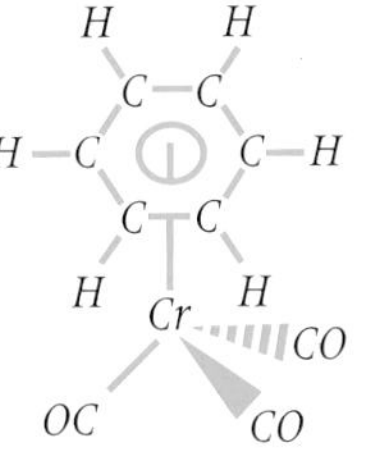

Figure 19. A chromium carbonyl attached to benzene changes its electronic character from electron-rich to electron-poor so that it reacts with negative rather than positive molecules.

Attaching a metal fragment based on chromium to benzene, to give the compound in Figure 19, sucks the electrons away from the benzene and changes its preference for positive molecules to one for negative molecules. So, by attaching metals to organic molecules, some of the traditional rules governing how organic molecules may be joined together are broken. This provides organic chemists with greater freedom when they design pathways to complex target molecules.

Organometallic complexes are also very useful as templates for fusing together several organic molecules, often to form cyclic compounds. A particularly spectacular and useful example of this is a reaction known as the Pauson-Khand reaction after Peter Pauson and his collaborator Ihsan Khand. Pauson and Khand, working at the University of Strathclyde in the 1970s, were interested in the reactivity of alkynes attached to metal fragments based on the metal cobalt (Figure 20a). (Alkynes contain two carbon atoms linked by a triple bond.) They discovered that on reacting these organometallic complexes with alkenes, they were able to generate five-membered ring structures known as cyclopentenones (Figure 20b).

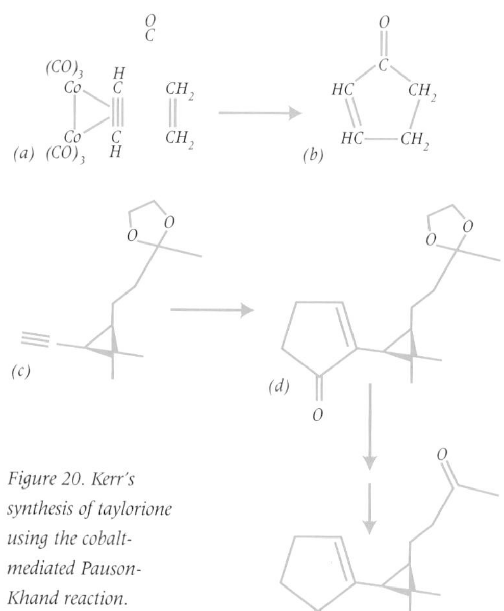

Figure 20. Kerr's synthesis of taylorione using the cobalt-mediated Pauson-Khand reaction.

These are formed from a molecule of alkyne, a molecule of alkene and a molecule of carbon monoxide. To date, however the precise details of how this occurs remain an unsolved mystery. Nevertheless cyclopentenones, or closely related structures are at the core of a wide range of interesting organic molecules and the reaction has found many applications in organic synthesis in recent years. For example, taylorione (Figure 20e), a natural product isolated from the common leafy liverwort *Mylia taylorii* found on acidic rocks in upland areas throughout the northern hemisphere, was synthesised by William Kerr's research group at the University of Strathclyde in the mid-1990s. The Pauson-Khand reaction was used to construct the five-membered ring in taylorione starting from the alkyne (Figure 20c); subsequent elaboration of the product of the Pauson-Khand reaction (Figure 20d) led to taylorione.

Arguably the most important role for organometallic compounds in organic chemistry both now and in the future, however, is as catalysts of organic reactions. The properties of organometallic compounds make them excellent in this role and the past 50 years have seen the discovery of many such catalysts. In order to appreciate their beauty fully, we can look in some detail at an example that has generated a lot of excitement in the past few years.

In the early 1990s, Bob Grubbs at the California Institute of Technology in Pasadena introduced the organic chemistry community to the *'alkene metathesis'* reaction, a reaction which for some time had been used for synthesising polymers but had been overlooked by organic chemists. Overall, the alkene metathesis reaction transforms two alkenes into two new alkenes. Alkenes are one of the most important building blocks of organic chemistry. They are found in many of Nature's organic chemicals such as vitamin A, quinine and chlorophyll, and many manmade organic chemicals such as pharmaceuticals, perfumes and flavourings. The alkene metathesis reaction may be catalysed by several

organometallic compounds such as the one in Figure 21a in which the metal ruthenium is joined to rather a lot of 'packing material' (two chlorine atoms and two phosphorus atoms carrying three six-carbon atom rings, Cy) plus a carbon atom carrying two hydrogen atoms. The carbon and ruthenium atoms are held together by a double bond, and they are fundamental to how the catalyst works. Thus our representation of the catalyst may be reduced to the version Figure 21b without any real loss of information.

(a) $Cl_2(PCy_3)_2Ru{=}CH_2$ ≡ (b) $[Ru]{=}CH_2$

Figure 21. The active catalyst of many alkene metathesis reactions Cy=cyclohexyl.

For organic chemists wanting to construct complicated molecules, the reaction is particularly useful if the two original alkenes are relatively simple and easily obtained (like alkenes I and II in Figure 22) which bear just one non-hydrogen group – R^1 and R^2, and one of the product alkenes is more complex, such as alkene III. Inspection of these structures suggests that a second product must be generated containing two carbon atoms and four hydrogen atoms. This is ethene, the simplest alkene of all. This conversion is an example of the alkene metathesis reaction and is catalysed by the ruthenium compound (Figure 21b).

How does the catalysis occur? Well the whole process is like a country dancing routine (Figure 22). First of all, the ruthenium and carbon atoms in the catalyst exchange partners with the two carbon atoms in alkene I. They do this through square complex 1 in which the two double bonds have reorganised themselves to form four single bonds. Breakdown of the ring and the taking of new partners generates the double bond of ethene and a new ruthenium-carbon double bond. Repetition of the sequence just described, but now with the alkene II, produces square complex II which collapses to the desired product alkene III and the original ruthenium catalyst. The catalyst now continues on round and round the catalytic cycle until all the reacting alkenes have been converted into product alkenes. At this point the catalyst can be separated from the products and used again.

The alkene metathesis reaction has captured the imagination of organic chemists and it is currently being exploited in many and varied contexts. As an example, epothilone A, a molecule with promising anticancer properties, has recently been constructed by a pathway that involves an alkene

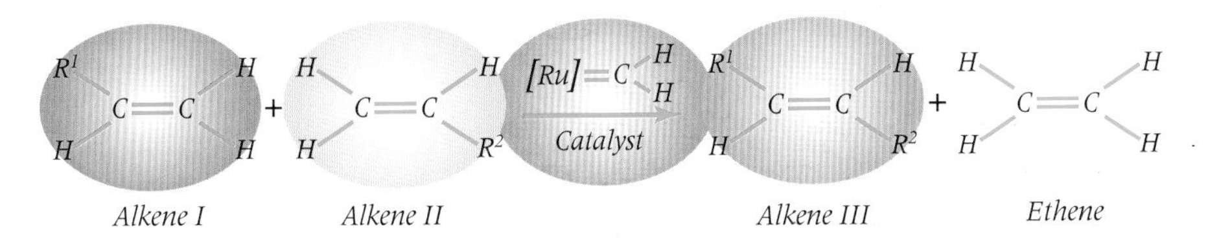

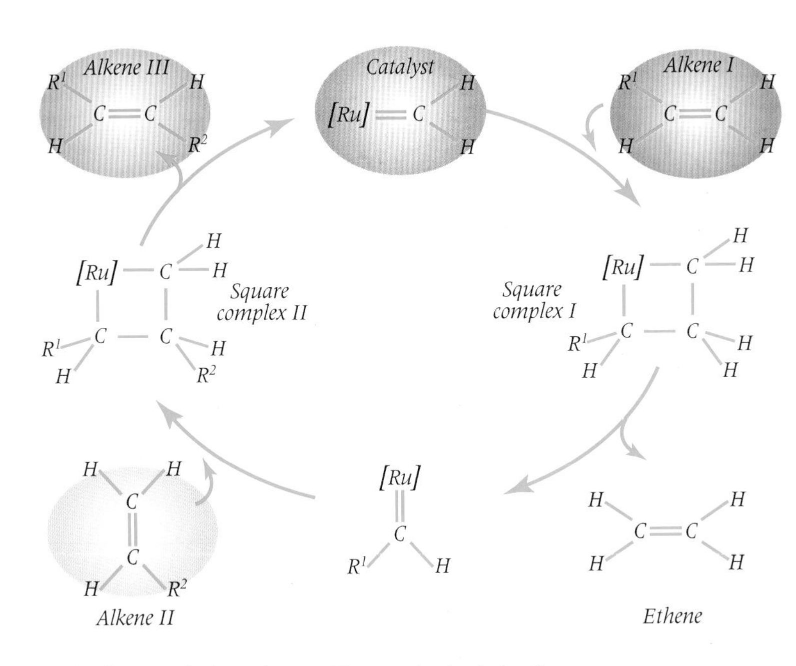

Figure 22. The alkene metathesis reaction resembles a country-dancing routine when viewed at a molecular level.

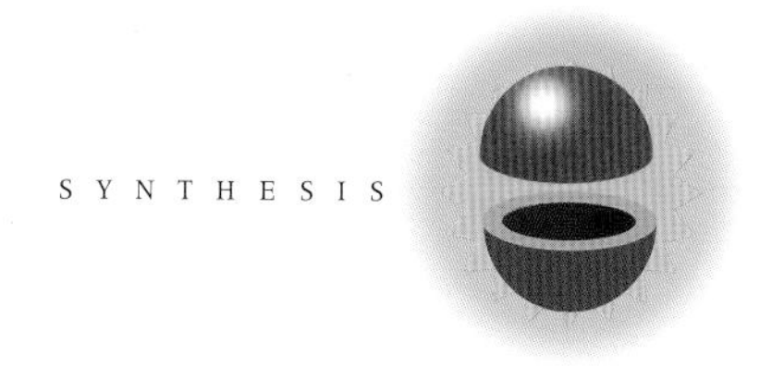

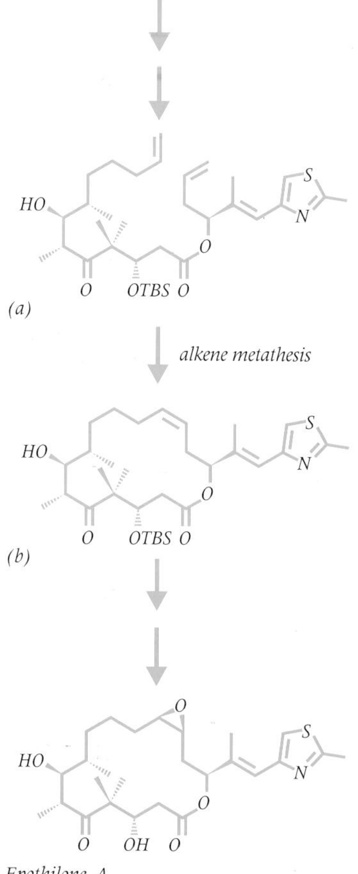

Figure 23. The alkene metathesis reaction has been used to make epothilone A, a molecule with promising anticancer properties.

metathesis reaction at an important point. Thus the first molecule (Figure 23a), which contains two alkenes – albeit with rather complicated R groups! – was converted into the second molecule (Figure 23b) using the alkene metathesis chemistry we have just discussed.

Combinatorial chemistry

So far we have seen that much of the development of organic synthesis has been concerned with constructing ingenious reactions and carefully thought-out reaction pathways in a highly-directed way to give the maximum amount of the required product. This approach can be extremely slow, and new organic syntheses can take months or years to perfect. Nevertheless, it may be appropriate if, for example, pharmaceutical companies, who are the main beneficiaries of organic syntheses, want to make a particular target molecule that they know has the required biological activity such as the anticancer drug, taxol, mentioned earlier. But searching for new drugs is also a hit-and-miss business if the molecular biology of the target disease is not yet well-understood. Although it is likely that there is a drug for every disease, the difficulty is always in seeking that rare needle in the vast haystack of possible structures. Drug companies therefore like to have large numbers of organic molecules available to test for biological activity. As a result they assemble vast libraries of compounds for future assessment.

The enormous diversity of possible chemical structures thus represents a huge challenge to the synthetic chemist seeking new drug molecules. In the past few years however, a new synthetic technique called *combinatorial chemistry* has come to the aid of the frustrated medicinal chemist. This is a method by which large numbers of compounds can be rapidly synthesised and screened for biological activity. Not only is combinatorial chemistry accelerating the previously slow process of drug discovery, it is also transforming the techniques used in organic synthesis creating a new revolution in the field.

The method has its origins in the work of immunologists who were investigating the way by which antigens – the proteins that cause an immune response in animals – were recognised by antibodies. Proteins consist of a long chain of amino acids, and in each protein antigen there is a short amino acid sequence (a peptide) called the *epitope*. This is the key region that is recognised by and binds to the

antibody. Chemists have tried to make all of the short sequences within an antigenic protein to pin down the epitope sequence, but this is a long and tedious process, often requiring the preparation of hundreds of individual peptides. Worse still, if the epitope is discontinuous, and comprised of several amino acids that are not in a continuous sequence, the number of potential epitope sequences increases exponentially and it is almost impossible to find the epitope.

In 1991, Richard Houghten at the Torrey Pines Institute for Molecular Studies in San Diego and Kit Lam at the Arizona Cancer Center, Tucson described how it is possible, through combinatorial chemistry, to make and test huge numbers of different peptide sequences. They used a synthesis method developed in the 1960s by the Nobel prize-winner Bruce Merrifield at the Rockefeller Institute in New York for making peptides. Here, the first amino acid is attached to a solid support and further amino acids are coupled onto it in a chain-like sequence with the excess reagents washed away after each step. At the end of the synthesis, the final peptide is removed from the support (Figure 24). The trick Houghton and Lam employed to give such a huge hike in productivity was to make the peptides on tiny polystyrene resin beads only 100 micrometres across, and to use a process of *'mix and split'* to make every possible combination of amino acids.

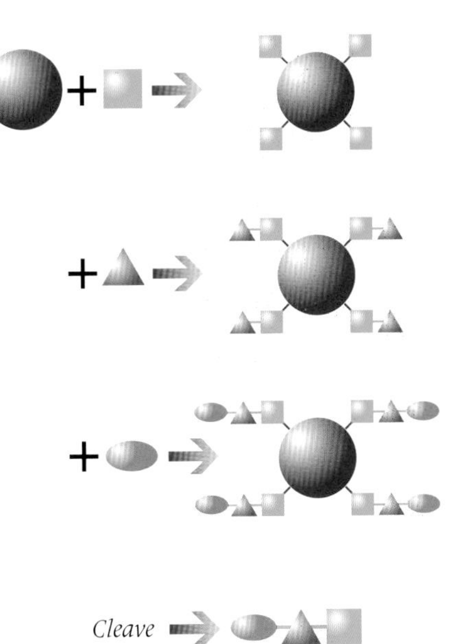

Figure 24. Solid-phase chemistry allows the synthesis of compounds attached to a resin bead. A final cleavage step generates the product in solution.

Mix and split

The mix and split process works as follows. Initially a number of amino acids are individually attached to quantities of resin beads. After the first coupling step, all of the resin beads are mixed, and then split into individual portions for the next synthetic step (see Figure 25). Each portion now contains beads with each of the amino acids attached and when reacted with further amino acids, every possible combination of dipeptide is generated. Further cycles of mix and split followed by amino acid coupling increases the total number of peptides.

The size of the peptide library generated was the geometric product of the number of amino acids used, to the power of the number of coupling steps. Thus, using 20 amino acids in each cycle permitted the production of 400 dipeptides, 8000 tripeptides or 160,000

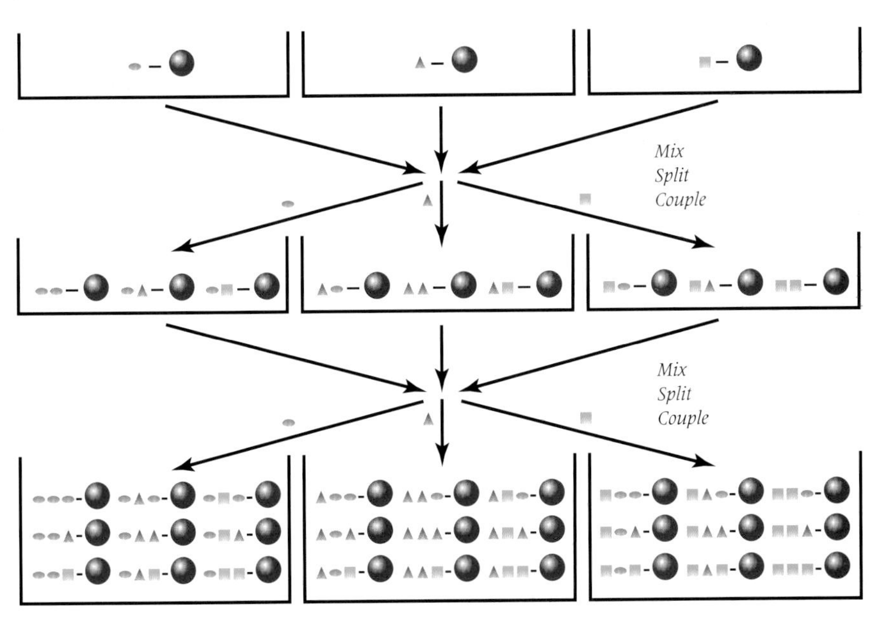

Figure 25. The mix-and-split process. Chemical building blocks (depicted here as an ellipse, triangle and square representing monomers such as amino acids used to make peptides) are attached to beads, which are mixed and split into three groups for a second coupling with the same three monomers. The procedure is repeated to give the 27-compound library shown here.

tetrapeptides. At the end of the synthesis, the combinatorial library consisted of a large number of resin beads, with each bead containing many molecules of only one peptide sequence. The combinatorial library process generates huge numbers of peptides, using only a few chemical transformations as the reactions that required addition of the same amino acid were efficiently combined into the same reaction flask. Furthermore, the use of this *solid-phase chemistry* approach allowed each reaction to be carried out in high yield by using an excess of the chemical reagents (Figure 24). Although this would normally lead to contamination of the products with excess reagents and unreacted starting materials, the resin beads containing the precious bound peptides could be isolated and cleaned-up by a simple filtration and wash. Although Houghten and Lam proposed different ways of screening the peptides to find epitope sequences, they were both successful in finding a potent antigenic sequence from amongst millions of peptides.

A typical automated laboratory.
Courtesy of Pfizer.

This seminal work initiated a range of techniques that have revolutionised organic chemistry and drug discovery in less than 10 years. Within a year of the original publications of Houghten and Lam, several other research groups had reported similar peptide library work either using resin beads or by preparing peptides on a range of other solid supports including paper sheets, polypropylene pins and even glass chips. Investigation of new solid-phases for combinatorial chemistry has also been complemented by a trend towards miniaturisation (for example, the use of microtitre plates – dishes with lots of tiny wells in which the reactions are carried out) and the robotic automation of the most repetitive of the manual tasks necessary in making these libraries.

The most significant development however has been the application of the technology to the synthesis of libraries of non-peptide molecules. Although biologically active, peptides do not make good drugs as they are very poorly absorbed when taken by mouth and are subject to rapid metabolism and excretion if they do enter the bloodstream. So in the early 1990s several farsighted organic chemists had branched away from peptides and begun to make real drug-like organic molecules on solid phase. One highly influential piece of research was by Jon Ellman at the University of California at Berkeley. He published a research paper in 1992 which demonstrated that the class of benzodiazepinone molecules, a structure commonly found in drug molecules such as anxiolytics (drugs that reduce anxiety), could be readily made on solid phase. Furthermore, the synthetic route offered sufficient flexibility such that a number of different reagents could be used to make a library of diverse benzodiazepinones, any of which might offer unprecedented pharmacological activity.

Jon Ellman.
Courtesy of Greg Butera, University of California, Berkeley.

The combination of rapid synthesis and the solid-phase generation of diverse organic molecules proved irresistible to medicinal chemists in the pharmaceutical industry, and today just about every major company has a group dedicated to the production of novel organic molecules using combinatorial chemistry. In addition, there has been a renaissance in the study and development of solid-phase chemistry, as numerous academic and industrial groups have investigated the range of chemistry that is possible on resin beads. Traditionally organic chemistry has been carried out by dissolving the reagents in a solvent so that they form a homogeneous mixture. We are now reaching a stage where many of the drug-like compounds favoured by medicinal chemists can successfully be prepared on a solid phase. Thus, libraries of compounds are now routinely prepared to add to the pharmaceutical company files for high throughput screening against new disease targets. The hope is that with a sufficient number of different compounds available to be tested, the screening will successfully give lead molecules that will kick-start a new drug discovery programme.

GlaxoWellcome

When designing the *combinatorial library* the chemist will consider using a range of commercially available small molecules that can be used as building blocks to be attached to a central molecular template. If there are a 100 possible building blocks in each of three positions on the template, the theoretical or 'virtual' library size is one million compounds. A chemist may decide to make all of these compounds using the mix-and-split approach, with the intention of testing the compounds in mixtures. Alternatively, he or she may wish to make just a representative selection of a few hundred compounds from the virtual library. These compounds can be prepared individually through a process of *parallel synthesis,* which is making many compounds simultaneously, possibly in an automated fashion. If any of the first set of compounds is active, a follow-up 'targeted' combinatorial library will readily be assembled and will aid the rapid optimisation of biological activity.

The list of compounds in a virtual library is a highly valuable resource to the medicinal chemist, and a range of computational techniques are being employed to select compounds which are maximally diverse and which allow the structural variation within the library to be expressed from just a few of the compounds. So influential have the techniques of combinatorial chemistry been, that they have instigated radical reappraisals of the ways by which molecules are prepared and synthesised. In particular, the most extreme of the methods now being explored actually challenges the need to make any compounds at all! A research group at Abbott Laboratories, Illinois, led by Stephen Fesik has developed a method that uses the building blocks of a virtual library to discover new drug molecules without actually making the library itself.

This technique is based on the following idea. Drug molecules work by binding to a target protein in the body such as

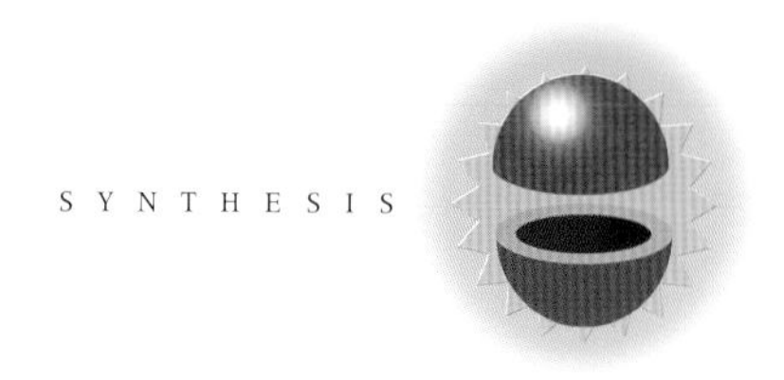

an enzyme or receptor, so it might be possible to identify the building blocks of a drug first by binding each one to the protein. A set of the building-block compounds that could be used in the preparation of a combinatorial library is individually incubated with a target biological protein (see Figure 26). Using nuclear magnetic resonance (NMR), a technique that can map the location of various atoms in a molecule (see *Analysis and structure of molecules*), building blocks that bind to the protein were detected by a change in the magnetic properties of the protein. Following the discovery of one compound that binds to the protein, the process was repeated with a new set of compounds in the presence of the first discovered compound. Changes in the NMR revealed that a second compound could also bind, at the same time and at a site distinct from the first compound. The final step in this innovative process requires a little computational work to devise ways by which the two

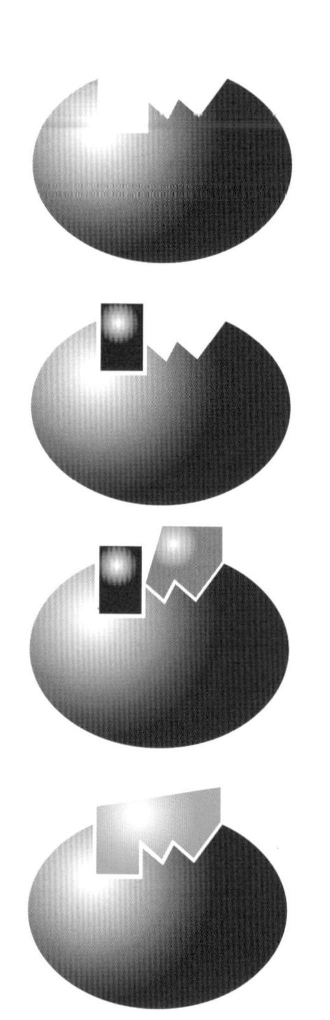

Figure 26. Novel compounds that bind to enzymes or receptors can be discovered by finding the best two fragments that bind and then synthetically combining them.

compounds could be joined such that the product molecule will still bind to the target protein, gaining the binding benefit of both molecules. At this stage a limited programme of synthetic chemistry generated a number of enjoined molecules, some of which had exceptional affinity for the target protein. Overall, this process allows the selection of one of the most potent compounds in a virtual library by the NMR analysis of the constituent monomers alone.

Combinatorial chemistry is revolutionising the way drug discovery is being done, as the enormous productivity it brings is allowing the exploration of large numbers of new molecules. As we have seen, the concepts are also initiating new ideas where it may not even be necessary to make more than a handful of compounds. With every major pharmaceutical company now using combinatorial techniques for drug discovery, we have seen a revolution in the way that chemical synthesis is practised. What is even more exciting is that combinatorial chemistry is now being used to make other groups of chemicals with required properties such as ceramics, catalysts, polymers and semiconductors.

Supramolecular chemistry

So far, we have seen the huge diversity of organic structures that can be made using synthetic methodologies developed in the 170 years since Wöhler made and identified the first synthetic organic compound. Much of the work of organic chemists has involved creating completely new compounds that don't exist in Nature. However, over the past 30 years, there has been increasing interest in preparing chemical systems that mimic the behaviour of natural ones. This interest has arisen, of course, because Nature carries out so many difficult chemical processes with such efficiency. Such an approach requires more than just the properties offered by the chemical functional groups and structures found in single organic molecules.

One of the crucial characteristics of biological molecules is that they often interact through weak electrostatic attractions, for example, the interaction between a receptor protein and a small drug molecule which provokes a

pharmacological response. This is called *molecular recognition* (Figure 27). Large biological molecules

Figure 27. The binding between drug and biological receptors through hydrogen bonding, shown by ''''''''' (A = acceptor sites, D = donor sites).

such as proteins are held in their characteristic three-dimensional shapes by the weak interaction of *hydrogen bonding* – the attractive force between positively charged hydrogen atoms and the negatively charged electrons on oxygen and nitrogen atoms. It is also the glue that holds DNA strands together through the exquisite complementarity in the positions of hydrogen bond *'acceptor'* and *'donor'* groups on nucleic acid base pairs (Figure 28). Without hydrogen bonding, life would not exist.

Understanding how these long-range, weak forces control the architecture and behaviour of large molecules and assemblies of molecules is now an exciting new field of chemistry. It is called *supramolecular chemistry* – that is, chemistry beyond the molecule. It is leading to new synthetic methods giving new types of structures with novel applications.

Figure 28. Hydrogen bonding between complementary nucleic acid bases guanine and cytosine.

Some of the first attempts to copy the supramolecular chemistry of biology were to create artificial receptors whereby alkali metals such as sodium and potassium bind into the centre of a cyclic organic compound in what is called a *host-guest complex*. In 1967 Charles Pedersen, then with the American chemical giant DuPont, made a cyclic polyether which consists of an 18-membered ring of 12 carbon atoms interspersed with six oxygen atoms (called 18-crown-6 because of its distinctive three-dimensional coronet shape). The slight negative charge associated with the oxygen atoms allows the molecule to bind weakly to a wide range of positively charged species. Shortly after, Jean-Marie Lehn of the Louis Pasteur University in Strasbourg, France and Donald Cram of the University of California, Los Angeles (UCLA) independently prepared more complex, three-dimensional host molecules called *cryptands* and *spherands* respectively (Figure 29). All three pioneers in this field won the Chemistry Nobel Prize in 1987.

Charles Pedersen.
Courtesy DuPont.

Jean-Marie Lehn.
©The Nobel Foundation.

Donald Cram.
Courtesy of Robin Robin and University of California, Los Angeles.

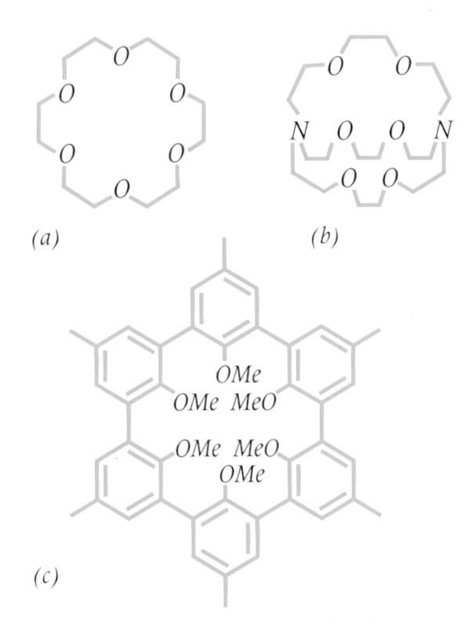

Figure 29. Three novel supramolecular structures: (a) 18-crown-6 ether made by Charles Pedersen, (b) [2.2.2] cryptand made by Jean-Marie Lehn and (c) a spherand made by Donald Cram.

Boxing clever

Cram emphasised the importance of 'preorganisation' in creating host structures with rigid geometries and of right size and shape to accommodate a desired guest molecule. A good example is the molecular prison he made to isolate the hyper-reactive cyclobutadiene molecule mentioned earlier, which eagerly reacts with itself. Cram had developed three-dimensional structures large enough to encapsulate small molecules rather than just metal ions, and

selected as his molecular prison, or *carcerand*, a molecule reminiscent of an empty Chinese lantern (Figure 30). At low temperatures, the bars of the

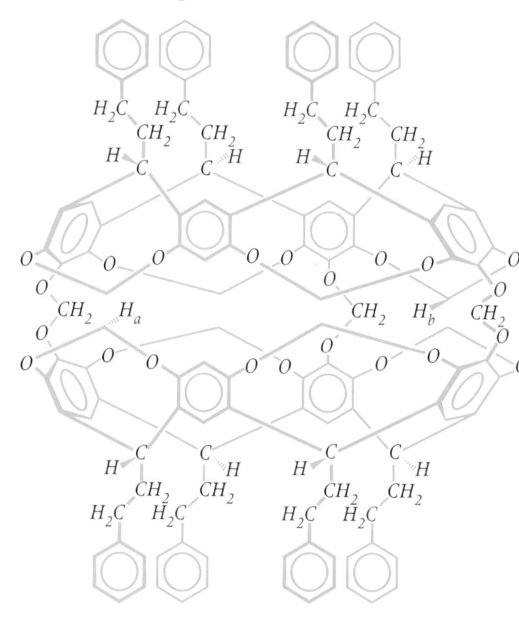

Figure 30. Cram's carcerand in which he imprisoned cyclobutadiene.

carcerand are closed and nothing can get in or out, but at temperatures greater than 130 °C, the bars bend and twist and, although it's a bit of a squeeze, small molecules can be forced in and out (Figure 31). By boiling up the carcerand in a mixture of chlorobenzene and α–pyrone the UCLA team managed to encapsulate α–pyrone in the

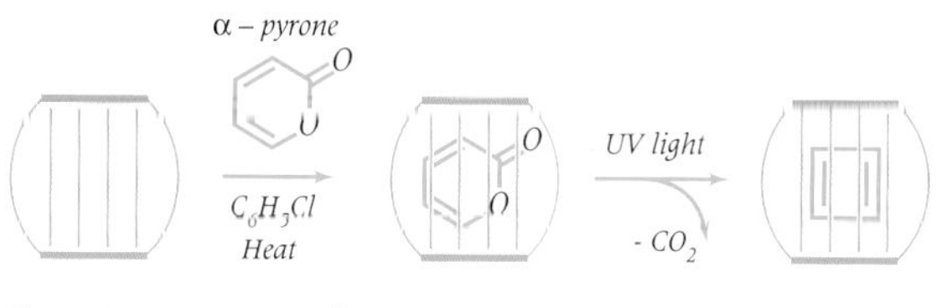

Figure 31. Incarceration of cyclobutadiene.

carcerand. By irradiating this material with ultraviolet light, the α–pyrone was transformed into cyclobutadiene. However, trapped within its protective shell, the cyclobutadiene was unable to reach other molecules of its type to react with and thus its characteristics and chemistry could be studied at leisure.

Now the same technique is being applied to make and study other highly reactive compounds and intermediates, for example, reactions with derivatives of benzene where one functional group attached to the benzene ring is substituted by another. Such reactions are extremely common in the fine chemicals, pharmaceutical and paints industries. Many of them are believed to proceed by a mechanism which postulates the existence of intermediate 'benzyne' containing a carbon-carbon triple bond (Figure 32).

This is so reactive it had only ever been seen trapped in solid argon at -200 °C! In 1997 benzyne was synthesised in one of Cram's molecular prisons and the details of its structure and reactivity revealed for the very first time.

$NaNH_2$, NH_3, $-HCl$ — 'Benzyne'

Figure 32. Benzyne, a proposed intermediate in reactions of benzene derivatives like the one shown above.

The ideas of host-guest molecules developed by Cram and the other pioneers have blossomed in many directions and been put to good practical use. The British chemist Andy Hamilton, in 1988 while at the University of Pittsburgh, made artificial receptors that bound to certain types of drugs. He noted that barbiturates (Figure 33a) possess four hydrogen-bond acceptors and two donor sites in a specific spatial array and designed a fully complementary partner (Figure 33b) bearing the opposite arrangement. When the two molecules were mixed they recognised each other and combined to form a host-guest complex with six inter-component hydrogen bonds (Figure 33). This synthetic molecule acts as a prototype chemical sensor, or artificial nose, for detecting barbiturates.

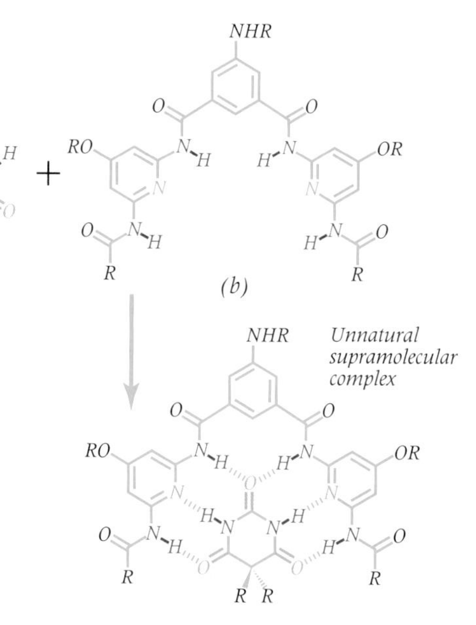

Figure 33. Recognition between barbiturates and a synthetic receptor – a prototype chemical sensor.

This sort of approach – designing and making molecules that selectively bind to others – has much wider application than just artificial sensors for drugs. Molecules that can selectively bind strongly to other species potentially could also be used to extract them from solution or gases. Areas of application of such systems include removing urea from blood (artificial kidney dialysis); isolating the vast quantities of uranium and gold naturally present in low concentrations in sea water; eliminating the cadmium, lead and other pollutants from industrial effluents; and even the extraction of caffeine from tea and coffee!

Molecular train sets

Chemists have now gone much further with the supramolecular concept. One of the most important characteristics of supramolecular biological systems is that they have the propensity to 'self-assemble' under the driving forces of hydrogen bonding and other weak chemical effects. These control many 'mechanical' biological processes such as protein-folding and the unwinding of DNA (see *The chemistry of life*). Chemists have now used this same approach to build assemblies of molecular units – interlocking individual molecular pieces, in much the same way as engineers assemble components for machines. Indeed, the ultimate components of devices may well eventually be individual molecules (which behave as wheels, cogs, spindles, shuttles, switches, and so on) which interact (a cog might spin a wheel, for example) in a carefully-designed way to create a 'nano' machine. Synthetic chemistry is now set to play a pivotal role in the exciting new area of *nanotechnology* (see *New science from new materials*).

Fraser Stoddart.
Courtesy of the University of California, Los Angeles.

One of the pioneers in this field is Fraser Stoddart, who moved from the University of Birmingham in 1997 to take over from Donald Cram who was retiring from his professorship at UCLA. Stoddart uses another supramolecular tool – the strong attraction between electron-rich and electron-poor aromatic (benzene) rings – in his approach to molecular engineering. In the late 1980s, he was making sticky ring-shaped molecules that bound to paraquat – a powerful weedkiller sold by ICI (now Zeneca). The paraquat ring system is positively charged – in other words, electron-deficient – and Stoddart took advantage of this feature to form a supramolecular complex between it and a large, doughnut-shaped crown ether containing a complementary pair of electron-rich aromatic rings. In fact, the fit between the two molecules was

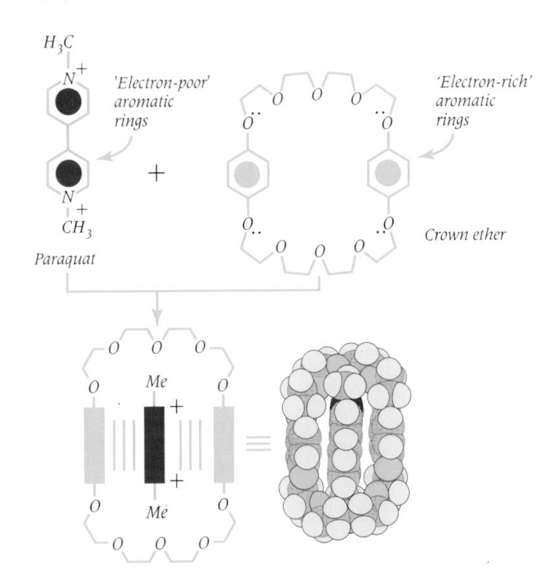

Figure 34. The threading of a paraquat molecule through an electron-rich crown-ether ring.
Adapted from Angew. Chem. Int. Ed. Engl., **1996**, *35, 1154-1196.*

Courtesy of Hornby Hobbies Ltd.

Box 2 Molecular rings that interlock

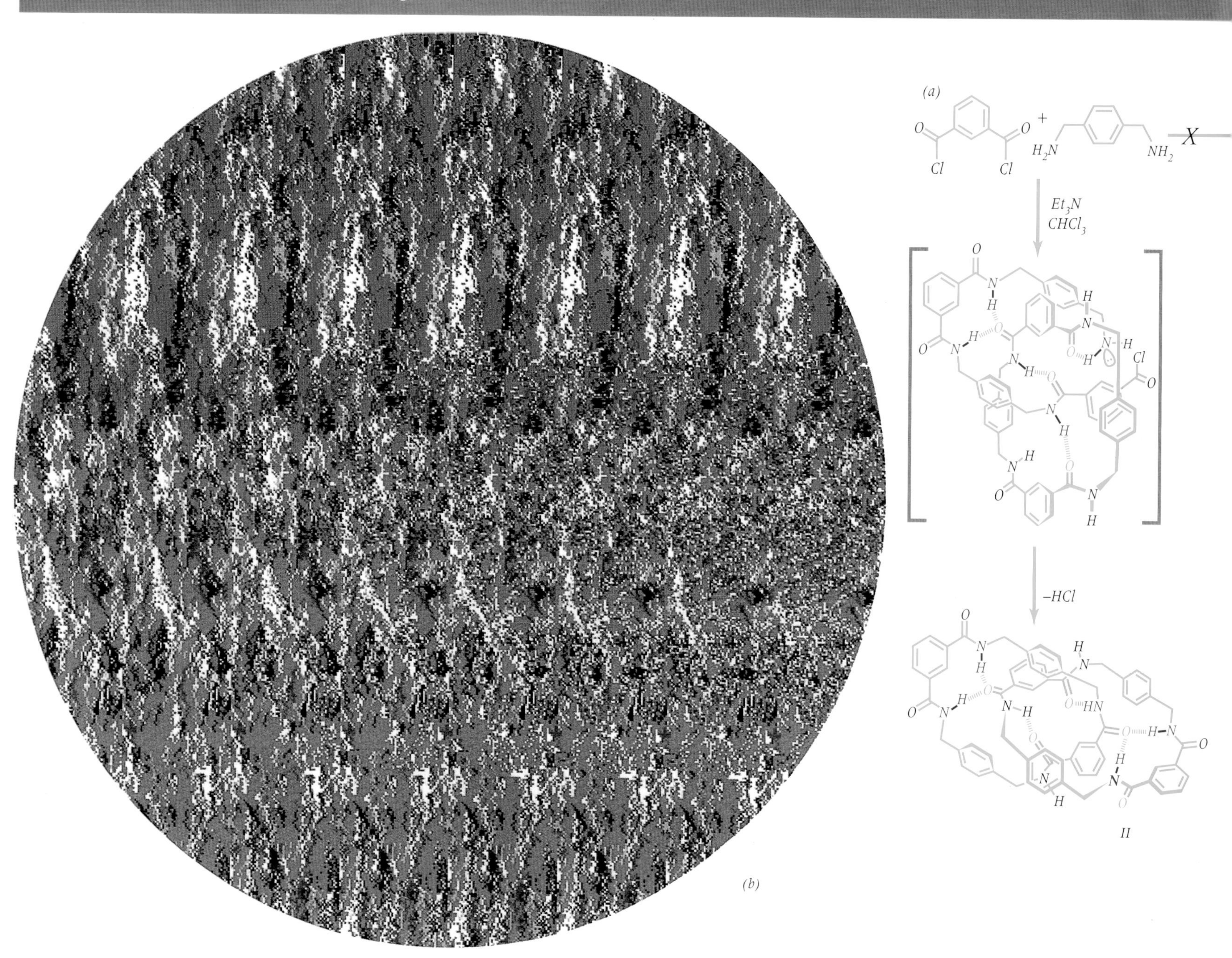

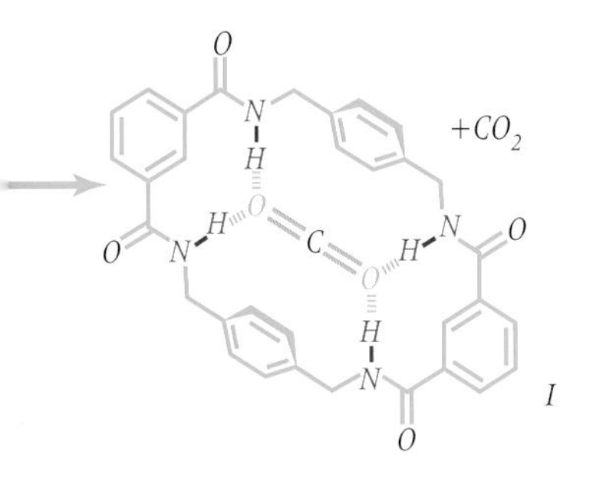

Other sorts of mechanically-interlocked systems have been synthesised purely by chance. The large cyclic compound I was designed to act as a chemical sensor for carbon dioxide through complementary hydrogen bonding interactions. However, the attempted synthesis led not to discrete molecules of I but rather to interlocked pairs of them, II. The formation of II is brought about by the same sort of inter-component hydrogen bonding that was designed to encourage II to bind to carbon dioxide.

(a) The serendipitous synthesis of hydrogen bond-assembled interlocked rings.
(b) The crystal form of (II) is shown in this autostereogram ('magic eye' picture). To see the image you must stare through the picture so your focal point is behind the page. Placing the picture behind a piece of glass and focusing on your reflection helps if you cannot see the picture straight away. Compound (II) is the tiniest interlocked ring system known – the autostereogram image is 2 x10^{8} times larger than the actual molecule itself!

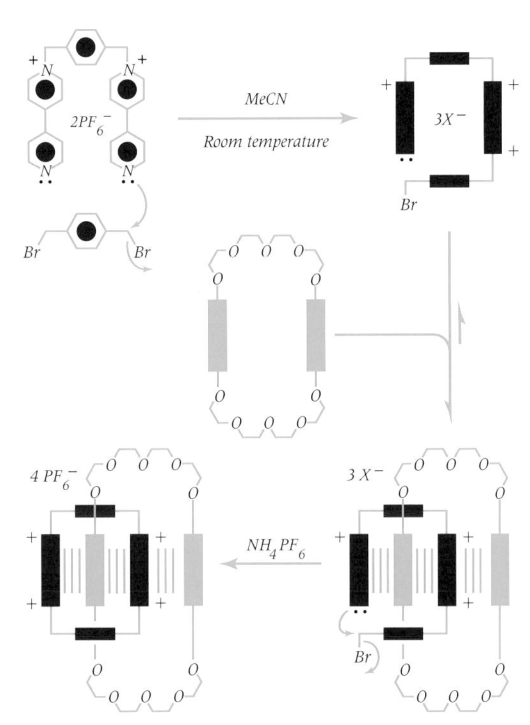

Figure 35. The mechanical interlocking of two molecular rings directed by the mutual recognition between appropriately located electron-rich (red) and electron-poor (blue) aromatic rings.
Adapted from Angew. Chem. Int. Ed. Engl., ***1996****, 35, 1154-1196.*

so good that the paraquat sat right in the middle of the doughnut, with the ends poking out each side (Figure 34).

Stoddart immediately realised that this arrangement was special, and he could do more with it than just make the two pieces stick together – he could interlock them permanently. He made a version of the same system where the ends of two paraquat units could be connected by aromatic rings also to form a cyclic structure. When this ring-forming reaction was carried out in the presence of the electron-rich doughnut molecule, the paraquat macrocycle formed around it, mechanically joining the two structures like links in a chain (Figure 35). The beauty of having mechanical linkages instead of connections with rigid chemical bonds is that the mechanically-interlocked components can move with respect to each other, just as required for the moving parts of machinery.

The Birmingham-UCLA team have also used the same strategy to prepare molecular 'shuttles' where a bead can be moved up and down a linear thread in response to signals with protons or electrons. They have even made a molecular 'train set', where a bead moves from station

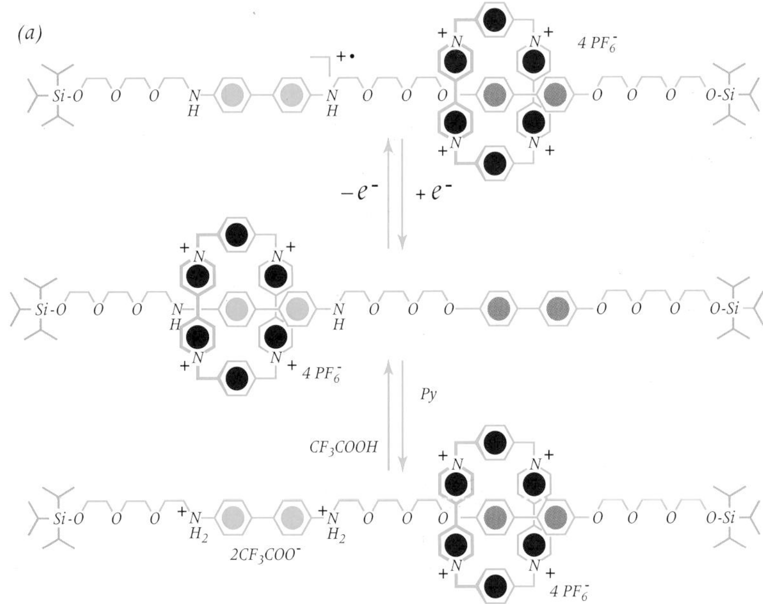

Figure 36 (a). Towards molecular machinery: a switchable molecular shuttle.
Adapted from Angew. Chem. Int. Ed. Engl., ***1996****, 35, 1154-1196.*

to station around a circular track (Figure 36). The trains are never late and, according to Stoddart, never crash into each other even when two or more are on the same track! Although practical information storage systems or *molecular computers* based on these ideas are still many years away,

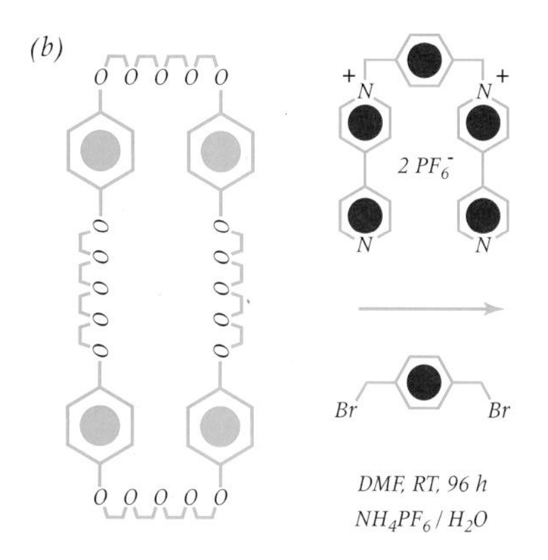

Figure 36 (b). A molecular train set.
Adapted from Angew. Chem. Int. Ed. Engl., **1996**, *35, 1154-1196.*

simpler molecular machines – where the properties of materials are controlled or switched through interlocked assemblies – are just around the corner. Supramolecular chemistry has fundamentally broadened the field on what it is possible to achieve through organic synthesis.

The future of synthesis

The past 150 years has witnessed extraordinary progress towards the creation of ever-increasing molecular complexity. Despite the advances seen, the synthesis of complex molecular targets still presents the organic chemists with enormous challenges.

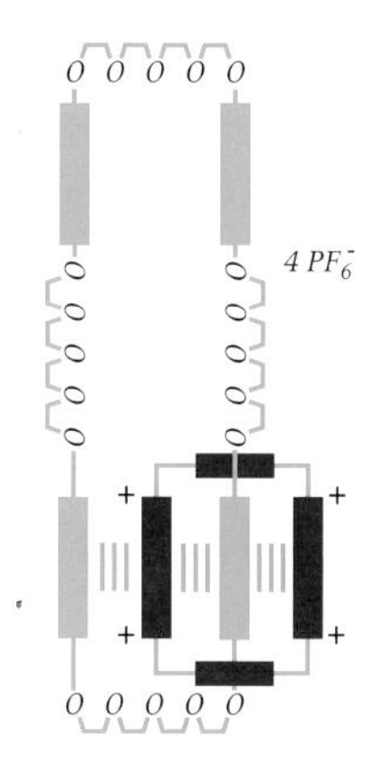

Preparing large naturally-occurring molecules is still regarded as a 'grand challenge'. Some of them have very unusual structures which make significant demands on the synthetic ingenuity of the chemist, highlighting gaps in currently available chemical technology. This not only stimulates the development of new synthetic reactions, but also tests these procedures to their limit. The chemical synthesis of a molecule can also help biologists who want to understand how a molecule interacts with its biological receptor, and may yield simplified analogues that could be potential drugs. Increasingly chemists try to copy the way Nature makes the molecule, using types of reactions that automatically prime the product for the next reaction step. This means that the reagents needed for several steps can be put all together in one reaction flask and then left to get on with the job – an approach called *one-pot synthesis*, or *cascade chemistry.*

Chemists are still discovering new, useful reactions, and stereochemical control will continue to be a significant feature in the search for increased efficiency and selectivity. The chemical industry is increasingly aware of the need to make best use of the world's limited raw materials and to keep waste to a minimum so as to protect the environment. As a result, chemists are always looking for reactions that exploit environmentally-friendly catalysts and can be carried out at room temperature in water rather than the traditional organic solvent. 'Green chemistry' is now very much the name of the game.

Synthetic chemistry will become ever more technologically sophisticated. Computation, miniaturisation and robotics will continue to have an impact on the way syntheses are done. Computer programs can be used to design routes to a compound retrosynthetically, and in the commercial world take into account economic aspects such the cost and availability of starting materials. For research purposes at least, we can expect to see experiments carried out on chips and plastic beads as well as in round-bottom flasks.

In years to come, it may be possible to design complex pieces of molecular architecture that automatically self-assemble from simple building blocks into a useful device. Although that view of chemistry is science fiction now, the 21st century is sure to reveal many such exciting advances in organic synthesis.

Further reading

1. *Classics in Total Synthesis*, K. C. Nicolaou and E. R. Sorenson, VCH, Weinheim: 1996.
2. *Organic Synthesis*, J. Fuhrhop and G. Penzlin, VCH, Weinheim: 1994.
3. *Principles of Asymmetric Synthesis,* R. E. Gawley and J. Aubé, Tetrahedron Organic Chemistry Series, Volume 14, Pergamon, Oxford: 1996.
4. *Asymmetric Synthesis,* R. A. Aitken and S. N. Kilényi, Blackie, London: 1992.
5. *Asymmetric Synthesis of Natural Products,* A. Koskinen, Wiley, New York: 1993.
6. *Stereochemistry of Organic Compounds*, E. L. Eliel and S. H. Wilen, Wiley, New York: 1994.
7. *Asymmetric Synthesis,* G. Procter, Oxford University Press, Oxford: 1996.
8. *Metathesis in Organic Synthesis,* A. Furstner (ed), Springer-Verlag, Heidelberg: 1998.
9. *Transition Metals in Organic Synthesis,* S. E. Gibson (ed), UOP, Oxford: 1997.
10. *Transition Metals in the Synthesis of Complex Organic Molecules,* L. S. Hegedus, University Science Books, Mill Valley: 1994.
11. *Combinatorial Chemistry,* N. K. Terrett, Oxford University Press, Oxford: 1998.
12. *Supramolecular Chemistry: Concepts and Perspectives,* J-M. Lehn, VCH, Cambridge: 1995.
13. *Supramolecular Chemistry,* F. Vogtle, Wiley, Chichester: 1991.

Glossary

Alkene metathesis A reaction that converts two alkenes into two new alkenes. It is catalysed by a wide range of metals, and in recent years it has been put to spectacular use in target-oriented organic synthesis.

Asymmetric synthesis The preparation of a compound (existing as 'handed', or chiral isomeric forms) enriched in one enantiomer – generally using an achiral starting material.

Cascade chemistry Reactions in which a programmed sequence of reactions occurs, usually involving multiple bond formation, to give a desired product in an efficient way.

Cavitand A molecule possessing a permanent rigid cavity into which certain other molecules or ions can bind.

Chiral auxiliary An enantiomerically (single-handed) pure chemical unit which is attached to an achiral molecule. The chiral auxiliary stereoselectively controls subsequent chemical reactions.

Chirality This refers to the situation when an object is not superimposible with its mirror image. This may refer to molecular structures or larger objects (for example, hands).

Chiral pool There are many enantiomerically pure compounds, which are readily available as building blocks in chemical synthesis. These are informally termed 'chiral pool' compounds.

Combinatorial chemistry A range of techniques, some possibly automated, that permit the rapid synthesis of large numbers of chemical compounds usually through the 'combination' of several different synthetic building blocks.

Conformation The particular three-dimensional array of atoms in a molecule of fixed constitution and configuration. Different conformations can be accessed by rotation about single bonds.

Conformational analysis The interpretation or prediction of the physical and chemical properties of compounds, based upon analysis of molecular conformations.

Crown ether A rather flat cyclic (crown-shaped) molecule with precisely arranged oxygen atoms which can bind to appropriate metal ions.

Cryptand A type of cavitand with a three-dimensional hollow inside the molecule into which suitable ions can fit and bind.

Enantiomer These are molecules which are isomeric by virtue of the fact that they are the mirror image of each other, but non-superimposible.

Enantiospecific synthesis The unambiguous conversion of an enantiomerically pure starting material into an enantiomerically pure product.

Epitope The amino acid sequence within a protein that is specifically recognised by, and can bind to, another protein such as an antibody.

Functional group A particular arrangement of atoms in a molecule which confers reactivity. Examples include the carbonyl (C=O) and primary amine (NH_2) groups, and also carbon-carbon multiple bonds, but not the unreactive saturated parts of molecules (of which alkanes are wholly composed).

Host-guest complex An assembly of two or more molecules (or a molecule and an ion) held together by 'weak' or non-covalent bonds. The 'host' (usually larger) molecule has convergent binding sites; the 'guest' has complementary divergent binding sites.

Hydrogen bonding An attractive, relatively weak interaction (about 2 to 5 per cent of the strength of a typical carbon-carbon covalent bond) between a positively polarised hydrogen atom and a negatively polarised (usually oxygen or nitrogen) atom. Networks of multiple hydrogen bonds are responsible for holding together DNA and regulating many biological processes.

Isomer Different compounds possessing the same molecular formula are known as isomers. A given compound may have an isomer that differs in constitution or stereochemistry, in the latter case the isomer is referred to as a stereoisomer.

Mix-and-split synthesis A synthetic process by which large numbers of peptides or non-peptide molecules can be efficiently made on a solid-phase support such as resin beads. The name comes from the process by which the resin beads are combined, mixed, then subdivided before each step in the synthesis.

Molecular recognition The selective affinity between two or more molecules with complementary binding sites.

One-pot synthesis A synthesis, normally involving more than one basic transformation, which can be conducted in a single reaction vessel.

Optical activity The ability of a compound to rotate a plane of polarised light.

Optical isomer See enantiomer (which is the recommended term).

Organic chemistry Often defined as the chemistry of carbon-containing compounds, including synthesis, analysis and theory.

Organometallic chemistry The study of compounds containing one or more carbon-metal bonds. Interest in transition metal organometallic chemistry has increased exponentially over the past 30 years, and this

work has provided many catalysts and reagents regularly used in complex organic synthesis.

Parallel synthesis The process by which many compounds can efficiently be made simultaneously, using either solution or solid-phase chemistry often with some degree of automation.

Physical organic chemistry The study of theoretical, physicochemical (and spectroscopic) aspects of carbon-containing compounds.

Racemic mixture A racemic mixture (or racemate) is a 50:50 mixture of two enantiomers.

Retrosynthetic analysis Also known as retrosynthesis, retrosynthetic analysis is the method of applying hypothetical dissection to a target molecule, such that each dissection constitutes the reverse of a viable synthetic reaction. In this way a complex target can be broken into simpler fragments until starting materials are identified, and the retrosynthetic operations then provide a guide as to how the fragments may actually be assembled to give the target molecule.

Solid-phase chemistry The synthesis of chemical compounds when one reagent or chemical building block is covalently attached to an insoluble support such as a polymeric resin bead.

Spherand A three-dimensional, ball-shaped molecule with a rigid binding cavity.

Stereochemistry Aspects of chemistry that involve three-dimensional considerations in the spatial arrangement of atoms.

Supramolecular chemistry This is chemistry *beyond* the molecule and relates to the structure, properties and characteristics of assemblies of two or more molecules held together by weak, non-covalent forces.

Total synthesis The science of constructing target molecules from simple fragments. Most often this refers to the synthesis of an organic molecule (which may be a natural product) from readily available starting materials.

Biographical details

Professor Sue Gibson is Daniell Professor of Chemistry at King's College London. Her research interests involve many aspects of organometallic chemistry including the use of palladium catalysts to make unusual amino acids, control of stereochemistry using chromium complexes of aromatic compounds, and exploitation of the alkene metathesis reaction in organic synthesis.

Professor Karl Hale is currently Professor of Chemistry at University College London. His research interests are in complex, pharmacologically-active, natural product total synthesis, and in the design and development of new asymmetric organic reactions. His group is also developing a research presence in chemical biology.

Professor David Leigh is an EPSRC Advanced Research Fellow and Professor of Synthetic Chemistry at the Centre for Supramolecular and Macromolecular Chemistry in the Department of Chemistry at the University of Warwick. His research interests cross the boundaries of biology and materials science, and focus on the use of novel supramolecular architectures in synthesis, medicine, materials and 'molecular machines'.

Dr Nick Terrett is manager of Lead Discovery Technologies at Pfizer in Sandwich, Kent. The department includes an active combinatorial chemistry team. He has worked on a range of different projects and is a co-inventor of the anti-impotence drug Viagra.

Professor Jonathan Williams is Professor of Organic Chemistry at the University of Bath. His research interests include the use of enzymes and transition metals in asymmetric synthesis.

Acknowledgements

Professor Sue Gibson wishes to thank Stef Biagini, Mark Peplow, Ellian Rahimian, Miguel Gama Goicochea, Alex Comely, Jerome Jones and Nicole Whitcombe for helpful and stimulating discussions.

Special thanks are due to Professor Nigel Simpkins for overseeing the compilation of this chapter.

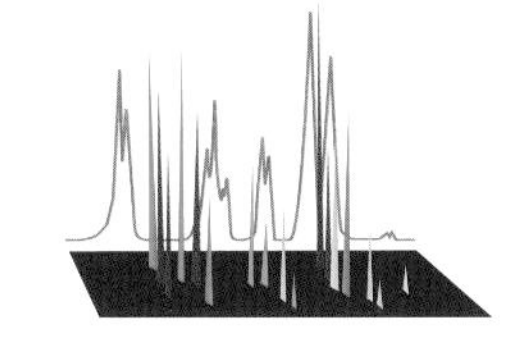

Analysis and structure of molecules

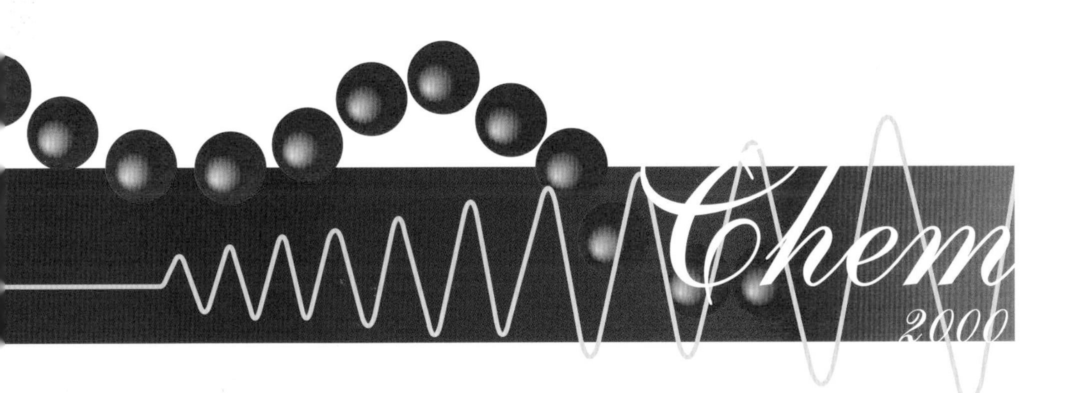

Analysis plays a vital part in our lives – all the products we use rely on quality control in their production. Nuclear magnetic resonance spectroscopy is an excellent source of information on the dynamic properties of molecules. The future applications of this and other techniques are endless.

Courtesy of the Laboratory of the Government Chemist.

Dr Melinda Duer
University of Cambridge
Dr Katherine Stott
University of Cambridge

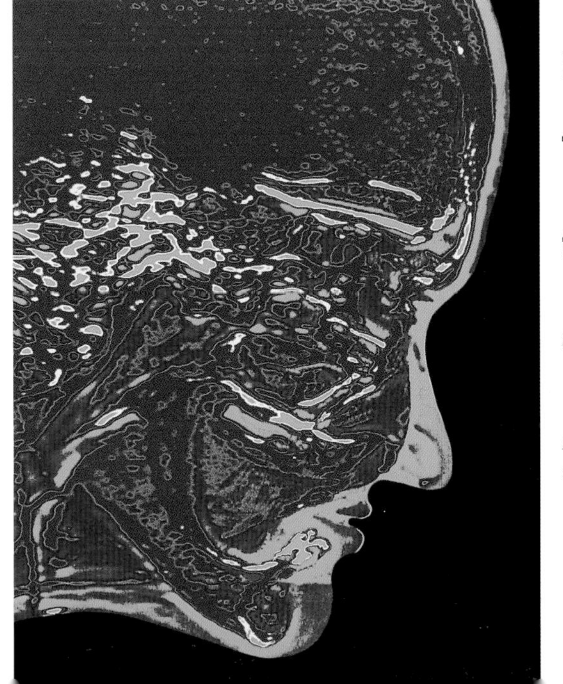

Mehau Kulyk/Science Photo Library.

Over the past century, chemists have accumulated a vast armoury of ingenious analytical methods to determine the structure of the chemical substances they produce. One of the most widely applicable is nuclear magnetic resonance.

Analysts undertaking analytical work on DNA.
Courtesy of the Laboratory of the Government Chemist.

Courtesy of West Midlands Police.

A portable gas chromatograph for environmental and occupational health and safety monitoring.
Courtesy of The Perkin Elmer Corporation.

Pick up any newspaper and it's bound to contain a news story in which analytical chemistry plays a part: a celebrity is accused of drink-driving or an athlete of taking drugs; toxic metals are found in a river, or terrorists have planted a bomb. The outcome of each case will involve identifying and measuring traces of a chemical – alcohol, a drug, a metal or an explosive. Mostly, the role of analysis is less obvious. When a new wonder drug is announced, its discovery and development will have depended on a whole battery of analytical tools. Sometimes the analytical requirement is exotic: how much iron is there in Moon dust?

Over recent decades, chemical analysis has become increasingly sophisticated, taking advantage of computer technology and robotics. It is now possible to detect a minute quantity of a material, which can pose problems of interpretation on the safety front. You may remember the scare over benzene in Perrier water in 1990. This arose only because a new analytical technique was so sensitive that it could detect minute amounts of the benzene (a carcinogen) – down to one part per billion! The amounts detected were between 7 and 20 parts per billion. There is now no detectable quantity of benzene in Perrier water.

Perrier water production halted

By Alan Friedman in New York

PERRIER, the French mineral water company, has halted production, and North American sales have been stopped, because of suspected contamination.

Perrier Group of America, Perrier's American importer, around $40m in lost sales, said Mr Ronald Davis, president of Perrier Group of America.

US sales of Perrier's regular and flavoured waters are $160m a year. Mr Davis said on Saturday that the search for the cause of the "chemical month, appeared to be "human error". He described it as "a very freak accident."

Tests of Perrier bottles in North Carolina and Georgia found the benzene level to be 12.3 to 19.9 parts per billion, well above the FDA's permissi-

Courtesy of Financial Times.

Chemical analysis plays an enormous part in our lives – all the products that we use – food, household products, materials of all kinds rely on quality control usually involving some kind of analysis. In industry, analysis is so important that most industrial chemists are in fact analytical chemists. All professional chemists whether working in industry or academia routinely do analysis.

What sort of analysis do chemists do? Before describing some of the amazing modern analytical methods, it is worth commenting on a little history.

Classical analytical chemistry

Analysis has traditionally depended on carrying out a chemical reaction on a substance – a test reaction – and observing a result. The simplest one that everybody knows is the litmus test for distinguishing between acids and bases. Older readers who were chemistry students may recall being required to analyse (qualitatively) a complicated mixture of compounds by the arduous and difficult *group analysis* scheme used to identify metal compounds. This was developed mostly by Carl Fresenius as early as 1840. He divided metals into six groups, each of which was distinguished by the fact that they could be precipitated by a particular reagent. The system includes, for example, a test in which certain metal salts give a characteristic precipitate when hydrogen sulfide is passed through an acid solution – cadmium sulfide, for example, is bright yellow.

Quantitative analysis depended on either *gravimetric* or *titrimetric* methods. One of the best ways of quantifying the nickel content of a mixture is to precipitate the nickel as the dimethylglyoxime salt and weigh the precipitate. Indicators such as phenolphthalein (discovered in 1877) and methyl orange (discovered in 1878) change colour when a solution changes from acid to alkaline. This led to *acid-base titrations*, whereby the concentration of an acid or base could be calculated from the relative amount of acid dripped onto a known amount of base needed to reach the 'neutral endpoint' as shown by the indicator. It is worth commenting that progress in quantitative analysis went hand-in-glove with the development of new apparatus such as reliable weighing balances and calibrated glassware such as burettes and pipettes.

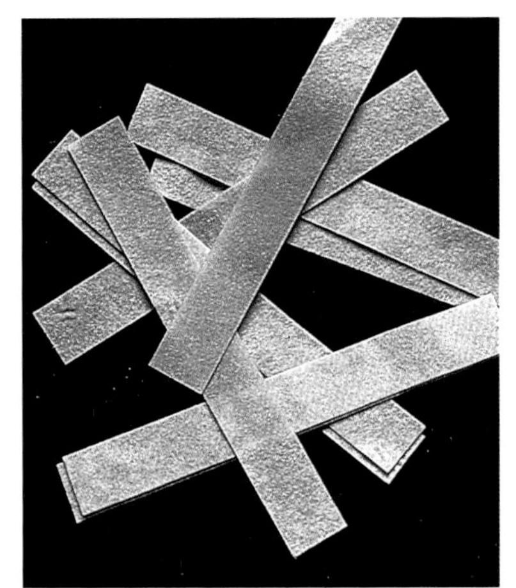

Litmus paper turns blue with bases and red with acids.
Charles D Winters/Science Photo Library

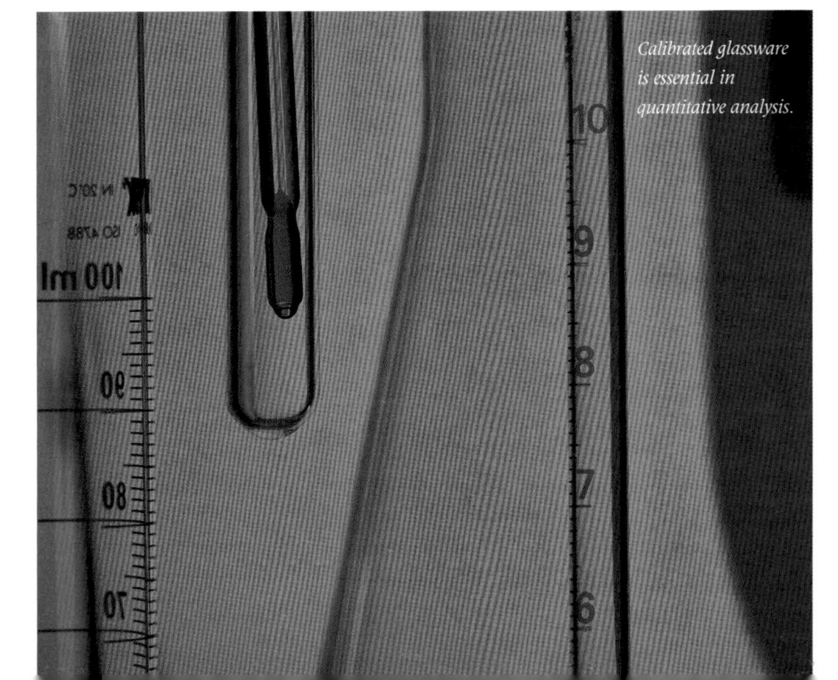

Calibrated glassware is essential in quantitative analysis.

During the 20th century, there have, of course, been huge advances in analytical techniques. Although much traditional 'wet' chemistry has been supplanted by physical methods, as we shall see later, careful quantitative analysis is still relevant in chemical research. To illustrate this, here is a modern story. It concerns the first 'high-temperature' superconductors which were discovered in 1986. These ceramic oxide materials were found to conduct electricity without any resistance at higher temperatures than any previous superconductor (see *New science from new materials*). Soon afterwards a superconducting temperature of 93K (or – 180 °C) was achieved with a compound made of yttrium, barium, copper and an unknown amount of oxygen with

Traditional titration carried out at school.
Tony Stone Images.

the general formula $YBa_2Cu_3O_x$. The value of x depends on the oxidation state of copper, particularly whether or not there is any Cu(III). The best way of determining this turns out to be *classical iodometry*, introduced by Robert Bunsen (of burner fame) in 1853. In a very simplified description, copper (either in oxidation state II or III) oxidises colourless iodide ions (I^-) to coloured iodine (I_2), while itself being reduced to copper in oxidation state I; the amount of iodine produced can be determined by its reaction with sodium thiosulphate. Thus, knowing the concentration of sodium thiosulphate leads to the oxidation state of copper in the original material, in other words, the value of x. The other method giving comparable accuracy is a traditional thermogravimetric analysis. The $YBa_2Cu_3O_x$ sample is heated under flowing hydrogen gas which reduces it to a mixture of yttrium oxide (Y_2O_3), barium oxide (BaO) and copper metal; x is found from the percentage mass loss.

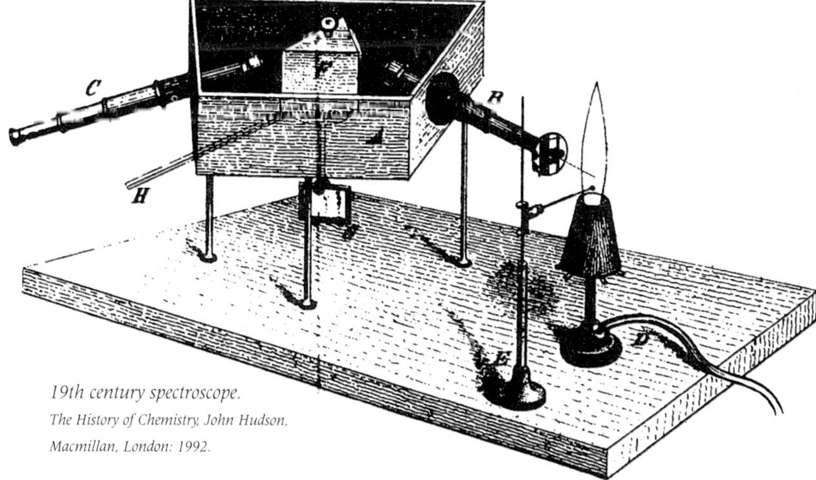

19th century spectroscope.
The History of Chemistry, John Hudson, Macmillan, London: 1992.

It is also worth pointing out that combustion analysis, originally introduced in the 19th century by Justus von Liebig, Jean-Baptiste-André Dumas and Johan Kjeldahl, and which determines the relative amounts of carbon, hydrogen, nitrogen and sulfur in a compound, is still a vital part of the analysis of organic compounds in, for example, the food industry.

The rise of physical methods

The most spectacular advances in analytical science, however, have followed the discovery of some new physical phenomenon – usually directly related to the electronic structure of atoms and molecules. An example with a long history is *atomic spectroscopy*. In the 19th century, scientists realised that when some compounds are heated in flames, they emit light with a colour characteristic of certain constituent elements. For instance, compounds containing sodium usually show a bright yellow flame (this is the origin of the sodium lamps used in street lighting). If the emitted light is then passed through a spectroscope, it is split into bands of colours – the spectrum. The bands, or lines, are extremely narrow and can be used to identify the elements. This atomic spectroscopy enormously enhanced analytical chemistry. At first, the method was not very quantitative but, today, there are several variants of the technique which when combined with modern technology give incredibly sensitive and precise results. For instance, when the source of light is an inductively coupled plasma

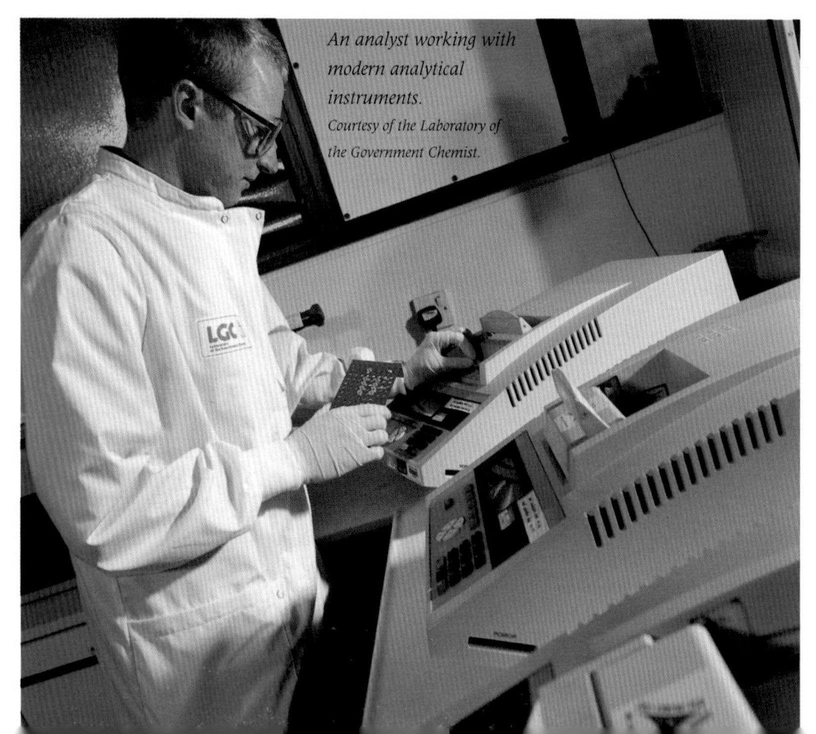

An analyst working with modern analytical instruments.
Courtesy of the Laboratory of the Government Chemist.

Box 1 Chromatography

Spill fruit juice on a white tee-shirt and you can often see chromatography at work. As the stain spreads it is clear that the original colour separates into different circles. In fact it was first noted by Friedrich Runge in the 19th century that if solutions of mixtures of coloured compounds are dropped onto a filter paper then concentric rings of the different colours form. The botanist Mikhail Semenovich Tswett coined the term 'chromatography' (coloured writing) in 1906. He passed a solution of plant pigments down a glass column packed with calcium carbonate particles, and then washed the column with solvent. Different coloured bands appeared as the components of the mixture were separated. As the washing continued the different coloured bands travelled separately down the column and out into the receiving glass vessel. Thus the components of the original mixture were separated.

Chromatography depends on the presence of a *stationary phase* and a *mobile phase*. In Tswett's experiments, the stationary phase was calcium carbonate and the mobile phase was the solution moving down the column. Suppose the mixture has only two components X and Y. If X adsorbs more strongly on calcium carbonate before being washed off, then it will pass more slowly through the column than Y. A powerful development of this technique is known as *high-performance liquid chromatography (HPLC)*. If the particles in the column are very small then separation can be very efficient, but the rate at which the solution passes through is very slow. However the solution can be forced through by applying high pressure. This results in extremely efficient separation.

Chromatography is not confined to solid stationary phase and liquid mobile phase. One of the major advances in chromatography was devised by two British chemists Archer Martin and Richard Synge who received the 1952 Nobel Prize for their work. This is *partition chromatography*, in which the mobile phase is once again a liquid, but the stationary phase is also a liquid which is held as a thin film on a solid support. The dissolved compounds partition to different extents between the solution and the liquid film. An important application of this, particularly to the separation of amino acids, was to use wet filter paper as the stationary phase.

Archer Martin.
©The Nobel Foundation.

Richard Synge.
©The Nobel Foundation.

Archer Martin and Richard Synge developed the technique of partition chromatography.

A very important type of chromatography uses gases as the mobile phase and either a liquid or a solid as the stationary phase in a long tube. *Gas chromatography* is often used in industry, for example, to analyse mixtures of volatile organic compounds such as perfumes and flavours in cosmetics and foods, and to monitor air pollution. To identify the components once separated, such chromatographical methods are often directly combined with an analytical technique such as mass spectrometry. GC-MS (gas chromatography combined with mass spectrometry) and LC-MS (liquid chromatography-mass spectrometry) equipment are to be found routinely in many R&D laboratories in the fine chemicals and consumer products sector.

(*ICP*), rather than a simple flame, it is possible to detect 0.08 millionth of a gram of magnesium in one litre of solution.

Very small amounts of substances can also be measured using electrochemistry (see *The power of electrochemistry*) – where an applied voltage causes an ion in solution to move to an electrode and undergo a chemical reaction. One clever use involves the *anion-selective electrode* where the voltage at an electrode is affected by one species of ion, therefore furnishing a direct and easy method of measuring its concentration. The crucial part of an ion-selective electrode is a membrane which permits diffusion of only one species. For example, the concentration of fluoride ions can be measured at billionths of a gram by employing a solid membrane of lanthanum fluoride (LaF_3) doped with europium fluoride (EuF_2) which lets through only fluoride ions. Another striking example is the glucose sensor, based on enzyme reactions, described in *The power of electrochemistry*. A more common system is the glass electrode which is universally used to measure the pH of solutions.

It is sometimes possible to analyse one compound in a mixture of many others. In this case analysis can be very easy. More often than not, it is impossible because the signal from the compound of interest is obscured by the signals from the other compounds; in this case it is essential to separate the target compound from the others before doing the analysis. There are several ways of doing this, but the most powerful method is *chromatography*, which depends on relative differences in diffusion of compounds through a given medium (Box 1).

The structure of compounds

The examples of analysis given so far offer no information about chemical structure, either in the simple sense of chemical groups, for example, in ethyl alcohol – CH_3CH_2OH, or in the sense of the three-dimensional atomic arrangement. A glance at almost any chapter in this book will reveal the vital importance to chemistry of knowing the molecular structure of either the simplest molecules containing a few atoms to the most intricate biomolecular systems containing thousands of atoms.

A chemist using a gas chromatography (gc) smelling machine (the gc set up is split so that output from the column is split - some gas goes to the detector to give the trace and the remaining gas output can be smelt by the chemist). Courtesy of Quest International.

Chemists have a vast armoury of methods to determine the structure of the chemical substances they produce. Most methods involve probing the molecule through some kind of energetic interaction which reveals the molecule's characteristics. The probe might be electromagnetic radiation or particles such as electrons or neutrons, and they may measure tiny variations in very subtle physical or chemical effects. Today there are dozens of different analytical techniques, each of which may offer a certain kind of information, or be aimed at specific kinds of molecules or particular types of structures and phases (such as crystals or surfaces). Analytical techniques used in catalysis, for example are given in *Chemical marriage-brokers*, and those used to study fast reactions are given in *Following chemical reactions*. Here, we mention a few of the main methods.

X-ray, neutron and electron diffraction

Perhaps the most straightforward way of examining the structure of a compound is to shine light on it and then examine the light scattered from it. (This is how our eyes see the shapes and surfaces of things.) But to discern structure at the level of molecular bonds, the radiation wavelength must be of comparable dimensions to interatomic distances, and that falls within the regime of X-rays. The simplest interaction of X-rays is with compounds in a crystalline form – through the phenomenon of *diffraction*. A crystal consists of repeating arrangements of molecules which can be viewed as regularly oriented planes of similar atoms. When X-ray waves hit the sample, they are reflected off consecutive planes in such a way that the waves interfere and reinforce each other periodically to produce a *diffraction pattern*. The pattern contains clues to the arrangement of atoms in the crystal given by a famous law discovered in 1912 by William and Lawrence Bragg. *Bragg's Law* shows the angle of the incident radiation (and thus the angle of diffraction) at a given wavelength is inversely related to the distance d between the layers of atoms: $\lambda = 2d\sin\vartheta$. In other words, the reinforcement occurs only if $2d\sin\vartheta$ is satisfied.

The European Synchrotron Radiation Facility in Grenoble produces intense beams of X-rays suitable for X-ray crystallography on complex molecules like proteins.
Courtesy of ARTECHNIQUE

The diffraction patterns are recorded as a series of peaks in X-ray intensity. The crystal is rotated through all angles to obtain diffraction patterns for all planes, and the patterns are then analysed. Because X-rays are a form of electromagnetic radiation they are interacting with the clouds of electrons in each atom so what we actually have is a record of the distribution of electron density in the compound. The conversion of the scattering data to structure is actually extremely complicated and involves a great deal of computation. However, increases in computing power over recent decades and new software techniques have greatly speeded up X-ray analysis such that quite complex structures can now be solved relatively quickly.

X-ray diffraction, or *X-ray crystallography*, provides the most complete structural picture of the molecule. Most university research laboratories have at least one X-ray instrument. X-ray analysis is routinely used particularly in organometallic and coordination chemistry (the study of compounds in which metals are bound or 'coordinated' to groups of atoms) to determine the structure of a newly synthesised compound. This requires a single crystal (although for some compounds it is possible to carry out X-ray diffraction analysis on powders).

X-ray diffraction pattern of lysozyme.
Phillippe Plailly/Eurelios/Science Photo Library.

X-ray diffraction has become a particularly powerful tool in molecular biology (see *The chemistry of life*). It is used widely and routinely to solve very large three-dimensional structures containing thousands of atoms such as proteins, RNA, DNA and other complex polymers – which is vital for our understanding of how these important biological molecules function. Even the molecular structures of very large assemblies such as viruses have been analysed using a particularly bright source of X-rays called *synchrotron radiation*. This is produced in large ring-shaped machines where electrons accelerated close to the speed of light circulate. The electrons lose energy by emitting extremely intense electromagnetic radiation at all wavelengths. Very narrow, coherent beams of X-rays of precise wavelength can be siphoned off at 'beam line' stations around the ring and used for all kinds of chemical analysis including high-resolution X-ray crystallography. UK chemists have access to two main synchrotron sources in Europe – the Synchrotron Radiation Source (SRS) at Daresbury in Cheshire and the very powerful European Synchrotron Radiation Facility (ESRF) in Grenoble, France.

Another related type of structural analysis is *neutron diffraction*, or neutron scattering, which works on the same principles of Bragg's law. Although neutrons are subatomic particles (found in the atomic nucleus), according to quantum mechanics, they also behave like waves, and their wavelength corresponds to interatomic and intermolecular distances. Neutrons being electrically neutral pass through the electron clouds of the atom and interact with the nucleus, which means they provide slightly different information about atomic positions. Although neutrons generally do not achieve the resolution of X-rays, they can see hydrogen atoms (protons), which usually do not offer enough electron density to be recorded via X-ray diffraction. What is more, it is possible to distinguish between selected hydrogen positions in a molecule by substituting with the hydrogen isotope deuterium, because the two isotopes scatter neutrons very differently. Neutrons are particularly useful for studying structures where water or intermolecular interactions play a role. Their wavelengths and energies are also ideal for studying the larger-scale structure and dynamics of polymers (long chain molecules made of similar units). Because neutron beams are produced from high-energy sources – nuclear reactors or particle accelerators – and neutron diffraction requires very large crystals, the technique is less used by chemists than X-rays.

Electrons also behave as waves which are diffracted by matter. If electrons are accelerated through a voltage of 50 000 volts, then their wavelength is about right for probing interatomic distances. Since electrons are charged particles, they cannot penetrate far into solids or liquids – although the technique of *electron diffraction* can be very important for probing surfaces. On the other hand, diffraction of a beam of electrons all at the same energy by a gas can give very detailed structural information on the gas. Each atom in the molecule scatters the incoming electron beam; the scattering from neighbouring atoms interfere with each other and so the end result is a complicated interference pattern. This can then be converted into a so-called 'radial distribution curve' which gives important interatomic distances. An example is shown in Figure 1.

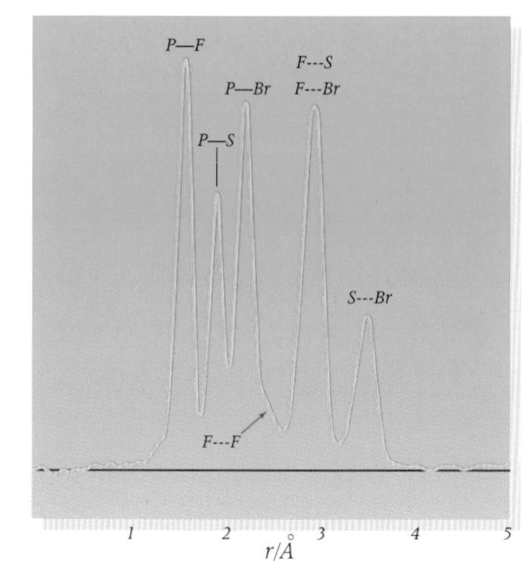

Figure 1. A radial distribution curve for $PBrF_2S$, where the bromine, fluorine and sulfur atoms are all bonded to the phosphorus atom.
Adapted from Structural Methods in Inorganic Chemistry, E. A. V. Ebsworth, D. W. H. Rankin and S. Cradock, Blackwell, Oxford, 1991.

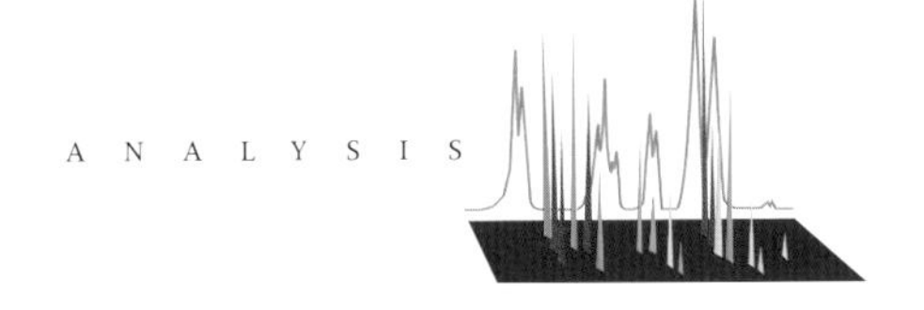

Box 2 Fourier transform – a useful mathematical trick

Analytical scientists are always trying to do two things: detect smaller and smaller amounts of material, and make measurements quicker and quicker. One of the most important advances enabling them to achieve this goal – and which became practical only with powerful computing – is a technique originating in the ideas of an 18th-century French mathematician, Jean-Baptiste Fourier, and later the PhD thesis in 1949 of a British physicist, Peter Fellgett. The technique is called Fourier transform. It can be used to analyse a complex oscillating signal into a series of simple sine and cosine waves (such as those associated with single wavelengths of light), a technique useful in many research areas. The beauty of Fourier transform in analysis is that it can be used with the aid of a computer to untangle individual signals in a spectrum from a jumble of data collected simultaneously, considerably speeding up the analysis and improving sensitivity.

Fourier analysis has now been applied very successfully to a wide range of analytical techniques, including infrared spectroscopy, mass spectrometry, microwave spectroscopy, and nuclear magnetic resonance (to be described later). As explained in the main text, these techniques rely on detecting some kind of signal that changes across a range of energies such as the electromagnetic spectrum. To show how Fourier transformation works, let's consider Figure a which shows a typical spectrum, in which the intensity of the signal is plotted (in this case) against frequency ν, where frequency is the velocity of light c divided by its wavelength, $\nu = c/\lambda$.

This example shows two narrow peaks corresponding, for instance, to blue and violet in the visible part of the spectrum. We can imagine the spectrum being recorded as the instrument scans over the whole range of frequency from the ultraviolet to the infrared. Most of the time, however, the instrument is not recording anything interesting, only background noise, thus wasting time. Clearly, it would be an advantage if all the information in the spectrum could be obtained at once.

It was Fellgett who first showed a practical way of 'interrogating' all wavelengths simultaneously, and it is known, therefore, as the 'Fellgett Advantage'. To see the principle we need to look at the information in the spectrum in another way. We know that light is a wave motion so if instead of plotting signal against wavelength, or frequency in our example, we record signal versus time, we would see two wave motions (actually sine waves), one corresponding to the blue light and one corresponding to the violet light. This is shown in Figure b. If we can record the individual wave motions it is a simple matter to redraw the spectrum versus wavelength. Because we are continuously collecting information about the two signals we are not wasting any time.

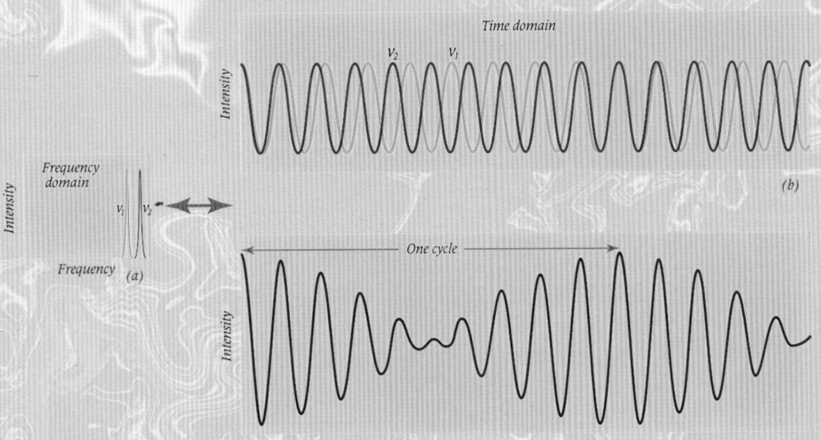

(a) Two frequencies in an electromagnetic spectrum at v1 and v2.
(b) The same two signals plotted as sine waves of intensity versus time
(c) The actual signal at the detector resulting from interference between the two waves.
Adapted from Principles of Instrumental Analysis, D. A. Skoog and D. M. West, 2nd edn, Saunders College, 1980.

Everyone is familiar with the coloured patterns of thin oil films on water. These result from interference between light waves (waves with crests and troughs in phase reinforce each other while those out of phase cancel out.) Thus in our example, the actual signal with time is the result of interference between the two wave motions as shown in Figure c. Thus the signal at the detector is rather complicated and it is not easy to extract the two individual waves from the final pattern. Clearly, if the spectrum contains many more individual signals the interference pattern is even more complicated.

Nevertheless, by applying Fourier's mathematical method, *Fourier analysis*, the complicated interferogram can be 'transformed' into a conventional spectrum. It requires special computer programs and very fast computers. We can thus collect information in the 'time domain' and convert it via Fourier Transform to the 'frequency domain' and make use of the Fellgett Advantage. We can take advantage of this Fourier Transform technique in two ways: first, the same spectral information can be collected in a much shorter period of time – one target of analytical scientists; secondly, the longer the time spent collecting the interferogram, the more accurate the information. Put differently, we can detect ever smaller signals against the background noise.

Light spectroscopy

We have mentioned earlier that atomic spectroscopy is a powerful analytical tool. It depends on electrons jumping from one energy level in the atom to another and emitting or absorbing a characteristic frequency of light in the process; some of these so-called transitions occur in the visible region of the spectrum and hence are easily seen. However *light spectroscopy* is much richer than this. In the first place, molecules can interact with light over the whole spectral range from the far ultraviolet to the far infrared and microwave regions. Secondly, the spectral pattern of the interaction can provide quite detailed information about the structure of the molecule.

In the ultraviolet and visible regions of the spectrum, the interaction is with the electronic energy levels of molecules (as for atomic spectroscopy). This can provide a ready means of identification (for example, in the case of transition metal coordination compounds and some organic compounds) and quantitative analysis. *Ultraviolet/visible spectroscopy* can be very sensitive if the 'absorbance' is large (for instance, the complex formed between iron and *o*-phenanthroline is highly coloured and provides a good method of estimating low concentrations of iron in natural waters). The ultraviolet spectrum of organic compounds can also

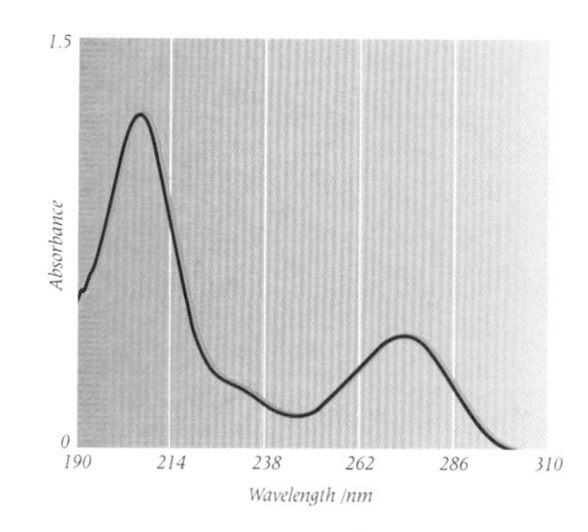

Figure 2. UV spectrum of caffeine.

show the presence of particular chemical groups, such as those containing double bonds. Much more sophisticated use of ultraviolet/visible spectroscopy can be applied to metal coordination compounds, where the position and intensity of the spectral bands can provide a great deal of information about the electronic structure of the molecule.

It is in the infrared region of the spectrum where most molecular information about chemical bonds and groups can be obtained. Molecules are constantly very rapidly stretching and shaking, and the energy levels corresponding to these motions occur mostly in the infrared region of the spectrum, from 4000 to 200 cm^{-1} (the common unit of the measurement of wavelength in the infrared region is the wave number – the number of waves per centimetre, cm^{-1}). *Infrared spectroscopy* probes these motions. Since the positions of particular bands in the infrared depends on the 'springiness' of individual bonds and on the masses of the appropriate atoms, the infrared spectrum can provide a 'fingerprint' of the various chemical groups in the molecule. This is very useful in organic chemistry, particularly if the infrared bands are very intense. For instance, an organic carbonyl (C=O) group gives a very intense band at about 1700 cm^{-1}.

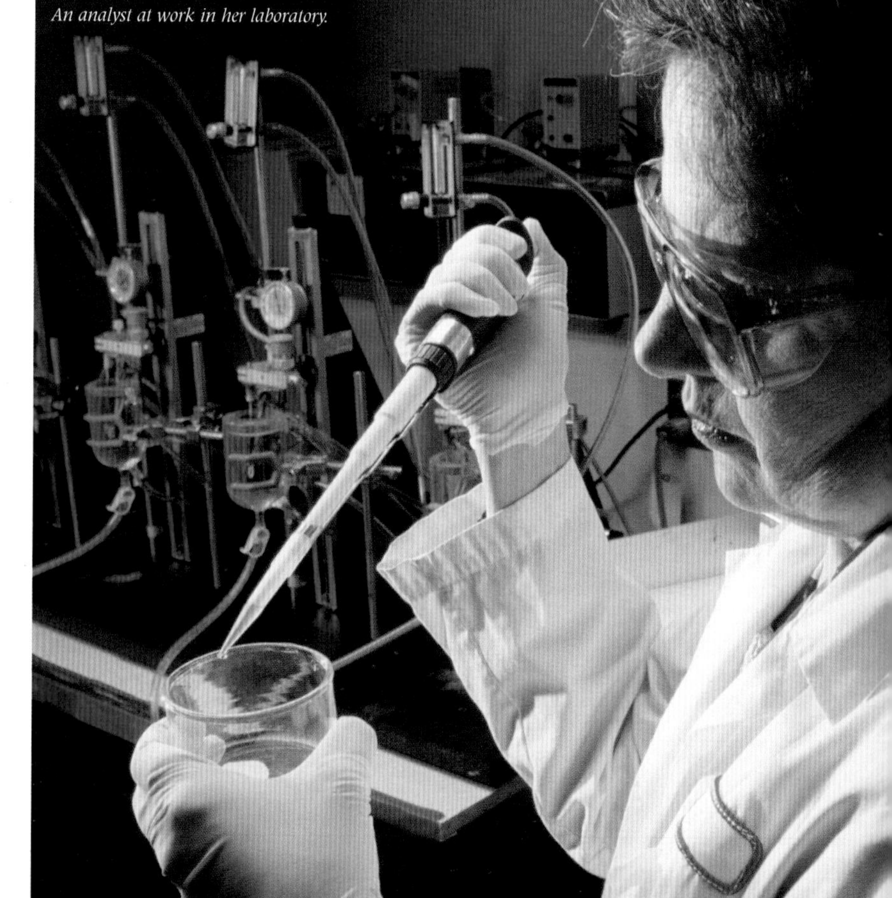

An analyst at work in her laboratory.

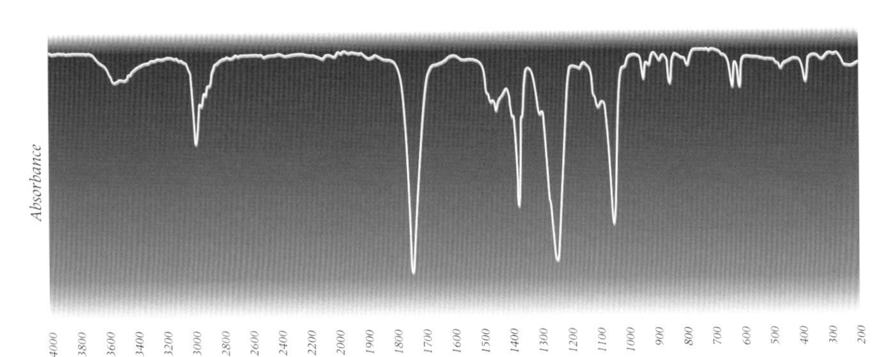

Figure 3. An infrared spectrum of ethyl acetate showing a C=O stretch between 1700-1800 cm^{-1}.

This type of analysis was very important in the Second World War, because it proved the best way of deducing the components of aircraft fuels used by both friends and foes. However, it was an extremely tedious and unreliable technique. The situation was revolutionised with the development of *Fourier transform* infrared spectrometers. The technique of Fourier analysis is so important in many areas of analytical science (Box 2).

In fact the situation in infrared spectroscopy is somewhat more complicated for two reasons: the infrared bands arise from vibrations of the whole molecule not just from isolated bits, and not all vibrations necessarily appear in the spectrum. These complications can be turned to advantage and provide subtle information about the symmetry of molecules. This is most strikingly demonstrated in the infrared spectra of transition metal complexes which contain carbonyl (CO) groups directly attached to the metal (some applications of this are given in *Following chemical reactions*). Such carbonyl groups give rise to very intense infrared bands, at around 2000 cm^{-1}. Suppose we have a compound, known from chemical analysis to be $L_4M(CO)_2$, where M is the transition metal, and L represents other groups attached to the metal (L for instance might be triphenyl phosphine, $P(C_6H_5)_3$). The question is whether the two carbonyl groups are positioned relative to one another at 90 degrees (the *cis* position) or at 180 degrees (the *trans* position). If they are *cis* there will be two infrared bands, but only one if they are *trans*. Such structural deductions have been extremely important in the development of organometallic chemistry. Although valuable for qualitative and structural analysis, infrared spectroscopy is seldom used for quantitative work, except in specialised areas, partly because it is difficult to get very accurate measurements of intensity.

An analyst working with modern analytical instruments.
Courtesy of the Laboratory of the Government Chemist.

As well as vibrating, molecules also rotate – especially in liquids and gases where the molecules can readily tumble around (although molecules can sometimes rotate in solids as well). This leads to energy levels associated with the rotation. In a liquid, the rotational energy levels are 'smeared out', resulting in broad absorption bands which all run into each other. In a gas, however, the levels are well-defined and separated by energies that depend on the moment of inertia of the molecule. This corresponds to light in the microwave part of the spectrum – hence *microwave spectroscopy*. Except under special circumstances (see *Epilogue*), the molecule must have a dipole moment to give a microwave spectrum, which means an asymmetrical distribution of electric charge is needed. However, where there is a dipole moment, microwave spectra can provide very accurate values for moments of inertia, bond lengths and, therefore, structure. An interesting application was to the molecule O_2F_2, which has an overall shape rather like hydrogen peroxide (H_2O_2, H-O-O-H). Microwave spectroscopy carried out by Robert Jackson at Harvard showed that the oxygen-oxygen bond in the fluorine compound is very short, a feature that has provided theoretical chemists with endless opportunities for speculation!

Mass spectrometry

Another technique that is regularly used by chemists is *mass spectrometry*. Here, the molecules in the sample are vaporised and broken into charged fragments (ions) with an electron beam, for example, then accelerated and separated by electric and magnetic fields before reaching a suitable detector. The ratio of the charge to the mass of the fragments is then recorded. This allows the mass of the molecule to be accurately determined as well as the nature of the atoms in the fragments. The pattern of fragmentation also depends on the most favourable energetic routes for decomposition and is again indicative of the structural components of the molecule.

Mass spectrometry is a technique that is rapidly growing in popularity because of improved methods of vaporising molecules. Until recently, it could not be applied to large molecules such as proteins because they would decompose, losing their identity completely. Now, however, there are two new methods for getting proteins into a sort of 'gaseous' state. One called *electrospray* involves spraying a solution of the protein through a syringe equipped with a needle at high electric potential. The electric field breaks up the liquid into a fog of charged droplets which then eject protein molecules carrying many charges. The resulting 'protein ions' can then be analysed in the mass spectrometer. In the other method, *laser desorption*, the sample protein is mixed with a matrix of smaller molecules such as urea. Then a laser pulse is fired with the right energy and wavelength to vaporise the matrix material without destroying the protein. The matrix and protein, which picks up charge from the matrix, are then propelled towards the detector. The mass of the charged protein is obtained from an accurate measurement of the time required for the protein to travel to the detector. Once the mass of the protein has been measured, it can be broken into fragments in a controlled way and the mass spectra measured. Substituting some of the atoms with their isotopes gives information about the amino acid sequence of the protein and its three-dimensional structure. This technique can be used on very small amounts of large biological molecules and has great promise for the future.

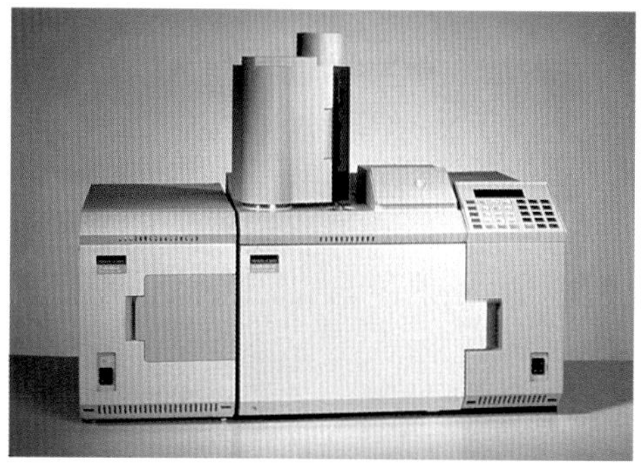

The TurboMass™ Mass Spectrometer can perform routine and complex analyses. Courtesy of The Perkin Elmer Corporation.

Nuclear magnetic resonance

Perhaps the chemist's favourite structural analytical tool is *nuclear magnetic resonance* (NMR) spectroscopy, and it is on this technique that we are going to concentrate for the rest of the chapter. It's a surprising method that depends on an unusual combination of nuclear physics

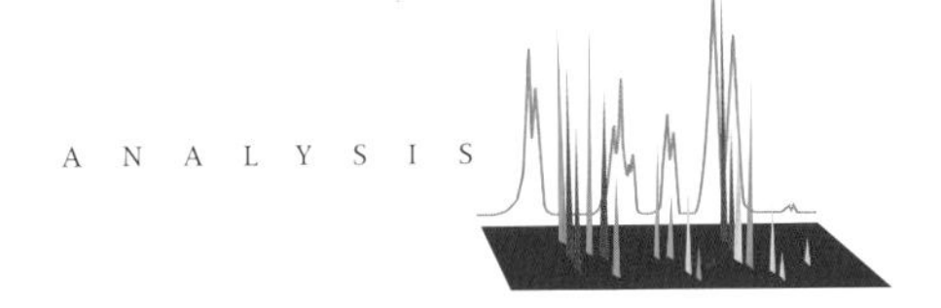

Box 3 The pioneers of NMR

The idea of nuclear magnetic resonance was first mooted by C. Gorter, a leading Dutch physicist. However, despite several attempts, he did not succeed in observing the phenomenon that he had predicted. Then in September 1937, he visited Rabi at Columbia University. Gorter thought that it would be easier to observe NMR from a gaseous beam of molecules, and Rabi was an expert on just this way of handling molecules. Rabi and his colleagues began work two days after Gorter left, and in January 1938 published a now famous research paper which describes the first observation of NMR: on lithium chloride molecules in a molecular beam. They acknowledged Gorter's original idea in their paper, but it was Rabi who was awarded the Nobel Prize in 1944 for the experiment.

Rabi and Gorter were both looking for a method of measuring magnetic moments of nuclei. The Second World War put the work on ice, as it was not fundamental to the war effort. Rabi moved to Massachusetts Institute of Technology (MIT) where he became the associate director of the Radiation Laboratory there, developing microwave sources for use in radar systems. Two other young physicists, Felix Bloch and Edward Purcell, who were later to make their mark in NMR, also spent the war years in the US working on radar. Although the War held up direct progress on NMR, the experience these physicists gained in microwave technology was to have a very positive impact on their development of NMR later.

The success of Rabi's initial experiments really enthused Gorter, and despite being in the middle of war-torn Europe, he tried again to observe NMR in bulk materials and again failed. Unfortunately for Gorter, it was to be another physicist who eventually chalked up this important success, and some 10 years after Gorter first started musing about it. In 1945, Bloch returned to his research laboratory at Stanford and began work on NMR. He had had an idea about how NMR could be observed in solids while sitting in a concert a year before, and he now teamed up with William Hansen, an electronics expert (also at Stanford), and Martin Packard, an enthusiastic graduate student, to put the idea into practice.

Meanwhile, Purcell was employed by MIT at the end of the War to write up the wartime research notes on the radar work that had taken place there. In September 1945, over lunch one day with his colleagues Henry Torrey and Robert Pound, Purcell raised the question of detecting NMR in solids. All three were employed writing books and were somewhat bored. The prospect of doing some real experiments was exciting. The experiments had to be done in their spare time – evenings and Sundays – since the main effort had to be on writing up the wartime notes. It took some time for the three to collect together and build the necessary apparatus, but finally on the evening of 13 December 1945, they attempted their first observation of NMR from a solid – a kilogram lump of paraffin wax. They observed nothing. They tried again the next Saturday. Still nothing. They were ready to give up, not least because it was very cold in the otherwise unused laboratory where they had set up their equipment. Just as they were about to leave the laboratory on that Saturday, one of them suggested turning up the current in the electromagnet. Now, finally, they saw what they had been looking for: the first observation of NMR from a solid.

Stanford University News Service.

Harvard University News Service.

Felix Bloch (left) and Edward Purcell (above) were the first to observe a nuclear magnetic resonance signal from a solid.

Shortly after, in January 1946, Bloch, Hansen and Packard observed an NMR signal from water using very different apparatus, though however, set up in an equally cold laboratory. They might have beaten Purcell's group to the observation, if Packard had not insisted upon visiting his parents in Oregon in December! Fortunately, the Nobel Prize committee recognised Purcell and Bloch's discoveries as being independent and simultaneous when they awarded the 1952 prize for physics to them jointly.

It was not until 1950, when NMR measurements were refined, that Warren Proctor and Fu Chun Yu at Stanford discovered that the compound ammonium nitrate (NH_4NO_3) gave two separate signals for the nitrogen nuclei, not one. The effect, attributed to the orbiting electrons, spelt the end of NMR for measuring nuclear magnetic moments and the start of its applications in chemistry.

Geoff Tompkinson/Science Photo Library

and chemistry, and as such could have been developed independently by either chemists or physicists; such is the way that science progresses. NMR depends on the magnetic properties of certain nuclei and the subtle influence of their chemical environment. Since the 1960s it has been used first by organic chemists to help solve the structure of relatively simple molecules, then as the methodology developed NMR has been applied to ever more complex structures including large biological molecules, complex solids and living tissue. Today it can be used to deduce the network of bonds in a molecule, identify atoms close in space, and probe the complicated way in which the molecule moves and flexes over time. In addition, NMR is not limited to a single state of matter, but is used on solids, liquid crystals, liquids and solutions. For all these reasons, NMR occupies a unique position in analytical chemistry.

The story of NMR begins in 1938, in a nuclear physics laboratory at Columbia University, New York (see Box 3). The threat of war was driving nuclear physics research at a fast pace. Physicists already knew that some nuclei have magnetic properties, and at this time the race was on to find a method for measuring the magnetic moment of a nucleus – the strength of its magnetic field. Isidor Rabi and his colleagues at Columbia showed that if a nucleus held in a strong magnetic field is bombarded with radio waves, the nucleus absorbs waves with a certain 'resonant' frequency – nuclear magnetic resonance. Their theory then went on to propose that the frequency of the absorbed radio waves was directly related to the desired nuclear magnetic moment. It was not until the postwar years that several leading nuclear physics laboratories could devote their attention to this idea. What they found however, was not quite what they had expected. The frequency of the waves absorbed by a nitrogen nucleus, for instance, varied depending on the compound the nucleus was contained in. The physicists were thwarted by the electrons that inevitably surround any nucleus in real life; their experiment would give them the data they wanted only if they could strip a nucleus of its accompanying electrons. The demise of the experiment seemed inevitable.

Isidor Rabi was the first scientist to demonstrate the phenomenon of nuclear magnetic resonance.

But electrons are the very stuff of chemistry. They are what glue atoms together into molecules; they are what move around in a chemical reaction. An experiment that depended on the influence of the electrons around a nucleus was just what analytical chemistry was waiting for. And so, an interesting physics experiment became the most powerful tool for examining molecules that the chemist has today.

The NMR experiment

The NMR experiment relies on the fact that some nuclei (depending on the arrangement of the constituent protons and neutrons) have intrinsic angular momentum, or spin. This means that like electrons, which also have spin, they are magnetic – the nucleus behaves like a tiny bar magnet when placed in a magnetic field. Of the many magnetic nuclei, the most important from the chemist's point of view are the hydrogen nucleus, which is a single proton, and the carbon-13 nucleus (carbon-12 is the normal carbon isotope). This is because the elements, hydrogen and carbon, are found in many molecules of interest to chemists.

An 800 MHz NMR spectrometer. Courtesy of Bruker.

In the absence of any magnetic fields, the nuclear bar magnets are all randomly oriented. However, when a sample is placed in a magnetic field, the nuclear bar magnets orient so that they have a constant component which is aligned with the field, with the net effect that the sample becomes magnetised. NMR spectroscopists talk about the *sample magnetisation* – this is the sum of all the tiny nuclear bar magnets in the sample. If the sample is now removed from the field, the magnetisation shrinks over time as the nuclei gradually assume random orientations once more – a process called *relaxation*. Eventually, the sample magnetisation returns to zero as the now randomly oriented nuclear bar magnets cancel each other out. Relaxation processes are generally exponential, and characterised by a constant known as the *relaxation time*. The relaxation time is highly dependent on the sample mobility.

The effect of the magnetic field is then to force the nuclear bar magnets to adopt certain orientations, such that they 'precess' like spinning tops. Now, magnetic resonance occurs if the nuclei are bathed in electromagnetic radiation (which is the equivalent of a rotating magnetic field) whose frequency matches the precessional frequency of the spinning nucleus. The nuclei can then readily absorb the electromagnetic energy, flipping from one energy state where the constant component of the nuclear bar magnets are oriented parallel to the magnetic field to one where they are antiparallel to it. The resonance frequency actually lies in the radio range (megahertz) of the electromagnetic spectrum, and depends on the strength of the applied field (as well as the nucleus being studied).

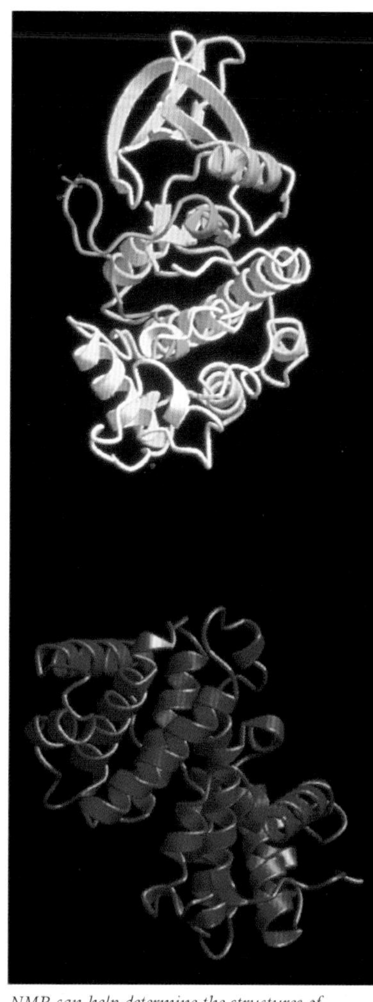

NMR can help determine the structures of complex molecules, such as the proteins shown.
Mol. Biophysics, Oxford Univ. Wellcome Trust Medical Photographic Library.

In early NMR spectroscopy, the signal was recorded as a function of applied magnetic field. The spectrum therefore consists of a series of peaks as the bar magnets in different environments, with different resonant frequencies, flip in turn when the energy gap for each spin hits the energy of the electro magnetic wave. This means that most of the time is wasted in scanning magnetic field strengths which don't hit a resonance for any nucleus in the sample, and in practice only the most sensitive nuclei could be studied. However, suppose a powerful radio frequency pulse is applied; if the frequency spread of the pulse is wide enough, all the bar magnets will flip together. The way the NMR signal is then recorded is rather ingenious; once the nuclear bar magnets have been pushed away from their equilibrium orientations, they rotate around the applied magnetic field in the NMR experiment, just like a gyroscope which is pushed off-'axis'. The rate of rotation is the resonance frequency of the respective nucleus. These rotating bar magnets then create an oscillating electrical current in a wire coil wound round the sample and this signal, known as the *free-induction decay* (FID) is recorded. This decaying signal will, in fact, show an interference pattern as the individual signals interfere with each other. We thus have a time domain spectrum which can be converted to the usual frequency domain signal by the Fourier transformation technique as described earlier in Box 2. This technique is called *pulse Fourier transform NMR* (pulse FT NMR). The pulse can be repeated over and over again – many thousands of times in fact – with the expected gain in the signal-to-noise ratio.

The chemical shift

The next key concept to take note of is that the chemical environment around the nuclei – the distribution of electrons – affects the resonance frequency. If the resonance frequency of all hydrogen nuclei (which are single protons), for example, were the same, then nuclear magnetic resonance would not be much use as an analytical tool since there would only ever be one peak. However, what happens is that in the presence of the applied magnetic field, the surrounding electrons set up their own magnetic field which usually opposes the external field. This effectively shields the nucleus, so that the nucleus sees a slightly different field from that of the applied field. Because the exact electronic environment varies from one proton to another in the same compound, their resonant frequencies will be different.

An NMR spectrum can be recorded with such a spectrometer for any magnetic nucleus. One of its strengths is that spectra for different nuclei can be recorded separately, so, for instance, the hydrogen (^{1}H) and carbon-13 (^{13}C) spectra of an organic compound are recorded in two separate experiments.

Each spectrum, therefore, consists of a series of peaks, or lines (Figure 4), corresponding to the nucleus in different electronic environments. The particular frequency of each line is known as the *chemical shift*, which is the shift in the spectrum from a designated reference standard. The intensity of each line is

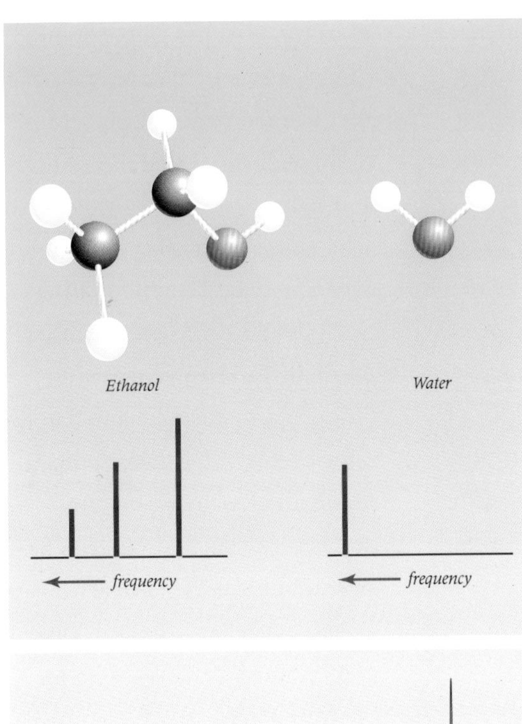

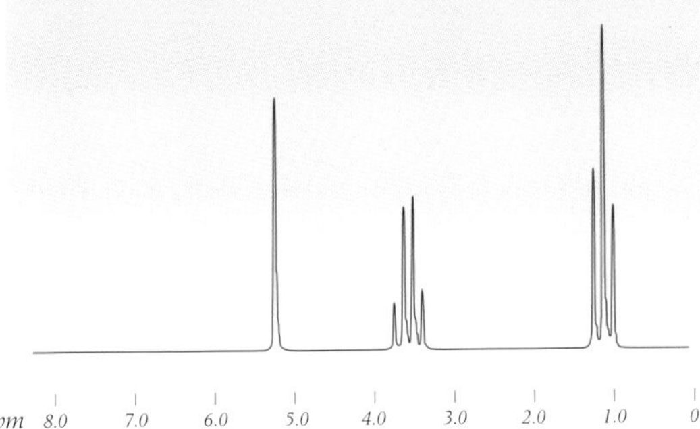

Ethanol, 95% neat; trace of conc. HCl

Figure 4. The proton NMR spectra for ethanol (CH_3CH_2OH) and water show up the different environments of hydrogen nuclei in the two compounds. The spectrum of ethanol gives three lines; one (relative intensity three) from the three hydrogens in the CH_3 group; one (relative intensity two) from the two hydrogens in the CH_2 group and one (relative intensity one) from the single hydrogen bonded to the oxygen. In water, on the other hand, both hydrogen atoms are equivalent, both bonded to oxygen with identical bonds, so only one line is seen. The bottom spectrum shows a higher resolution proton NMR spectrum of ethanol, illustrating how the lines are split by coupling between two sets of hydrogen nuclei.

proportional to the number of nuclei in the sample in that particular environment. So for instance, a proton-NMR spectrum of ethanol (CH_3CH_2OH) gives three signals; one from the three hydrogens in the terminal methyl (CH_3) group (relative intensity 3); one from the two hydrogens in the methylene (CH_2) group (relative intensity 2) and one from the single hydrogen bonded to the oxygen (relative intensity 1). In water on the other hand, both hydrogen atoms are the same, both bonded to oxygen with identical bonds, so only one line is seen.

If we look more closely at the ethanol spectrum, we see that the CH_3 and CH_2 signals are, in fact, not single lines, but split into three (triplet) and four (quartet) components respectively. This is because the spins, or magnetic moments, of the hydrogen nuclei 'couple' with each other. The effective magnetic field seen at one proton will depend on whether a nearby proton magnetic moment is lying parallel or antiparallel to the applied field, so will give rise to two lines, one either side of where the uncoupled peak would have been. This is called *spin-spin coupling*. The distance between the peaks is constant and given the symbol *J*. If the protons are equivalent, that is, with the same chemical shift we don't see the signal splitting. But in the case of two or more protons that are not equivalent, the spin-spin coupling is additive producing clusters of lines depending on number of ways the nearby protons can affect the net field. This is the scenario for the methyl hydrogens in ethanol, each being coupled to the two methylene (CH_2) hydrogens. The methyl hydrogen sees three possible net fields, one where both methylene hydrogen magnetic moments are parallel to the external field (↑↑), one where they are both antiparallel (↓↓), and one where one methylene hydrogen magnetic moment is parallel and the other antiparallel (↑↓) or (↓↑). Over the whole sample, all three possible net fields occur, ensuring three lines for the methyl resonance on the ethanol proton-NMR spectrum.

It might be asked why the proton on the hydroxyl group shows just a single line; it appears to be coupled to the CH_2 protons in exactly the same way as the CH_3 protons and hence might be expected also to show a 1:2:1 triplet. Furthermore, coupling between the hydroxyl proton and the CH_2 would be expected to have an effect on the CH_2 protons, which does not happen. The reason is that the hydroxyl proton is very rapidly exchanging with protons in the solvent. Looking at it from the point of view of the CH_2 protons, they get 'dizzy' and are not sure whether

Box 4 NMR and the Periodic Table

To most organic chemists, NMR spectra from hydrogen and carbon-13 nuclei provide all the structural information needed. However, inorganic chemists view the whole Periodic Table as their stamping ground. There are, in fact, many nuclei that can behave like little bar magnets. There are two important problems, however. The first is that only certain isotopes of an element display magnetism; for instance, carbon-12 gives no NMR signal. The sensitivity of the NMR experiment, therefore, depends on the relative abundances of isotopes. Thus, whereas fluorine-19, which is magnetic, is the natural isotope occurring with 100 per cent abundance, iron-57 is present in only 2.2 per cent in naturally occurring iron.

The second problem is that the inherent sensitivity of the technique varies from nucleus to nucleus; for instance, on a scale with the hydrogen nucleus as unity, the sensitivity for iron-57 is 0.000034. Thus in an NMR experiment these two factors make iron-57 only 0.00000074 times as sensitive as hydrogen! Because hydrogen-1 is the most sensitive, accessible nucleus, it is not surprising that, with some notable exceptions such as fluorine-19, phosphorus-31 and thallium-205, proper exploitation of the Periodic Table (including carbon-13) mostly had to wait for the improved sensitivity of the NMR experiment brought by the development of Fourier transform NMR and pulse techniques. With these difficulties out of the way, what sorts of information might we get from the NMR of some other nuclei?

A straightforward example comes from an investigation of the reaction between tungsten hexafluoride (WF_6) and tungsten hexachloride (WCl_6). Excluding (WCl_6) there are nine possible isomers. Looking at the fluorine-19 NMR spectra with a knowledge of the chemical-shift properties of fluorine-19 nuclei permitted the nine isomers to be assigned, and therefore it was possible to follow the reaction in detail.

The nine possible isomers of the tungsten complex WCl_nF_{6-n} were assigned using fluorine-19 NMR.

Adapted from Structural Methods in Inorganic Chemistry, E. A. V. Ebsworth, D. W. H. Rankin and S. Cradock, Blackwell, Oxford, 1991.

the hydroxyl proton magnetic moment is 'up' or 'down' and hence the coupling is smeared out. Similarly the hydroxyl proton shows only one line. This apparent problem can be turned around. Whether or not NMR signals are smeared out depends on the speed of the chemical exchange; in fact as the speed of exchange increases the NMR spectrum can change gradually from many lines to fewer lines. From the changing pattern of the NMR spectrum with temperature, a great deal can be learned about the kinetics of the exchange process. This has been an enormously important technique in chemistry and biochemistry.

Like infrared spectroscopy, NMR can build up a picture of the structure of a compound from the characteristic absorption peaks, in this case from chemical shifts and coupling constants.

Once the chemical shift had been discovered, chemists were very quick to realise the potential of the NMR phenomenon. However, even the pioneering chemists could not have realised the eventual power of the technique. The first real applications of NMR to problem-solving in structural chemistry began in 1953. The information NMR provided was particularly useful in deciding between different structures having the same chemical formula. Some examples of simple structural problem-solving by NMR are shown in Figure 5. These examples are typical of the kind of questions solved by NMR in the early days.

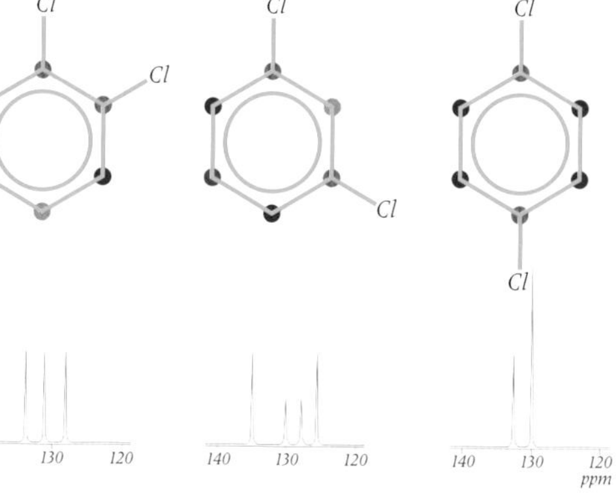

Figure 5. 1,2-, 1,3- and 1,4-dichlorobenzene have the same chemical formula, but they are easily distinguished in the carbon-13 NMR spectrum since 1,2-dichlorobenzene has three peaks, 1,3-dichlorobenzene has four peaks, but 1,4-dichlorobenzene has only two peaks, due to the intrinsic symmetry present in the molecules. The intensity reflects the number of nuclei in each of the different chemical environments.

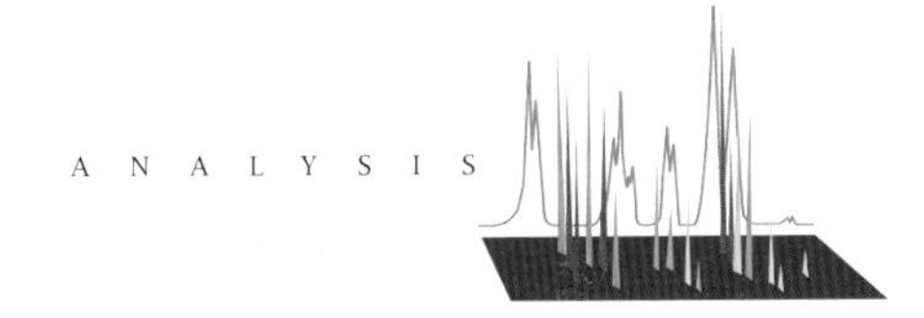

By the 1960s, NMR was starting to be used as a routine research tool in organic chemistry labs to solve the structures of fairly simple molecules. However, during the following decades the applications that NMR could be used for was to expand rapidly in a way that even the NMR specialists of the time would never have dared to predict. Indeed, NMR has now been widely applied to other elements than carbon and hydrogen (see Box 4).

More complex molecules

The information provided by the simple NMR spectra in Figure 5 relies on being able to identify peaks as belonging to certain nuclei on the basis of their chemical shift alone. Unfortunately, our current ability to use the chemical shift is such that it is useful only for pigeonholing certain types of nuclei, as, for example, the protons in hydroxyl (OH) or methyl (CH_3) groups in the spectrum of ethanol shown earlier. What happens if you have more than one methyl group in

Box 5 The technology of NMR

The processing of the experimental data from an NMR spectrometer into a frequency spectrum is performed by today's modern computers in the twinkling of an eye. However, in 1964, when a postdoctoral scientist, Richard Ernst, working with Varian Associates in Palo Alto, first implemented this way of recording NMR spectra, computers were somewhat slower. The digital data from the experiment came out on paper tape and then had to be parcelled up and posted to IBM to be punched onto computer cards, then put onto magnetic tape, and only then finally fed into the most powerful computer then available. The results were posted back a week later but if the computer operator tore the tape or dropped the cards, months of work were lost! To say the least, developments in computers have been essential to the advancement of NMR.

Richard Ernst is regarded as the father of modern NMR spectroscopy. He revolutionised the way that the spectra are recorded and developed two-dimensional NMR spectroscopy. Without these advances, NMR would never have found its huge range of applications in chemistry. Accordingly, Richard Ernst was awarded the Nobel Prize for Chemistry in 1991.
Courtesy of ETH, Zurich.

Magnet technology also had to develop before many of today's applications of NMR could be realised. Even a simple molecule may yield NMR spectra with a veritable forest of lines, with neighbouring lines overlapping extensively. The separation of these lines in the spectrum increases linearly with the strength of the magnetic field used in the NMR experiment. Not only this, but the sensitivity of the experiment increases rapidly as well. Consequently, there has been a relentless drive towards ever stronger and better quality magnets since the advent of NMR.

One of the most important developments here has been the *superconducting magnets* which operate at the temperature of liquid helium. Superconducting materials (see *New science from new materials*) pass an electric current without any resistance and the electromagnets made from these materials are incredibly strong. These magnets were developed and sold around the world by a UK company, Oxford Instruments. In 1945, Bloch's group were using a magnetic field strength of 0.2 Tesla produced by a cast-off electromagnet. Today, superconducting magnets producing field strengths of 18.8 Tesla are available for NMR experiments, with a 23 Tesla magnet on the drawing board. These magnets come at a price of course, around £2 million for a magnet at this field strength. They are also somewhat larger than the magnets used by the early researchers in NMR: an 18.8 Tesla magnet stands some 4 metres high, the height of a small house!

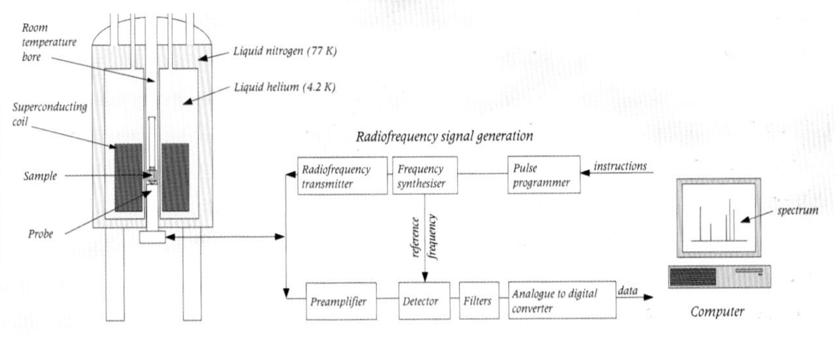

The layout of a modern NMR spectrometer.

It isn't only the attainment of higher magnetic field strengths which improves the quality of NMR data. The design of the radiofrequency probes and hardware is continually being upgraded, with a new era of ultra-low noise cooled 'cryoprobes' and amplifiers on the horizon.

your molecule and, therefore, more than one methyl resonance in the proton NMR spectrum? Clearly, the more peaks there are in the spectrum, the less easy it is to assign them to a particular nucleus. The information is there, but how do we extract it?

There is another problem encountered with larger molecules, because there is a limit to the number of peaks that can physically fit into a particular region of the spectrum, before they start overlapping. In severe cases, the peaks merge to a broad envelope in the crowded regions. How do we disentangle the spectrum? Both these problems have been overcome, and solving each of them represented a huge leap forward in the usefulness of NMR. Generally, so long as one peak can be unequivocally assigned to a certain nucleus, the rest of the peaks can be deduced, by the techniques described below.

New NMR techniques

We described how a powerful radio frequency pulse will flip all the nuclear bar magnets together, so that a free-induction decay can be recorded, and converted via Fourier transformation to the usual frequency spectrum, and how the main benefit of pulse-FT techniques is sensitivity. The other major advantage of pulse-FT NMR is that it opens up the possibility of a vast number of new experiments in which the spectroscopist controls the information contained in the free-induction decay using sequences of pulses designed to excite the spins in a specific manner. The details of this procedure are complicated, often delving deep into quantum mechanics. Suffice to say the resulting spectrum is often reminiscent of the simple 'one-pulse' spectrum, but has been edited in some way so that only nuclei in specific chemical environments show up. The editing is performed by applying a whole array of radiofrequency pulses to the sample, rather than one single one as in the simple NMR experiment previously described. There are a huge number of variations on this 'multiple pulse' theme, with exotic names such as 'NOE difference' and INADEQUATE. An application of the NOE difference pulse sequence is shown in Figure 6.

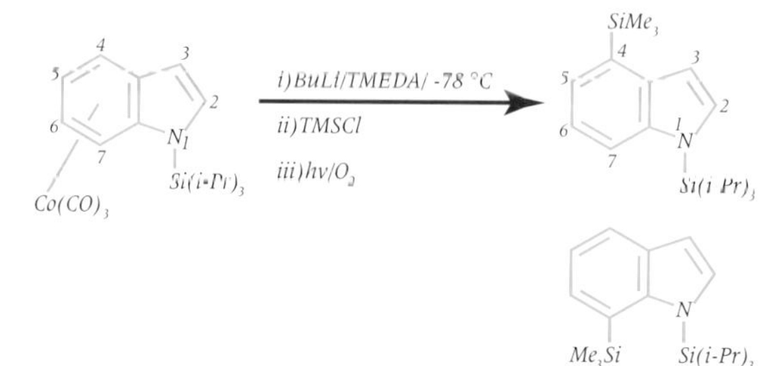

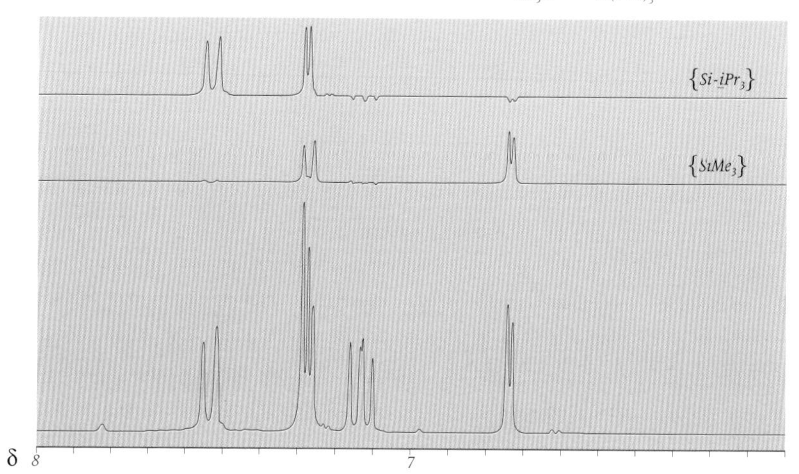

Figure 6. The reaction shown has two possible products, but NMR can tell the difference. The pulse sequence used to obtain this spectrum was designed to reveal atoms which are close in space to the target atom (in this case nitrogen), via an effect known as the Nuclear Overhauser Effect, or NOE. The spectra show that the trimethyl silyl ($SiMe_3$) group is present at the 4 position of the indole rather than the 7 position.

Another way to resolve more complex spectra is to present them in two or more dimensions. *Two-dimensional NMR* is a particularly subtle application of pulse sequences. It was first conceived in 1971 by the Belgian physicist Jean Jeener, and first demonstrated in practice in 1974 by Richard Ernst. Suppose that the pulse sequence has a regularly increasing interval between pulses. After each application of the pulse sequence, the free-induction decay can be Fourier-transformed to give a conventional spectrum of intensity as a function of frequency. Thus a series of

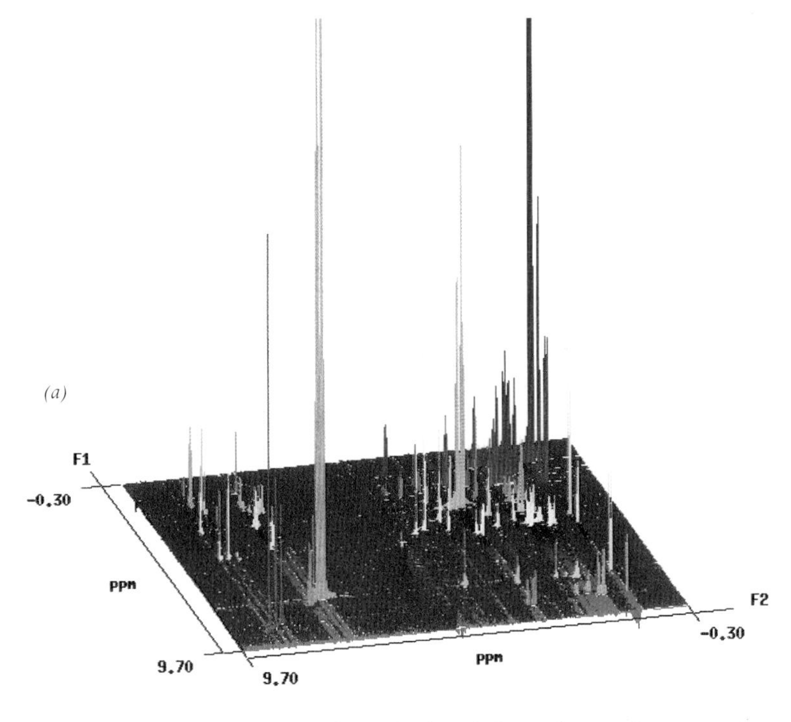

Figure 7. A two-dimensional spectrum of a peptide, shown both as peaks (a) and as contours (b). Unlike the one-dimensional spectrum, the two-dimensional spectrum reveals cross-peaks connecting hydrogens which are close in the bonding network (generally only hydrogens which have between two and five intervening chemical bonds). To those familiar with NMR spectra, it is easy to identify one uniquely placed peak as belonging to a particular nucleus. From this point it is possible to walk round the structure, via the cross peak 'stepping stones', to reveal the rest of the bonding network.

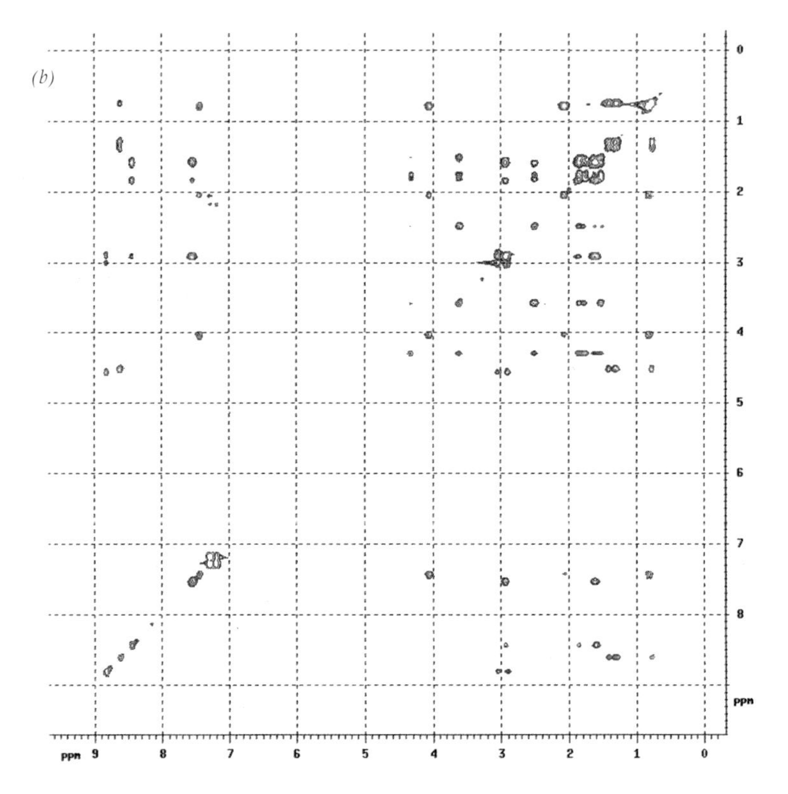

spectra, each referring to a particular delay interval between pulses, is obtained. The series of intensities at any particular frequency can then be Fourier-transformed again, this time with respect to the delay interval. If this is done for each spectral frequency, a two-dimensional (2-D) spectrum results, with each intensity point being a function of two distinct frequencies (Figure 7).

What does a 2-D spectrum tell us? The answer is, it depends crucially on the particular sequence of pulses used to obtain it. However, the interpretation is usually based on spotting peaks which have different frequency coordinates in the two dimensions. These peaks are called *cross-peaks*. The frequency coordinates at which the cross-peaks appear correspond to two different frequencies, and therefore to two different nuclear environments. Depending on the type of 2-D spectrum, the presence of a cross-peak indicates that the nuclei are either close in space (closer than 50 nanometres), connected by a defined number of chemical bonds, or chemically exchanging. 2-D spectra can have a very large number of cross-peaks, and so are extremely rich in information compared with their 1-D counterparts. The 2-D spectrum in Figure 7 has been set up to utilise the scalar coupling to reveal which nuclei are connected by chemical bonds.

We can see that two-dimensional NMR solves two problems at once. Not only is it possible to spread out a complex and overlapping one-dimensional spectrum into a two-dimensional spectrum of separated cross-peaks, but the cross-peaks themselves are a vital source of information about the molecule, because they map out relationships, such as chemical bonds or short interatomic distances, between pairs of nuclei within the structure.

With increasingly complex molecular systems, it is still possible that the two-dimensional spectrum, however spread out, may suffer from overlap problems, just as its one-dimensional

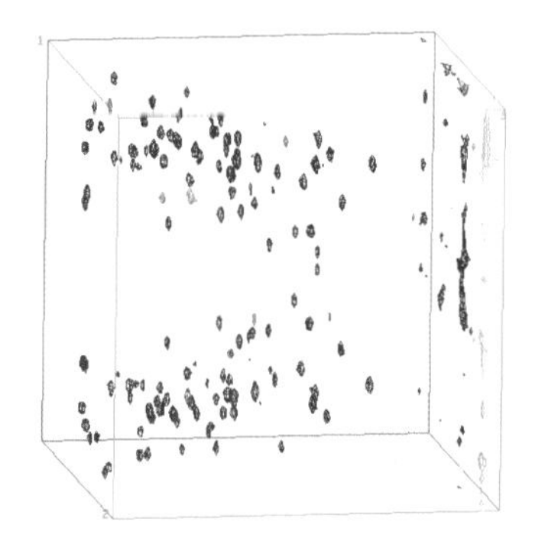

Figure 8. A three-dimensional NMR spectrum. It is easiest to visualise this spectrum as a cube in 3D frequency space.

counterpart does. In addition, the cross peaks in a two-dimensional spectrum connect only two nuclei; extra information could be gleaned if the cross peaks connected more nuclei. The philosophy behind two-dimensional spectroscopy can be extended to three and even four dimensions. A four-dimensional experiment can take up over a week of valuable time on an NMR spectrometer.

It is not as difficult as it may seem to visualise a multi-dimensional spectrum, since the spectra are not really related to physical space – this is just the way we choose to display them (Figure 8). In a two-dimensional spectrum, one cross-peak has two frequency coordinates, and therefore connects two different nuclei. Likewise, in a three-dimensional spectrum the cross peaks connect three nuclei, and in a four-dimensional spectrum, the cross-peaks connect four nuclei.

NMR in biochemistry

Since 1953, NMR was almost exclusively the province of chemists, until the 1970s, when two-dimensional NMR opened the doors to studies of more complex biomolecules.

Proteins consist of a long chain of amino acids, which in the biologically functional form fold up into a complex three-dimensional structure. Until 1984, X-ray crystallography was the only technique to provide 3D structural information, and then only for proteins of which a single crystal could be grown. In 1984, the first 3D structure of a protein, BUSI (bull seminal plasma inhibitor) was performed by NMR, in the pioneering laboratory of the Swiss chemist, Kurt Wüthrich. The structure was solved by 2D NMR techniques, in which the distances between various parts of the molecule could be measured, using the intensity of the cross-peaks. When fed into a computer, these measured distances formed the basis of a highly involved calculation to determine which structure satisfied the distances

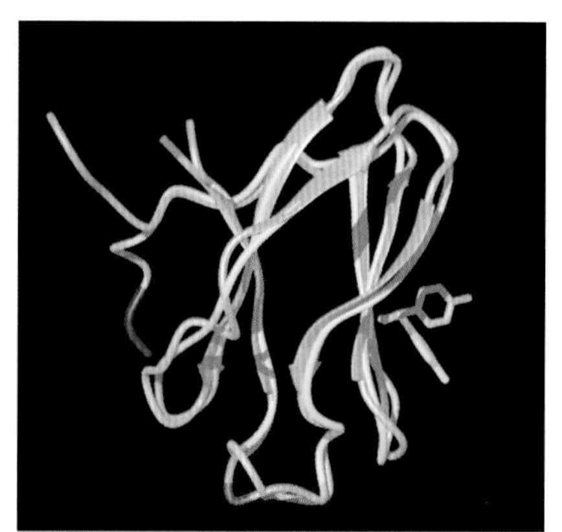

Figure 9. The structure of the protein Tendamistat, with the NMR and X-ray structures superimposed.

obtained by NMR. The scientific community was very sceptical of the method, since the 3D crystal structures of similar proteins were already well known, and therefore could have been used to obtain the NMR structure. However, the matter was resolved when the structure of a completely new protein, Tendamistat, was worked on simultaneously by NMR spectroscopists and X-ray crystallographers. Virtually identical structures were obtained (Figure 9), and a new subject was born.

BUSI and Tendamistat are both relatively small proteins with a molecular weight of around 7 kilodaltons (7000 amu). With the development of new ingenious pulse sequences, better hardware and stronger magnetic fields, it is possible to study much more complex biological molecules; currently the limit is around the 40 kilodaltons. One of the most crucial advances in biomolecular NMR was the development of techniques in which naturally abundant nuclei, for example, nitrogen-14 and/or carbon-12 (N-14 has 'inconvenient' spin and C-12 has no spin) can be replaced by the NMR-active isotopes nitrogen-15 and carbon-13, present only at very low levels in Nature. Proteins are made in the laboratory by inserting the gene that makes the protein into the DNA of an organism which can be easily grown quickly and easily in large numbers in laboratory conditions, for example, the bacterium *E. coli*. The bacteria are then forced to make the protein, which is extracted by breaking open the cell, thereby releasing the contents for purification. To make protein enriched in isotopes nitrogen-15 or carbon-13, the bacteria are grown in media in which the starting materials for making the protein, ammonium chloride and glucose, are substituted with these isotopes, respectively.

Given the amino acid sequence that makes up a protein chain, the major challenge in biomolecular NMR is knowing both which peak belongs to which nucleus, and the unit it comes from. Often this process of 'assignment' takes longer than the NMR experiments and structure calculations combined. Labelling with nitrogen-15 and carbon-13 can make this process easier. The protein 'backbone' is made up of -NH-CHR-CO- repeat units, in which the 'R' chemical group defines the type of amino acid making up the unit. By labelling the nitrogen and carbon nuclei in the protein, a great deal of information can then be obtained using 3D experiments in which the presence of a cross-peak links a hydrogen to a carbon and a nitrogen. The carbon or nitrogen need not be in the same amino acid unit, but can be one step forwards or backwards along the backbone, depending on the way the experiment is set up. That way it is possible to 'walk down' the chain, using the cross-peaks as stepping stones, assigning each one in turn. These experiments are called *triple resonance* NMR techniques, because the three frequency axes correspond to the three different nuclei: the hydrogen proton, carbon-13 and nitrogen-15.

Nowadays, determining the structures of biological molecules by NMR is a highly competitive field, and many laboratories around the world are dedicated to this area. In fact, biochemists have been the driving force behind many of the recent technical developments in NMR.

Aside from structure determination, NMR is an excellent source of information on the dynamic properties of biomolecules (as it is for other chemical compounds), and is a useful probe of transient interactions, processes where atoms such as hydrogens are exchanged, and the mechanisms of biochemical reactions. Using NMR, it is possible to gain an understanding of how biological molecules interact in Nature, for example, how proteins bind to DNA and how enzymes recognise their substrates. These interactions are not only interesting at a purely scientific level: with this information, we can understand diseases at the molecular level, and accordingly, better drugs can be designed. Another useful feature of NMR is that it can be done in 'real time', enabling, for example, studies of the complex way in which proteins fold up (see *The chemistry of life*). All these subjects are currently active and competitive areas of research.

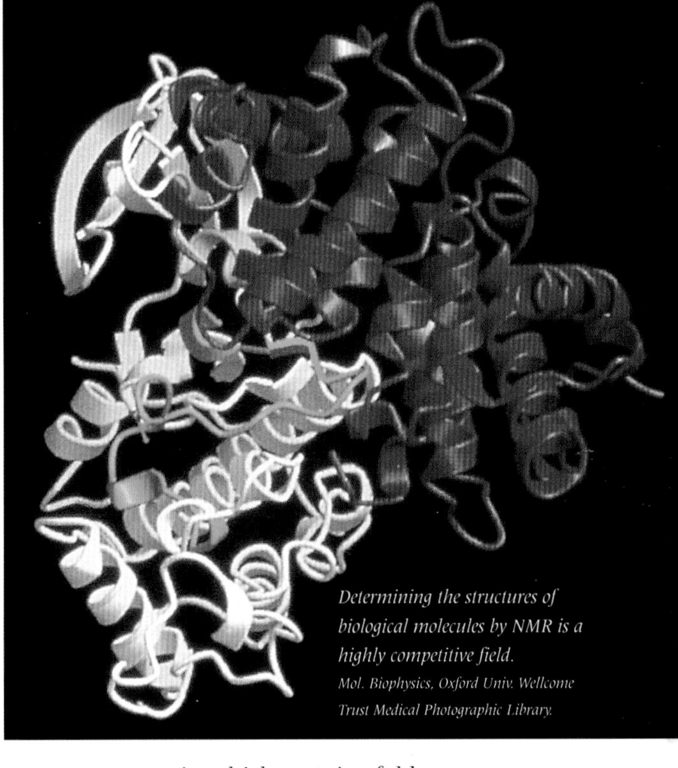

Determining the structures of biological molecules by NMR is a highly competitive field.
Mol. Biophysics, Oxford Univ. Wellcome Trust Medical Photographic Library.

However, there is a practical limit to the size of molecule it is possible to study by NMR, because with increasing molecular size, molecular tumbling slows down, and the relaxation processes get faster. Beyond a certain point, relaxation occurs so fast that there simply isn't enough time to perform the pulse sequence and record the spectrum. It is possible to replace some of the hydrogens by deuterium, slowing down the relaxation to a certain extent, and there are pulse sequences currently being developed which promise to extend the size limit a little further at very high fields. However, in this respect, crystallographic techniques may always win out in the study of larger biomolecules. Despite this limitation, there is no doubt that NMR is a rich and unique source of information, much of which is still poorly understood, and there remains a great deal of research to be done in order to extract this untapped information. What is certain is that biomolecular NMR has an exciting future.

NMR of solids

In all the above examples, the samples are dissolved in solution. Even though some of the earliest NMR experiments were on solid samples, NMR became, in its development during the 1950s, a technique for use solely on solution-state samples. The myth grew up that it was 'impossible' to do NMR on solids. The reason was simply that strong magnetic interactions between the nuclear bar magnets meant that the NMR spectra from them were composed of very broad lines, so broad that lines from different chemical sites could not begin to be distinguished. The very advantage of NMR, the clear distinction of different chemical sites, appeared to be lost. This was all changed by Raymond Andrew. Working in Bangor in the late 1950s, Andrew recognised that spectra of solution-state samples did not suffer the line-broadening inherent for solid samples because natural molecular tumbling in solutions averaged away the effects of the nuclear interactions responsible. He realised that he could mimic this effect for solids by simply spinning the whole sample at a particular angle, the so-called magic-angle, relative to the static magnetic field in the NMR experiment, and so reduce a spectrum of broad lines to a solution-like one of sharp, narrow lines.

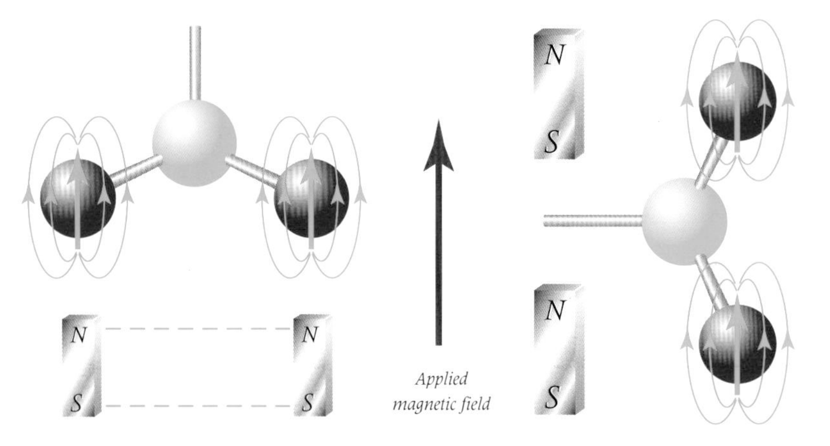

Figure 10. This diagram illustrates the interaction between nuclear bar magnets (the so-called dipole-dipole interaction) which affects the resonance frequency of both nuclei. The nuclear bar magnets are oriented by the external magnetic field in the NMR experiment and are held in their spatial position within the molecule by chemical bonds. The strength of the interaction depends on the relative position of the nuclei, which in turn depends on the orientation of the molecule, for instance, in the orientation of the molecule on the left, the nuclear bar magnets are repelling each other (high energy interaction); on the right they tend to attract each other (low energy interaction). The energies of the resonances of each nucleus are modified by the energy of the interaction between the nuclei; every molecule orientation gives (in general) a different strength of interaction, and so if a sample consists of many crystallites (ie many different molecular orientations present), a broad line in the NMR spectrum results.

With this new technique of *magic-angle spinning*, modern *solid-state NMR* was born. Chemists now appreciate that NMR for solids is just as feasible as for solutions, and furthermore, that it has probably an even greater range of applications. The nuclear interactions that can make the spectra of solids so broad actually render solid-state NMR extremely useful. Those interactions contain information about the structure and the movement of molecules within the solid and the skill of solid-state NMR spectroscopists since the 1950s has been in inventing techniques and methods by which this information can be drawn out of NMR spectra.

One of the most interesting uses of solid-state NMR is in the study of molecular dynamics in solids. The traditional view of solids is that they are static, rigid entities; very little motion of their respective molecules being possible with the tight molecular packing that was supposed to exist in solids. Before the advent of NMR, there was no method for studying molecular motion in solids, and so this view remained unopposed amongst chemists; indeed, the rigid nature of bulk solids seemed to agree with it. Scientists were then amazed when NMR experiments showed that rotational motions of parts of molecules and even whole molecules is not just possible, it is positively common in molecular solids. This is an extremely important area because we are now realising that macroscopic properties of a solid, such as its brittleness, flexibility and so on, are dependent on how the molecules in the solid can twist and bend. For instance, the brittleness of a solid (such as a plastic) is determined by the degree to which the molecules in it can move to absorb an impact. If the energy from the impact can be absorbed into harmless rotations, for instance, the solid is more likely to survive the impact. If not, since the energy has to go somewhere, it drives the molecules apart and the material shatters. We can also use studies of molecular motion to probe experimentally the intermolecular forces that hold molecular solids together. Thus, the variation in the molecular motion with changes in temperature gives valuable information about how strongly the neighbouring molecules are bound together in the solid. This type of work aims to advance our fundamental understanding of the forces responsible for forming solids.

NMR can monitor molecular rotations because the positions of the lines in an NMR spectrum depend upon the orientation of the molecule being examined. As a molecule rotates and its orientation changes, so do the positions of the lines in its NMR spectrum. If the molecule moves on the timescale of the recording of the NMR signal, the overall effect is a characteristic 'distortion' of the NMR spectrum, caused by the changing positions of the NMR lines as the spectrum is recorded.

In recent years, much of the attention has been on polymers (molecules made of chains of similar units), and the correlation between molecular motion within polymers and their overall material properties. Many polymer materials are semicrystalline; that is they are composed of crystalline regions where the polymer chains are well-ordered and intervening amorphous regions where the polymer chains are almost randomly coiled. As Figure 11 shows, a single polymer chain may dip in and out of crystalline regions, looping round in amorphous regions to reenter the crystalline region (see *The age of plastics*).

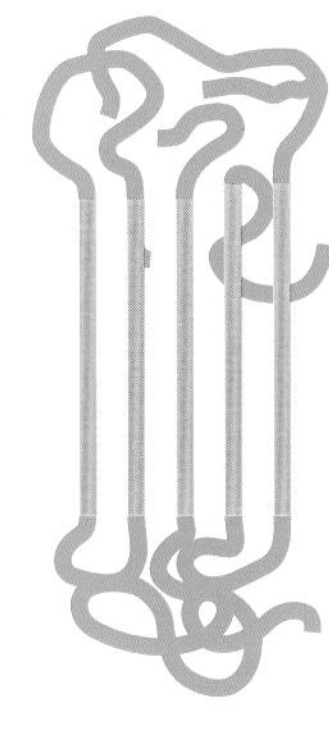

Figure 11. Many polymers are semi-crystalline, being composed of crystalline (yellow) and amorphous (green) regions. A single polymer chain will be involved in both regions, looping round in the amorphous region to re-enter the crystalline region.

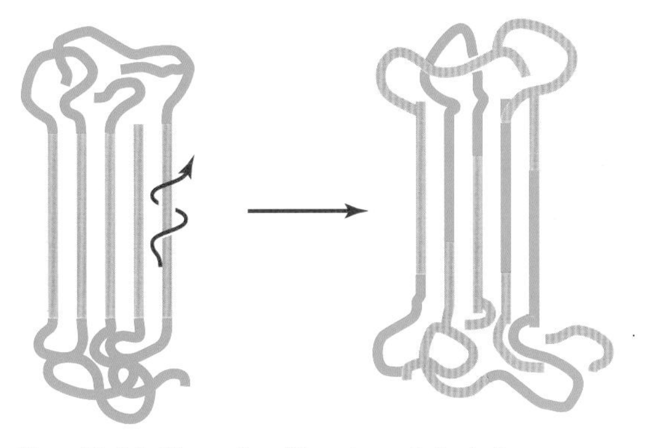

Figure 12. Spiralling motion of the polymer chains in the crystalline regions (yellow) of polyethylene leads to diffusion of the chain in and out of the amorphous region (green). The amorphous and crystalline regions of the polymer can be distinguished in a proton-carbon-13 correlation spectrum.

NMR has shown that the chains in the amorphous regions of most polymers are highly mobile, twisting and flexing at kilohertz rates. A detailed study of polyethene in 1991 by Hans Spiess and colleagues in Mainz revealed that chains in the crystalline regions (Figure 12) undergo a spiralling motion which results in a given region of a chain diffusing in and out of the amorphous region of the material. This process almost certainly accounts for the flexibility of the material.

Figure 13. Many polymers contain phenylene rings (blue). NMR has shown that these rings can rotate to absorb the energy from mechanical stresses on the material.

Many polymeric materials contain phenylene rings – hexagonal rings of carbon atoms – at intervals down the length of the polymer chain (Figure 13). When mechanical stresses are applied to a material, the mechanical energy has to be absorbed in some way. NMR studies on several polymeric materials containing phenylene rings have shown that these rings are able to rotate, and in doing so can absorb mechanical energy. The motion leaves the essential structure of the material intact, and so renders it tough to mechanical stress.

These examples provide just a flavour of some of the recent work on polymers. There are now increasingly detailed studies of the molecular dynamics in widely diverse areas, for instance, in biological solids, such as cell membranes; in organometallic compounds, where the motion of organic groups bonded to metals is of great interest; and in important catalytic systems, such as zeolites.

In the future, improvements in resolution from larger magnetic fields, faster magic angle spinning (up to 40 kilohertz is currently available) and more powerful radiofrequency pulses will undoubtedly widen the applications of solid-state NMR still further. Current technology already means that structure determinations analogous to those for solution state NMR are now possible. In just the past few years this field has moved on apace and certainly has not yet reached its full potential.

Magnetic resonance imaging (MRI)

Soon after the first successful NMR experiments at Stanford University, Bloch obtained a strong proton NMR signal when he positioned his finger in the radiofrequency coil of his spectrometer. Most of the signal would have come from water, which makes up 70 per cent of the human body. However, water in the body is in a very different environment to free water, in fact MRI applied to medicine depends on the protons in water in different environments having different properties. Water molecules diffuse and tumble differently inside, for example, tumour cells and normal cells. This motion can be quantified by measuring the relaxation time or diffusion rate of the water protons. Different types of tissue will yield different values for these parameters, and so can be discriminated.

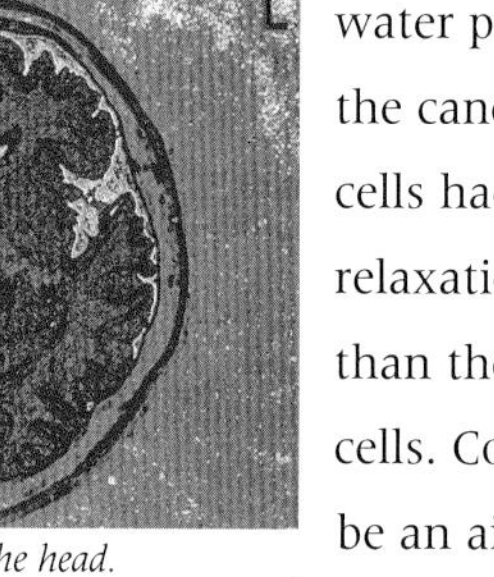

A cross-sectional image of the head.
BSIP Ducloux/Science Photo Library

Paul Lauterbur who invented magnetic resonance imaging (MRI).
Courtesy of the University of Illinois at Chicago.

In 1971, Raymond Damadian, a doctor at the State University of New York, found that he could discriminate between normal cells and cancer cells in excised rat tissue by measuring the relaxation time of the NMR signal from the water protons within the cells. The water protons in the cancerous cells had a longer relaxation time than the normal cells. Could this be an aid to the diagnosis of cancer? What did it tell us about the nature of the cancerous cells? It was the observation of similar experiments which inspired Paul

Lauterbur, a chemist working at a company called NMR Specialties (which he had helped to set up while at the Mellon Institute in Pittsburgh), to invent magnetic resonance imaging (MRI) in 1972.

While watching a graduate student perform NMR on rat tissue, Lauterbur was struck at first by how little information NMR provided compared to the conventional means of looking at cells – the optical microscope. However, having watched the student first sacrifice and dissect the rats, he realised that unlike optical techniques, the NMR information could have been obtained directly from the intact living organism. Pondering the experiments again that evening, he arrived at a broader question: was there any way to locate an NMR signal from within a three-dimensional object, without opening it up?

Lauterbur soon realised that it was possible to create an 'internal picture' of an object by NMR, and he quickly wrote down his ideas in a notebook and had them witnessed by a colleague. His ideas were based on the use of a magnetic field that varies through space – a magnetic field gradient. While Lauterbur performed his preliminary experiments, Peter Mansfield at the University of Nottingham had a more ambitious aim. He had no knowledge of Lauterbur's work, but he had an idea of how he might get an NMR picture of a crystal, similar to an X-ray crystal structure. However, a colleague pointed out that in order to obtain atomic resolution the magnetic field gradients would have to be immense, and also that even with the best techniques available at the time the linewidths from the solid crystal would be too broad to get the required resolution. So instead he settled for layers of camphor crystals separated by cardboard, and achieved a resolution of less than one millimetre. Lauterbur began imaging small living objects, such as clams 4 millimetres across, while Mansfield and his colleagues continued with *NMR microscopy* on solids, and in 1976 produced the first picture from a live human subject – a cross-section through a finger, revealing true anatomical detail. Mansfield continued to be a pioneer in the field, developing better and better methods for imaging larger parts of live human subjects and for advancing the resolution of micro-imaging techniques well past the sub-cellular level. Nowadays, NMR microscopy is a worthy rival of optical microscopy.

How does NMR imaging work? Imagine you are observing an NMR signal from within an object, for example, a drop of water trapped deep inside a sponge. The exact frequency of the line corresponds to the chemical shift of the protons in water, but it is also directly proportional to the strength of the applied field. If you could make the field stronger across the part of the sponge containing the drop of water, the line would shift to a higher frequency, by an amount proportional to the increase in

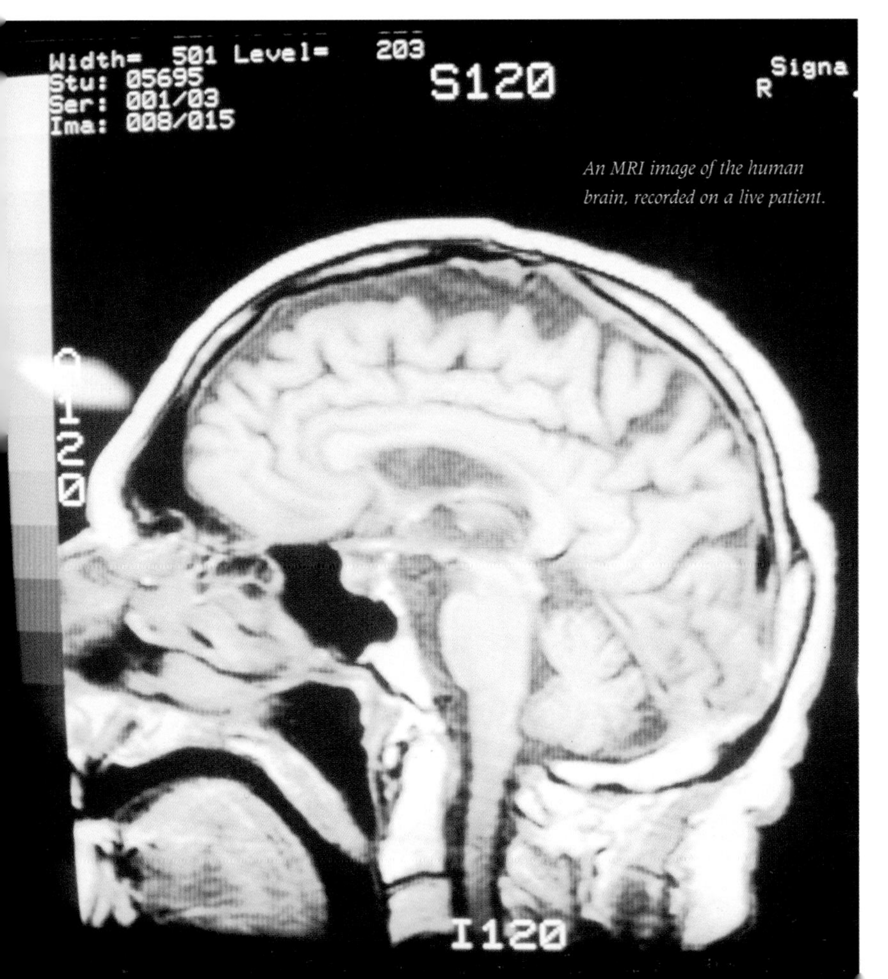

An MRI image of the human brain, recorded on a live patient.

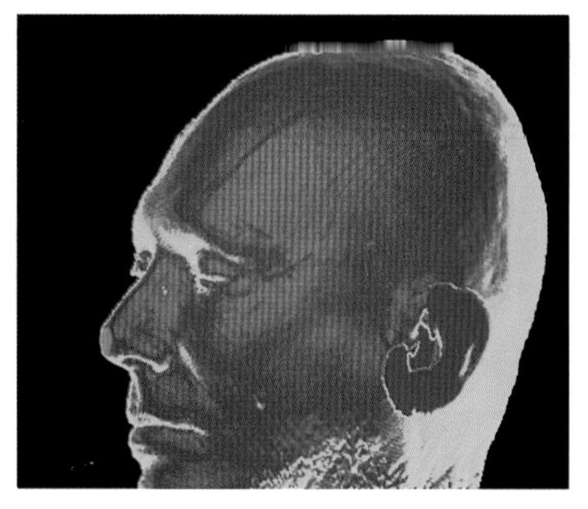

Coloured 3D MRI scan of a human head.
GJLP-CNRI/Science Photo Library.

field strength. A more sophisticated way to locate the drop might be to apply a magnetic field which increased in a steady fashion across the sponge. That way the exact frequency of the NMR signal would fix the position of the drop to within a 2D slice of the 3D sponge. Further measurements in different directions would fix the drop first to a line and then to a point. This is the basis of *magnetic resonance imaging* (MRI); a magnetic field gradient turns NMR frequencies into spatial coordinates.

Being able to take a 'picture' of a water droplet is one thing, but how would you go about differentiating between, for example, water in tumour cells and normal cells? If the water protons relax faster in the normal cells than in the tumour cells, then after a certain period of time the signal arising from the normal cells is less intense than from the tumour cells. A computer can display different intensities as different shades of grey, or as different colours.

In, for example, a human head, different tissues have different relaxation properties, hence tissues such as grey and white matter give rise to different pixel intensities within the computer-generated image. The contrast can be improved by enhancing any relaxation time difference between the water protons in the two environments. Developments in this field hinge on quite unrelated observations of years ago. For a typical proton NMR spectrum of say an organic compound, the signals from the different protons cover a narrow range. However researchers discovered that, in the presence of small concentrations of paramagnetic compounds (compounds containing unpaired electrons), the proton spectrum can be widely spread and, at the same time, the relaxation properties of the different protons made very different. The reason for this effect is that the protons are in instantaneous contact with the paramagnetic molecules, and the unpaired electrons on this compound interact with the protons on the organic compound, the interaction being different for different protons. Early paramagnetic compounds which show this effect were usually complexes of europium or praseodymium. There is now an intense effort to synthesise new complexes with appropriate relaxation effects which will target one type of cell but not another.

A typical MRI scanner.

One of the first technical hurdles in MRI was to increase the magnet gap to allow the study of biological samples larger than the 4-millimetre live clam used by Lauterbur to demonstrate imaging of living animals. Nowadays, the magnet gap is large enough to slide in a whole human body. A typical clinical whole-body magnet has a 100-centimetre bore with a smooth-enough magnetic field going down the cavity to image the whole length of the human spine. Scientists have devised better ways of recording the image. For instance, it is generally more time-efficient to generate several parallel 2D slices of the 3D image, rather than the whole 3D image. The current clinical standard for brain imaging is to take 20 to 30 slices, each with a resolution of 256 x 256 pixels – this takes only 8 minutes.

Magnetic resonance imaging is one of the most important tools available in biology and medicine. For obvious public relations reasons, the prefix 'nuclear' is omitted, since the word has the unhelpful connotations of radioactivity, despite being completely safe. The key is that the method is noninvasive, and unlike the medical use of X-rays, is extremely sensitive to differences in parts of the body with a high water content. Thus, although broken bones remain the province of X-rays, MRI is used for the study of abnormalities in tissues, muscles, and other soft parts of the body, and is also increasingly useful for studies of blood flow. As a relatively young technique, it has a rich future. Increasingly, MRI has been used to study physiological processes such as diffusion and perfusion. For example, MRI which is sensitive to diffusion in brain tissue, is used to detect those areas of the brain that are at risk of damage in

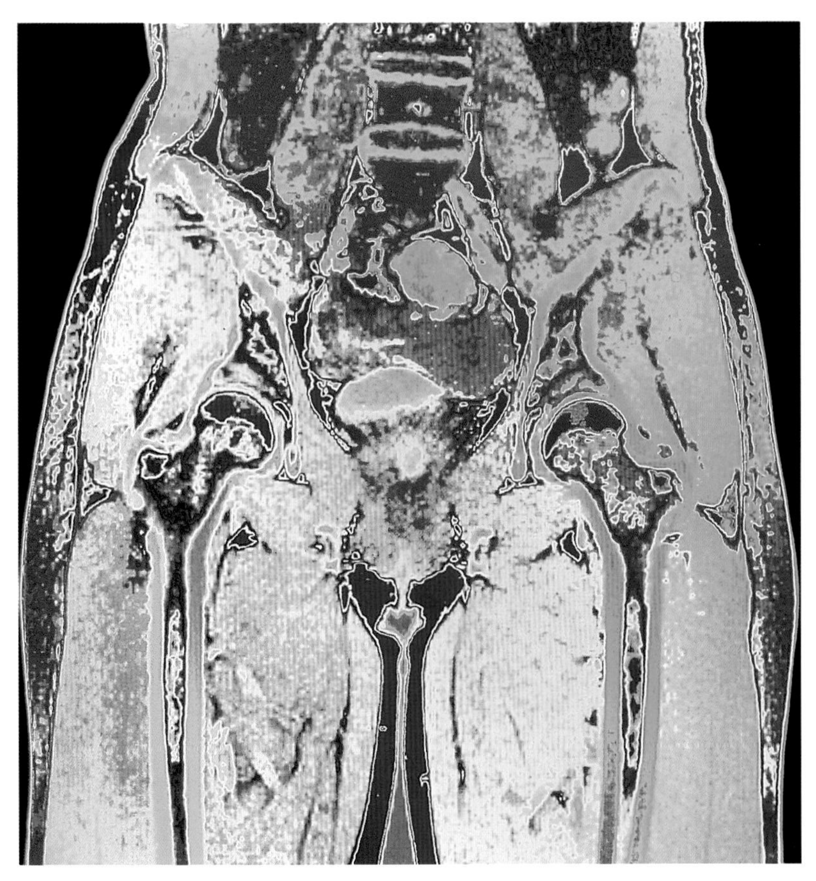

An MRI scan of a section through the hips of a 20 year old woman.
Simon Fraser/Science Photo Library.

MRI scans on a light box.
Geoff Tompkinson/Science Photo Library.

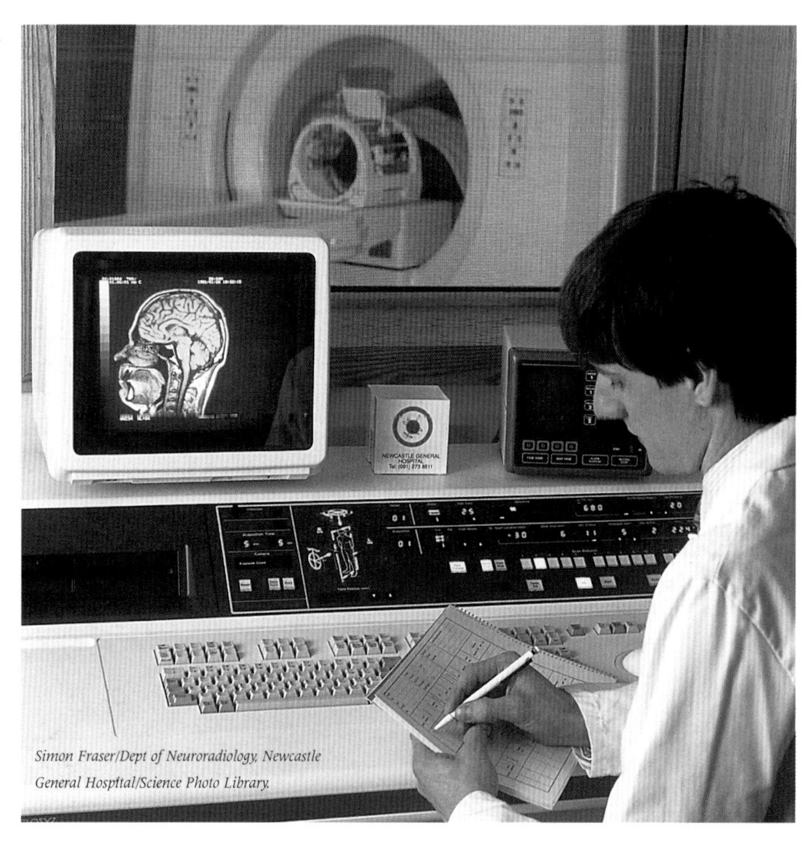

Simon Fraser/Dept of Neuroradiology, Newcastle General Hospital/Science Photo Library.

the hours following a stroke. Other nonmedical applications of MRI include imaging of materials, and even 'real-time' changes to these materials, for example, the processes of drying cement, and the uptake of solvents by polymers.

As with the onset of biochemical applications to solution-state NMR, the technical demands of MRI have driven the NMR field quickly forwards over the past two decades, most notably in the development of strong, rapidly-switchable magnetic fieldgradients, which have in turn revolutionised traditional NMR spectroscopy. Interestingly, the resources expended on MRI now dwarf those dedicated to traditional NMR spectroscopy, which is also reflected in the increased public awareness of magnetic resonance techniques.

The future

Technological advances are likely to increase further the range of materials suitable for study by NMR spectroscopy and MRI. Recently, Paul Callaghan and his research group at Massey University in New Zealand performed NMR experiments on samples of ice in Antarctica one metre long, using the Earth's own magnetic field as the magnetic field in the experiment; further *in situ* studies are planned for the future. This may well be the first of many future experiments where large samples are able to be examined by NMR.

Scientists in the US have recently begun combining atomic force microscopy and NMR. In atomic force microscopy, an atomic-length probe scans the surface of the material under study, ultimately producing an image of the atoms there. The idea behind the new experiment is to excite and detect nuclear magnetic resonances from individual nuclei at the surface of the material. Although it may yet be many years before such experiments are practicable, the implications for such experiments for studying catalysts for instance are enormous and well worth waiting for.

There are still many developments to be made in MRI and solid-state NMR in particular. In MRI, the current spatial resolution is of the order of micrometres at best. If this were to be improved so that resolution on molecular scales were possible, a whole new area would be opened up. Imagine being able to routinely image molecules inside a solid! The applications would be endless.

Further reading

1. *The History of Chemistry*, John Hudson, Macmillan, Basingstoke and London: 1992.
2. *Principles and Practice of Analytical Chemistry*, F. W. Fifield and D. Kealey, 4th edn, Blackie, Glasgow: 1995.
3. *Principles of Instrumental Analysis*, by D. A. Skoog and D. M. West, 2nd edition, Saunders College, Philadelphia: 1980.

(These three books were used as reference in the preparation of the general introduction, for which the authors are grateful.)

4. *The Development of NMR, The Encyclopedia of NMR*, Vol 1, D. M. Grant and R. K. Harris (eds.),Wiley.
5. *Nuclear Magnetic Resonance Spectroscopy*, Robin K. Harris, Longman, Harlow: 1986.
6. *Exploring Chemical Analysis*, Daniel C. Harris, W. H. Freeman and Company, New York: 1996.

Glossary

Acid-base titration A method by which the concentration of an acid or base is determined by reaction with a solution of known concentration of base or acid.

Atomic spectroscopy A method of identifying elements by heating the sample (traditionally in a flame) so that they emit (or absorb) characteristic wavelengths of radiation.

Bragg's law The relationship defining the conditions for the diffraction of radiation by a crystal whereby the angle of the incident radiation (and thus the angle of diffraction) at a given wavelength is inversely related to the distance *d* between the layers of atoms: $\lambda = 2d\sin\vartheta$.

Chemical shift The frequency difference between a standard reference and a given line on an NMR spectrum – the 'chemical shift' of the line.

Chromatography A method of separating compounds by exploiting differences in diffusion through a given medium (see *Stationary phase* and *Mobile phase*).

Diffraction A phenomenon whereby reflected light waves constructively and destructively interfere with each other to produce a diffraction pattern.

Electron diffraction A method of obtaining structural information about surfaces or gases by passing electrons through the sample and measuring the diffraction pattern obtained (see *Diffraction* and *Bragg's Law*).

Electrospray mass spectrometry A novel method of getting a high-molecular weight compound into the 'vapour state' for mass spectrometry by spraying it through a syringe with a needle at high electrical potential which then imparts a mutually repulsive charge to the molecules (see *Mass spectrometry*).

Fourier transform A mathematical technique which can analyse a complex oscillating signal into a series of simple sine waves. It is used in combination with analytical techniques such as infrared spectroscopy and NMR, with the aid of a computer, to unscramble individual signals in a complex spectrum from data collected simultaneously, considerably speeding up the analysis and improving sensitivity.

Free induction decay The oscillating electrical signal which is measured directly in a Fourier transform nuclear magnetic resonance (FT-NMR) experiment; Fourier transformation of this then leads to the frequency spectrum (see *Nuclear magnetic resonance*).

Gas chromatography A form of chromatography in which the mobile phase is a gas. It is often used in the food industry and for environmental analysis (see *Chromatography* and *Mobile phase*).

Gravimetric analysis An analytical technique used to quantify the amount of an element or a compound of the element in a mixture by isolating and weighing it in as pure a form as possible.

Group analysis An analytical scheme used to identify metal compounds by reacting them with a particular agent to give a coloured precipitate.

High performance liquid chromatography (HPLC) A highly efficient form of chromatography using a stationary phase of very small particles packed in a narrow column through which the mobile phase moves very slowly under high pressure (see *Chromatography*).

Inductively coupled plasma (ICP) A source used in atomic spectroscopy to excite the atoms in a sample. It is produced by ionising argon gas with radiofrequency induction coil. The sample is introduced as a fine mist and ionises in the plasma.

Infrared spectroscopy An important analytical method for determining the structure of compounds. Infrared radiation is absorbed by molecules undergoing bond stretching, bending and twisting to give a characteristic pattern of wavelengths.

Iodometry A type of titration used to quantify metals and compounds in particular oxidation states. An oxidising analyte is added to excess iodide to liberate iodine which is titrated against sodium thiosulfate (see *Titrimetric analysis*).

Ion-selective electrode An electrode whose voltage is selectively affected by an ionic species.

Laser desorption mass spectrometry A method of getting a high-molecular weight compound into the 'gaseous state' for mass spectrometry by mixing it with a matrix of smaller molecules that are then vaporised with a specially tuned laser (see *Mass spectrometry*).

Light spectroscopy An important method of analysing molecular structure by the interaction of light with a compound. Characteristic wavelengths of light are absorbed or emitted.

Magic angle spinning A technique to produce solution-like NMR spectra from solid samples which involves rotating the sample rapidly about a particular angle (the magic angle of 54.74°) to remove the parts of the spectrum which depend on molecular orientation.

Magnetic resonance imaging The use of an applied magnetic field which varies over space so that the chemical shift of a given line depends on its position in space, allowing spatial mapping of materials (see *Chemical shift*).

Mass spectrometry A method of analysing the structure of a compound by vaporising and breaking it up into ions whose mass and charge are then measured. The pattern of fragmentation also provides information about structure.

Microwave spectroscopy An analytical method for determining the structure of gaseous compounds that depends on the absorption of microwave radiation by the rotation of a molecule with a dipole moment (an asymmetric distribution of charge).
Mobile phase The phase in chromatography in which the mixture of compounds to be separated is dissolved, and which moves through the stationary phase (see *Chromatography* and *Stationary phase*).
Neutron diffraction (scattering) A method of obtaining structural and dynamic information about materials by passing neutrons through the sample crystal and measuring the diffraction pattern obtained (see *Diffraction* and *Bragg's Law*).
Nuclear magnetic resonance
A spectroscopic technique which uses radio frequency electromagnetic waves to stimulate energy transitions associated with nuclei of atoms. The experiment is conducted with the sample held in a magnetic field and the transitions correspond to changes in the orientation of the nuclear magnetic moment relative to the direction of this field.
Nuclear magnetic resonance microscopy
The use of magnetic resonance imaging to examine the spatial structure of materials (see *Magnetic resonance imaging*).
Partition chromatography A form of chromatography in which the mobile and stationary phases are liquids, the latter being held as a film on a solid support. The mixture of compounds 'partitions' to different extents between the mobile phase and the liquid film (see *Chromatography*, *Mobile phase*, and *Stationary phase)*.
Pulse Fourier transform NMR An NMR experiment in which the transitions are excited by one or more radiofrequency pulses (rather than continuous irradiation) and the resulting signal (the free induction decay – FID) Fourier-transformed to produce a frequency spectrum (see *Fourier transform* and *Free induction decay*).
Relaxation time The characteristic time constant associated with nuclear magnetic moments returning to their equilibrium orientations once a sample is removed from the magnetic field used in the NMR experiment.
Sample magnetisation The net magnetic field arising from all the nuclear magnetic moments in a sample which are partially oriented by the magnetic field used in the NMR experiment.
Solid-state NMR Nuclear magnetic resonance applied to solid samples; most NMR is still performed on solution samples where the spectral resolution is usually better.
Spin-spin coupling An interaction between nuclear magnetic moments which results in NMR lines being split into series of closely spaced lines.
Stationary phase The phase in chromatography through which the mixture of compounds to be separated moves (see *Chromatography* and *Mobile phase*).
Synchrotron radiation Electromagnetic radiation of all wavelengths produced when charged particles travelling close to the speed of light pass through a magnetic field. With the appropriate set-up, synchrotron radiation is an important analytical tool.
Titrimetric (volumetric) analysis
A quantitative method of analysis in which the volume of a solution of accurately known concentration is made to react quantitatively with a solution of a substance being determined.
Triple resonance An NMR experiment in which three different types of nuclei, for example, nitrogen-15, carbon-13, hydrogen-1, are irradiated (or excited) during the experiment. This usually leads to a multidimensional experiment which enables the spatial connections between the different types of nuclei to be determined.
Two-dimensional NMR NMR experiments designed to produce frequency spectra in which each line is a function of two frequency variables, rather than one which is more usual in spectroscopy.
Ultraviolet/visible spectroscopy The absorption or emission of ultraviolet or visible radiation by a compound to give a characteristic pattern of wavelengths that depends on the compound's electronic structure.
X-ray crystallography (diffraction) An important method for determining the structure of molecules by passing X-rays through the sample crystal and measuring the diffraction pattern obtained (see *Diffraction* and *Bragg's Law*).

Biographical details

Dr Melinda Duer is a university lecturer in the Department of Chemistry at the University of Cambridge. Current research interests include the use of NMR to study the structure and dynamics of glasses, polymers and biomaterials.

Dr Katherine Stott is a postdoctoral research associate in Biomolecular NMR in the Department of Biochemistry at the University of Cambridge. Current research interests include NMR method development and new techniques for the magnetic alignment of biomolecules.

Acknowledgements

The authors would like to thank Professors Robin Harris, Peter Sadler, Jim Turner and Ken Packer for help with the preparation of the text.

Chemical marriage-brokers

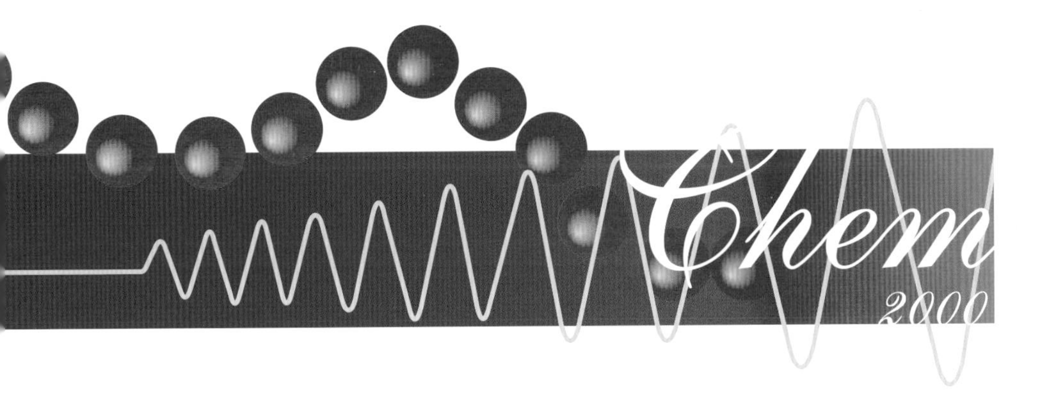

Dr Phil R Davies
Cardiff University
Dr Anthony Haynes
University of Sheffield

Modern catalysis plays a huge part in our lives. Many synthetic materials found in our homes as well as medicines and agricultural chemicals are made using catalysts. Plastics, the antifreeze in the radiators of our cars, and even fermentation are also dependent on catalysts.

Catalysis is commonly used in the brewing industry.

Catalytic converters are now in widespread use throughout the motor industry, reducing exhaust emissions, giving a cleaner environment.
Courtesy of Rover Group.

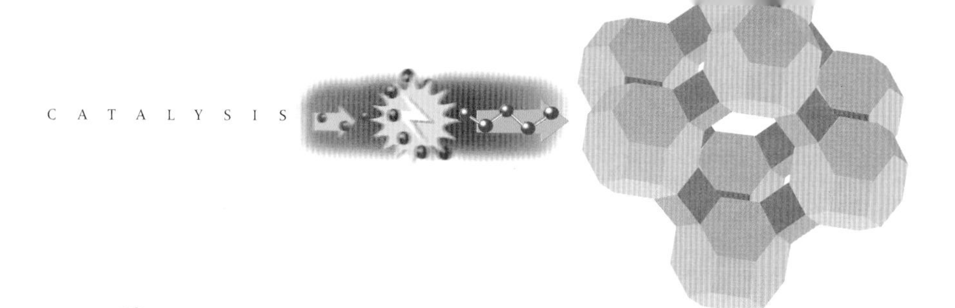

The majority of industrial chemical processes used to make the materials that Society relies on depends crucially upon substances called catalysts which speed up often impossibly slow reactions without themselves being changed. Similarly, most molecular processes in living organisms are mediated by biological catalysts called enzymes. Catalysis is thus vitally important to people's everyday lives. It is an extremely active area of research, as chemists search for more efficient catalysts to create energy-saving, more environmentally-friendly processes for the chemical industry.

Aspartame is used widely in soft drinks.
Source: NutraSweet® Information Centre.

One of the earliest and perhaps most pleasurable pieces of industrial chemistry practised by humans was fermenting fruit and grains to make wine and beer. The natural sugars are converted to ethanol by the mediation of complex molecules called enzymes released by yeasts. Enzymes, in fact, act as *catalysts* in making sluggish reactions go considerably faster. Today, most processes underpinning the chemical industries depend on *catalysis* to carry out chemical reactions that would otherwise be impracticably slow.

Modern catalysts play a huge part in our lives. Take the car, for example: its fuel – a carefully designed mixture of medium-weight hydrocarbons is produced by catalytic processing of crude oil; the plastics which are increasingly used in vehicle components rely on catalysed polymerisation processes; and the antifreeze in the radiator is made by a selective catalytic oxidation process. Even the exhaust gases from the engine are passed over a catalyst to convert them into more environmentally-friendly molecules before being released into the atmosphere.

Similar arguments apply to the many synthetic materials found in our homes, as well as to medicines and agricultural chemicals. Nature also depends on catalysts. Fermentation is just one of thousands of biological processes depending on enzymes.

The concept of catalysis was first formulated by the Swedish chemist Jöns Jacob Berzelius – the same man who developed the notation used for the elements of the Periodic Table. In his annual report to the Royal Swedish Academy of Sciences in 1835, Berzelius drew together the disparate experiments of several scientists of the time, including such eminent figures as Johann Wolfgang Döbereiner, Humphry Davy and Michael Faraday, and concluded that the experiments: "...constitute a sufficient number of examples to establish the existence of catalytic power." Berzelius described this catalytic power as substances: "...able to awaken affinities which are asleep

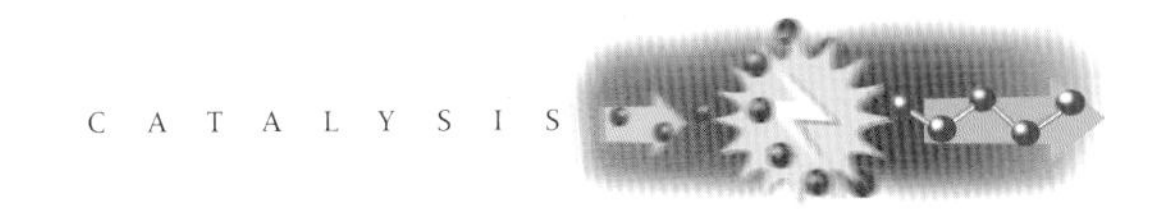

Jöns Jacob Berzelius, at the age of 47.
Reproduced courtesy of Library and Information Centre, The Royal Society of Chemistry.

at this temperature by their mere presence and not by their own affinity." This description has a curious echo today in the use, by the Chinese and Japanese, of the same ideograph for the word 'catalyst' as for 'marriage broker'. Considering the poor understanding of chemical theory in Berzelius' time, (the concept of the rate of a reaction, for example, was not developed until the end of the 19th century), his description of catalysis is very good. Substitute the word reactivity' for the word affinity' and you arrive at something not very far away from our current understanding of a catalyst as: "A substance that participates in a chemical reaction and increases its rate without a net change in the amount of the catalyst in the system." In particular, Berzelius' definition includes the important observation that catalysts help reactions to occur at lower temperatures than would normally be needed. This of course is why catalysts are so important to industry – a reaction at lower temperatures costs less money to run.

Rise of the use of catalysts in industry

The exploitation of catalysis on a major industrial scale began late in the 19th century with the production of sulfuric acid over a platinum catalyst. The process was patented in 1831 by Peregrine Phillips, a vinegar manufacturer in Bristol, but it was not until 1875 that Rudolph Messel, working for the W.S. Squire company, developed a full-scale industrial process. In the 19th century, the main customer for sulfuric acid was the dye industry. Nowadays, sulfuric acid (which is still made using the Contact process developed by Phillips and Messel, though a vanadium pentoxide catalyst, V_2O_5, has replaced the platinum one) is such a vital component of so many industrial processes, the amount of acid used by a nation is sometimes used as a measure of its technological development!

An example of acid production earlier this century. Reproduced courtesy of Public Record Office MUN 5/297/626

Despite this long history, catalysis is still one of the most active areas of research in chemistry, though its aims have changed significantly. In these environmentally conscious days, the ability of a catalyst to promote a reaction with great *selectivity* (in other words to produce the desired product with the minimum amount of waste products) is just as important as being able to drive the reaction with lower energy costs.

How does a catalyst work?

There is still considerable debate over the detailed mechanisms of nearly all catalysed processes but the overall effect of a catalyst is understood – in every case a catalyst makes a reaction go faster by providing a reaction pathway requiring less energy than those available in the absence of the catalyst. At the start of most reactions there is an energy barrier called the *activation energy*. All catalysts reduce the size of

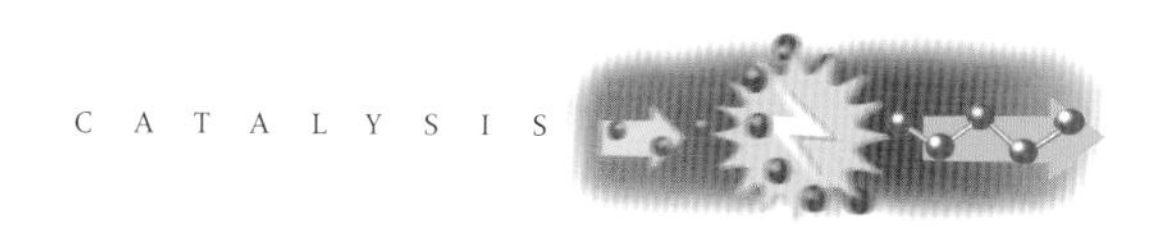

this barrier, usually by providing an alternative pathway.

The situation is similar to the problem of travelling over a mountain. When Hannibal marched on Rome, for example, he had to take his elephants over the Alps. Nowadays he could have taken them through one of the tunnels, expending considerably less energy in the process. In the case of catalysis, the alternative route for the reaction usually involves the formation of a transitionary intermediate compound of the reacting molecules and the catalyst. Once the product molecule forms, the catalyst is regenerated so that it can go on to catalyse reactions between further molecules and so on in a continuous *catalytic cycle*.

The synthesis of ammonia (NH_3) from nitrogen (N_2) and hydrogen (H_2) over an iron catalyst is the classic example. The nitrogen molecule is held together with a tough triple bond that requires an enormous amount of activation energy to break. In the gas phase, the only feasible way of doing it is with an electric discharge (a spark) and, except where electricity is cheap, that is a very expensive procedure. By 1898, more than 300,000 tons of nitrogen were being used to make ammonia for fertiliser, and it was clear that unless a cheap way could be found to 'fix' nitrogen from the air, all other known sources of the element would be exhausted by the mid-20th century. Chemists around the world worked on the problem until 1905, when a German chemist, Fritz Haber, made a breakthrough – he discovered that iron acted as a catalyst for the reaction between nitrogen and hydrogen to make ammonia:

$$N_2 + 3H_2 \longrightarrow 2NH_3$$

His discovery, which earned him a Nobel prize, was developed into a full-scale industrial process (still used today) for the Badische Anilin und Sodafabrik company (BASF) by Carl Bosch in 1914. It's often claimed that the *'Haber-Bosch'* process actually saved the world from starvation by making the ammonia for fertiliser.

The role of the iron catalyst is to provide an easy way for the strong nitrogen-nitrogen bond to break by forming new bonds to the nitrogen atoms. The nitrogen atoms are then free to combine with hydrogen atoms at the catalyst surface in a step-wise fashion to form a series of intermediate structures – first an imide (NH), then an amide (NH_2) and finally, ammonia (NH_3). The ammonia readily departs from the metal, regenerating the clean iron surface. The intermediates and overall energy changes for the catalysed and uncatalysed reaction pathways are illustrated in Figure 1.

The red line traces the path of the uncatalysed reaction, and the blue line follows the situation when a catalyst is introduced. It's clear that the transformation of reactants to products happens much more easily if it follows the catalysed path. Another important point shown by Figure 1 is that the starting and finishing points of the reaction are not changed by

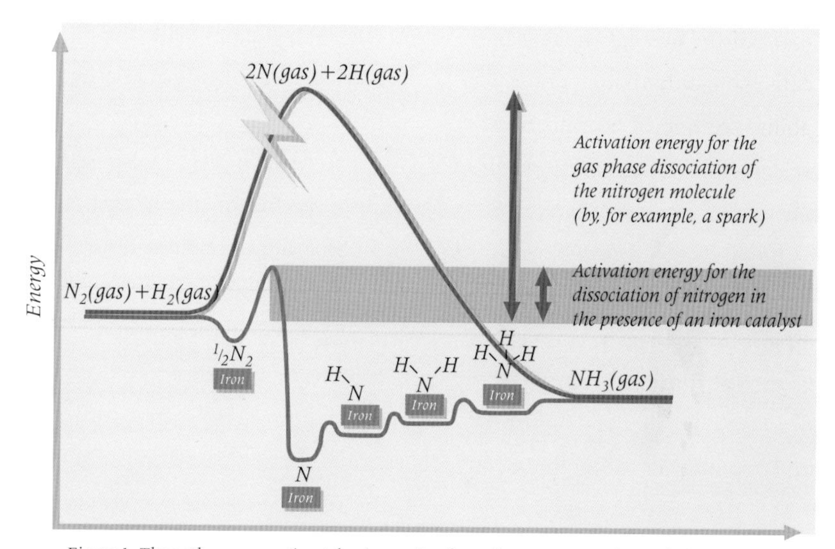

Figure 1. The pathway a reaction takes in moving from the reactants to the products can be represented in terms of the energy of the system. Most reactions must cross an energy barrier (called an activation energy) before they can happen. This, for example, is why graphite is stable in air though it burns very fiercely when heated. Catalysts make reactions go faster by lowering the activation energy barrier.

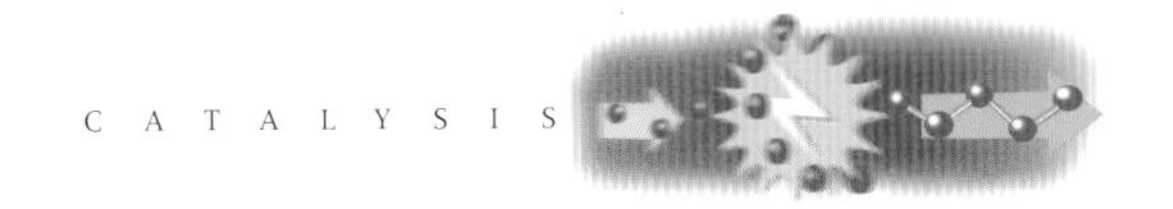

the catalyst. The position of 'equilibrium' for a reaction – in other words, how much of the products form relative to the reactants – is determined by the difference in energy between the reactants and products. What this means is that a catalyst can't make more of a product than would be produced (eventually) without it, it just helps make it faster. Nevertheless, when there are two or more competing reactions producing different products from the same reactants (which is often the case) a catalyst can produce more of the desired product by catalysing only that reaction. The ability to activate one reaction selectively is an increasingly important requirement of industrial catalysts.

What makes a good catalyst

As seen from the early use of catalysts, these materials are often metals or metal oxides particularly those of the transition elements (such as platinum, iron, cobalt and rhodium). Transition elements have complex, highly versatile electronic structures which participate in chemical reactions to form variable numbers of differing types of bonds with many kinds of atoms and molecules. It is this ability that allows them to offer alternative low-energy pathways for reacting molecules. In some instances the metal may be attached to other atomic or molecular units called *ligands* to form a *complex*. Rather than presenting a solid surface to which gaseous or liquid reactants have access – so-called *heterogeneous catalysis* (the catalyst is in a different phase from the reactants), metal complexes usually react in solution – *homogeneous catalysis* – because they are in the same phase as the reactants (Figure 2).

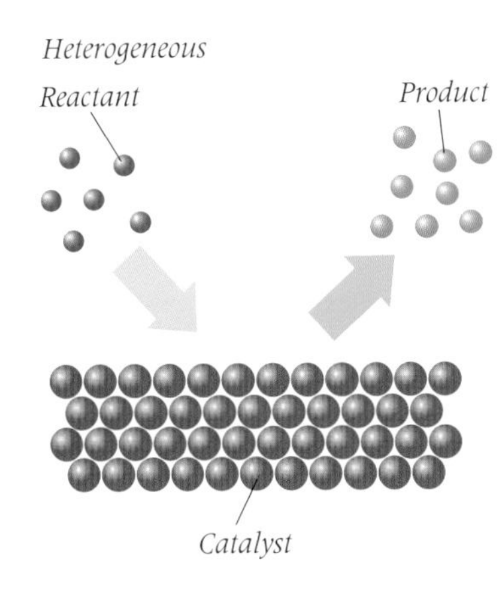

Figure 2. Heterogeneous and homogeneous catalysis.

Other materials also behave as catalysts. Acids can catalyse a wide variety of reactions, and some of the new mineral catalysts used in industry are environmentally-friendly acid catalysts. Another important group of catalysts are the enzymes mentioned earlier, which are large biological molecules.

In all cases, for catalysts to work efficiently, the *active site* – the location on the catalyst at which the reaction takes place – must first have the right electronic characteristics to form some kind of bond with at least one of the reactants leading to a lower energy reaction pathway, and these bonds must also break readily after the reaction. Secondly, the active sites must have the appropriate geometry to give access to the reactants. Indeed the configuration around the active site often determines the selectivity – which of a number of possible molecules are produced. Finally, there needs to be a large number of active sites which the molecules involved in the reaction can readily reach.

In heterogeneous catalysis, the number of active sites on a solid surface is limited (only about 1 in 10^8 atoms of a typical solid are at the surface). For a catalyst to have any effect, the reactants must collide with the right site on the catalyst surface. The number of collisions between the reactants and the catalyst active sites is one million-millionth (10^{-12}) of that between the reactants in the catalyst's absence. This means that for a reaction to go any faster, the activation energy must be reduced by as much as half. In contrast, since homogeneous catalysts are

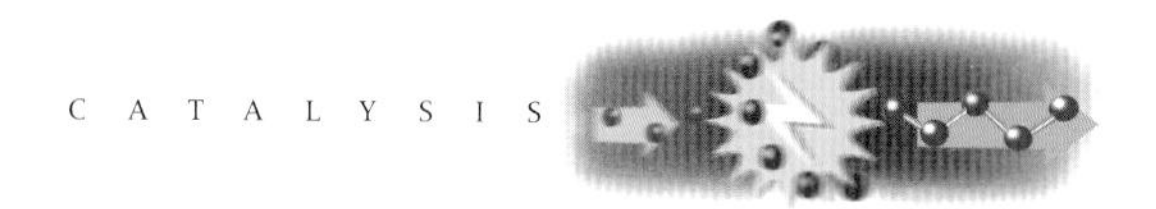

dissolved in solution, every catalyst atom can be an active site so that quite small changes in activation energy can be of benefit. Heterogeneous catalysts do, however, have one outstanding advantage – being in a different phase they are easily filtered off from the reaction mixture. Homogeneous catalysis, on the other hand, requires expensive separation techniques.

Heterogeneous catalysis and fuels

Because heterogeneous catalysts are easier to handle, they are used in 85 per cent of all catalysed chemical processes. Their economic and social role is revealed very clearly in one particular sector, that of providing the world's fuel.

Today, most of the energy needed to drive our cars and buses comes from crude oil, a complex mixture of hydrocarbons which includes long-chain alkanes, branched alkanes, alkenes (which contain carbon-carbon double bonds) and aromatic hydrocarbons (which are made of carbon-rich ring-like structures). To make crude oil suitable for use, however, it is 'refined' by several vitally important processes. The two major ones are *reforming*, in which linear alkanes are converted into branched or cyclic versions, and *cracking*, which is breaking up the heavier hydrocarbons into lighter ones. Both processes rely on heterogeneous catalysis.

How catalytic reforming won the War

"Never in the field of human conflict has so much been owed by so many to so few."

The honoured "few" of Winston Churchill's famous tribute to the RAF's victory over the Luftwaffe in the Battle of Britain, should also include a group of scientists whose feet never left the ground. These were the chemists who, shortly before the battle, developed a new reforming catalyst capable of producing 100-octane fuel from the distillate produced in the oil refinery. The new fuel helped allies' planes perform so much better that they were able to defeat an enemy who had beaten them in the same aircraft just a few months previously in the skies over France.

Reforming is essential because linear alkanes explode in petrol engines causing the phenomenon of *knocking*. Branched chain alkanes and cycloalkanes (produced by reorganising the carbon-carbon bonds in the hydrocarbon chains – a process called *isomerisation*) or aromatic compounds such as benzene and toluene (made by removing some of the hydrogen atoms – *dehydrogenation*) burn more smoothly. The success of the process is indicated by a fuel's octane number which is a measure of its anti-knocking properties. The Table gives some octane numbers of typical hydrocarbons.

Hydrocarbon		
Name:	*Structure:*	*Octane No.*
n-Hexane		19
n-Heptane		0
n-Octane		-19
2,2-Dimethylpentane		89
2,2,3-Trimethylbutane		113
Cyclohexane		110
Benzene		99
Toluene		124

Octane numbers for typical hydrocarbons found in petroleum.

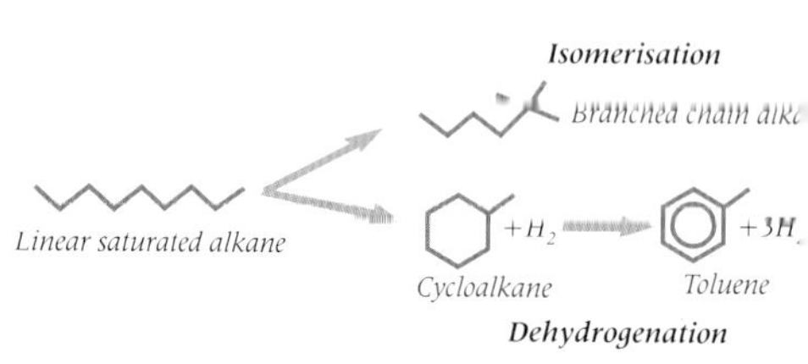

Figure 3. Some examples of the reactions that happen during catalytic reforming of fuel.

Oil companies are highly secretive about the catalysts they use because their business is so competitive. However, many of them involve small platinum particles, 2 nanometres across, supported on a base of aluminum oxide. This is a typical, though by no means the only, form for a heterogeneous catalyst. The support is there to provide a high

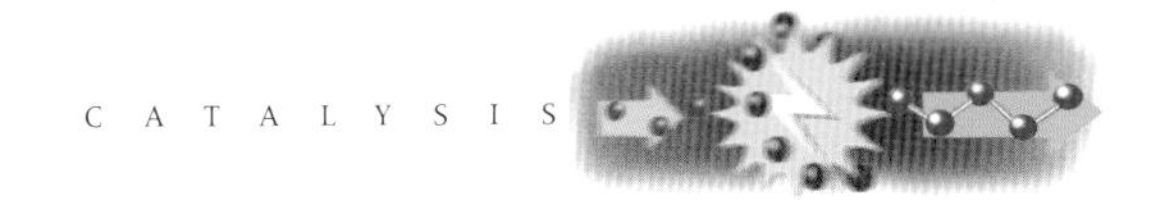

surface area (to maximise the number of active sites). It is usually inert, giving mechanical and thermal stability for the active component – although in the case of reforming catalysts the oxide support also plays an active role, catalysing the isomerisation reactions while the platinum is responsible for hydrogenation and dehydrogenation steps.

Zeolites – cracking good catalysts

The larger molecules in crude oil burn too slowly to be used in a petrol engine. So to squeeze more petrol out of crude oil, these heavier hydrocarbons can be split into smaller ones. This is a demanding exercise because it involves breaking relatively strong carbon-carbon bonds and so requires high temperatures. Catalysts such as platinum on aluminum oxide could not withstand such temperatures, instead the cracking catalysts of choice are *zeolites*. These natural minerals were discovered by the Axel Fredric Cronstedt, in 1756, a member of the Stockholm Academy of Sciences. (He also discovered nickel, another element widely used as a catalyst.)

The name zeolite derives from the Greek words *zein*, meaning 'to boil', and *lithos*, meaning 'stone'. It was chosen by Cronstedt because he found stilbite – the first zeolite he discovered – released large amounts of water when heated, giving the impression that the substance was boiling. In fact, that is the clue to why zeolites are such valuable catalysts. They are riddled with tiny interconnecting channels, or pores, which provide a huge absorbing surface area anything up to 500 square metres per gram. That's about the area of two tennis courts in an amount of catalyst you can pick up with a teaspoon! The channels, which are between 0.3 and 1 nanometre wide, are constructed such that almost every atom of the catalyst's crystal lattice is at a surface and therefore available as an active site. Zeolites are thus sometimes referred to as three-dimensional surfaces. The channels are formed by stacking rings made up from a lattice of tetrahedral units of silica (SiO_4) that link together at the corners. The rings can involve 4, 6, 8, 10 or 12 tetrahedra.

The natural mineral, stilbite, discovered by Axel Fredric Cronstedt whilst walking his dog. The story that Cronsted named the mineral after his dog is probably wrong. The sample shown was obtained from Lonaula, Pune, Maharashtra, India and is part of the collection in National Museum of Wales.

Photograph reprinted with permission from the National Museum of Wales. NMW 94.23G.M.2

Figure 4. The building blocks of zeolites. Zeolites are constructed from tetrahedral units such as the SiO_4 unit shown in (a) which link together through the oxygen atoms as shown in (b). The linked tetrahedra can form rings which stack together to form planes (c) and fold to form 'cages' such as that shown in (d). The cages can stack together giving linked channels as in (e).

There are about 30 naturally occurring zeolites which are made up primarily of silica units with some replaced by alumina (AlO_4) ones. The latter require an extra charge-balancing cation such as sodium or calcium, because the

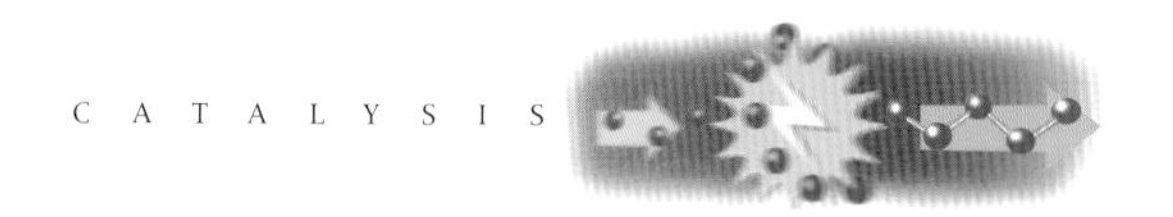

highest oxidation state for aluminium is 3+ compared with 4+ for silicon. Since Cronstedt's discovery, chemists have also made more than 150 synthetic zeolites, some of them incorporating non-natural building blocks, for example, zinc (ZnO_4), germanium (GeO_4) and phosphorus (PO_4).

Zeolites were first introduced into the chemical industry in the 1940s as absorbents and filters rather than catalysts. However, in the 1960s Jule A. Rabo and colleagues working in the Union Carbide laboratories showed that the new synthetic zeolites also had very good catalytic properties. Zeolites have two characteristics which caused particular excitement. The first is their high, but tunable acidity – the ability to donate hydrogen ions, or protons *(Brønsted acidity)* to a reactant, or accept pairs of electrons from it *(Lewis acidity)* to make a new chemical bond. This is important because it is the acidic sites that are catalytically active. They are formed where an aluminum atom replaces a silicon. Aluminum has a valency of only three, whereas silicon has a valency of four which leaves one of the oxygen atoms in the lattice with one bonding site too few. The oxygen atom either picks up a hydrogen atom to form a hydroxyl (OH) group (a Brønsted acid site because it can later donate the hydrogen ion) or leaves the lattice altogether leaving a positive charge on a silicon atom, (a Lewis acid site which can accept negative electrons). The total number of acid sites therefore depends on the aluminum concentration in the zeolite. The type of acid site (Lewis or Brønsted) can be controlled by heating (which causes Brønsted sites to lose hydroxyl groups to give Lewis acid sites) or by reaction with proton donors or acceptors such as water and pyridine. Different reactions happen at the differing acid sites, so controlling the zeolite's acidity allows chemists to decide which reactions occur.

Brønsted acid sites

Heat

Lewis acid site

$+H_2O$

Figure 5. Acid sites are formed on the surface of a zeolite where an aluminum atom replaces a silicon atom. One of the associated oxygen atoms loses a bonding site and therefore a hydroxyl group is formed. These hydroxyl groups have strong Brønsted acidity. If the zeolite is heated, the hydroxyl groups can combine to eliminate water leaving oxygen vacancies in the lattice. These vacancies are strong Lewis acid sites.

The second useful aspect of zeolite structure is that you can control, even design, the precise shape of the pores. The size and shape of the channels affect how easily particular molecules diffuse through them and thus what products form. For example, the cracking of large hydrocarbon molecules in some zeolites produces almost exclusively *para*-xylene. Xylenes are benzene rings with two methyl groups each attached at one of three possible relative positions around the ring (see Figure 6). You can see that the *ortho* and *meta* versions of xylene are too wide to escape from the catalyst pores, while the *para*-xylene can squeeze out quite easily. The undesirable products are therefore forced to rearrange themselves to *para*-xylene before escaping. In this way zeolites can select the products of the reaction.

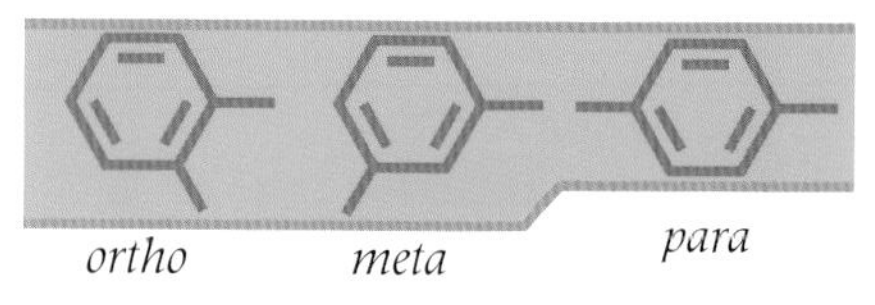

Figure 6. The relative dimensions of the xylene isomers. Para-xylene can squeeze through much narrower channels than either of the other two isomers of xylene.

Tailored catalysts

In recent years, chemists have been designing and making ever more sophisticated zeolite catalysts with channels of precise dimensions. Some chemists have been using organic molecules such as *N,N*-diisopropylethylamine and cyclohexylamine to control the growth of a zeolite crystal from solution. The molecules act as templates – they each carry a positive charge which pulls the crystallising negative silicate ions around it in such way as to produce cavities and channels of the required dimensions. However, the problem is to predict

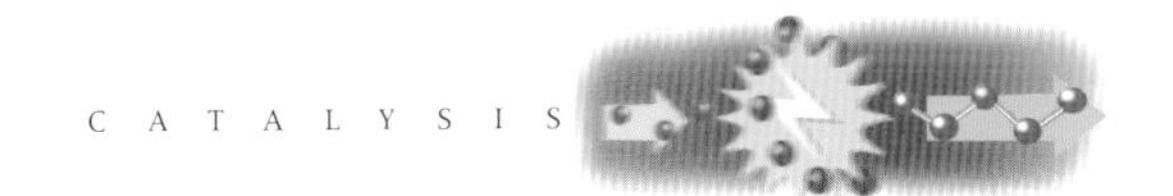

which template will give the required structure. Dewi Lewis, Sir John Meurig Thomas and colleagues at the Royal Institution in London and David Willock at Cardiff University have tackled this problem with a computer program called ZEBEDDE (zeolites by evolutionary de-novo design) which can predict suitable templates to produce particular zeolite structures.

One drawback of the majority of zeolites is their small pore size (mostly less than 1 nanometre) which hinders diffusion of larger molecules. Recently, however, C.T. Kresge and his colleagues in the Mobil Research and Development Corporation in New Jersey have succeeded in growing silica frameworks with much larger pore sizes, ranging from 1 nanometre (a microporous solid) to as much as 10 nanometres (a *mesoporous* solid) with surface areas as large as 1000 square metres per gram. The researchers use carefully chosen liquid-crystal templates to control growth in solution and obtain the required pore dimensions.

These large custom-built pores are ideal substrates on which to 'graft' other molecules. This offers the tantalising prospect of combining the best of the homogeneous catalysts – high selectivity and activity at low temperatures – with the benefits of heterogeneous catalysts – ease of recovery of the catalysts and durability. Sir John Meurig Thomas and his colleagues' work in the Royal Institution is an excellent example of this approach. They have produced a hybrid heterogenous/ homogeneous catalyst with exceptionally good activity and selectivity for the selective oxidation of large molecules such as cyclohexene, by grafting titanocene (an organic compound in which a titanium atom is

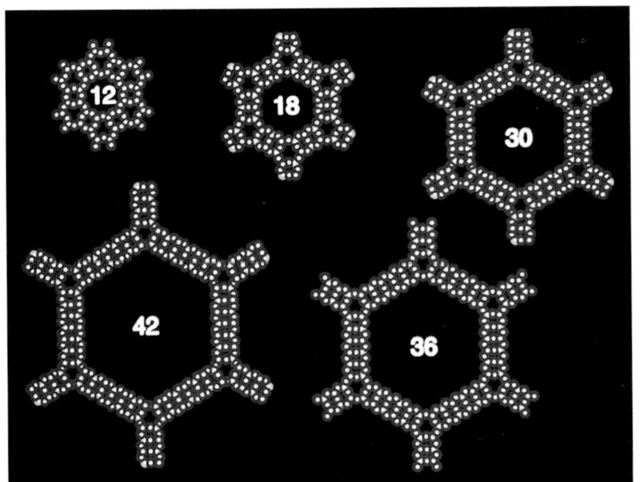

Figure 7. The maximum size of channels in synthetic zeolites has increased dramatically in recent years from a maximum of 1 nanometre to about 10 nanometres. The number of tetrahedral units in the walls of the channels of the new mesoporous' zeolites are shown in the graphic. (Reproduced from Faraday Discussion 105 with permission of the authors.)

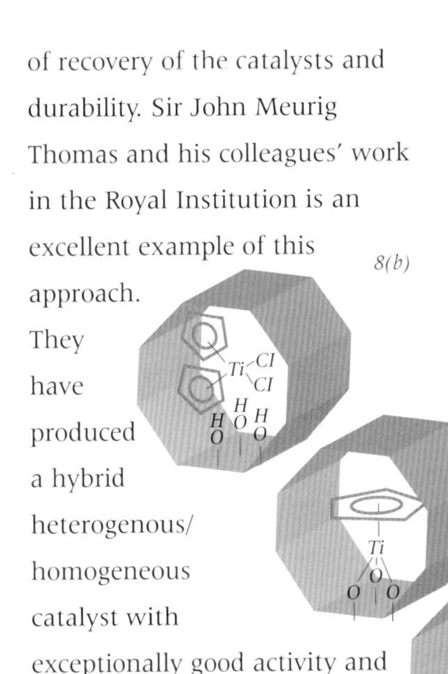

Figure 8. (a) A computer-generated image of the diffusion of titanocene through the pores of a mesoporous solid (reproduced from Faraday Discussion 105 with permission of the authors.); (b) a schematic representation of the steps involved in the reaction of the titanocene with the walls of the pores of the zeolite. The titanocene complex reacts with hydroxyl groups at the surface to form an adsorbed titanium-containing complex. On heating in oxygen, this complex decomposes to give the catalytically active centre. This approach is designed to make the most of the strengths of both heterogeneous (ease of catalysts processing) and homogeneous catalysis (high selectivity).

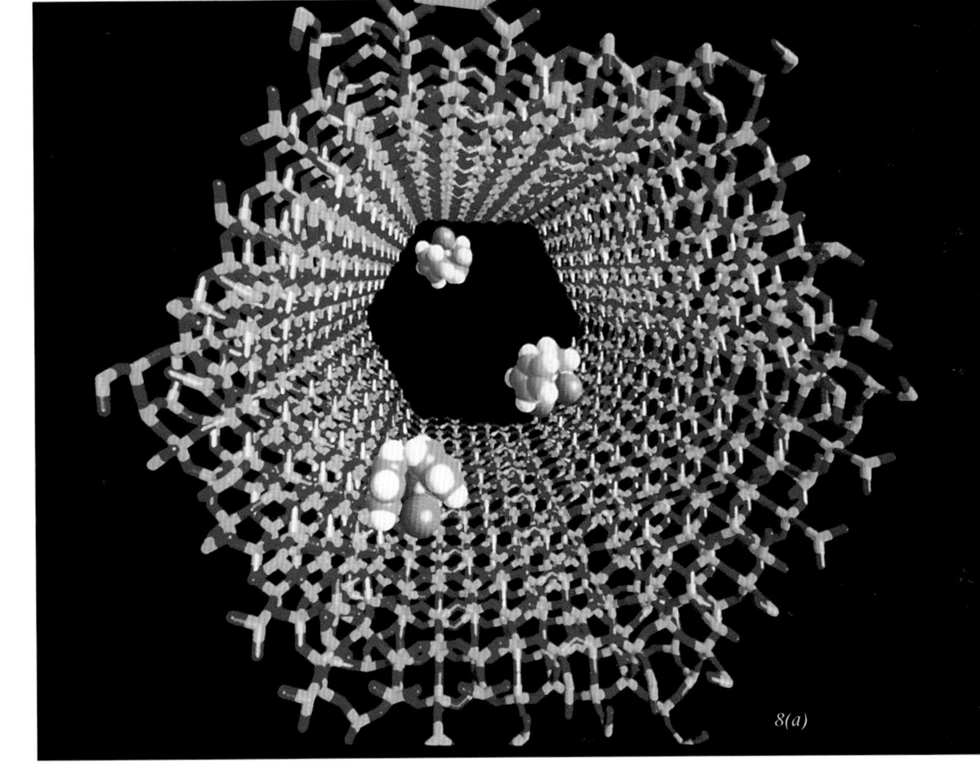

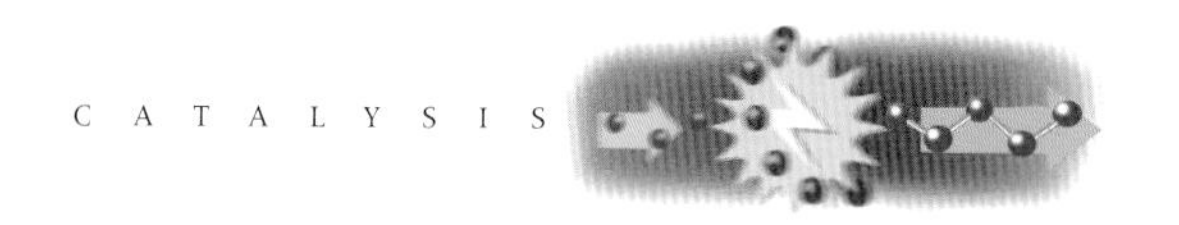

sandwiched between five-membered carbon rings – see Box 3) onto the walls of one of the new mesoporous solids. Because of the large pore sizes the titanocene can spread rapidly and evenly throughout the catalyst surface to produce a large number of identical reaction sites. Once the titanocene has reacted with the walls it is converted to catalytically active titanium (Ti^{4+}) sites by heating in oxygen.

Fuels for the future

Ultimately the world's oil reserves will run out. Whether it takes 50 or 100 years, we will need alternative supplies of energy. People have already suggested that oil, which contains an enormous variety of complex, difficult-to-synthesise chemicals, is far too valuable a resource to be burnt in petrol engines. Many of the solutions to this problem depend on catalysis and some, for example, *Fischer-Tropsch* catalysis, are already in use.

The Fischer-Tropsch reaction makes use of the world's rich coal reserves which should last much longer than oil. Coal is first broken down into a mixture of hydrogen and carbon monoxide – so-called synthesis gas, or *syn-gas* – using a process called *coal gasification*:

$$C + H_2O \longrightarrow CO + H_2$$

The carbon monoxide and the hydrogen are then recombined (over metal catalysts such as cobalt, iron, nickel and ruthenium) to form useful long-chain carbon molecules. The Fischer-Tropsch reaction was used extensively in the past but eventually became uneconomical – except for those countries without access to good supplies of oil such as Germany during the Second World War and South Africa during its years of isolation. In the early 1970s (when oil prices were increasing rapidly) there was considerable interest in how Fischer-Tropsch catalysts worked. The hope was that a better understanding of the reaction mechanism would help to develop more efficient catalysts. The key question was whether or not the carbon monoxide dissociates into carbon and oxygen atoms while adsorbed on the catalyst surface. If the molecule remains intact, then the reaction probably proceeds through a hydroxymethylene intermediate as shown in Figure 9.

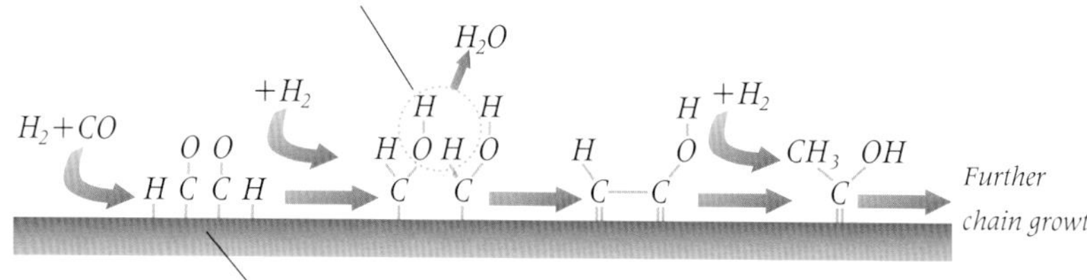

Figure 9. A suggested mechanism for the Fischer-Tropsch reaction in which carbon monoxide remains in a molecular state at the catalyst surface.

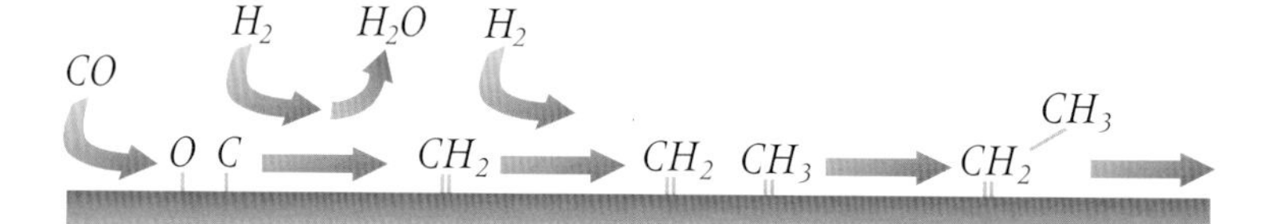

Figure 10. An alternative mechanism for the Fischer-Tropsch reaction in which carbon monoxide dissociates at the catalyst surface.

If, on the other hand, the carbon monoxide breaks up, the reaction must involve an adsorbed carbon atom which picks up hydrogen atoms to form adsorbed CH_2 groups. These subsequently combine to form longer-chain alkanes.

To answer this question, chemists studied carbon monoxide at the surfaces of a series of metals, using a technique known as photoelectron spectroscopy in which molecules at the surface are ionised with photons of light (see reference 3 for further details). The energies of the ejected electrons depend upon the bonding characteristics of the molecule from which they originate. The spectra revealed the presence of molecular carbon monoxide on poor Fischer-Tropsch catalysts but none on effective catalysts, suggesting that carbon monoxide does dissociate during the reaction.

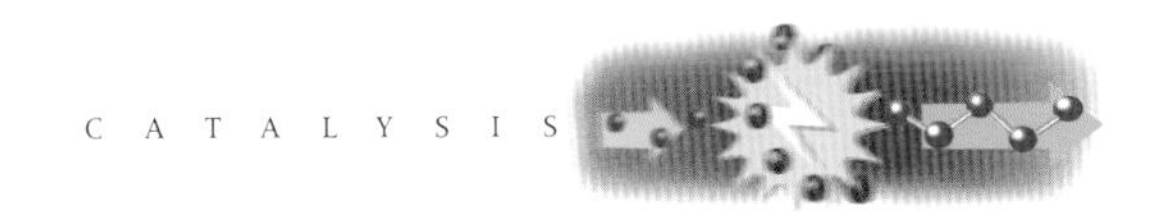

Methanol – a fuel for the future?

Another future fuel is methanol (CH_3OH). More than 25 million tonnes of methanol are manufactured each year using a catalyst of copper supported on zinc oxide. As well as being an important feedstock for the chemical industry (it is used both as a solvent and as a reactant), methanol can also be burnt directly as a fuel or as a source of hydrogen for fuel cells – electrochemical cells which convert chemical energy to electrical energy with very high efficiencies (see *The power of electrochemistry*).

Like the Fischer-Tropsch process, making methanol involves syn-gas although a different mixture of gases is used (typically one part carbon dioxide to one part carbon monoxide and eight parts hydrogen, and a few other minor components depending on its source). The copper/zinc oxide catalyst is both highly selective and very active, that is to say, it produces virtually pure methanol, in good yield at

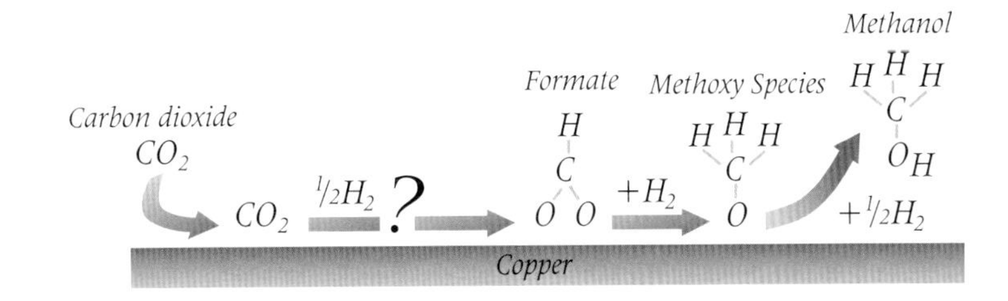

Figure 11. Known intermediates in the synthesis of methanol from carbon dioxide and hydrogen over a copper catalyst. Debate is now centred on how the carbon dioxide is activated at the copper surface, since experiments suggest that carbon dioxide on its own does not react with copper fast enough to support the synthesis of methanol.

the relatively low temperature of 200 to 300 °C and at pressures between 50 and 100 atmospheres.

$$CO/CO_2/H_2 \longrightarrow CH_3OH$$

There is, nevertheless, still vigorous debate about how the catalyst works. There are many clues – for example, experiments where the reacting gases are 'labelled' with radioactive versions of the constituent elements have shown that all the carbon in the methanol comes from the carbon dioxide rather than carbon monoxide. We also know that the reaction happens on the copper component of the catalyst not the zinc oxide.

Another clue is that spectroscopy reveals the presence of a particular molecular species – a formate (HCO_2) – on the catalyst surface. Formate reacts with hydrogen to form methanol and so could be a key intermediate.

Chemists have found it difficult to glean further information about the reaction because it is carried out at such high temperatures and high pressures. Instead they have turned to model studies that look at the reactions of the individual components. The key problem is that carbon dioxide does not react with copper fast enough to explain the rate at which methanol is produced over the catalyst. Since the overall reaction cannot be faster than its slowest step the activation of carbon dioxide has become a question of profound interest.

The clue that may hold the answer to the mystery of the unreactive carbon dioxide actually has come from studying oxygen. Researchers had assumed that any oxygen on the catalyst would be in the form of an unreactive oxide (O^{2-}). Work over the past 10 years in the chemistry department at Cardiff University, however, (see reference 4) has shown that oxygen can exist at metal surfaces in several different chemical states, all of which are far more reactive than the oxide. The first of these states is the dioxygen (O_2) molecule itself. This remains stable until it acquires the two electrons necessary to break the O-O bond, but that can be long enough for it to react with other materials adsorbed onto the catalyst surface.

At very low temperatures (-269 °C) molecular oxygen can have a lifetime on copper surfaces of several hours and this has allowed chemists to use a devastatingly powerful new family of techniques called *scanning probe microscopy* (SPM) actually to see the molecules.

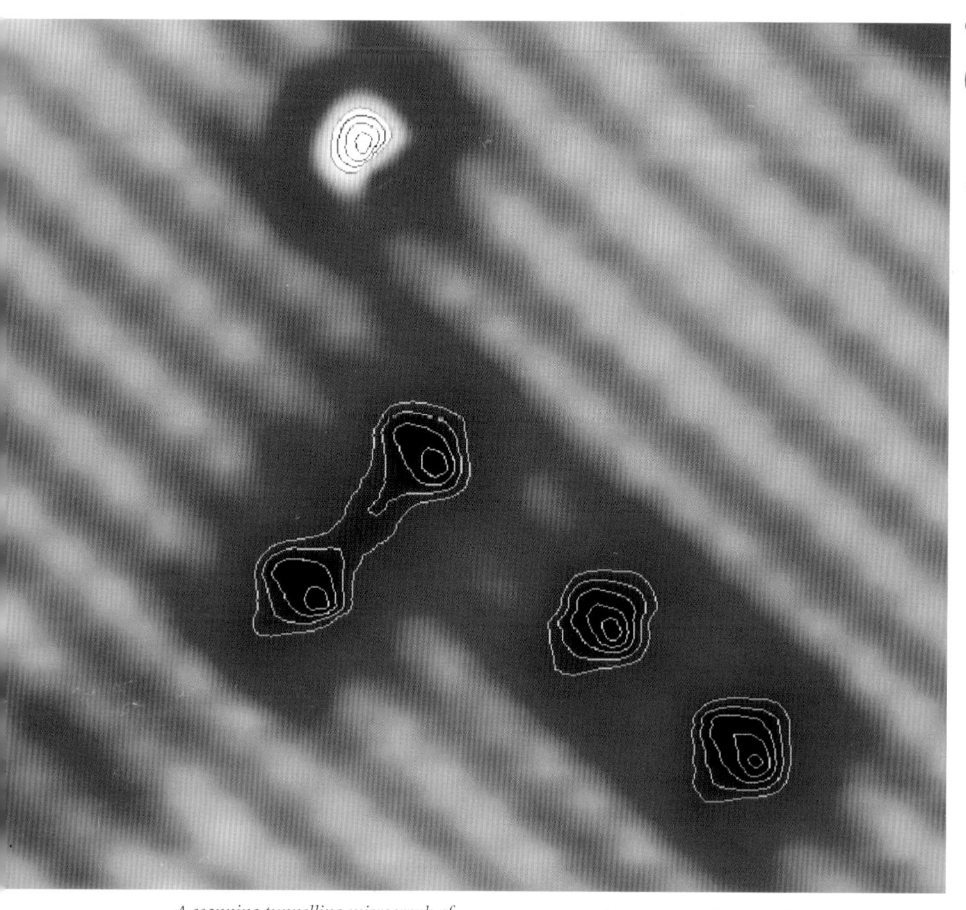

A scanning tunnelling micrograph of molecular oxygen on a copper surface showing clearly the individual atoms.
B.G. Briner, M.Doering, H.P. Rust and A.M. Bradshaw, Phys. Rev. Lett. **78**, 1516 (1997)

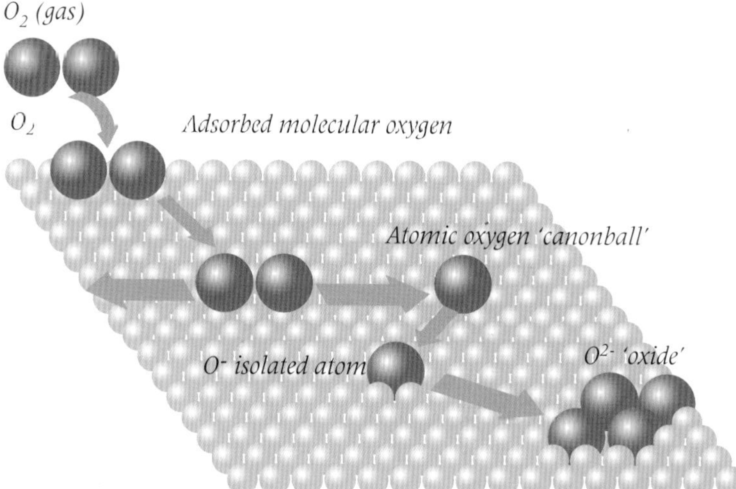

Figure 12. A pictorial representation of the different states of oxygen at metal surfaces. Adsorbed molecular oxygen, which has been observed at surfaces by many techniques including scanning tunnelling microscopy, dissociates to produce very energetic oxygen atoms which fly up to 4 nanometres across the surface before finding stable adsorption sites. Once adsorbed the atoms remain very reactive until islands of oxygen build up around them.

Seeing atoms and molecules at surfaces

The SPM techniques do not rely, as traditional microscopy does, upon the reflection of light to create an image, rather they register the shape of the surface using a needle tip that can be positioned by piezoelectric drives to within one hundred-billionth of a metre, in other words, to a distance that is smaller than the width of a single atom! The needle tip is tracked back and forward over the surface detecting atoms by measuring either the number of electrons that 'tunnel' from the surface to the needle (*scanning tunnelling microscopy* – STM) when an electric potential is applied to the needle, or the repulsive force between the sample and the tip (*atomic force microscopy* – AFM). SPM allows us to see individual atoms and watch reactions in progress. Since its invention about 15 years ago it has revolutionised our understanding of catalysis.

When oxygen molecules dissociate on the surface, STM images show that the oxygen atoms can end up as much as 8 nanometres apart, even though the oxygen-oxygen bond length is only 0.132 nanometres. This implies that when the bond breaks the oxygen atoms fly apart like cannonballs. Even after the oxygen atoms have found stable sites, however, as long as they remain isolated they are highly reactive, it is not until islands of oxygens develop that the unreactive oxide develops. Recent work has shown that when carbon dioxide encounters the reactive oxygen atoms rather than the oxide islands it reacts very readily to form a carbonate. This suggests a different mechanism for methanol synthesis in which carbonate is hydrogenated to

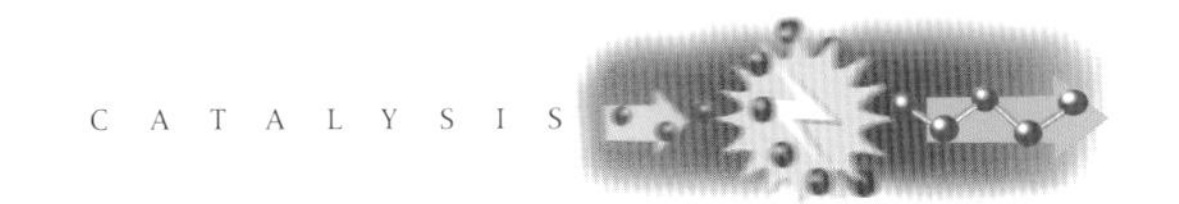

Box 1 The importance of surface studies to catalysis

The way that the atoms are arranged on the surface of a single crystal depends upon the angle at which the surface cuts through the crystal lattice. In 1982 Gabor Somorjai at the University of California at Berkeley and colleagues showed that the exact structure of the surface can have a profound effect on its catalytic properties. He studied the synthesis of ammonia on different iron surfaces with different structures and found that some surfaces were 418 times more catalytically active than others. This example nicely illustrates the importance of the surface to catalysis. Since 99.999999 per cent of the atoms of a catalyst are not at the surface they have no contact with the reactants. In fact, of the 1 in 10^8 atoms that are at the surface, it is thought that very few have the right combination of geometry and electronic structure to act as a catalyst. Identifying these active sites is one of the holy grails of catalytic research.

Studying surface chemistry with electrons

Obviously studying the chemistry of the surface is essential if we are to understand the processes that go on in heterogeneous catalysis. Unfortunately this is not as easy as it seems; most spectroscopic techniques sample all of the atoms of the catalyst and the contribution to the signal from the surface atoms is swamped by the atoms in the bulk material. Chemists have therefore had to search for techniques that will probe only the surface region.

Electrons provide one possible solution to the problem of studying surfaces because they can travel only short distances through a solid before they lose energy. If we study only those electrons with characteristic energies we can be sure we are looking at the surface region.

Over the past 30 to 40 years, a wide range of spectroscopic methods based on electron spectroscopy has been developed, giving a comprehensive picture of the surface. Some of these are listed below:

Technique:	*Information obtained:*
Low energy electron diffraction (LEED) Because electrons can behave like waves they can be diffracted like light waves. This can be used to give information about the arrangement of atoms at a surface in a similar fashion to the way X-ray diffraction tells us about the structure of the bulk.	*The structure of the surface, ie the geometric arrangement of atoms at the surface*
Photoelectron spectroscopy Electrons are knocked out, or ionised from a sample using photons of light, the energy of the electrons that emerge depends upon the atom from which they are ejected and its precise chemical environment. Photoelectron spectroscopy can therefore give quantitative information on the elements and molecules at a surface.	*Elemental composition and information on the chemical environment.*
Auger electron spectroscopy After an atom has been ionised, it is left in a highly excited state and needs to get rid of some of the energy it possesses. One way it can do this is to eject another electron, a so-called 'Auger' electron, named after the Frenchman, Pierre Auger, who discovered them. The energy of the Auger electron also depends upon the atom from which it is ejected and so Auger spectroscopy can also give quantitative information on the elemental composition of a surface.	*Elemental composition of the surface.*
Electron energy loss spectroscopy If an electron collides with a molecule at a surface it can give some of its energy to cause vibrations in the molecule. By measuring the amount of energy lost by electrons after they have collided with a surface a vibrational spectrum (very similar to that provided by infrared spectroscopy for bulk samples) can be obtained. In many cases this spectrum, in addition to identifying molecules at the surface, can also tell us something about their orientation.	*Molecular species present; information on the surface orientation of molecules is also sometimes available.*

No single technique is sufficient in itself and the most important advances in surface chemistry have usually come from combinations of techniques. In recent years, a variety of new methods have emerged to complement the electron spectroscopies, including using reflected infrared radiation, X-ray techniques using synchrotron radiation, and in particular scanning probe microscopy methods which are capable of imaging individual atoms at surfaces.

The importance of surface chemistry stretches far beyond even the broad topic of catalysis. Corrosion and adhesion for example are surface phenomena, the fabrication of semiconductor devices requires exquisite understanding and control of surface processes and the astonishing synthesis of diamond films which is currently revolutionising materials chemistry occurs via a series of surface reactions. Catalysis has been the driving force for much of the surface chemistry that is studied but its effects stretch to much broader horizons.

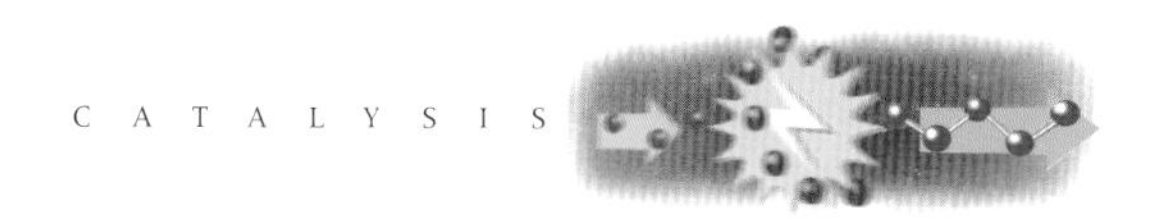

formate (Figure 13). Debate, however, continues to rage over how methanol is formed on copper.

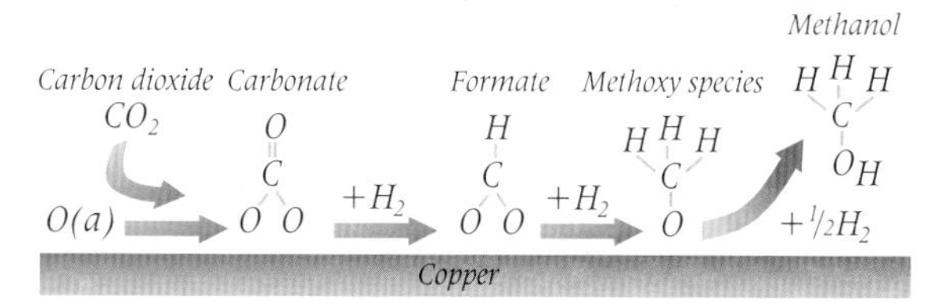

Figure 13. A suggested reaction mechanism for the synthesis of methanol at copper surfaces from hydrogen and carbon dioxide. Trace quantities of oxygen at the copper surface react with the carbon dioxide to give a carbonate which is subsequently hydrogenated to a formate. The activation of carbon dioxide by oxygen transients is significantly faster than the reaction of carbon dioxide with the clean copper surface.

Such surface studies are crucial to a better understanding of heterogeneous catalysis. Some of the methods used are listed in Box 1.

Homogeneous catalysis

Although most catalytic processes in commercial use are heterogeneous, some reactions used both in the laboratory and by industry rely on homogeneous catalysis, particularly to make complex organic compounds – *fine chemicals* – rather than fuels and simple molecules used as feedstocks to manufacture other chemicals. As explained earlier, homogeneous catalysts, which are mostly transition metal complexes, are often more active because all the metal ions are dissolved in solution where they can perform their job. They therefore work under milder conditions with greater selectivity. Furthermore, since the discrete molecules of a homogeneous catalyst can often be isolated, researchers can study their chemistry in detail and make subtle changes to tune the properties of the catalyst.

Nature's catalysts

Another set of homogeneous catalysts are those used by Nature – *enzymes*. They are proteins, large molecules, mostly soluble in a watery environment, whose complex shapes contain active sites – 'clefts' of precise geometrical shape designed to 'recognise' and hold in place a particular molecule while it reacts. A simple way to imagine the way in which an enzyme works is the 'lock and key' model shown in Figure 14. In some enzymes, a metal ion lies at the heart of the active site where it plays a key role in the catalytic activity. Carbonic anhydrase, for example, is a zinc-containing enzyme which catalyses the reaction of water with carbon dioxide in living organisms. The metalloenzyme speeds up the rate of the reaction 1 billion (10^9) times, but if the zinc ion is stripped away, the enzyme is no longer active.

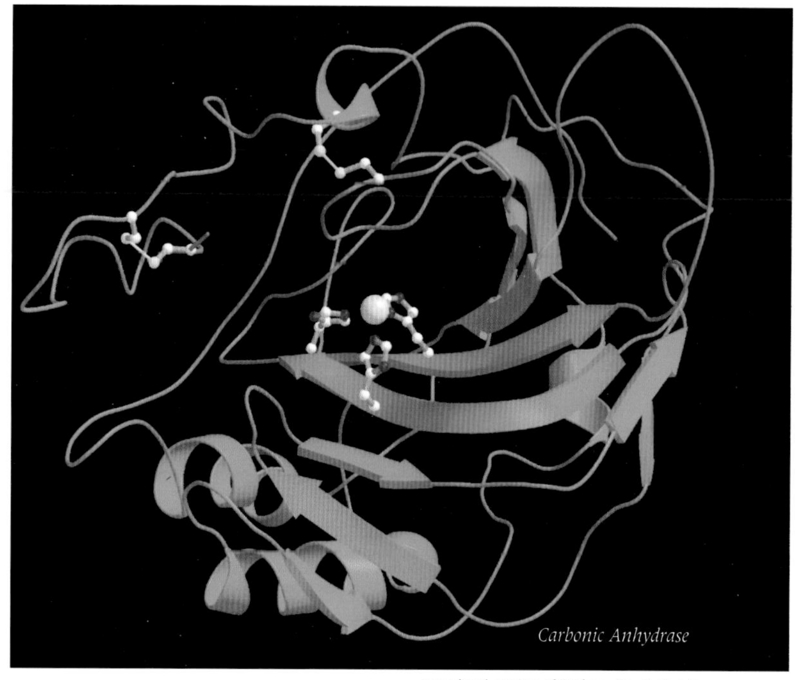

Reproduced courtesy of Professor David Christianson

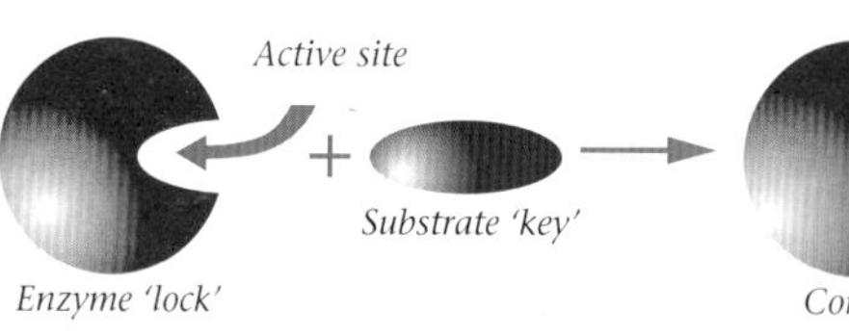

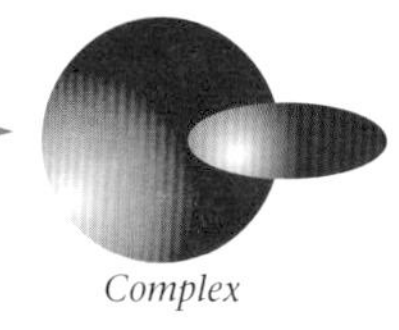

Figure 14. The enzyme has an active site, or 'lock', which recognises and binds the substrate 'key'. The chemical reaction occurs within the resulting complex.

Enzymes are used in industry for specific reactions (Box 2). They are highly selective and work under very mild conditions. In some ways they are too selective, usually acting on one particular molecule, or substrate. Many chemists would like to devise catalysts based on transition metal complexes that work as effectively as enzymes but that act on a wide variety of substrates.

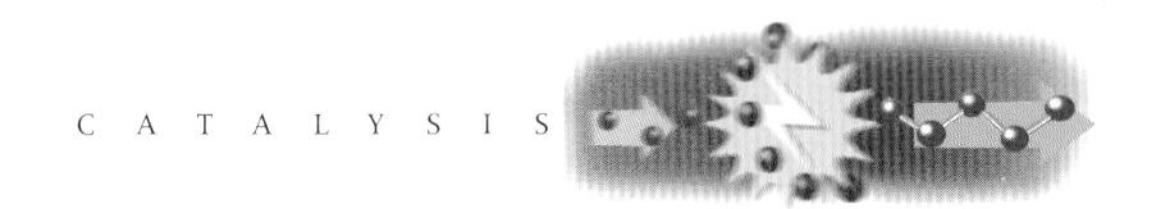

Box 2 Enzymes in industry

Enzymes, the catalysts of Nature, are present in all living matter. This family of high-molecular weight polypeptides (proteins) promotes a wide variety of chemical transformations. Typical reactions catalysed by enzymes include the hydrolysis of esters and amides – in other words, the break-up of esters or amides into the organic acids and the respective alcohols or amines from which they were derived. Such enzymes promote reactions extremely efficiently, under very mild conditions, for example, at temperatures just above room temperature, a pH near 7.0 and under atmospheric pressure. This efficient methodology is very appealing to the chemical industry.

The natural catalysts are extremely selective in their actions; thus enzymes that hydrolyse esters (called esterases) do not affect amides. Conversely amide-hydrolysing enzymes (amidases) do not cleave esters. Not only are enzymes selective they also display exquisite stereoselectivity (see Synthesis). This means that they distinguish between compounds that have the same structure but are differently arranged in three-dimensional space, such as chiral compounds, which are mirror images of each other – so-called enantiomers. For example an esterase, when given the opportunity to hydrolyse a mixture of a chiral esters, ($R_1R_2CHCO_2CH_3$), will often hydrolyse one enantiomer more rapidly than its mirror image, giving a chiral acid ($R_1R_2CHCO_2H$) and a recovered chiral ester. This process is important in the production of non-steroidal anti-inflammatory agents.

Other enzymes catalyse reactions other than hydrolysis. Reductase enzymes catalyse reduction reactions, for example, the reduction of a ketone (R_1COR_2) into a secondary alcohol (R_1CHOHR_2). Once again, one enantiomer of such a chiral alcohol will be formed preferentially, in many cases. The reductase enzymes are more complicated than the hydrolase enzymes. This is because the hydrogen atoms in the secondary alcohol R_1CHOHR_2 are not derived from the enzyme but from an associated *cofactor* – and solvent. (Enzymes often require other molecules, cofactors, to work.) Because of the need to have an enzyme and cofactor together to effect the reduction, reductase enzymes are often employed in their natural habitat, in other words, in a whole cell such as that of a bacterium or fungus. Thus bakers' yeast (a fungus) has a set of reductases which, with their cofactor(s), reduce many ketones stereoselectively to the corresponding alcohols, important components in many pharmaceutical products.

It is not uncommon to use whole cells in biotransformations. Not only are the necessary cofactors in place but unstable enzymes can be used without being damaged by extraction procedures. This is particularly true in the area of oxidative transformations. Interesting enzyme-catalysed oxidations include reactions that add a hydroxy (OH) group to a molecule. Some of these hydroxylations are particularly important in the pharmaceutical industry; the whole-cell catalysed conversion of progesterone into 11-hydroxyprogesterone opened the way to the simple preparation of a wide range of anti-inflammatory steroids.

It has long been recognised that enzymes will perform transformations on natural starting materials. In recent years, it has become increasingly apparent that many enzymes will equally well catalyse their particular reactions on non-natural substrates, while retaining chemical and stereo-selectivity. Even if the enzyme is not exactly right for a particular transformation, its structure can be altered by genetic engineering to confer better properties. For example, soap powders contain an amidase (subtilisin) which degrades protein. The enzyme has been adapted by controlled mutation to be effective at different temperatures, and is added to various washing powders depending on whether they are going to be used in a 40 °C, 50 °C or 60 °C wash.

Even a microorganism can be mutated to include a new enzyme to allow a desirable cascade of reactions to take place. For example the microorganism *Erwinia herbicola* was known to convert (D)-glucose into 2,5-diketo-(D)-gluconate. Equally it was known that a different microorganism *Corynebacterium sp* contained a reductase which transformed 2,5-diketo-(D)-gluconate into 2-keto-(L)-gulonate, which itself is easily dehydrated to form vitamin-C. By transferring the *Corynebacterium* reductase gene into *Erwinia* a recombinant organism was produced capable of converting (D)-glucose into 2-keto-(L)-gulonate, thus simplifying the preparation of vitamin C, a food additive.

As outlined above, enzymes are having an increasingly important role to play in the chemical industry, particularly the food, cleaning and pharmaceutical sectors.

Professor Stan Roberts
University of Liverpool

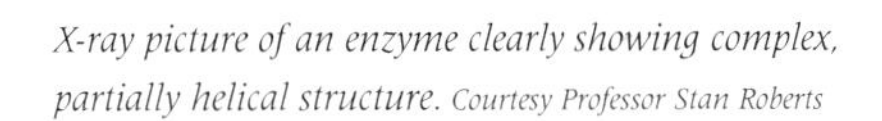

X-ray picture of an enzyme clearly showing complex, partially helical structure. Courtesy Professor Stan Roberts

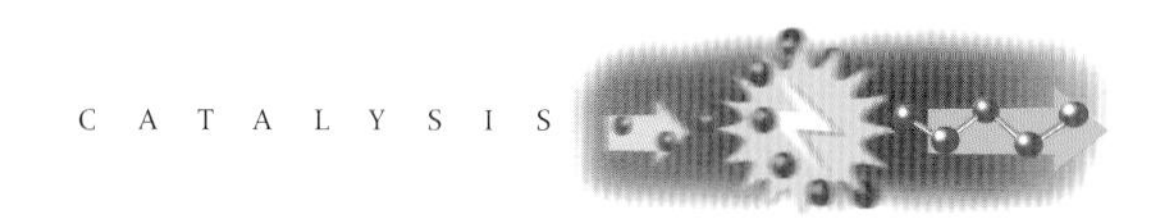

Versatile transition metal complexes

Transition metal complexes are very effective catalysts because the core metal atoms can bind to a wide variety of molecules or ions (see *Make me a molecule*). Although they may remain bound throughout the catalytic process without being consumed or directly involved, these *spectator ligands* are of key importance because they can be carefully chosen to tune the reactivity of the metal so as to modify its catalytic properties. Potential ligands come in all shapes and sizes. They can be simple ions such as chloride (Cl^-), small neutral molecules like water (H_2O), ammonia (NH_3) or carbon monoxide (CO) or larger more complicated structures.

One class of ligands frequently used in catalysts are the *phosphines*. These consist of hydrocarbon units linked to a phosphorus atom which then binds to the metal (Figure 15). Chemists have synthesised many different phosphines by subtly altering the hydrocarbon groups in order to tailor the ligand for the job in hand. Sometimes two phosphorus atoms are linked together by a bridging unit to form a 'claw' which binds even more strongly to a metal.

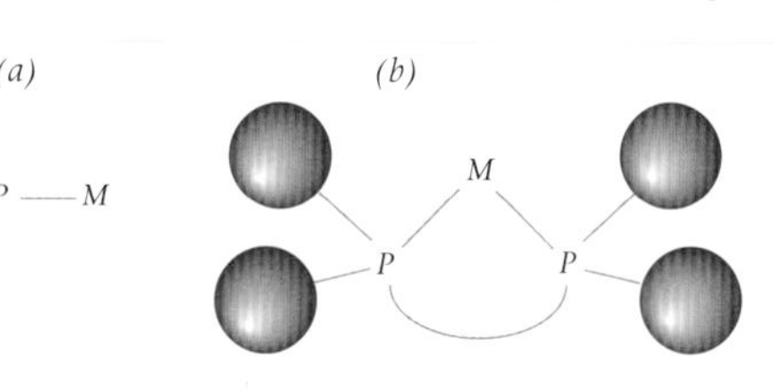

Figure 15. Phosphine ligands (a) contain a phosphorus atom (P) linked to three other chemical groups (usually hydrocarbons). Diphosphines (b) contain two phosphorus atoms linked by a bridge. Each of the phosphorus atoms can bond to a metal (M).

To bring about a catalytic reaction, the metal must also be able to bind to the substrates. This happens at an active site whose geometry, as for enzymes, is profoundly influenced by the size and shape of the ligands. Since the choice of ligands is almost infinite, the chemist has great scope for designing a catalyst with the required activity and selectivity. Sometimes an active site must be generated by detaching one or more ligands from the metal catalyst, providing a 'gap' where the substrate can bind. This sort of reaction can often be promoted thermally (by increasing the temperature) or photochemically (by shining light on the reaction mixture). Altering the metal at the heart of the complex can also lead to dramatic changes in catalytic behaviour, as can more subtle tinkering, such as changing its electrical charge (oxidation state).

The precise pathway, or mechanism, by which a homogeneous catalyst works is generally easier to study than that for a heterogeneous catalyst. This is because the discrete molecules involved in homogeneous catalysis can often be defined precisely using the battery of modern analytical techniques such as nuclear magnetic resonance (NMR) (see *Analysis and structure of molecules*), infrared, and ultraviolet spectroscopy, along with crystallography using X-ray diffraction. Such studies are much more difficult for a heterogeneous catalyst where the reaction occurs only on the surface, which cannot be distinguished using the above techniques from the bulk atoms that outnumber them by a factor of 10^8.

The catalytic cycle

For many homogeneous processes, the catalytic cycle

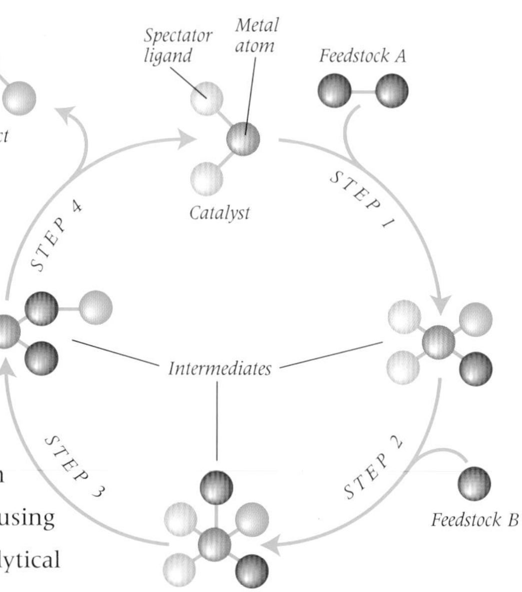

Figure 16. A catalytic cycle for an insertion reaction. In step 1 (an addition reaction) the bond linking the two components of feedstock A is broken by reacting with the metal catalyst. In Step 2, feedstock B binds to the metal. A migration reaction then occurs (step 3) and finally the product is released (step 4) regenerating the catalyst.

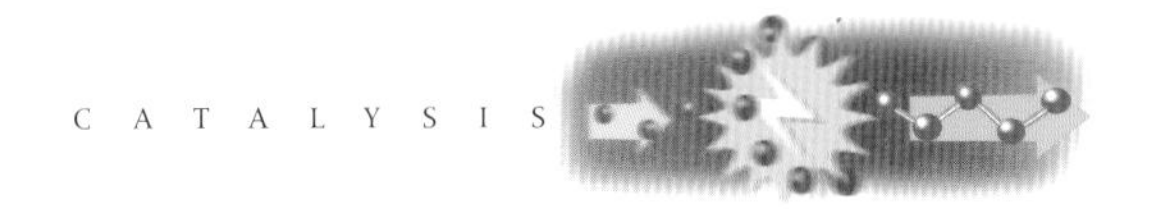

linking reactants to products through a series of reactions is thus easier to define. In many cases the intermediates can be studied individually. The diagram in Figure 16 illustrates the principle of a catalytic cycle. You can think of it as a factory production line with individual parts being added at different stages of the assembly process with the catalyst acting as the robot assembler. The cycle shown represents a type of reaction called an insertion. The net effect is to insert an atom or molecule (represented by the pink ball) between two parts of a substrate (the red and green balls). This overall process is achieved by a

Box 3 Sandwich compounds and their use as catalysts

The first 'sandwich' compound to be discovered was ferrocene – $Fe(C_5H_5)_2$ – which forms orange crystals. The unusual molecular structure of this compound was arrived at separately in 1954 by Sir Geoffrey Wilkinson, then at Harvard University and by Ernst Otto Fischer at Munich University. It contains an iron atom sandwiched between two planar rings which consist of five carbon atoms, each bonded to a hydrogen atom. These C_5H_5 rings are represented by blue disks in the diagrams below. The chemical reactivity of the rings is similar to that of benzene (C_6H_6) a purely organic molecule.

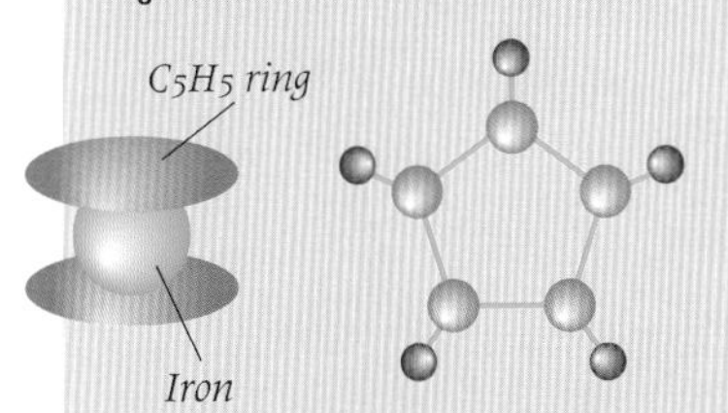

(a) The structure of ferrocene. An iron atom is sandwiched between two C_5H_5 hydrocarbon rings, represented by the blue discs.

We now know of a whole class of sandwich compounds which contain various transition metals as the 'filling'. The 'bread' of the sandwich can also come in various flavours, where the hydrogen atoms of the C_5H_5 rings are replaced by more complex chemical structures. Chemists in both academia and industry have recently become very interested in these so-called metallocenes, primarily because of the discovery that they can be used as catalysts for making valuable polymers such as polyethylene and polypropylene, which are manufactured on a huge scale, worldwide. Polymers of this type have long been made using heterogeneous catalysts based on titanium and aluminium. These were discovered by Karl Ziegler and Giulio Natta, who were awarded the Nobel Prize in 1963.

In the late 1970s Walter Kaminsky's research group at the University of Hamburg found that metallocenes containing the metals titanium, zirconium or hafnium were extremely good polymerisation catalysts, when they added certain aluminium compounds to the reaction mixture as promoters. Subsequent work, notably by Hans Brintzinger at the University of Constance has helped us to understand how these catalysts work.

Growing polymer chain

Monomer (Ethylene, C_2H_4)

(b) The mechanism for polymerisation of ethylene by a metallocene catalyst. The long-chain polymer molecule grows from an active site on the metal (zirconium in this case) by incorporating ethylene (C_2H_4) monomer units sequentially.

In metallocene polymerisation catalysts, the rings tilt to give a 'bent sandwich'. This creates an active site on the metal atom where a polymer chain can grow by incorporating monomer units, one by one. In the example shown , the monomer is ethylene (or ethene, C_2H_4) and the resulting polymer is polyethylene (or polythene) (see *The age of plastics*). A range of other polymers can be made using similar catalysts. For example, in polypropylene, one of the red hydrogens in Figure b on each monomer unit is replaced by a methyl (CH_3) group. The way these 'side-groups' are positioned along the polymer chain affects the physical properties of the resulting plastic. If all the methyl groups are attached to one side of the polymer backbone, the plastic is more rigid, whereas an alternate up-down arrangement gives a more transparent material. The outcome can be controlled by making subtle changes to the rings attached to the metal, which influence the size and shape of the active site. The metallocene catalysts shown below each produce a polymer with different properties. Such control is less easily achieved for heterogeneous Ziegler-Natta catalysts in which the active sites are less well defined.

(c) Metallocene polymerisation catalysts. Chemists can change the shape of the active site by altering the structures of the organic rings sandwiching the metal atom.

These homogeneous metallocene catalysts have been commercialised by a number of companies and are beginning to compete with the established Ziegler-Natta catalysts in certain areas of the plastics market.

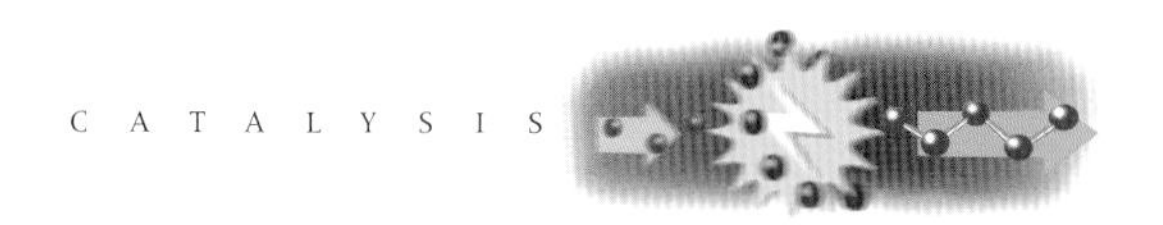

Sir Geoffrey Wilkinson (left) and Ernst Otto Fischer discovered the structure of ferrocene which is now an important homogeneous catalyst.

series of reactions which form a loop. In each step chemical bonds are either broken or made. The catalyst itself, made up of a metal ion and its surrounding ligands, continuously travels around this loop, producing one molecule of product on each lap. The overall speed of the catalytic process is governed by the slowest reaction in the cycle, often referred to as the *rate determining step*. If a more active catalyst is to be designed, a way must be found to accelerate this step.

A whole branch of chemistry – *organometallic chemistry* – has grown up around the need to understand such reactions of metal compounds with organic molecules. This area received a huge impetus with the discovery of the 'sandwich' structure of ferrocene in 1954 (Box 3). The late Sir Geoffrey Wilkinson of Imperial College, London shared the Nobel prize in 1973 with a German chemist, Ernst Otto Fischer, for their independent pioneering work in this field. The chemistry of the so-called 'sandwich' compounds is still an area of huge importance, and has led to the development of a new generation of extremely efficient catalysts for the manufacture of polymers like polythene (Box 3).

As well as the discovery of sandwich compounds, Wilkinson was also instrumental in the design of other transition metal catalysts. One of these is now known universally as Wilkinson's catalyst, and is used to catalyse the reactions of hydrogen (H_2) with organic molecules containing carbon-carbon double bonds (C=C). Wilkinson's catalyst, contains the precious metal, rhodium, surrounded by three phosphine ligands and a chloride ligand. The catalytic cycle which operates is shown in Figure 17. It is thought that for Wilkinson's catalyst to become active, one of the phosphine ligands must first detach from the rhodium (step 1). This creates an active site where a molecule of hydrogen can react with the central rhodium atom (step 2). The alkene then binds to the complex (step 3) and each of the hydrogens can hop in turn onto the alkene to give the product alkane (steps 4 and 5). The whole cycle can then repeat.

Figure 17. Catalytic cycle for hydrogenation of a carbon-carbon double bond using Wilkinson's catalyst. The reactions are described in the main text.

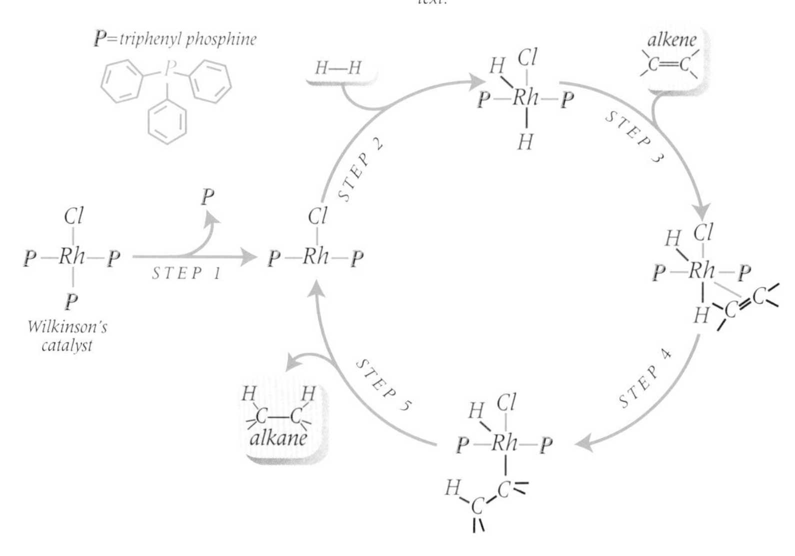

A whole range of catalysts, many also containing rhodium, have been developed to catalyse these kinds of reactions. An everyday example is the hydrogenation of vegetable oils to make butter substitutes. Another important application is in the manufacture of pharmaceuticals such as L-DOPA, a drug used in the treatment of Parkinson's disease. (Apparently, the production of L-DOPA ceased in 1986 because the catalytic synthesis was so efficient that it led to stockpiles of the drug that could meet demand for several years!) The drug can exist in two chiral, or mirror-image forms, only one of which is active for the treatment. It is therefore vital to

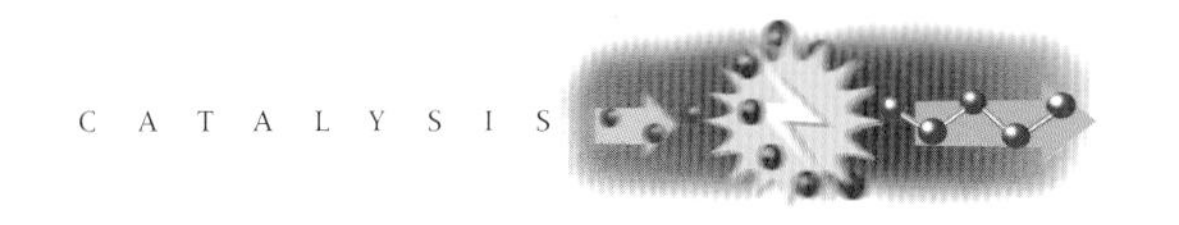

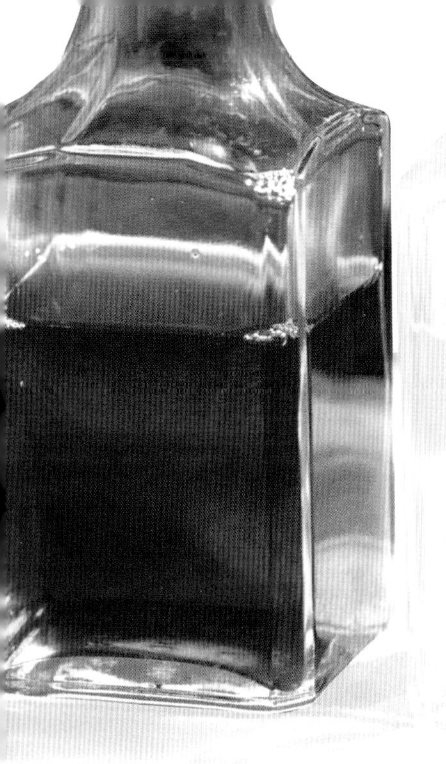

find a synthetic route to the drug which gives the correct mirror image. (This approach is called asymmetric synthesis, for more information see *Make me a molecule*.) The American company, Monsanto, developed a rhodium catalyst containing ligands designed to create an active site with just the right shape to produce the desired version of the drug. So-called asymmetric catalytic hydrogenation is also used in the synthesis of the low calorie sweetener, aspartame (commonly known as NutraSweet®) and *S*-Naproxen, an anti-inflammatory drug. In the latter case the mirror-image molecule, *R*-Naproxen, is a liver toxin, showing just how important it is to obtain the drug in the correct form.

A case history: a homogeneous route to acetic acid

If you look at the bottles of vinegar, the next time you go to the supermarket, you may well see that there are brands, invariably cheaper, bearing the name 'non-brewed condiment'. You might wonder what the difference is between the brewed and non-brewed products. Traditionally, vinegar is made from alcohol, itself produced by fermentation. The oxidation of the alcohol gives acetic (or ethanoic) acid – the chief constituent of vinegar – and gives it the familiar bitter taste. The same oxidation process occurs when you leave a bottle of wine open to the air for too long. Pure acetic acid (CH_3COOH) is actually a colourless liquid, like water, and the normal colour of vinegar is provided by other ingredients. As we will see, using a metal catalyst, acetic acid can now be made in a much more efficient way than by the old fermentation method. It is made on a very large scale as a bulk commodity chemical which has a whole range of uses, apart from cheaper vinegar (Figure 18).

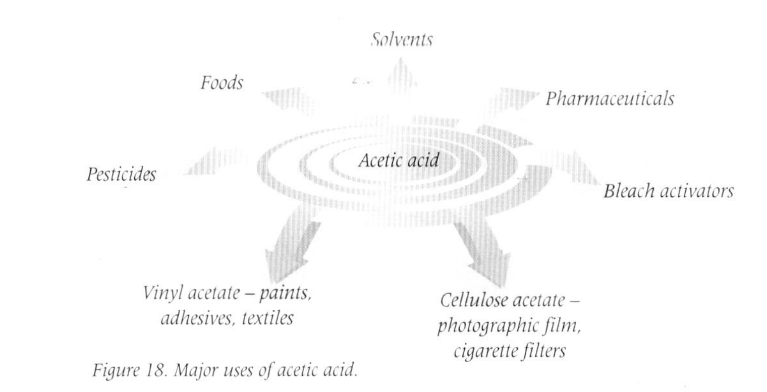

Figure 18. Major uses of acetic acid.

The global demand for acetic acid has thus risen rapidly during the 20th century. In response, the chemical industry has evolved a series of processes to manufacture the product, each one more efficient than the last. The older processes used ethylene (C_2H_4, ethene) or acetylene (C_2H_2, ethyne) as their hydrocarbon feedstocks. Figure 19 shows a number of routes to acetic acid, all involving the use of catalysts.

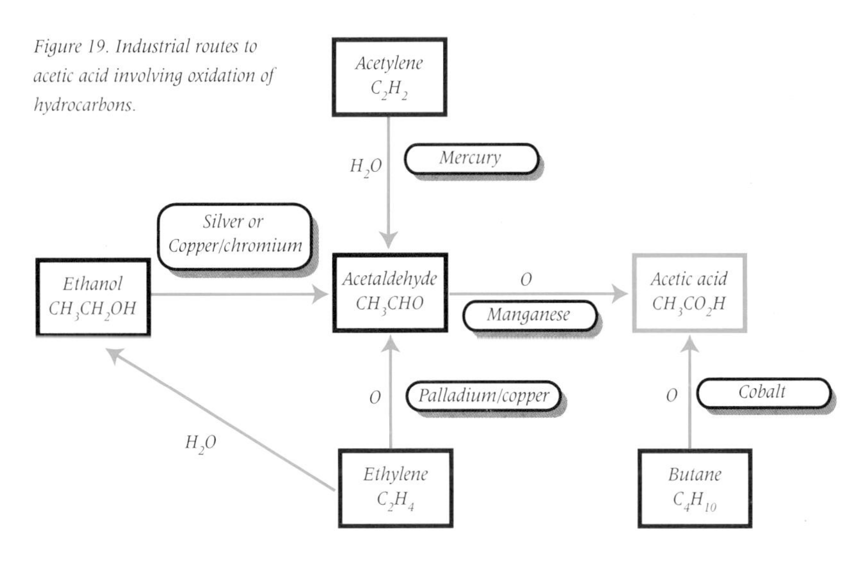

Figure 19. Industrial routes to acetic acid involving oxidation of hydrocarbons.

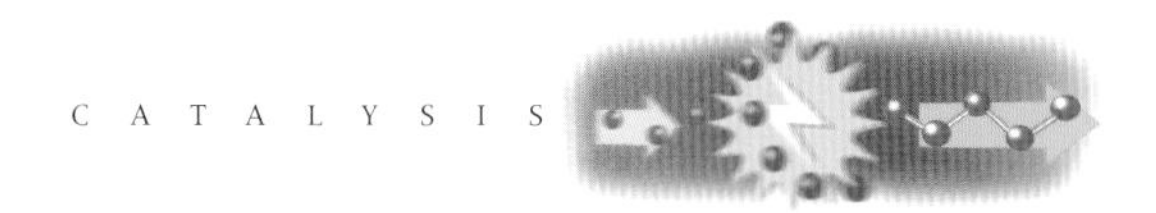

©The British Petroleum plc 1999

After the Second World War, the industrial production of acetic acid took change of direction, with butane (C_4H_{10}) becoming the preferred feedstock. In 1952, Celanese introduced a process at its plant in Pampa, Texas, to manufacture acetic acid by the oxidation of butane using a homogeneous cobalt catalyst. By 1973 this technology accounted for 40 per cent of total acetic acid capacity with plants operated by a number of companies, including Union Carbide. During the 1960s, however, new developments had been taking place which would eventually supersede the oxidation reactions. The old processes suffered from a lack of selectivity, so a substantial price is paid to separate acetic acid from the various byproducts. A new approach, using a more controlled catalytic reaction overcame these problems, and it is now regarded as one of the foremost success stories of homogeneous catalysis.

The story began in the early 1960s when BASF in Germany developed a new process for making acetic acid which involved the reaction of methanol with carbon monoxide. The reaction is termed an 'insertion', because a molecule of carbon monoxide is inserted into the carbon-oxygen bond of methanol (CH_3OH). The homogeneous catalyst used by BASF comprised of cobalt with an iodide promoter, which improved the catalytic activity. One of the great advantages of this new process was that both starting materials, methanol and carbon monoxide, can be made from natural gas (Figure 20) or alternatively, coal. The economics of the process therefore no longer depended on the price of oil which was subject to large fluctuations.

Despite this advance, the BASF process was soon improved upon by researchers at Monsanto in St Louis, Missouri, and in Texas City. They developed a similar catalyst based on the precious metal, rhodium (which lies directly beneath cobalt in Group 9 of the Periodic Table). In 1968, Monsanto announced its new process, which operated at lower temperature and pressure than the cobalt system. The rhodium catalyst converted methanol into acetic acid with remarkable selectivity, with less than 1% of unwanted byproducts. Many plants based on the Monsanto

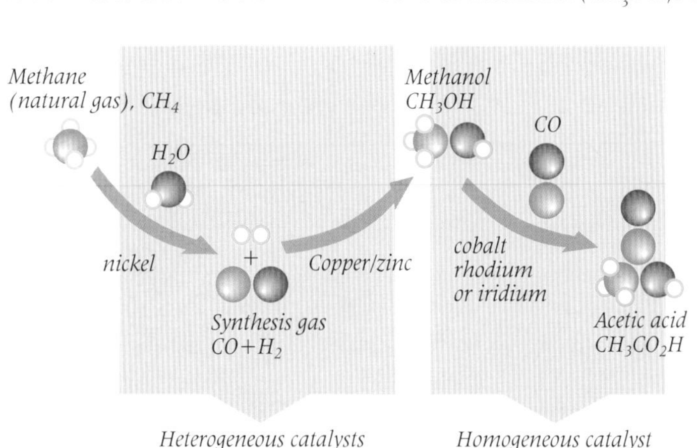

Figure 20. The route to acetic acid from natural gas, using both heterogeneous and homogeneous catalysts.

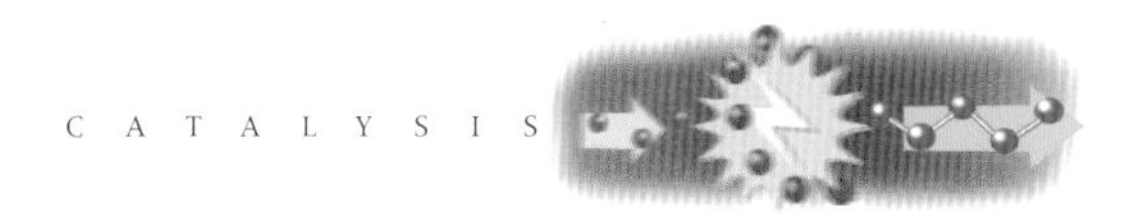

technology were built around the world, and in 1993, 60% (more than 3 million tonnes annually) of the world acetic acid capacity used the rhodium catalyst system.

Many research man-hours have been put into understanding exactly how the Monsanto acetic-acid process works. Denis Forster and his colleagues at Monsanto initially identified the precise chemical structures of a number of the rhodium compounds involved in the catalytic cycle, similar to those shown in Figure 17 for hydrogenation using Wilkinson catalyst. Using special equipment, they were able to measure the infrared spectrum of a real catalytic reaction at high temperature and pressure. The spectrum of the rhodium compound which they observed under the extreme conditions of this experiment was identical to one they could isolate in the laboratory. By studying the chemistry of this compound, they could build up a series of reaction steps which fitted together to form a catalytic cycle.

Although Forster and his colleagues discovered much of the chemistry underlying Monsanto's acetic acid process, the studies have continued to this day. In the early 1990s, chemists at the University of Sheffield proved the existence of one of the molecules which had eluded Forster. The discovery, using modern infrared and NMR spectroscopic methods, helped to explain the extremely high selectivity of the catalytic process.

The story is not quite finished. In 1986, the Monsanto technology was acquired by BP Chemicals, which has continued to work to improve the process. A sustained effort by BP's research chemists and engineers came to fruition with the announcement, in 1996, of a new catalyst that works even better than rhodium. The journey down Group 9 of the Periodic Table has continued to iridium. This element (another precious metal) was originally tested by Monsanto, but under the conditions that they used, it did not work as well as rhodium. BP Chemicals has now discovered additives that promote the iridium catalyst, and can improve its activity to such an extent that it out-performs rhodium. The new catalyst has been given the name Cativa, and is already installed on a plant in the US, and on BP's UK plant in Hull. The company also plans to build new plants to satisfy the growing market in the Far East.

Improvement of the iridium catalyst, to the point where it out-performed rhodium, depended on finding promoters that could speed up the slowest reaction in the catalytic cycle to give a faster output of product. Chemists at the BP Research Centre in Hull and at the University of Sheffield think that they have unravelled

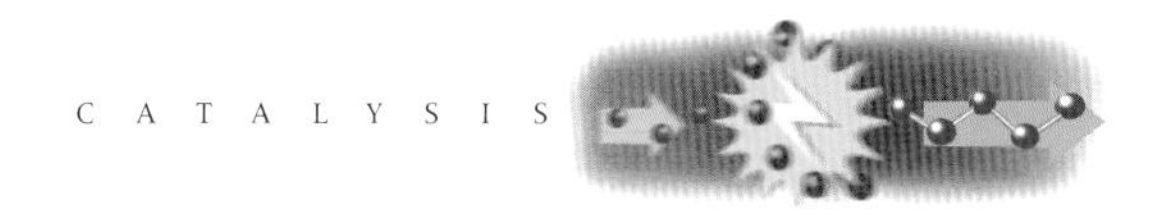

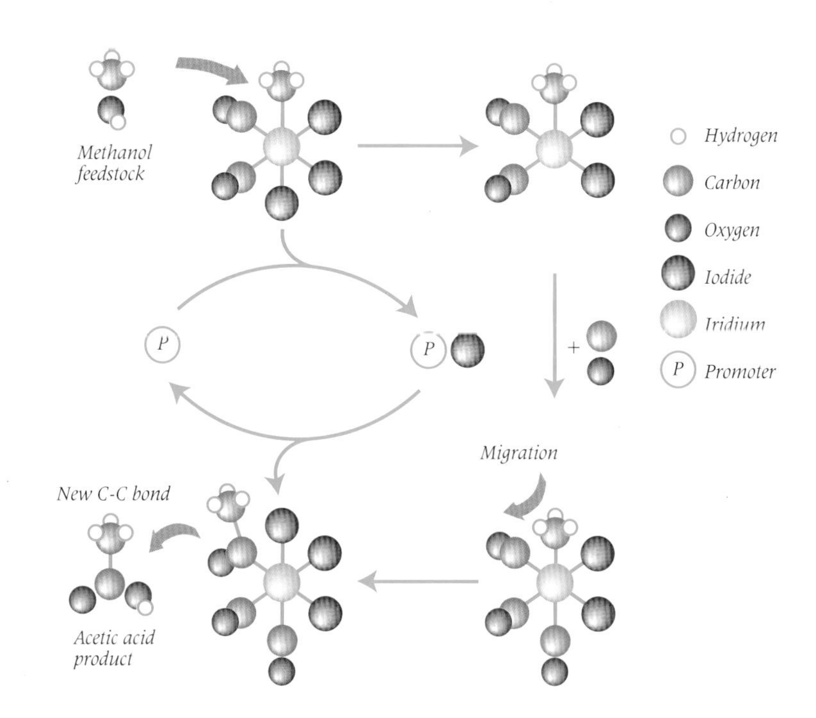

how the promoters work. The trick is to encourage a methyl (CH_3) group to combine with a molecule of carbon monoxide (CO) while both are bonded to the iridium catalyst. To do this it appears to be necessary to remove one of the three iodide ligands from the central iridium atom. The job of the promoter is to keep a hold of this iodide until the bond between the CH_3 and CO groups has formed (Figure 21). A carbon-carbon bond is formed in this sequence of steps, giving an acetyl ($COCH_3$) group which is a fundamental unit of the final product, acetic acid (CH_3COOH).

Figure 21. The role of a promoter in the iridium catalyst for the manufacture of acetic acid. Starting from the top-left, the promoter helps to remove one of the iodide ligands from the iridium catalyst. This creates a space where a molecule of carbon monoxide can bind. A new carbon-carbon bond then forms by migration of a methyl (CH_3) ligand (which originates from the methanol feedstock) onto a carbon monoxide (CO) ligand. The iodide can then hop back to the iridium so that the promoter is free to repeat its task.

Future challenges

The case history above shows how a catalytic process can evolve as ways are found to improve upon the catalyst's properties. The search continues for more active and selective catalysts in all parts of the chemical industry. One long-term aim is the selective conversion of hydrocarbons into more valuable chemicals, for example, via selective oxidation. The biggest advances are expected to involve completely new processes, particularly replacements for current non-catalytic reactions such as the conversion of propene ($CH_3CH{=}CH_2$) to propene oxide (CH_3CHOCH_2). Catalytic routes for these processes would be cheaper and much more environmentally friendly.

One of the world's most abundant resources is natural gas, the main constituent of which is methane (CH_4). As well as being burnt to provide energy, methane is a source of carbon which can be used as a building-block for larger organic chemicals. Currently, this is achieved by a rather circuitous route, via synthesis gas. Chemists are exploring a number of avenues by which methane and similar hydrocarbons could be converted directly into other compounds like methanol.

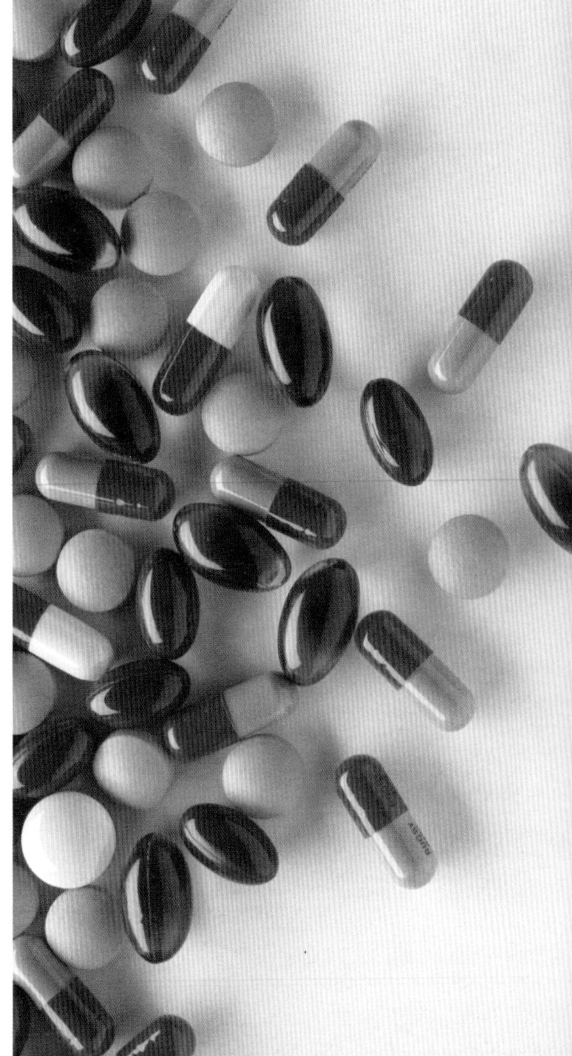

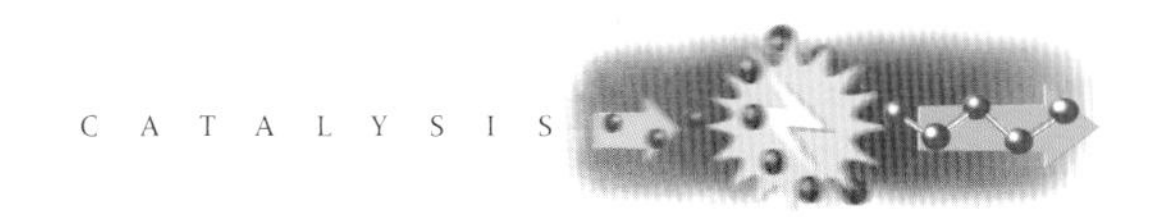

Mirror image catalysis

Another key goal, mentioned earlier in this chapter, is so-called *asymmetric catalysis*, where a chiral product is made in preference to its mirror-image partner. These compounds are of vital importance for pharmaceuticals and other biologically-related products. The processes used at the moment are often not catalytic, and if they are they are usually homogeneous. As we explained, the chemical industry prefers heterogeneous catalysts, so researchers are currently trying to develop chiral catalysts that work heterogeneously. One approach is to tether chiral catalysts onto a substrate or even to the internal surface of a zeolite, as described earlier. Another method is to modify the surface of a catalyst with an adsorbed organic molecule which acts as a template, controlling the configurations of the intermediates to produce only one chiral molecule.

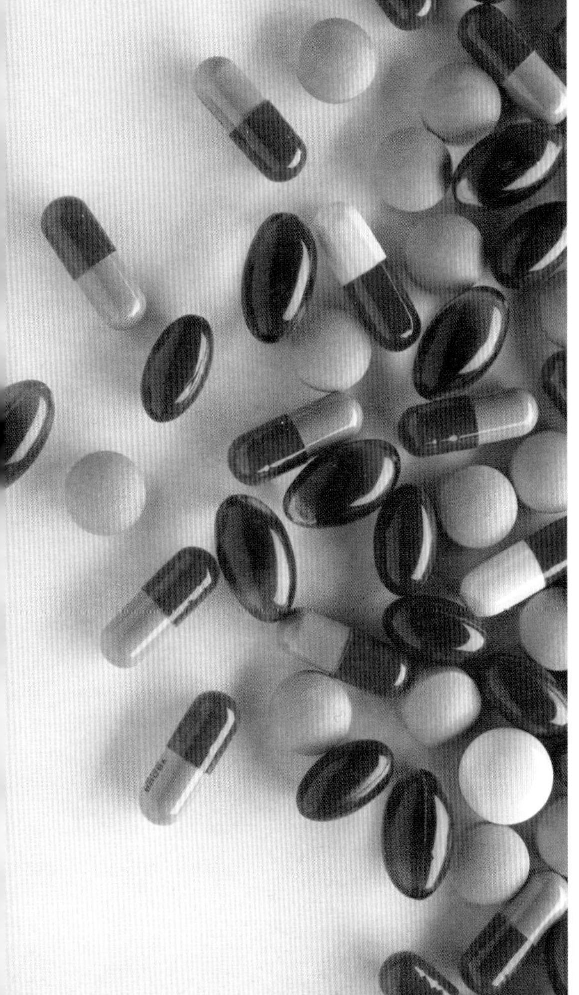

More generally, the chemical industry requires highly selective, energy-efficient processes with minimal adverse environmental consequences. The role of the chemist is to explore and understand the behaviour of catalytic systems and to use this knowledge to help design the next generation of catalysts.

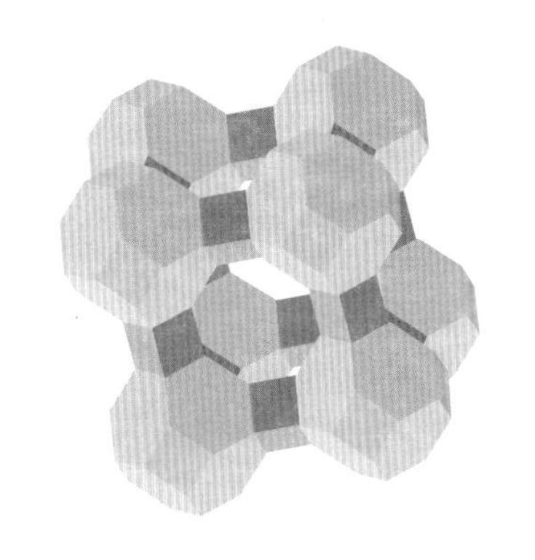

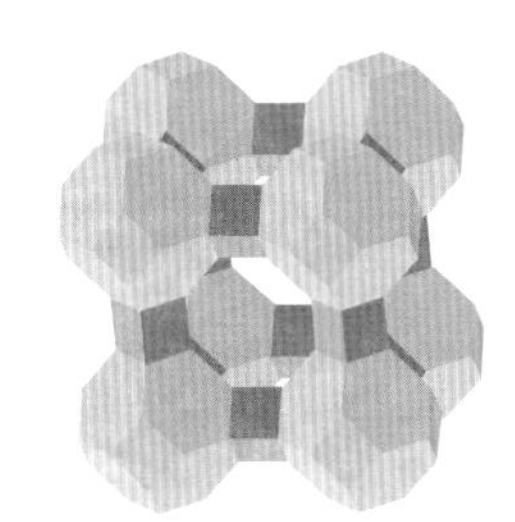

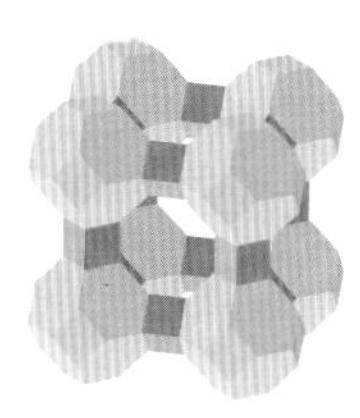

Further reading

1. *Principles and Practice of Heterogeneous Catalysis*, J. M. Thomas and W. J. Thomas, VCH Publishers, Weinham, FRG and New York: 1997.
2. *Heterogeneous Catalysis: principles and applications*, G. C. Bond, Clarendon Press, Oxford: 1974.
3. "Photo-electron spectroscopy and surface-chemistry", M. W. Roberts, *Chemistry in Britain*, 1981, **7**, 510.
4. "The role of short-lived oxygen transients and precursor states in the mechanisms of surface reactions; A different view of surface catalysis", M. W. Roberts, *Chemical Society Reviews*, 1996, **25**, 437.
5. "The Magic of Catalysts", *New Scientist, Inside Science*, No 35 (23 June 1990).
6. *Homogeneous Catalysis – The Applications and Chemistry of Catalysis by Soluble Transition Metal Complexes*, G. W. Parshall and S. D. Ittel, 2nd edn, Wiley, New York: 1992.
7. *Acetic Acid and its Derivatives*, V. H. Agreda and J. R. Zoeller eds, Marcel Dekker, New York: 1993.

World Wide Web sites of interest

1. Directory of UK academic interests in applied catalysis http://biotite.xtl.ox.ac.uk/iac/dirindex.html
2. An index of pages with SPM information and pictures http://www.ifm.liu.se/Applphys/spm/links.html
3. IBM's page of scanning tunnelling microscopy images http://www.almaden.ibm.com/vis/stm/stm.html
4. The homepage of surface science and catalysis in Cardiff http://www.cf.ac.uk/uwcc/chemy/surfsci/homepage.html
5. ECTOC3 – An Electronic Conference on Organometallic Chemistry http://www.ch.ic.ac.uk/ectoc/ectoc-3/
6. Organometallic Chemistry page http://www.scc.um.es/gi/gqo/
7. Catalytic Mechanisms Research Group, University of Sheffield http://www.shef.ac.uk/uni/projects/cmrg/
8. Details of Cativa, BP Chemicals new acetic acid process http://ci.mond.org/9613/961305.html

Glossary

Activation energy The amount of energy needed to get a reaction started.

Active site The position on the catalyst where the catalytic reaction happens.

Catalytic cycle A sequence of chemical reactions that can be linked to form a loop. The catalyst cycles round the loop converting feedstocks to products.

Complex A molecule formed when *ligands* bond to a metal atom or ion.

Feedstock A compound (substrate) fed as starting material to the catalytic reaction.

Heterogeneous catalysis The reaction of the substrate (a gas or liquid) takes place at the surface of the solid catalyst.

Homogeneous catalysis The catalyst and substrate are both dissolved in a solvent where the reaction occurs.

Ligand An atom or group of atoms which are bonded to a metal. The type and number of ligands can tune' the catalytic properties of a metal.

Monomer A unit from which a polymer is built.

Organometallic chemistry The chemistry of molecules containing bonds between a metal and carbon.

Polymer A long-chain molecule made by the stepwise linking of many small monomer units.

Promoter An additive which improves the efficiency of a catalyst.

Selectivity The ability of a catalyst to produce the desired product with the minimum amount of unwanted by-products.

Substrate A molecule on which the catalyst acts.

Transition metal A metallic element found in the central block of the Periodic Table, usually with variable numbers of electrons available for bonding.

Biographical details

Dr Phil Davies is a lecturer in physical chemistry at Cardiff University. He researches into the mechanisms of the fundamental surface reactions that operate in heterogeneous catalysis and other surface-mediated processes such as the activation of carbon dioxide at metal surfaces and the decomposition of methanol at copper surfaces.

Dr Tony Haynes is currently BP Chemicals Lecturer in Homogeneous Catalysis at the University of Sheffield researching into the mechanisms of homogeneous transition metal-catalysed reactions, particularly those involving carbon monoxide.

Acknowledgements

The authors would like to thank Professor Peter Maitlis, Professor Wyn Roberts and Dr Glenn Sunley for their assistance with this chapter. They would also like to thank Professor Sir John Meurig Thomas and Professor Alex Bradshaw for their kindness in providing copies of some of the pictures used.

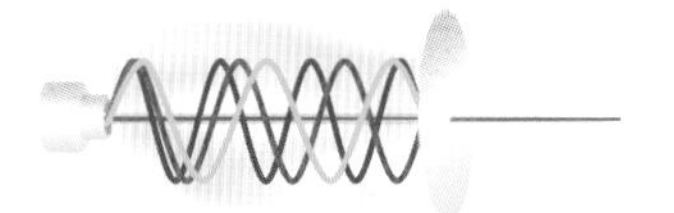

Following chemical reactions

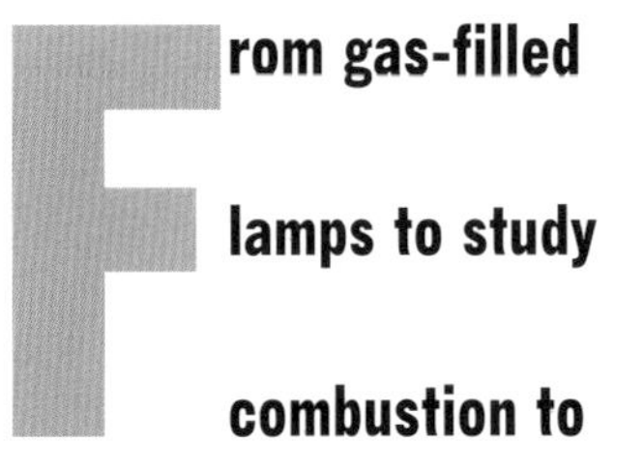

From gas-filled lamps to study combustion to

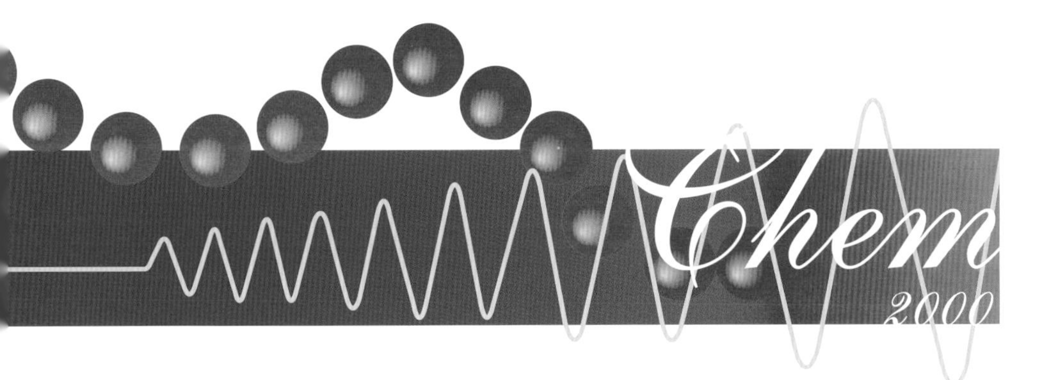

state-of-the-art-lasers, chemists continue to find out what really happens when molecules interact with one another.

Dr Helen Fielding
King's College London

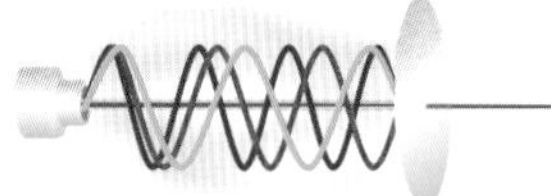

Chemistry is all about reactions between atoms and molecules. Thanks to a new generation of lasers, chemists can now watch the minute details of what happens during a chemical reaction and learn how and why certain molecules interact with one another. This allows us to begin to understand what is going on in important chemical processes such as combustion, atmospheric ozone depletion, photosynthesis, and even how the first molecules formed in the Universe.

In the past few decades, researchers have made such tremendous progress in the field of *reaction dynamics* that they can follow even the fastest chemical reactions and actually observe the intermediate stages from reactants to products. By determining the *transition state* – the point of 'no return' through which the reacting molecules or atoms pass – chemists can learn the fine details of how the reacting molecules approach each other, how bonds are broken and made, and how the product molecules move apart afterwards (Figure 1a). In some cases, an actual, highly reactive *intermediate* compound is formed during the reaction (Figure 1b).

The most obvious way to study a chemical reaction is to probe the energy changes that always accompany it. The laws of nature predict that any system, whether physical or chemical, will tend towards the most stable arrangement where its potential energy is lowest. A chemical reaction is a manifestation of this basic law: the internal, or potential, energy of the products is lower than that

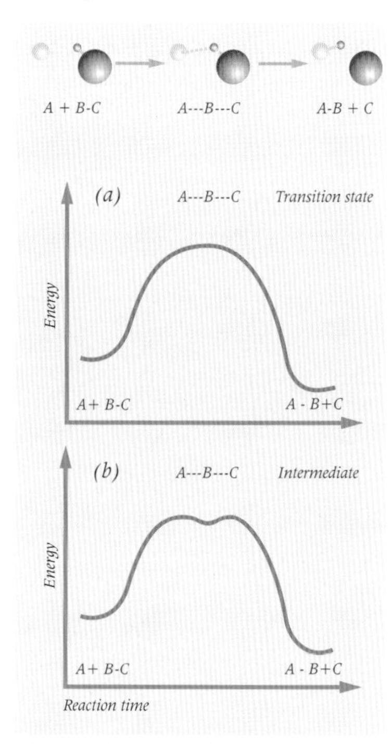

Figure 1. The transition from reactants to products in a typical reaction $A + BC \longrightarrow AB + C$ *showing the bonds being made and broken. Figure 1a represents an energy level diagram for a reaction occurring via a transition state (for example,* $H + H_2 \longrightarrow [H \cdots H \cdots H] \longrightarrow H_2 + H$*); there is a fairly large energy barrier in going from reactants to products and the intermediate exists at the maximum point on the energy curve. Figure 1b represents a reaction which occurs via an intermediate (for example,* $I + HI \longrightarrow [I \cdots H \cdots I] \longrightarrow IH + I$*); there is an energy barrier in going from reactants to products but the intermediate lies in a dip in the top of the energy curve.*

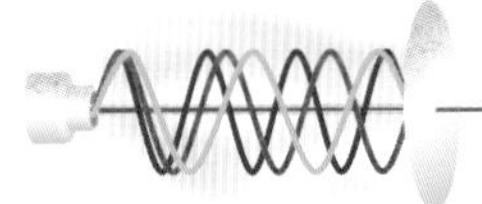

of the reactants. However, in many cases, for a reaction to happen, there is an energy barrier to overcome because the transition state has a higher energy than either the reactants or products. If an intermediate forms, then there is a slight dip in the energy barrier (see *Chemical marriage-brokers*). Many reactions therefore require the input of energy, say, in the form of heat or electromagnetic radiation *(eg* light). How this energy interacts with the reactants and what happens to the energy afterwards are two of the key questions asked when studying the reaction path.

The internal energy of a molecule is distributed in several ways: in the arrangement of electrons in the orbitals of the molecule, and in the vibrations and rotations of the chemical bonds. According to quantum mechanics – the theory that explains the behaviour of very small objects – these energies can have only certain values. These give rise to a series of increasing *electronic*, *vibrational* and *rotational energy levels*, the difference between adjacent energy levels being equivalent to a quantum of light energy given by its frequency times Planck's constant ($E = h\nu$). Each electronic level has a series of vibrational and rotational levels associated with it (Figure 2). Usually a molecule (AB) sits in its lowest energy level, or *ground state* but can be 'excited' to a higher level (AB*) by absorbing a *photon* of light with the appropriate wavelength. Ultraviolet and visible wavelengths are needed to excite electronic levels, while vibrational and rotational levels absorb the longer-wavelength, lower-energy infrared and microwave radiation. If the molecule is already in an *excited state* it may 'relax' back to a lower energy level or the ground state, emitting characteristic radiation in the process. In fact, because the pattern of energy levels in a molecule are highly individual, the resulting *absorption* or *emission spectrum* obtained by probing them with radiation provides an indispensable way of identifying the chemical structure. Indeed modern *spectroscopy* is a key analytical technique in following chemical reactions.

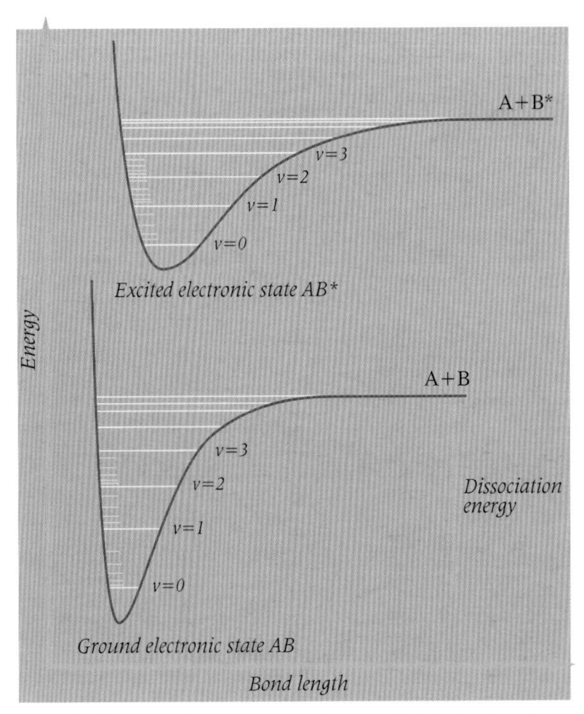

Figure 2. The potential energy curves for the ground electronic state of a typical diatomic molecule AB and an excited state AB. The vibrational energy levels are labelled v = 0, 1, 2 etc and are shown all the way up to the point where they converge at the dissociation limit. The rotational energy levels associated with the lowest vibrational states of the molecule are also shown on the far left of the potential energy curves.*

To undergo a reaction the molecule has to absorb enough energy not only to excite the molecules to higher energy levels but also to cause them to break up, or dissociate. The resulting transfer of electrons brings about a rearrangement of bonding and thus the formation of new molecules. Spectroscopy can, therefore, also be used to identify the transitionary steps of this process. If carried out over a period of time, it can also monitor the changing concentrations of the chemical species involved in the reaction, thus probing how fast it goes – in other words, the *reaction kinetics*.

Highlighting reactions with flash photolysis

Chemists have been measuring the kinetics of reactions for more than a century by merely mixing reactants together in a container and measuring the concentrations of the reactants and products. Although the short-lived transition states and intermediates could not be

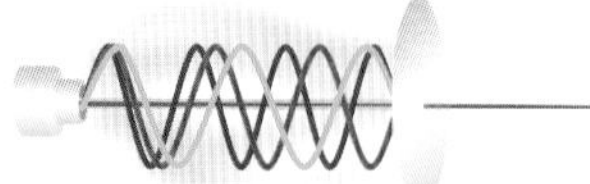

observed directly, their existence was inferred by monitoring the formation of products and trying to deduce a mechanism. Only fairly slow reactions could be tackled. It was not until just after the Second World War that very fast reactions triggered by visible light could be observed, as a result of a revolutionary experimental development. In 1949, Ronald Norrish and George Porter, at the University of Cambridge, developed an exciting technique called *flash photolysis*. As its name implies, it involves breaking up molecules using light. This novel tool could measure chemical reactions lasting just a few milliseconds or even tens of microseconds and earned the two scientists the 1967 Nobel Prize for Chemistry. In their first experiments, Norrish and Porter used batteries of condensers from war-surplus stores to pass huge electrical currents of short duration through flash lamps containing noble gases (similar to those used in modern flash photography). These short, intense pulses of light were then used to 'pump' large numbers of atoms or molecules in a gas into highly excited electronic states whose subsequent growth or decay could be monitored over a period of time.

Ronald Norrish (left) and George Porter (below) developed the technique of flash photolysis for studying fast reactions.

Courtesy ICI

Figure 3 provides a schematic diagram of the apparatus used in a typical flash photolysis experiment. The short light pulse – the photolysis flash – is directed into a glass bulb containing the reactants. A significant number of the reactant molecules absorb this light to form excited versions of themselves, or they break up to create free radicals (highly reactive molecules with single, unpaired electrons) and other short-lived fragments. Now, here is the clever part. By firing a second 'probe' flash through the reaction vessel over a series of carefully timed intervals after the photolysis flash, the absorption spectra of the species present at that instant can be obtained. In those early days the visible spectrum was recorded on photographic film; today, an array of electronic photodetectors is used. If one of the species in the reaction vessel absorbs light of a particular wavelength, it will be observed as a dark line in the spectrum of the probe pulse.

Norrish, Porter and their Cambridge colleagues, exploited this flash technique to study a huge variety of problems as wide-ranging as the explosion reaction of hydrogen and oxygen, the recombination of iodine atoms to form molecular iodine and the photolysis of organic molecules, acetaldehyde and acetone, at high intensities of light. One particularly interesting experiment, carried out by Porter in the 1950s, which has important implications in our present-day understanding of the chemistry of our atmosphere, was the photolysis of chlorine dioxide.

Porter and his colleagues recorded a series of absorption

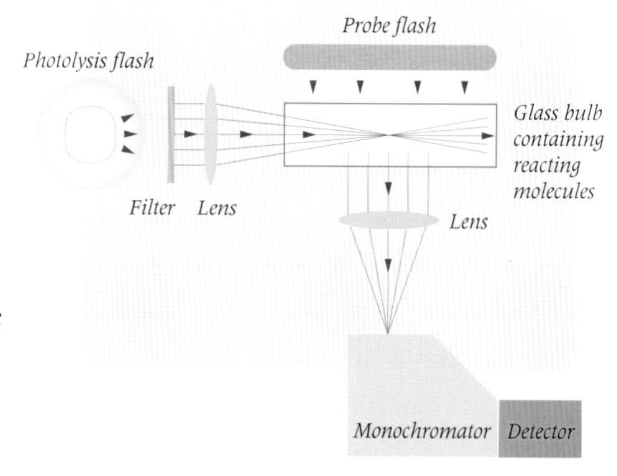

Figure 3. A schematic layout of the apparatus used in a typical flash photolysis experiment. The photolysis and probe flashes are directed into a glass bulb containing the reactant molecules: the photolysis pulse creates the excited species of interest and the probe pulse is used to record absorption spectra of the species present at specific times after photolysis. If a particular molecule absorbs a particular frequency of light then this will appear as a dark line in the spectrum of the probe pulse.

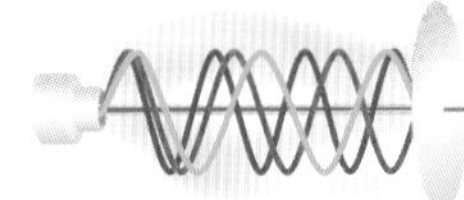

spectra in separate experiments with increasing delays on a microsecond timescale between the photolysis flash and the probe flash. Each chlorine dioxide molecule present before the photolysis flash dissociates to form the diatomic radical – chlorine monoxide (ClO), and atomic oxygen.

$$ClO_2 + h\nu \longrightarrow ClO + O$$

The absorption spectra of the ClO fragments were visible after the photolysis flash and their intensities were seen to decay with time (Figure 4). Their concentrations decrease as they react to form molecular chlorine and oxygen.

$$O + ClO_2 \longrightarrow ClO + O_2$$
$$ClO + ClO \longrightarrow Cl_2 + O_2$$

It turns out that, because of its stability, chlorine monoxide is of immense importance in photochemical reactions where chlorine is incorporated into a molecule in the presence of oxygen. However, it was not until 30 years after these pioneering experiments that people found they were of any practical significance. It turns out that the chlorine monoxide radical plays a significant role in the formation of the 'ozone hole' above Antarctica. Chlorofluorocarbons from aerosols and refrigerant compounds decompose in the upper atmosphere releasing chlorine atoms which react with ozone (O_3) to generate the chlorine monoxide radical, thus removing the ozone.

In addition to their work on radicals in atmospheric chemistry, Norrish and his students also put a lot of effort into understanding the kinetics of radicals involved in combustion processes – in particular, the role of antiknock. Controlling the combustion of a mixture of fuel and air inside the cylinder of an internal combustion engine is an important engineering exercise. If the mixture burns too quickly, the piston receives a hard jerk rather than a smooth push, which is known as knocking. As well as producing an unpleasant sound, knocking decreases the efficiency of the engine. Fuels are usually rated by their octane number which provides a measure of their tendency to cause knocking. The higher the octane number of the fuel, the better its performance. It turns out that by adding certain antiknocking agents such as tetraethyllead to fuels, it is possible to increase the octane rating (though there are now more environmentally-friendly ways of doing this). An example of one of Norrish's experiments to demonstrate the role of antiknock in the combustion of alkanes is shown in Figure 5. The sharp peaks

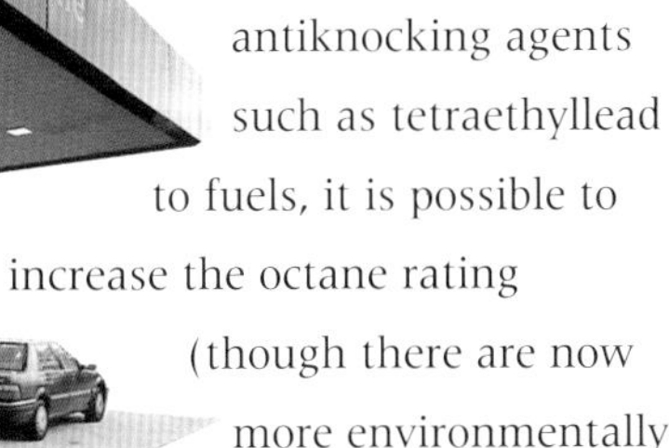

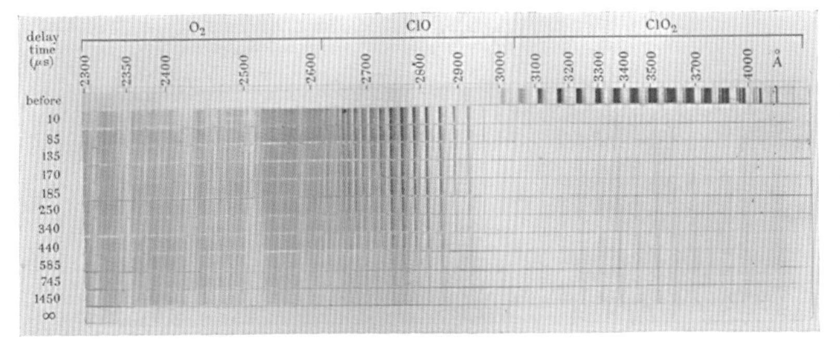

Figure 4. Flash photolysis spectra of the chlorine dioxide radical. Notice how chlorine dioxide is completely fragmented by the photolysis flash to the chlorine monoxide radical and the oxygen molecule in an excited state. The concentrations of these species decrease with time as they decay to form chlorine and oxygen in their ground states. Reproduced with permission from R Norrish, Proc R Soc, Plate 4, 1-37, A301, 1967, The Royal Society.

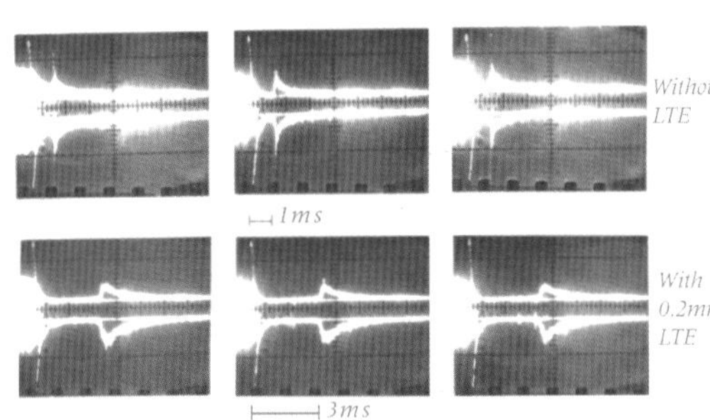

Figure 5. Effect of adding tetraethyllead on a hexane-oxygen explosion reaction. The sharp peaks observed in the oscilloscope traces in the top row indicate that the mixture has detonated. Notice how these are dramatically suppressed in the traces shown in the bottom row due to the addition of tetraethyllead.

'Fast reactions and primary processes in chemical kinetics', Proceedings of the 5th Nobel symposium.

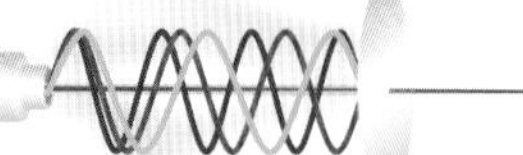

characterise detonation of a mixture of hexane and oxygen and can be seen to be dramatically suppressed by the addition of tetraethyllead.

Although chemists were at this time limited by the timescale of the flash lamps available, a race to study faster and faster reactions was just beginning. At a Chemical Society Faraday Discussions meeting (The Chemical Society was one of the forerunners of The Royal Society of Chemistry) in 1954 on fast reactions, the German chemist Manfred Eigen who later shared the Nobel Prize with Norrish and Porter, asked Oxford don Ronnie Bell how the English language would describe reactions that were "faster than fast". Ronnie Bell replied: "Damn fast reactions, Manfred, and if they get faster than that, the English language will not fail you, you can call them 'Damn fast reactions indeed'!"

Manfred Eigen
©The Nobel Foundation

Enter the laser

Fortunately, while the chemists were becoming more excited about the prospect of looking at faster reactions, the physicists were busy developing a new light source which was to revolutionise the field of chemical reaction dynamics. This new light source was the laser. The word laser is an acronym for 'light amplification by stimulated emission of radiation'. It produces a highly focused, or collimated beam of radiation which is not only of a single wavelength but is also *coherent* (the undulations of light all coincide, or are 'in phase' with one another). Figure 6 shows the principle of laser action. The first step in generating laser radiation is to create a *population inversion* in a material, that is to pump the constituent atoms or molecules up to excited states, usually by another light source. Next, a few of these excited species may release energy spontaneously by emitting photons. Each of these photons then interacts with other excited species and stimulates the emission of an additional photon with the same wavelength which travels in the same direction and, most importantly, is in phase with the first. This process continues throughout the *laser cavity* generating an avalanche of photons. Mirrors placed either end of the laser reflect the beam backwards and forwards and amplify the intensity still further. By arranging that one of the mirrors is only partially reflecting, laser light will be emitted in a well-collimated beam from one end of the cavity.

Today, a huge variety of lasers are available, exploiting various gases, solutions of organic materials, inorganic crystals and semiconductors as lasing materials that emit over a wide range of wavelengths from the extreme ultraviolet to microwaves. As a result they are extremely useful tools not only in the laboratory but also in everyday applications such as CD players, supermarket checkout scanners and

1. Initially the laser medium is in thermal equilibrium and most of the atoms or molecules are in their ground states.

2. The atoms or molecules are then pumped up to their excited states to create a population inversion.

3. A few of these excited species may emit photons spontaneously. These photons then go on to stimulate emission of additional photons with the same wavelength, direction of travel and phase.

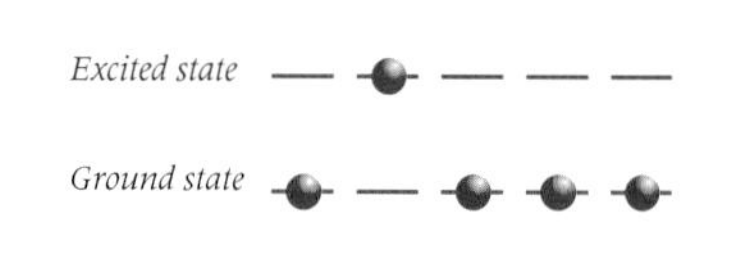

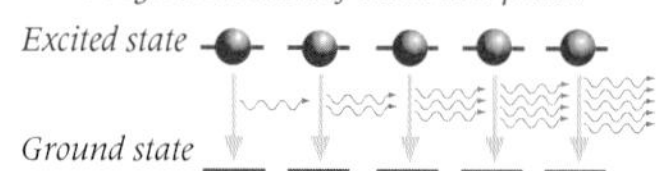

Figure 6. The principle of laser action: ground-state molecules are first pumped to an excited state to create a population inversion. Then, just one photon emitted spontaneously is sufficient to stimulate an avalanche of coherent photons, hence the acronym laser (light amplification by stimulated emission of radiation).

telecommunications. Lasers are also used routinely in ophthalmic surgery for removing cataracts and correcting poor eyesight and in the highly lucrative plastic surgery business for removing birthmarks. During the US Star Wars' era in the 1980s, millions of dollars were spent developing lasers for defence such as the dual-purpose Stingray laser, which generates a wide beam to scan for reflections from enemy sensors and an intense narrow beam to destroy night-vision equipment and gun sights.

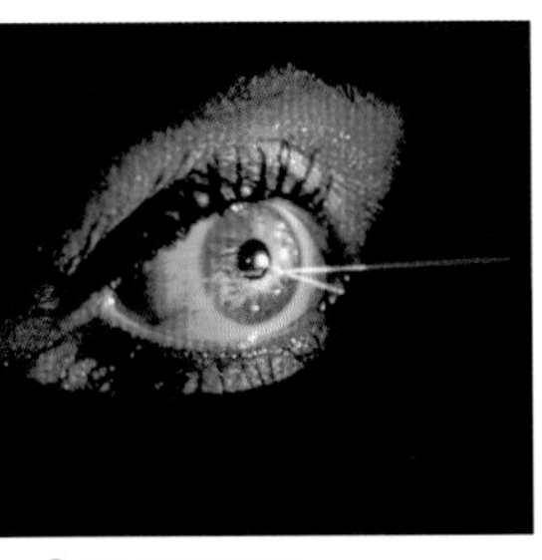

Laser eye surgery.
Alexander Tsiaras/Science Photo Library

Since the energy levels of laser materials are often fixed, the output of most lasers occurs at fixed frequencies. However, for some lasers such as those using dyes as lasing materials, it is possible to use different energy levels and hence 'tune' the output of the laser. Tunable dye lasers are very popular in research laboratories and have had a dramatic impact on chemical spectroscopy and photochemistry studies. One commonly used organic dye is rhodamine 6G, a large molecule built of several carbon rings. When such large aromatic organic molecules are dissolved in solution their electronic spectra look quite smeared out. This is because there are so many vibrational absorption levels for each electronic level due to the large number of ways in which the molecule can vibrate. The absorption and emission spectra of rhodamine 6G are shown in Figure 7. This particular molecule absorbs green light and can therefore be pumped to an excited state using either flash lamps or intense green laser light. The molecules then relax quickly and efficiently to various lower vibrational levels of the excited state. Stimulated emission may then occur from any of them, generating laser radiation with a wide range of wavelengths. By placing something like a prism or a grating in the laser cavity it is possible to select a specific wavelength.

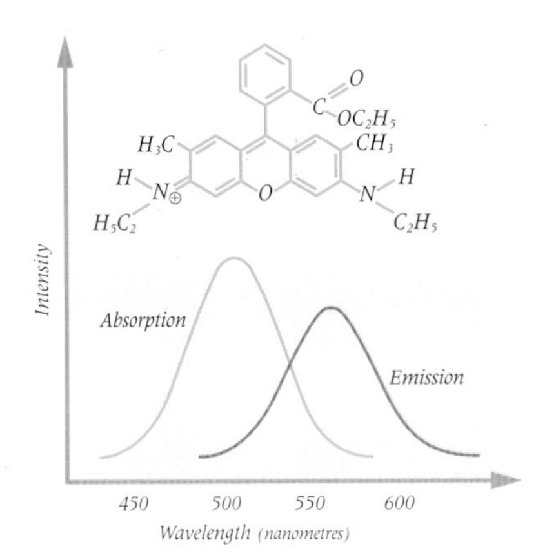

Figure 7. Absorption and emission spectra of the organic dye, rhodamine 6G.

The first commercial lasers produced continuous beams of light. However, it was not long before it was possible to produce short bursts of radiation in pulses lasting only a few nanoseconds (billionths of a second), about one-thousandth shorter than the pulses of light generated using flash lamps. These short pulses of radiation are produced by a process known as *Q-switching*. By temporarily stopping the flow of photons backwards and forwards in the laser cavity (for example, by rotating the end-mirror of a laser cavity), and preventing stimulated emission taking place, a huge population inversion is allowed to build up. Once this has happened, the photons are suddenly allowed to flow (by bringing the mirror back into position). A huge avalanche of photons then builds up and a giant pulse of light is emitted from the laser cavity.

Michael Topp and George Porter were among the first chemists to exploit these new nanosecond lasers and perform faster flash photolysis experiments at the Royal Institution in London in the mid-1960s. The major difficulty they had to overcome was the need to generate two pulses separated by a few nanoseconds, the photolysis and probe pulses. Their solution to this problem avoided electronics completely by using an *optical delay unit* (Figure 8), a method still used today in even

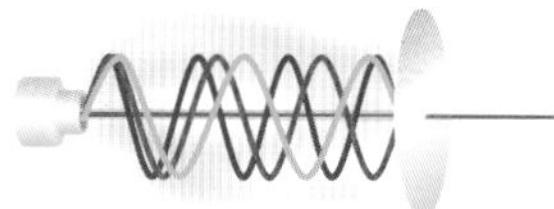

faster time-resolved experiments. A pulse of laser light is sent through a beam-splitter which, as its name suggests, splits the light beam into two portions. Some of the light is reflected directly into the reaction vessel forming the first photolysis pulse, while the rest of the light passes straight through the beam-splitter towards a mirror which is mounted on a moveable stage. At the mirror, the light is reflected directly back onto the beam-splitter where it is then directed into a vessel containing a liquid referred to as the scintillation solution. The large molecules in this solution behave in much the same way as the organic dyes used in dye lasers. A huge number of these molecules are pumped up to excited states which relax very quickly back down to low energy states by emitting light over quite a wide range of wavelengths. This *fluorescence* provides the nanosecond white light, or continuum probe pulse for monitoring the absorption spectra of the excited species in the reaction vessel. Recall that this part of the experiment is much the same as the earlier experiments based on flash lamp excitation. Varying the distance between the beam-splitter and mirror changes the distance that the probe light pulse must travel and determines the time-delay between the photolysis and probe pulses. It is quite straightforward to change the total light path to generate delays in the nanosecond regime since a path of 3 metres corresponds to a delay of 10 nanoseconds.

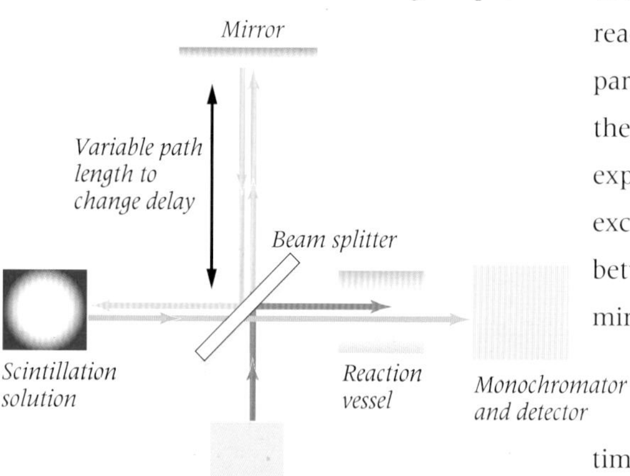

Figure 8. A nanosecond flash photolysis experiment. The magenta line shows the path of the photolysis pulse, the pale magenta line shows the path of the delayed probe pulse from the beam splitter to the scintillation solution and the green line shows the path of the white light or continuum probe pulse.

Molecules in solution

By this time, photochemists had moved on from studying radicals in the gas phase and were beginning to understand the importance of the excited states of larger molecules in solution. Chemists were quick to exploit the availability of laser technology over the whole spectral range. For large molecules in solution, spectra in the visible region are often broad and contain very little detail. However, spectra that provide information about the vibrations of molecules, for example in the infrared region, have proved powerful analytical tools in observing reactions (see Box 1).

Nevertheless one important early result using visible lasers involved the excited states of large organic molecules. Some of these excited states have all the electrons arranged so that their spins are paired and these states are referred to as singlet states. In other excited states, the electrons are organised so that overall there are two unpaired electrons with the same spin and these are known as triplet states. While flash photolysis techniques had been limited to the microsecond timescale it had only been possible to observe the absorption spectra of the triplet states of molecules, which are relatively long-lived (typical lifetimes are milliseconds to several seconds). However, with the advent of nanosecond flash spectroscopy came experiments to observe the much shorter-lived singlet state and, in 1970, Topp and Porter used their nanosecond apparatus to measure the lifetime of an excited singlet state of triphenylene which was just 45 nanoseconds (Figure 9).

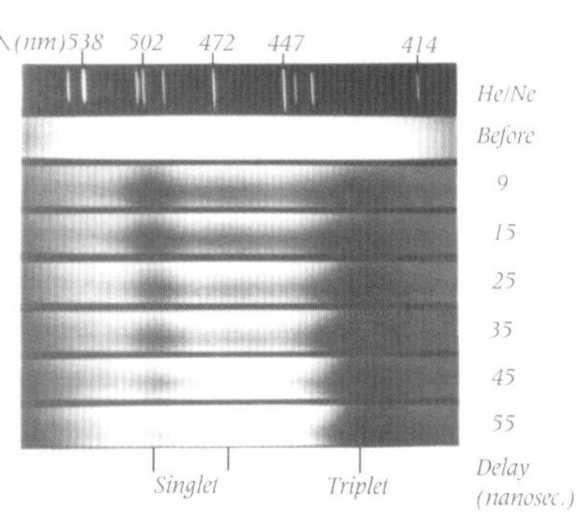

Figure 9. Absorption spectra of the excited states of triphenylene. Notice how the singlet state spectrum disappears within a few tens of nanoseconds whereas the spectrum of the much longer lived triplet state is clearly visible for much longer. Reproduced with permission from G Porter and M R Topp, Proc R Soc, Figure 4, 163-184, A315, 1970, The Royal Society.

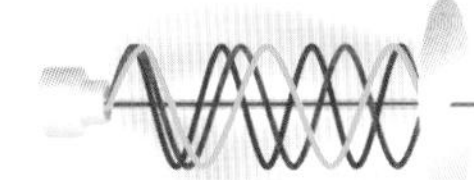

Box 1 Following reactions with resonance Raman and infrared spectroscopy

Although much of the work on reactions is done with visible light which interrogates electronic energy levels, chemists also use *resonance Raman spectroscopy* and *infrared spectroscopy* to look at the vibrational spectra of short-lived intermediates in solution.

Resonance Raman spectroscopy is a particular form of vibrational spectroscopy in which the probe visible laser is tuned to a frequency that matches (or is resonant with) the absorption of the sample. A portion of this light is scattered and most of it has the same frequency as the incoming light. However, some of the scattered light has a frequency that differs by an amount corresponding to the frequency of vibration of some of the bonds in the molecule – thus providing information about its vibrational properties. The experiment can be done in two stages; first a short-lived intermediate is generated by a powerful pulsed laser (flash photolysis) and this is followed by interrogation by a second probe laser. In this way structural information about the intermediate is obtained.This technique is called time-resolved resonance Raman (TR^3) spectroscopy, and is particularly suited to biological molecules.

Even though biological molecules may be huge, we are often interested in what is happening in one particular location in the molecule. For instance, the large protein haemoglobin, which carries oxygen in the blood, contains a flat ring-like structure, a porphyrin, with a central iron atom. It is well known from X-ray crystallography that the iron atom lies nearly in the plane of the porphyrin when oxygen is attached (as in the oxygenated form oxyhaemoglobin) but out of the plane when oxygen is not attached (Figure a). Moreover in the oxy form, all the electrons are paired up in the iron, and the molecule is 'diamagnetic', whereas in the de-oxy form the electrons are unpaired and the molecule is 'paramagnetic'. The interpretation is that the paramagnetic form of the iron atom is a bit bigger than the diamagnetic form, and is thus squeezed out of the fixed porphyrin ring.

How does this local change happen as the oxygen comes on and off? Resonance Raman spectroscopy homes in on this immediate environment and, ignoring the rest of the protein molecule, shows that the instant the oxygen leaves the iron, the iron becomes paramagnetic and thus bigger, stretching the porphyrin ring in the process. Having accomplished this, it then moves out of the plane in a more leisurely – albeit still very fast – manner. This ability of resonance Raman spectroscopy to home in on significant chemical behaviour is being exploited in all manner of biological systems.

(a) Representation of the change in structure on loss of oxygen from oxyhaemoglobin. In oxyhaemoglobin the iron atom is nearly in the plane of the porphyrin ring – represented by the ellipse, with the four lines representing four iron-nitrogen bonds – and is bound to the oxygen molecule as shown. When the oxygen leaves, the magnetic state of the iron changes, it increases in size and stretches the ring. The iron atom then moves out of the plane, and the ring adjusts to its original size.

When the other form of vibrational spectroscopy, infrared spectroscopy, is applied to large molecules, it usually produces very complicated spectra, which are difficult to make sense of. However, under some circumstances, it can, like resonance Raman spectroscopy, home in on certain parts of the molecule. This is particularly true for transition-metal compounds which contain carbonyl (CO) or cyanide (CN) groups bonded to the metal. Suppose that a simple metal carbonyl such as the chromium compound $Cr(CO)_6$ is subjected to a burst of ultraviolet laser light; one carbonyl is removed. Very fast infrared spectroscopy can show that the remaining short-lived species, $Cr(CO)_5$, has a shape like a square pyramid, and it can follow this intermediate's reaction with other molecules. Molecules containing carbonyl groups are often important in homogeneous catalysis and the technique presents one way of understanding the mode of reaction.

Both resonance Raman and infrared spectroscopy have also proved to be powerful methods of looking at excited states. There is a big search for molecules that can perform electronic 'tricks'. One approach is to make large, carefully-tailored molecules that contain transition metals, and where electrons can be persuaded, usually by shining light on the molecule, to hop from one metal atom to another. We then need to understand what the electron is doing, and this involves examining what kind of excited state follows the exposure to light. As a simple example, one classic application of TR^3 to excited states is William Woodruff's experiment, carried out at the University of Texas at Austin, on the ruthenium complex ion $[Ru(II)(bpy)_3]^{2+}$ (bpy stands for 2,2'-bipyridyl). Flash photolysis, followed by probing with visible spectroscopy, suggested that in the excited state the electron moves from the ruthenium atom to the bipyridyl group(s); the question was whether the electron is 'spread' over all three groups or sitting on just one. In Woodruff's experiment the pattern of the vibrational bands in the resonance Raman spectrum of the excited state showed clearly that the molecule can be described as $[Ru(III)(bpy)_2(bpy)^{-}]^{2+}$ – in other words, that the electron is indeed localised on one of the bipyridyl molecules. Such experiments, combined with corresponding infrared experiments, are proving invaluable in unravelling the complex behaviour of these important excited states.

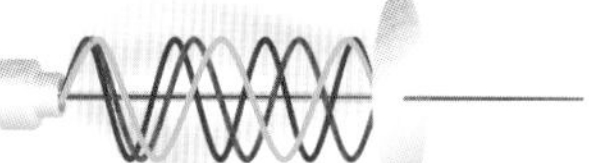

Box 2 Modelocking

In a laser, the light waves must 'fit' into the cavity. That is, the length of the cavity between the two mirrors must be a whole number of half wavelengths of light (see Figure a). A particular *laser cavity* will produce light with all these different wavelengths which are known as cavity modes. Normally, the light waves belonging to two different cavity modes are not in phase with one another, although of course two light waves belonging to the same cavity mode have to be in phase with one another because they are, by nature, coherent.

As its name suggests, modelocking involves locking the phases of the various cavity modes together so they can interfere with one another in an orderly manner. The best way to imagine this is to picture all the different cavity modes having a maximum at a particular point in space, at the same time (see Figure b). These waves will interfere constructively at this point in the cavity but destructively elsewhere, resulting in bursts of constructive interference separated by regions of destructive interference.

There is only one pulse in the cavity at any one time and each pulse lasts only a few picoseconds or femtoseconds depending on the number of cavity modes which are superimposed. As more and more modes are added to the superposition, they will interfere constructively for shorter and shorter periods of time, and so superimposing more cavity modes generates a shorter pulse. The shortest pulses which have been measured today are around 5 femtoseconds in duration and are obtained using a titanium sapphire crystal as the laser medium.

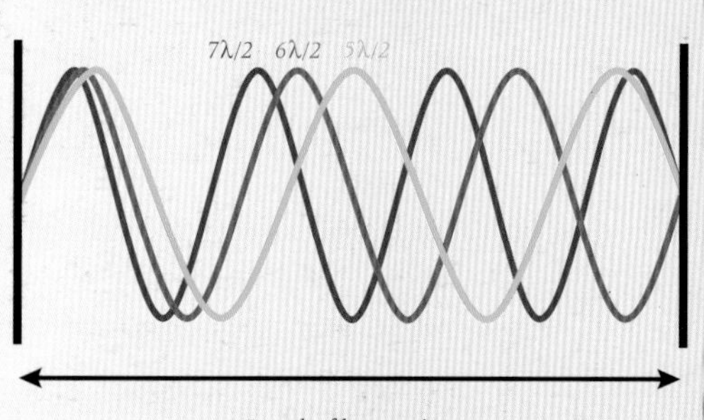

Figure (a) For a light wave to 'fit' into a laser cavity there must be a whole number of half wavelengths (wavelength = λ) between the two end mirrors.

Figure (b) The various cavity modes are only in phase with one another in a small region of space (pink area).

Picosecond lasers

From this time on, faster and faster chemical reactions became accessible as the laser physicists provided shorter and shorter laser pulses. By using a technique called 'modelocking' (see Box 2) it soon became possible to generate pulses of only a few picoseconds (10^{-12} seconds) duration and, in fact, the shortest modelocked pulses available today last just a few femtoseconds (10^{-15} seconds).

Peter Rentzepis, an American scientist working in New Jersey, exploited this new picosecond light source to measure various so-called radiationless processes in liquids in a series of elegant experiments. Usually, a molecule possessing vibrational energy in an excited electronic state relaxes quickly back to a vibrationless state by colliding with other molecules. This process is a radiationless process since the vibrational energy is lost as heat during collisions rather than by the emission of light. Following this radiationless transition, the molecule may either drop back down to its ground electronic state by emitting a photon or it may cross over to a different excited state in another radiationless transition. During this transfer from one electronic state to another, which is known as *intersystem crossing*, the molecule may twist or bend or the atoms may even rearrange themselves within the molecule.

In one particularly ingenious experiment, Rentzepis measured the rate of vibrational relaxation of the first excited singlet state of the organic molecule, azulene (Figure 10). Pulses of green light from a modelocked laser were used to excite high vibrational levels of the first excited singlet state of azulene from its ground state. These pulses of green light are created by combining two photons of infrared radiation in a birefringent crystal (see *The world of liquid crystals*). The infrared radiation itself does not have sufficient energy to excite the first

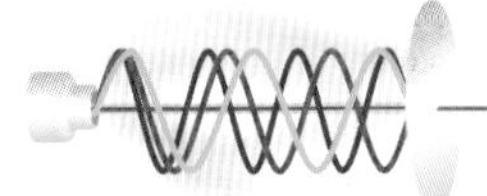

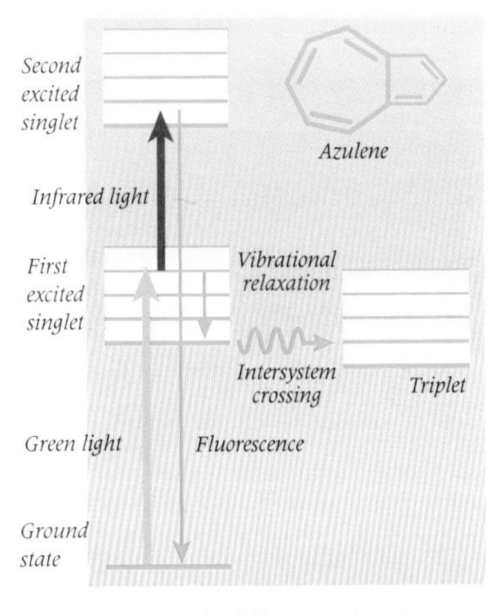

Figure 10. Energy level diagram of azulene.

excited singlet state from the ground state, however, it does have sufficient energy to lift the molecule from the high vibrational levels of this first excited state to the second excited singlet state. Unlike the first state, the second fluoresces rather strongly back down to the ground level. By photographing this fluorescence when both the green pulse and the infrared pulse are focused into a cell of azulene, Rentzepis deduced that the lifetime of the relaxation process was just 7.5 picoseconds. Because the lifetime is longer than the duration of the laser pulses he used, even after the green pulse has passed through the cell of azulene, the vibrationally excited levels of the first excited singlet state are still hanging around and so can be pushed up to the second excited singlet state when the infrared pulse comes along (Figure 11).

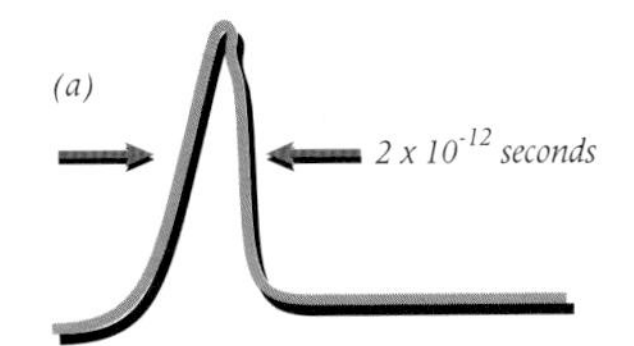

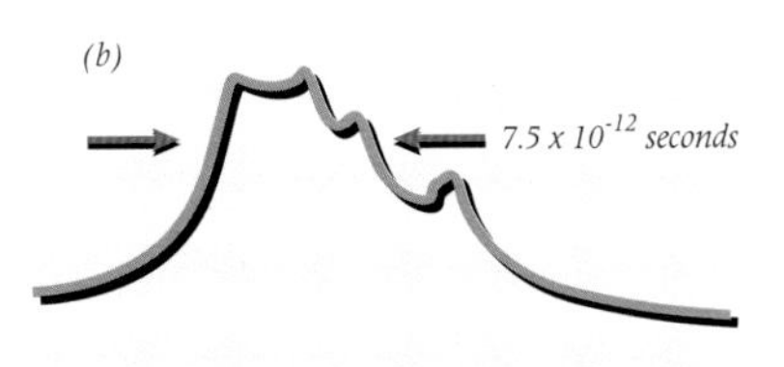

Figure 11. Time-resolved traces of (a) a laser pulse of two picoseconds duration and (b) the fluorescence from the second excited singlet state of azulene; its duration provides a direct measure of the lifetime of the first excited state of the molecule (just 7.5 picoseconds).
Adapted from 'Picosecond spectroscopy'. T. L. Netzel, W. S. Strive and P. M. Rentzepis.

Since most reactions carried out by chemists take place in solution, one of the main questions that needs to be answered is how do solvents affect a chemical reaction? One important role that a solvent plays is to stop a molecule breaking up by forming a cage from which the fragments cannot escape. This is appropriately named the cage effect and, although it was first thought of by James Franck and Eugene Rabinowitch at Göttingen in 1934, it has only become possible to observe it directly since the development of picosecond lasers. In the early 1970s Tung Chang, Geoff Hoffman and Kenneth Eisenthal carried out a series of original experiments at IBM in San Jose, California to investigate how various solvents affected the dissociation of molecular iodine into atomic iodine. Their somewhat controversial results stimulated a long-lasting debate about what they had actually seen and set the stage for a huge burst of activity in the theoretical and experimental understanding of the cage effect.

Eisenthal and his research group used the green light from a modelocked laser to pump molecular iodine up to an excited state which could decay producing two iodine atoms. By using an optical delay line similar to that designed by Porter and Topp (recall that the timing of the second, probe laser pulse depends on the distance it has to travel, which is controlled by a moveable mirror), they could fire a second pulse into the solution at a series of later times and monitor its absorbance by the iodine atoms. It is worth pointing out that picosecond delays are also relatively simple to generate since a delay of 10 picoseconds corresponds to a path length of 3 millimetres.

Figure 12 shows how the absorbance of the probe light pulse varies with the time-delay between the two pulses for a solution of iodine in carbon tetrachloride. The chemists learnt a great deal of new information concerning the flow of energy between the solvent and solute from these pioneering experiments. However, later work on this system has shown that the chemistry is actually rather more complicated than was appreciated at the time and involves an incredibly large

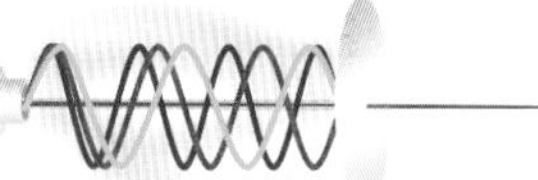

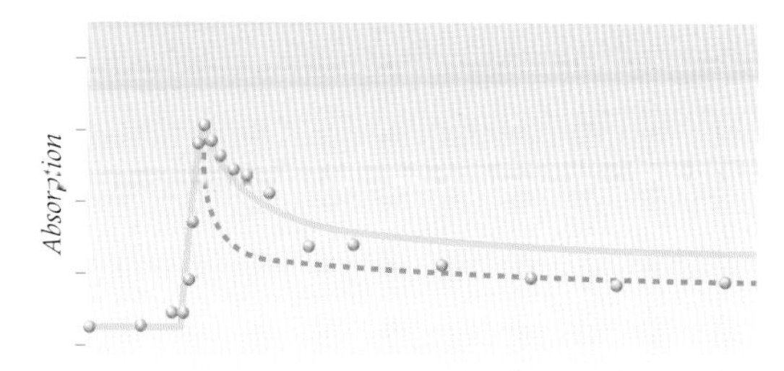

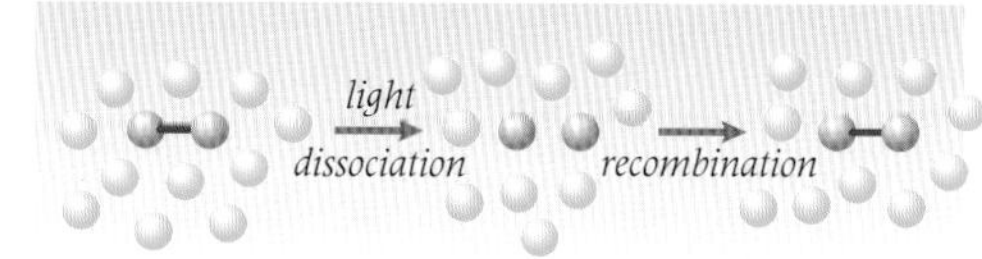

Figure 12. The results of a picosecond laser experiment to monitor the dissociation of iodine molecules in a solution of carbon tetrachloride and subsequent recombination. The graph shows how the absorption of picosecond light by dissociated iodine atoms changes with time. Notice how the number of iodine atoms rises to a maximum around 50 picoseconds after photolysis and subsequently decays as they recombine to form iodine molecules.

Adapted from Chemical Physics Letters, 25 (2), 201, (1974), with permission from Elsevier Science.

number of excited states of iodine. It has only recently been possible to develop a complete description of all the processes taking place.

Another important role that a solvent plays in a chemical reaction is to exert friction. British chemist, Graham Fleming, and his research team at the University of Chicago, used picosecond spectroscopy to investigate how changing the friction influences the speed with which the molecule, stilbene, can twist about its central bond (Figure 13). This molecule looks like a bent dumb-bell with two benzene rings joined by a carbon-carbon double bond. This molecule can exist in two forms, or isomers. When the benzene rings are arranged on opposite sides of the carbon-carbon double bond (so they appear above and below it) the molecule is known as *trans*-stilbene, whereas when they are on the same side it is known as *cis*-stilbene. Although there is a huge energy barrier preventing one isomer from changing to another (isomerisation) in the ground state of the molecule, this is not so for the excited state, and experiments to observe the rate of isomerisation of the *trans* excited state have been especially rewarding. It turns out that in the gas phase at low pressures, increasing the friction increases the rate of isomerisation. On the other hand, at higher pressures the opposite is true and the rate decreases as the friction is increased. The rate of the reaction passes through a maximum at the point where the system changes from being a gas to a liquid. Although various theories have been proposed to explain how friction influences a chemical reaction, there is still some controversy concerning its precise role in the liquid phase when the friction is very high.

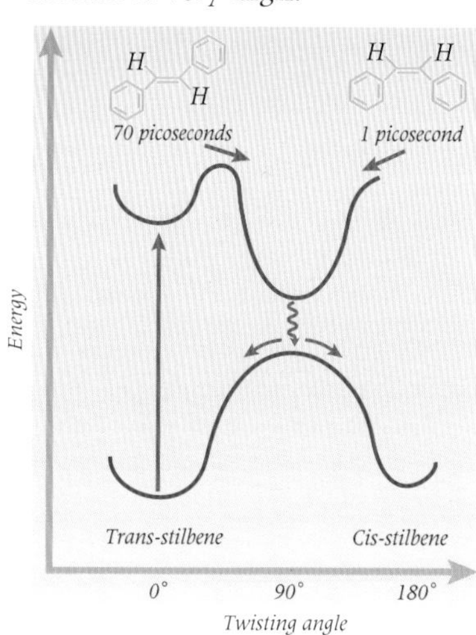

Figure 13. An energy level diagram of the ground and electronically excited states of stilbene. In the ground state there is a very large energy barrier preventing trans-stilbene converting into cis-stilbene and vice versa, although in the excited state this is not the case. Although isomerisation of the trans isomer is hindered by a small barrier and happens on a timescale of approximately 100 picoseconds, isomerisation of the cis isomer is barrierless and occurs on a 1 picosecond timescale. The vertical arrow on the diagram shows how the trans isomer would be created in its excited state from the ground state.

Photosynthesis

Currently, one of the most rapidly expanding areas of chemical dynamics in solution is photosynthesis. The transfer of solar energy between molecules in a leaf membrane, and the subsequent movement of electrons and protons across the membrane, is very rapid and occurs in less than a picosecond (see *The power of electrochemistry*). George Porter and his colleagues at Imperial College, London are among a growing number of groups employing state-of-the-art femtosecond lasers to unravel these primary processes in photosynthesis. At the moment, the system known as Photosystem II, which is also found in photosynthesising bacteria, is receiving most attention. The reaction centre of Photosystem II is the simplest unit in a green plant that takes energy from sunlight to remove electrons from water and release oxygen. Recently, Porter, together with

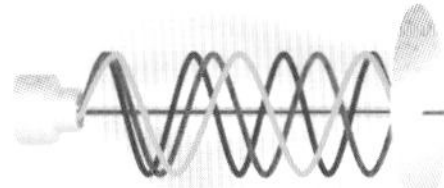

David Klug, James Durrant and their colleagues, have carried out some particularly ingenious experiments to pump two excited states on opposite sides of the reaction centre of Photosystem II. The energy then flows backwards and forwards between these two sides until equilibrium is reached and it turns out that this redistribution process occurs in just 600 femtoseconds.

Supersonic beams

Although most chemical reactions do take place in solution, in order to understand fully the chemistry of large and complicated molecular systems, it is often necessary to go back and look at each individual fundamental step. The best way to do this is to start with a small isolated molecule and to understand how it behaves on its own and how it interacts with other molecules depending on whether it is rotating, vibrating or travelling at great speed. The advent of the supersonic molecular beam as a source of cool molecules in the 1970s paved the way for chemists to observe the reactions of isolated molecules in well-defined energy states. It marked the beginning of an exciting new era in chemical dynamics. Although beam sources using nozzles were first developed in the early 1950s by Arthur Kantrowitz and Jerry Grey at Cornell University, it was 20 years before Richard Smalley, Lennard Wharton and Donald Levy at Chicago University first exploited the cold temperatures of supersonic molecular beams to record high-resolution spectra of molecules. Then, in 1986, Dudley Herschbach, Yuan Lee and John Polanyi shared the Nobel Prize for Chemistry for

Dudley Herschbach

Yuan Lee

John Polanyi

Pictures ©The Nobel Foundation

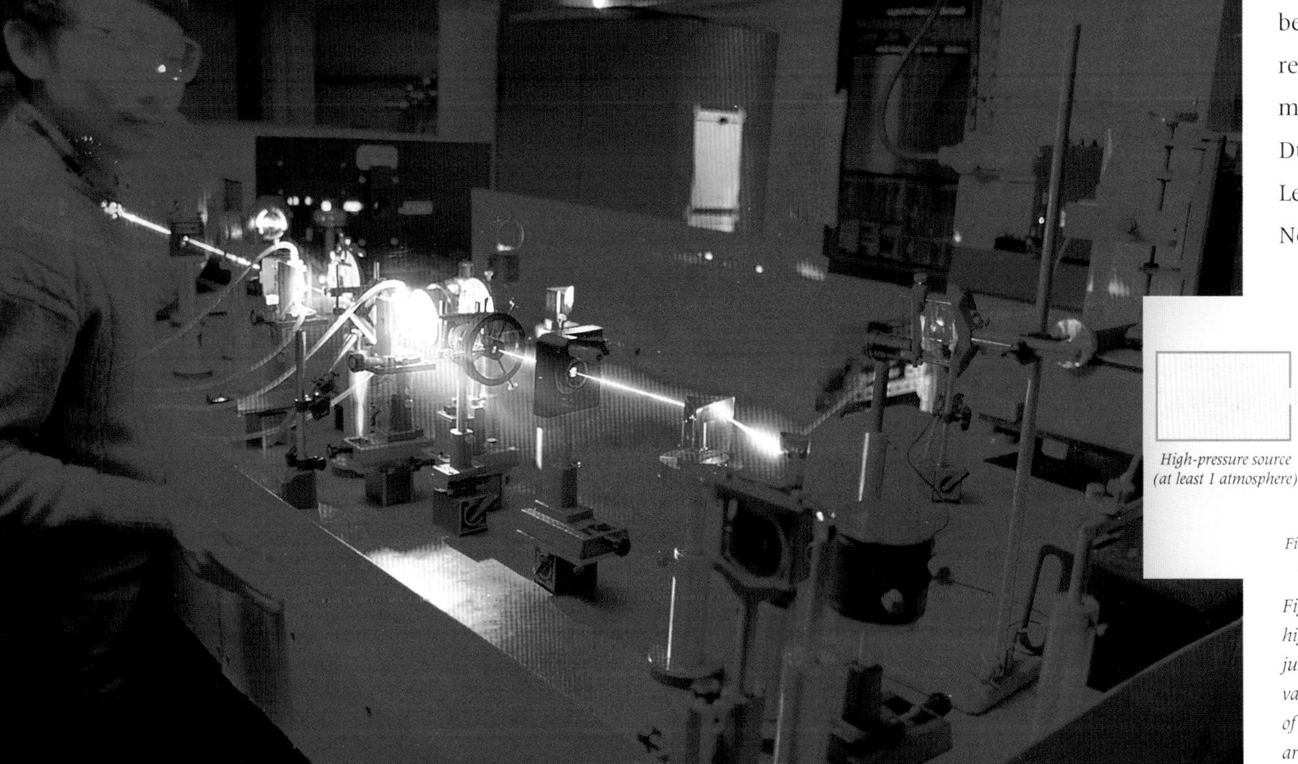

Physics Dept., Imperial College/Science Photo Library.

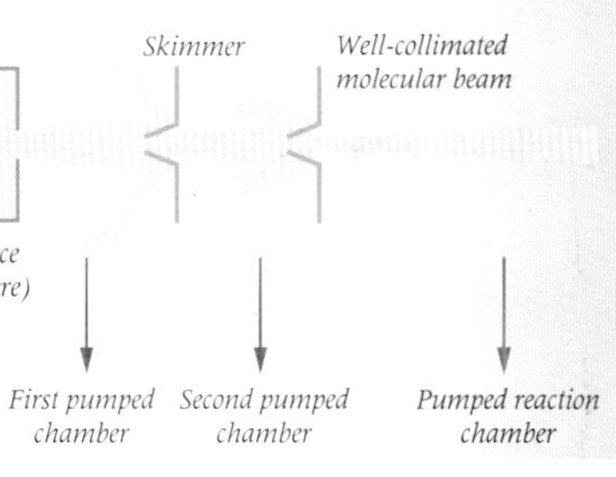

Figure 14. A diagram of a molecular beam. A high-pressure gas is forced through a tiny hole, just a fraction of a millimetre in diameter, into a vacuum chamber. As it expands it cools. A series of skimmers (rather like upside down funnels) are placed along the path of the beam to 'skim' off the edges and make it very well-collimated.

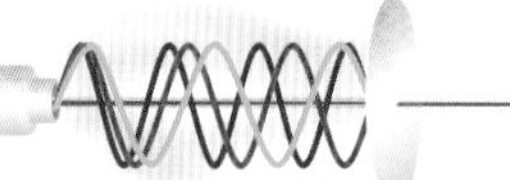

their pioneering work on the dynamics of elementary chemical reactions using molecular beam techniques.

A supersonic molecular beam is created by expanding a gas from a high pressure chamber through a tiny pinhole into a vacuum chamber (Figure 14). In the nozzle itself, the gas is compressed and behaves rather like a liquid. The thermal energy causing the high-pressure gas molecules to move with a wide range of speeds in all directions in the chamber is converted into translational motion in one direction as the gas 'flows' through the nozzle into the vacuum beyond. As the gas bursts into the vacuum chamber it expands and cools, in much the same way as drops of hairspray released in an aerosol feel cold as they land on your skin. In the expanded beam there are very few collisions and so almost all the molecules travel with the same speed in the same direction. By placing a series of 'skimmers' along the beam path, the molecular beam can be very well collimated indeed. A skimmer is rather like a small funnel but in reverse, with the beam going in through the small hole so that any molecules not travelling exactly along the axis of the funnel are 'skimmed' off the edges of the beam.

The study of chemical reactions using molecular beams has allowed chemists to understand how two molecules or atoms react to form new molecules and begin to understand something about the transition state. For example, in a classic experiment, Yuan Lee and his group at the University of California at Berkeley have studied the reaction between fluorine atoms and the deuterium molecule to form deuterium fluoride and a deuterium atom (deuterium is a heavier version of hydrogen).

$$F + D_2 \longrightarrow DF + D$$

The experiment works by firing two supersonic beams of reactants at right angles towards one another and observing the

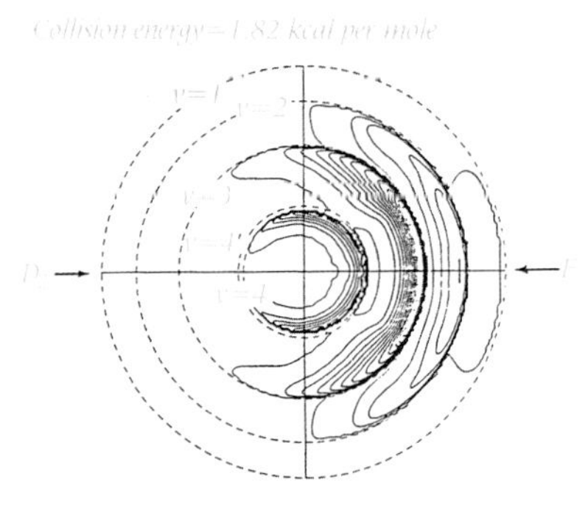

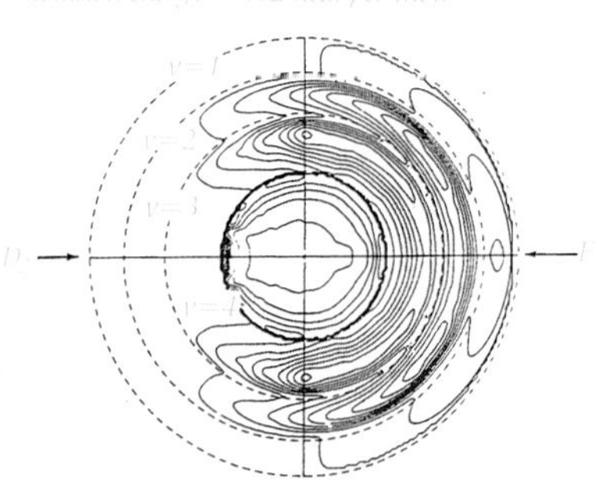

Figure 15. The reaction between fluorine atoms and the deuterium molecule to form deuterium fluoride. The circles denote the maximum velocity possible for a deuterium fluoride molecule in a specific vibrational state, $\underline{v}$ = 0, 1, 2 and so on. The contours associated with a specific circle show the speed and direction of a deuterium fluoride molecule within that vibrational state. From this diagram it is apparent that most of the deuterium fluoride is scattered in the same direction as that of the incoming deuterium molecule.
Adapted with permission from D. M. Neumark, A. M. Wodtke, G. N. Robinson, C. C. Hayden, K. Shobatake, R. K. Sparks, T. P. Schafer, and Y. T. Lee. Journal of Chemical Physics 82, 3067 (1985). ©1999 American Institute of Physics.

direction in which the products move apart and their speeds. It turns out that it is most convenient to see what is going on if we use something called a centre-of-mass frame. That is, if we pretend that we are sitting at the centre of mass of the two reactants so it appears to us that they actually collide head on. (Choosing the right reference frame is also important to aircraft pilots. Whilst in the air, it is the speed of the plane relative to the air that is important, although when landing it is the speed of the plane relative to the ground which is significant. The conversion between air speed and ground speed is carried out routinely.) The contour map in Figure 15 shows where the product molecule, deuterium fluoride, ends up. After the collision, the majority of the deuterium fluoride molecules move away in the same direction that the deuterium molecule was travelling in. This indicates that the transition complex is rather short-lived since it does not have sufficient time to rotate before the product breaks away. If the product was scattered over a wider range of angles, it would indicate that the transition complex had a longer lifetime

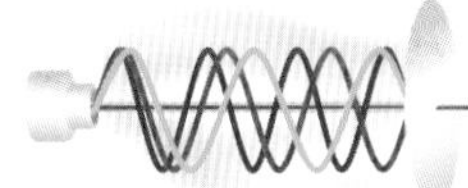

because it would have had time to rotate before the deuterium fluoride molecule was ejected.

In order to develop a really complete picture of how reactant molecules approach one another and how the products move apart it is necessary to determine the orientation of the reactants and products in space as well as the angles of approach and departure. Dick Zare, Andrew Orr-Ewing and colleagues at Stanford University in California have recently developed an elegant method called core extraction for determining the complete stereochemistry of the reaction between a chlorine atom and a vibrating methane molecule to form hydrogen chloride and the methyl radical.

Dick Zare.

$$Cl + CH_4 \longrightarrow HCl + CH_3$$

In these experiments a mixture of chlorine and methane is expanded through a pulsed nozzle (like a machine gun) and across a beam of ultraviolet laser light to dissociate the chlorine molecule. One of the carbon-hydrogen bonds in the methane molecule is then stretched either parallel or perpendicular to the velocity of the chlorine atom by exciting it with plane-polarised infrared radiation (with all the light waves oscillating in the same plane). Changing the polarisation of the laser light causes the orientation of the stretched carbon-hydrogen bond to change. After the chlorine atom has collided with the methane molecule, it removes a hydrogen atom and forms hydrogen chloride and the methyl radical. The hydrogen chloride molecule may then continue to travel in the same direction as the chlorine atom (known as forward scattering) or it may travel backwards in the opposite direction (backward scattering) or somewhere in between (sideways scattering). The speed of the hydrogen chloride product and its direction of travel are measured by ionising the molecule using lasers and timing the distance for the ions to travel to a detector. This technique is known as time-of-flight mass-spectrometry. By placing a mask with a small hole along the axis of the time-of-flight tube, only ions travelling exactly along this axis are collected. The Stanford team have named this method core extraction since it is rather like coring an apple.

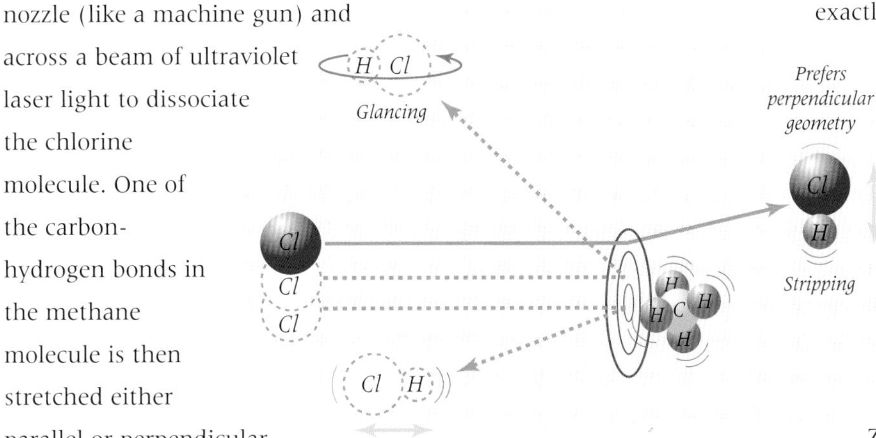

Figure 16. A model for the reaction between atomic chlorine reacting with vibrationally excited methane. When the hydrogen chloride molecule is forward-scattered it comes away rotating like an aeroplane propeller, whereas the backward-scattered molecule comes away rotating like a frisbee.

Adapted with permission from W. R. Simpson, T. P. Rakitzis, S. A. Kandel and A. J. Orr-Ewing. Journal of Chemical Physics 103 (17), 7313, (1995). ©1999 American Institute of Physics.

By the conservation of energy and momentum of the reactants and products, Zare and his team have managed to deduce that the hydrogen chloride molecule is most likely to be backward scattered when the stretched C-H bond points directly towards the chlorine atom, whereas for forward scattering to occur, the stretched C-H bond must be perpendicular to the direction of travel of the chlorine atom.

In a more refined version of their experiment, the Stanford group then used polarised laser light to determine the orientation of the hydrogen chloride molecule as it moves away from the collision complex. It turns out that when the hydrogen chloride molecule is forward scattered it comes away rotating around an axis which is parallel to its

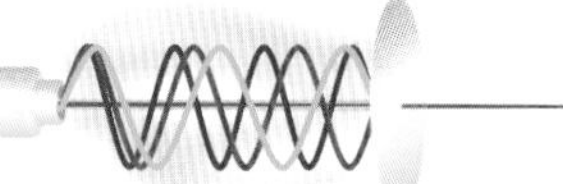

direction of travel, rather like the propeller of an aeroplane. On the other hand, when the molecule is backward scattered it comes away rotating around an axis which is perpendicular to its line of travel, rather like a frisbee.

Very recently, John Simons, Mark Brouard and colleagues at the University of Oxford have developed a powerful new approach, employing a technique known as Doppler-resolved laser spectroscopy, to obtain a complete three-dimensional picture of the reaction between an oxygen atom and a hydrogen molecule to form the hydroxy molecule and a hydrogen atom.

$$O + H_2 \longrightarrow OH + H$$

In these experiments, the researchers filled an enclosed container with a mixture of dinitrogen oxide (N_2O) and hydrogen, and then used a pulse of polarised ultraviolet laser light to break up the dinitrogen oxide so that oxygen atoms formed. The electric field of the polarised light wave causes the N-O bond to vibrate in a specific direction and consequently, when it has so much energy that it actually falls apart, all the oxygen atoms end up travelling in the same direction. The oxygen atoms collide with the hydrogen molecules to form hydroxy molecules and hydrogen atoms. The direction in which the products are scattered, their orientation and their speed are then deduced by spectroscopy. A second polarised laser beam pumps the hydroxy molecules into an excited state which then decays back to the ground-state by fluorescence.

The amount of fluorescence depends on how many hydroxy molecules absorb the radiation which in turn depends on the direction in which the molecules are travelling. A molecule travelling in the same direction as the second laser beam will 'see' the light wave as having a longer wavelength than it actually has, and so the laser will need to be tuned to slightly shorter wavelengths to excite the molecule. On the other hand, a molecule travelling in the opposite direction to the laser beam will 'see' the light wave as having a shorter wavelength than it actually has and so the laser will need to be tuned to slightly longer wavelengths to excite the molecule. This change in observed wavelength is the well-known Doppler effect, which is also responsible for the change in note heard as a train sounding its horn passes a stationary commuter on a station platform. By analysing the Doppler profiles of the fluorescence, John Simons and his group could determine the directions of the products of the reaction and their velocities. It turns out that if the hydroxy molecule is not vibrating, then those molecules that are rotating relatively slowly tend to be scattered backwards whereas those that come away rotating more quickly tend to be scattered in a forwards direction. On the other hand, if the hydroxy molecule formed is vibrating, the molecules seem to be scattered more or less equally both in the forward and backward directions.

Chemistry between the stars

Using a combination of molecular beam methods and lasers, it is also possible to study reactions between ions and molecules. Ion-molecule reactions are extremely important in the gas clouds that lie in the vast cold regions of space between the stars. In fact, the most common ion-molecule reaction in the Universe is that between a neutral hydrogen molecule and the charged hydrogen molecular ion to produce a neutral hydrogen atom and the triatomic hydrogen ion.

$$H_2 + H_2^+ \longrightarrow H + H_3^+$$

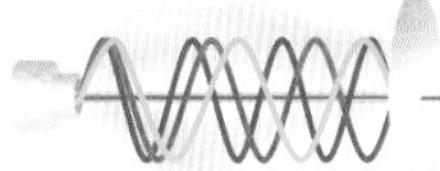

The Great Nebula in Orion.
©1982 PPARC, Royal Observatory Edinburgh and Anglo-Australian Telescope Board.

Tim Softley and his research group, also at Oxford, have recently developed a novel experimental method to study reactions such as this one. By using a combination of pulsed ultraviolet laser light and electric fields they are able to generate molecular ions with particularly well-defined vibrational and rotational energy. These molecular ions are then allowed to flow through a supersonic jet of neutral molecules where the reaction occurs. It appears that the end-over-end rotation of the hydrogen molecular ions hinders their reaction with the neutral species. As the molecules cool, their rotational energy decreases and consequently they become more reactive. This effect is particularly significant in the low temperatures of interstellar gas clouds.

Some of the most significant contributions to chemical reaction dynamics using ultra-fast lasers and molecular beams are currently being made by Ahmed Zewail at the California Institute of Technology. Zewail is the pioneer of an incredibly exciting new area of chemistry which has become known as femtochemistry. It involves using state-of-the-art femtosecond lasers to probe chemical reactions on the same timescale as the molecular vibrations which control their rates. The elegance and potential of the technique were rather beautifully illustrated in 1988 in an experiment to observe the break-up of isolated molecules of the ionic compound sodium iodide in real time.

Ahmed Zewail.
Courtesy of the Egyptian National Postal Organization.

The potential energy diagram for the dissociation of sodium iodide is shown in Figure 17. There are two potential energy curves (like those in Figure 2). There is a low-energy ionic state (orange curve), consisting of positively charged sodium ions and negatively charged iodide ions bound together by the attractive force of their opposite charges, and a high-energy covalent state (blue curve), consisting of neutral sodium and iodine atoms. These two potential energy curves are

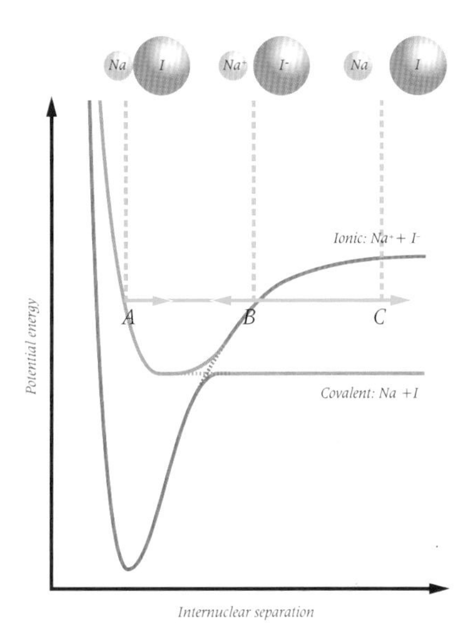

Figure 17. (a) A potential energy level diagram for the sodium iodide molecule. The orange curve is that of the ionic molecule whilst the blue curve is the higher energy covalent molecule NaI. Where these curves would be expected to cross (dashed lines) they actually interact with one another to form two different curves (solid lines) and each one has a mixture of ionic (orange) and covalent (blue) character. The horizontal black line shows what happens to a molecule which is pumped up to the high energy curve. It is formed at point A where it can be thought of as being covalent and it has a compressed bond. The bond then stretches and it becomes a molecule which can be thought of as being ionic (point B). Here, there is a 90-per-cent chance that the bond will simply compress back to point A, although there is a 10 per cent chance that the molecule which actually fall apart to form two neutral atoms (point C).

Courtesy of Professor M. J. Pilling, University of Leeds.

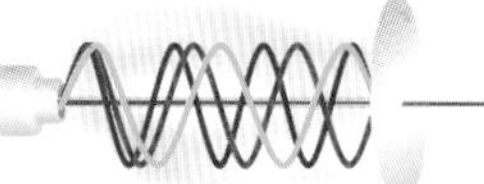

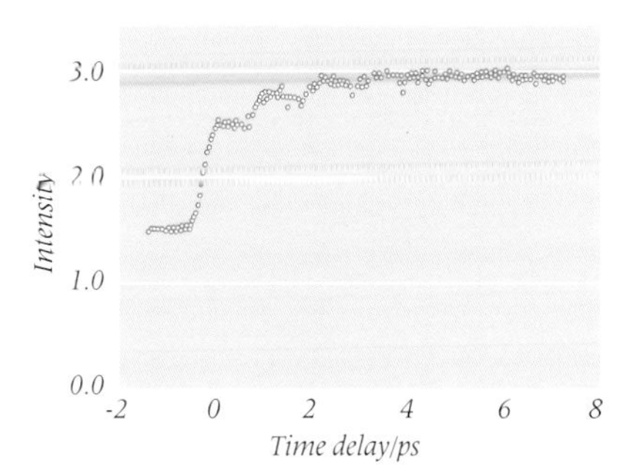

(b) The fluorescence of dissociated Na atoms (at point C) which have been pumped to an excited state, Na, with a delayed probe pulse. Notice how the fluorescence increases in steps as the molecule vibrates back and forth between points A and B.*

Courtesy of Professor M. J. Pilling, University of Leeds.

interleaved and, at the point where they would be expected to cross (dashed lines in the Figure), they actually interact and repel one another. This results in two potential wells of mixed character. The lower well is ionic at short bond lengths and covalent at long bond lengths whilst the upper well is covalent at short bond lengths and ionic at longer bond lengths.

The experimental technique is essentially a faster version of the nanosecond and picosecond flash photolysis experiments. There is one important technical difference: to create a time delay of 10 femtoseconds requires a path difference of only 3 micrometres, and so a rather more sophisticated translation stage is necessary. The pump pulse arrives first to pump the molecule from the ground state to the excited electronic state of the molecule. At the precise moment of its creation, the electronically excited molecule is compressed, that is the two atoms are closer together than the equilibrium position. The potential energy stored in the compressed molecular bond pushes the atoms apart and causes the molecule to vibrate back and forth through its equilibrium position. In terms of the potential energy diagram, the molecule is bouncing backwards and forwards between the inner (covalent) and outer (ionic) wall of the upper well. Each time the molecule passes the point where the two potential energy curves interact, there is a 10% chance that it might 'leak out' of the upper well and dissociate into neutral atoms. The number of neutral sodium atoms (Na) that have been created by dissociation at any particular time is monitored by pumping them up to an excited state (Na*) using a probe laser pulse. These excited atoms eventually release their energy in the form of light (fluorescence) and the dissociation reaction can be followed by monitoring the total amount of fluorescence as a function of the delay between the pump and probe pulses. Figure 17 shows how this fluorescence builds up in a series of steps with the time between the steps being equal to the time that it takes for

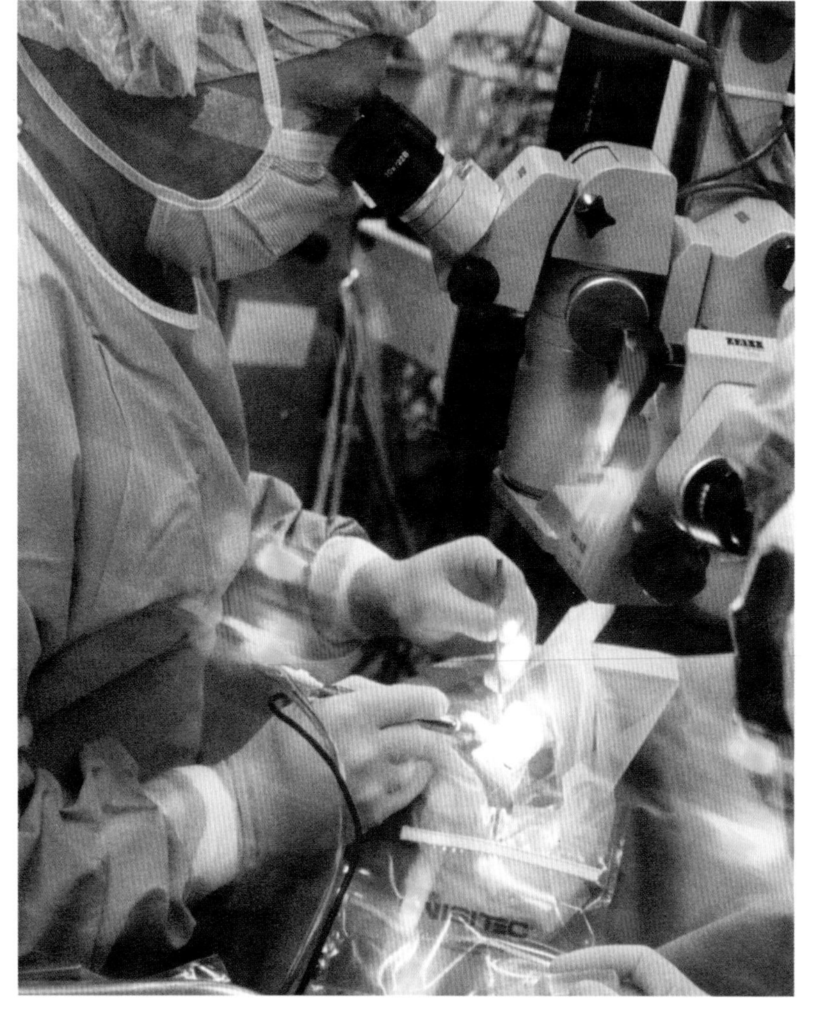

Laser surgery

the vibrating sodium iodide molecule to oscillate backwards and forwards in the upper well.

Since the early days of flash photolysis, when George Porter and Ronald Norrish were using gas-filled flash lamps in Cambridge to study combustion and the photochemistry of the triplet state in solution, an improvement in time resolution of one thousand million has revolutionalised the field of chemical dynamics. Now, chemists such as Zewail are using state-of-the-art femtosecond lasers to watch isolated molecules fall apart or rearrange themselves. Those working at the boundary between chemistry and biology, such as Porter and his colleagues at Imperial College, London and Godfrey Beddard at the University of Leeds, are trying to understand the primary processes in photosynthesis. Those on the boundary between chemistry and physics, such as Bart Noordam in Amsterdam at the FOM-Institute for Atomic and Molecular Physics and Helen Fielding at King's College London, are using picosecond techniques to watch electrons ionise from atoms and molecules. A combination of pulsed laser and molecular beam methods are allowing scientists such as John Simons and Tim Softley at Oxford to understand the fine details of reactions between two molecules or a molecule and an ion.

One of the ultimate goals of the chemist is to control chemistry at the molecular level. Scientists including Moshe Shapiro at the Weizmann Institute of Science in Israel and Paul Brumer at the University of Toronto in Canada have recently shown that lasers can be used to control the pathway of a chemical reaction. However, this work still has a long way to go before it is of any practical use. There are still, nevertheless, a great number of questions to be answered about basic molecular dynamics and when the simple systems are understood it will be possible to move towards a greater understanding of more complex processes, particularly those that occur in biology.

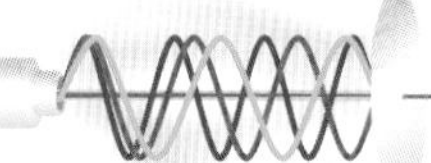

Further reading

1. "Forty years of Photochemistry. A summary by George Porter", George Porter, *J. Chem. Soc., Faraday Trans. 2*, 1986, **82**, 2445.
2. "Real-time laser femtochemistry. Viewing the transition from reagents to products", Ahmed Zewail and Richard Bernstein, *Chemical and Engineering News*, 1988, **66**, 24.

Glossary

Absorption spectrum The characteristic pattern of wavelengths absorbed by a particular molecule when excited from lower to higher energy levels.

Coherent The property of light waves being in phase.

Dye laser A popular source of tunable coherent radiation.

Electronic energy levels The energies of the various arrangements of electrons within a molecule.

Emission spectrum The characteristic pattern of wavelengths emitted by a particular molecule when it drops from higher to lower energy levels.

Excited state A molecule existing in higher energy levels (electronic, vibrational, rotational) than the ground state.

Femtochemistry The study of the dynamics of very fast chemical reactions which occur after just a few molecular vibrations.

Flash photolysis The technique of using a very short pulse of radiation to create molecules in short-lived states followed by spectroscopy to monitor the change in concentrations of the various species with time.

Fluorescence The light emitted by a molecule as it drops from a high energy electronic state to a low energy electronic state with the same number of unpaired electrons.

Ground state A molecule in its lowest energy levels (electronic, vibrational, rotational).

Infrared spectroscopy A technique in which light in the infrared part of the spectrum is absorbed at particular frequencies by a sample. These frequencies are characteristic of molecular motions (vibrations) in the sample, and provide information about its molecular structure.

Intermediate A stable molecule formed during the course of a chemical reaction.

Intersystem crossing The transfer of a molecule from one electronic state to another with a different number of unpaired electrons and without the emission of radiation.

Laser cavity The parts of a laser between the two end mirrors.

Optical delay unit An arrangement for creating a short time delay between two laser pulses by making them travel along paths of different lengths (used in studies of reaction kinetics).

Photon A quantum of light.

Population inversion The situation of more molecules being in a higher energy state than in a lower energy state.

Raman spectroscopy A technique, discovered by the Indian scientist C V Raman, in which visible light is scattered from samples at frequencies that differ from the impinging light. (When the scattered frequency is the same as the impinging light it is called Rayleigh scattering).The differences between impinging and scattered light correspond to molecular motions (vibrations) and the technique is thus complementary to infrared spectroscopy.

Reaction dynamics The study of a chemical reaction: how the reactants approach one another and the products move apart, and the time it takes for the reaction to occur.

Reaction kinetics The speed of a reaction.

Resonance Raman spectroscopy (see Raman spectroscopy) Normal Raman spectroscopy is a very weak effect. However if the impinging light matches an absorption band in the sample, under certain circumstances, the scattered Raman light can be extremely intense.

Rotational energy levels The allowed energies of a rotating molecule.

Spectroscopy The study of the interaction of radiation with molecules to deduce their energy levels and hence their structure.

Supersonic molecular beam A well-collimated beam of molecules which almost all travel at the same speed in one direction and consequently has a very low translational temperature.

Transition state A very short-lived, energetically unstable species formed during a chemical reaction.

Vibrational energy levels The allowed energies of a particular vibrating bond within a molecule.

Biographical details

Dr Helen Fielding is a reader in physical chemistry at King's College London. She carries out experiments and calculations to investigate electron processes in atoms and molecules.

Acknowledgements

I should like to thank Professors John Simons and Mike Pilling for providing useful advice and interesting anecdotes, and Professor Jim Turner for his description of Raman spectroscopy.

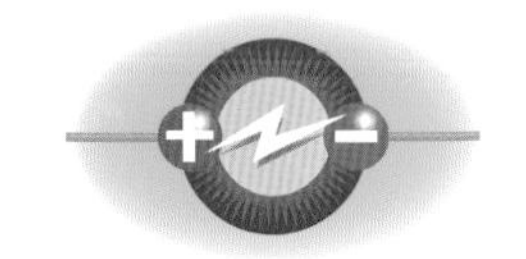

The power of electrochemistry

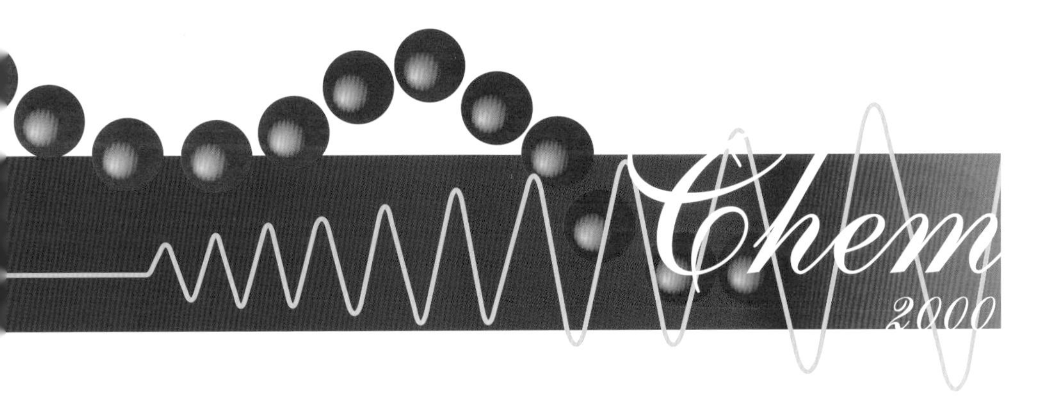

Electrochemistry is an important and fast-growing field. Electric vehicles, batteries, biological sensors and new electrochemical industrial processes – some important applications ensuring that electrochemistry has a bright future into the 21st century and beyond.

Dr Peter Birkin
University of Southampton

Courtesy Solectria Corporation/Photo by Sam Ogden.

Electrochemistry – the study of how electrons move between atoms and molecules under an electric potential – is today an important and fast-developing field. As well as providing insights into electron transfer processes and related chemical behaviour, applications of electrochemistry are helping to clean up the environment and provide new, more sensitive methods of chemical analysis.

Cell rooms, Lostock Northwich. Courtesy of ICI.

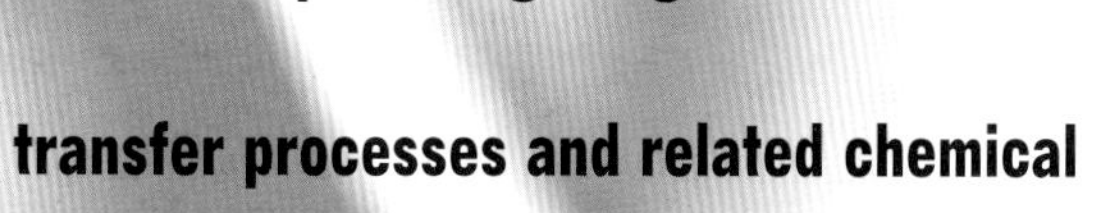

Courtesy Solectria Corporation/Photo by Sam Ogden.

Chemistry is largely about the interaction and exchange of electrons among atoms and molecules. *Electrochemistry* involves the study of these processes when electrons are transferred between molecules or *ions* (charged atoms or molecules) often at the surface of an electrically conducting material, an *electrode*, under the influence of a suitable driving force – in most cases an *electrical potential*, or voltage. Such electron transfer processes are found everywhere including the phenomenon of rusting and other types of corrosion, and in living cells. Nature depends on electron transfer processes to drive chemical reactions which, for example, convert food into chemical energy for use throughout an organism. Many enzymes (large protein molecules with the ability to catalyse a particular reaction, see *Chemical marriage-brokers*) operate by donating or removing electrons from molecules.

Over the past 200 years, researchers have exploited electrochemistry to perform useful tasks. These include extracting metals from their ores, synthesising chemicals and converting chemical energy into electrical power. As we approach the 21st century, electrochemistry is becoming increasingly important in developing environmentally-friendly industrial processes and sources of power, in cleaning contaminated soil and in detecting trace amounts of biologically important compounds.

As its name implies, electrochemistry combines electricity with chemical processes. The simplest *electrochemical cell* consists of a solution of ions (an *electrolyte*) in which are placed two electrodes (made of metals, conducting forms of carbon or semiconductors) connected to an outside voltage source. Applying a large enough voltage to the electrodes causes molecules or ions close to their surfaces to lose or gain electrons – processes called *oxidation* and *reduction* respectively. To maintain overall

electrical neutrality, oxidation at one electrode is always complemented by reduction at the other and *vice versa* – the electrons liberated from one electrode appear to move around the external circuit to be used at the other electrode. To complete the circuit, ions correspondingly move through the solution. The sign and magnitude of the current indicate what kind of chemistry is happening at each electrode and how fast. The level of voltage applied not only governs whether an oxidation or reduction occurs but also the speed of the process, thus giving control over the chemistry going on.

Oxidation and reduction are among the most important reactions in chemistry and reveal much about the electronic behaviour of the molecules concerned. Electrochemistry is thus a powerful probe of chemical reactivity. The ability to donate or gain electrons can be described by the term *redox potential*. A substance with a high or low redox potential will have a strong affinity to gain or donate electrons respectively. For instance, the element, fluorine, has one of the highest redox potentials known and subsequently gains electrons from other materials. It is therefore extremely reactive.

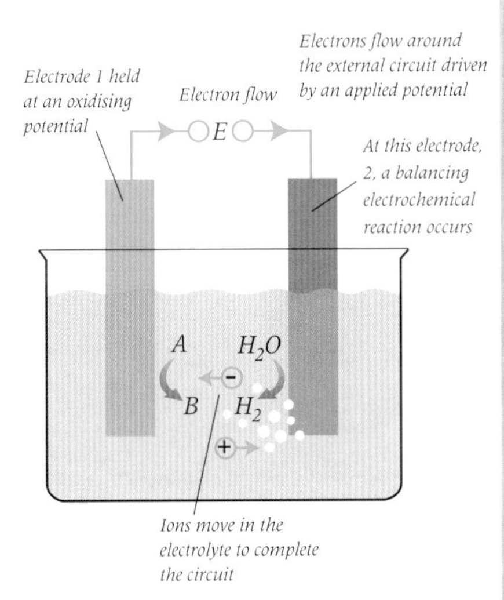

Figure 1. A simple electrochemical cell. The dissolved atom or molecule, A, can be oxidised at electrode 1 and converted to B. Electrons flow around the external circuit to electrode 2 where they can reduce water molecules. The circuit is completed by positive and negative ions from the electrolyte moving through the solution, but in opposite directions. In the figure above, curved arrows denote an oxidation or reduction.

Figure 1 shows a typical circuit. If two electrodes are placed within a water solution containing a molecule or ion 'A' and a potential is applied so that 'A' is oxidised to 'B' at electrode 1,

Box 1 The electrochemistry of rusting

Corrosion of metals, such as the rust that appears inexorably on cars, can be explained in terms of a simple electrochemical cell. The process consists of two reactions occurring perhaps at different sites on the surface of the metal. The metal is oxidised by atmospheric oxygen, which is itself reduced to water. At the point of corrosion, the metal itself dissolves.

$$\tfrac{1}{2}O_2 + H^+ + 2e \longrightarrow H_2O$$
$$M \longrightarrow M^{n+} + ne$$

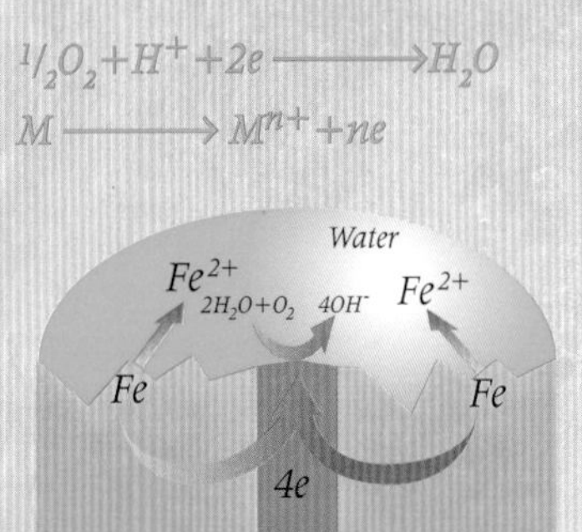

A Corrosion cell. Corrosion of a metallic surface takes place at two sites. At the first site, metal atoms dissolve to form ions. The electrons librated by this process travel through the metal where they can reduce oxygen gas.

The electrons generated flow through the metal between sites while ions on the surface of the metal move to complete the circuit. The two reactions are thus in balance and so equally affect the rate of corrosion. The type of metal also dramatically affects the rate. Some metals, such as aluminium, produce an oxide layer which is impervious to water and thus protects the metal from corrosion unless scratched. Other metals such as iron, produce an oxide that flakes away from the metal surface and so continually corrode.

In most cases, the corrosion rate is low but various chemical or biological factors can speed it up. To stop corrosion people employ many strategies. These include painting or coating the metal with a sacrificial metal (for example, zinc galvanising) which preferentially corrodes instead of the protected metal. It is also possible to protect a metal object electrochemically by maintaining its electrical potential so that the corrosion rate is virtually zero. Of course, this can be performed only on suitable systems (such as ships hulls, oil rigs or the steel inside reinforced concrete structures) but it is an effective way to protect a metal structure.

The cathodic protection of an oil rig. A rig's structure can be protected by placing platinum-coated titanium blocks on it. These become the anode while the rest of the structure constitutes the cathode. Because corrosion involves oxidation of the iron at an anodic site this protects the rig by allowing oxidation to occur only on the platinum electrodes. The iron is therefore cathodically protected.

Photo Courtesy of Trident Alloys Limited.

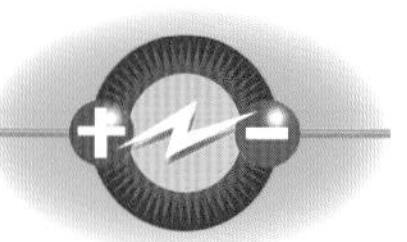

then a reduction process must occur at the other electrode to balance out the first process. This could be, for example, the reduction of water itself. The electrons lost by species 'A' at electrode 1 flow around the external circuit to electrode 2 where they can reduce water. The entire process is completed by the movement of ions in the solution. A simple example of a real electrochemical cell is the surface of a corroding metal such as iron (see Box 1).

Early progress in electrochemistry

The first serious investigations into electrochemistry date back to the end of the 18th century, when people first discovered ions, and studied how solutions of ions conduct electricity and the structure of the interface between an electrode and an electrolyte. Of the many researchers studying electrochemistry, the great 19th century scientist, Michael Faraday, is perhaps the most famous. He assigned – with a little help from his friends, Whitlock Nicholl and

Michael Faraday.
Reproduced courtesy of the Library and Information Centre, Royal Society of Chemistry.

William Whewell – the names of the components of an electrochemical system we still use today. These included 'electrode' (meaning 'electron way'), *'electrolysis'* (meaning 'electron splitting'), *'anode'* (positive electrode), *'cathode'* (negative electrode), *'anion'* (negative ion), and *'cation'* (positive ion).

Faraday showed that all forms of electricity were identical and helped to develop the laws of electrolysis. These laws showed that the mass of materials deposited on, or released by, an electrode would depend on the product of the current and the time it was passed for, the molecular mass of the material and the number of electrons involved in the process. Faraday's laws had far-reaching significance for the understanding of solutions containing ions because they implied that electricity itself was

Luigi Galvani 1737-1798.
Reprinted by permission of McGraw-Hill, F J Moore, A History of Chemistry,© 1939.

not continuous but was related closely to the then new ideas of elemental substances being made up of atoms and containing positive and negative particles.

Voltaic pile (battery) given by Count Alessandro Volta to Michael Faraday in 1814. The Royal Institution, London, UK Bridgeman Art Library London/New York.

Power from chemistry

The first practical use of electrochemistry was to generate and store electrical power and this is still a very active area of research today. Even as early as the beginning of the 18th century, people had noticed that animals such as the electric eel produced electricity. In 1780, Luigi Galvani discovered that

Alessandro Volta 1745-1827
Reprinted by permission of McGraw-Hill, F J Moore, A History of Chemistry,© 1939.

placing two different metals in contact with frog muscle produced an electrical current. Later his friend, Alessandro Volta, professor of natural philosophy at the University of Padua, demonstrated that metals alone could generate electricity. In 1800, he showed that a series of zinc and silver disks separated by layers of filter paper soaked in salt water could act as a source of electricity. This was in effect the first electrical *battery*.

The arrangement of metal plates and solution are the basis upon which all batteries operate.

Anatomy of a battery

Strictly speaking, a battery is a collection of several electrochemical cells from which electrical energy is obtained (although the stick batteries used in most personal stereos are actually single cells.) The battery converts stored chemical energy into useful electrical energy to power portable objects like laptop computers, mobile telephones, radios and so on.

Figure 2 shows a schematic representation of one cell of a battery. This is known as the *Daniell cell* after the English chemist, John Daniell, who invented it in 1836. It consists of a zinc electrode immersed in a zinc sulfate solution and a copper electrode immersed in copper sulfate, with the two electrolytes separated by a porous partition. Zinc metal dissolves from the zinc anode (the electrode that releases electrons on discharge) as zinc ions, giving up two electrons in the process. These two electrons flow around the external circuit and do useful work such as powering a radio, before entering the copper cathode (the electrode that consumes electrons on discharge) where they combine with copper ions from the surrounding copper sulfate to form copper metal.

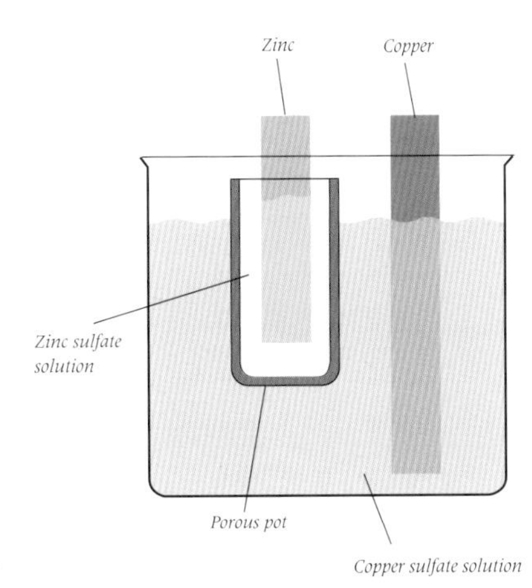

Figure 2. The Daniell Cell. In this simple battery, a copper electrode is immersed in a copper(II) sulfate solution and a zinc electrode in a zinc(II) sulfate solution. The two systems are separated by a porous pot which allows ions to move.

The process happens because the two metals have different *electrochemical activities* (the measure of a material's ability to oxidise or reduce another material). When combined, they generate a potential difference and thus a voltage between the terminals of the cell because one material will tend to give up electrons while the other consumes them.

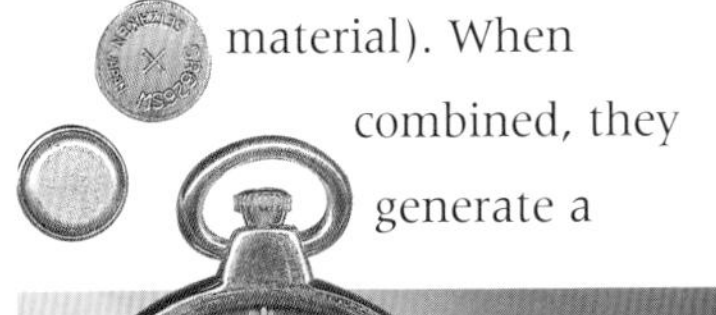

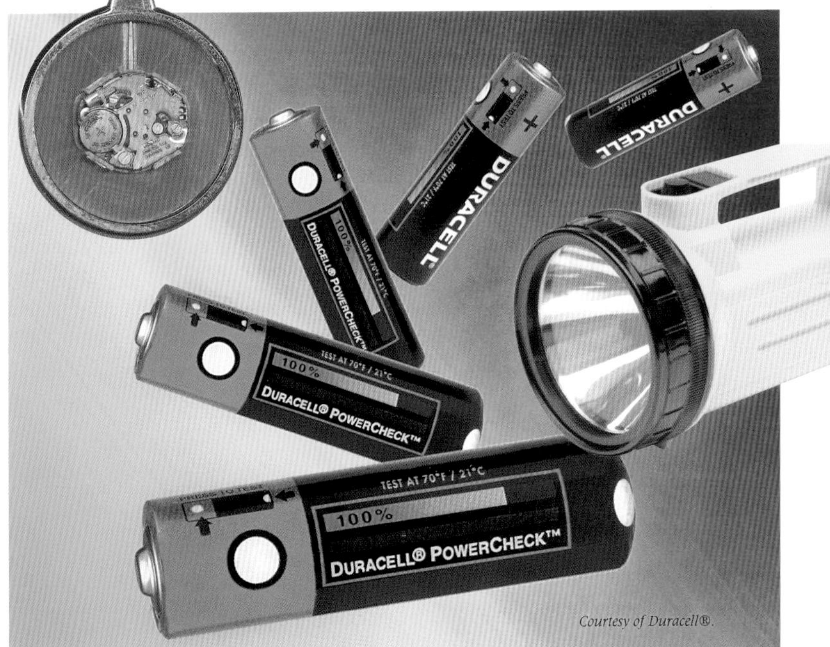

Courtesy of Duracell®.

Rechargeable batteries

Batteries have advanced significantly since Volta's discovery and are still constantly under further development and improvement. One of the mainstays of the battery industry is the *lead-acid battery* which is still used for heavy power applications. It is used extensively in vehicles for either starting purposes (the internal combustion engine) or as a power source for electric vehicles such as the milk float. Known as a *'secondary'* battery, it can be recharged and

then reused many times – as opposed to most batteries used in watches, which cannot be recharged and are known as *'primary'* batteries. The lead-acid battery was first discovered by G. Planté in 1860. It relies on the ability of lead to exist in two stable oxidation states, Pb(II) which has lost two electrons, and Pb(IV), which has lost four. When the battery is being used, or discharged, then lead metal is oxidised at one electrode to form insoluble lead(II) sulfate.

$$Pb(s) + SO_4^{2-} \rightarrow PbSO_4(s) + 2e$$

These electrons travel around the external circuit to the other electrode where lead(IV) dioxide is reduced to lead(II) sulfate.

$$PbO_2 + 4H_3O^+ + SO_4^{2-} + 2e \longrightarrow PbSO_4 + 6H_2O$$

The overall process removes sulfuric acid from the solution and converts lead and lead(IV) oxide to lead(II) sulfate. When the battery is recharged, the process is reversed. The lead-acid cell can produce short bursts of high current and has been difficult to replace. However, its relatively high mass compared with the overall amount of energy stored in the battery is a disadvantage, particularly in weight-sensitive applications such as powering electric vehicles and portable electronic devices, and has prompted research into batteries with higher energy densities. Some of the candidates include *lithium* batteries, *nickel-metal hydride*, *sodium-sulfur* and *zebra* batteries. However, to compete with the lead-acid battery, their performance needs to improve significantly to offset the low cost of the lead acid battery.

Sony lithium-ion battery module.
Courtesy of Electric Vehicle Association of Canada.

Lithium-based technology has been the most extensively researched, with the result that both non-rechargeable and rechargeable lithium batteries are now available. Their advantage is that, theoretically, they can store a huge amount of charge per kilogram of battery mass (295 amp hours per kilogram), which is 3 times more than the theoretical capacity of a lead-acid cell (86 amp hours per kilogram). This calculation is based on the materials used in the cell reaction only. This vast improvement relates to two key characteristics: first, lithium metal is very light; second, the electrochemical activity of lithium is much higher than that of many other materials used in battery technology. Unfortunately this second characteristic also creates a major snag. Lithium's high activity means that it is readily oxidised, causing passivating films to form over its surface which partially insulate the metal from the electrolyte.

Researchers have, however, overcome this problem by developing a rather clever idea – *insertion electrodes* which can accommodate the small lithium ion within their matrices, thus removing the need to use pure lithium metal. Two kinds of electrodes are used: one made of a metal compound (MA_2), such as

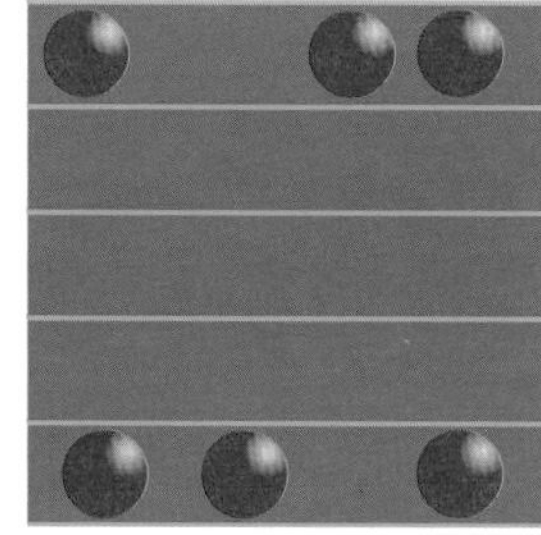

Lithium ion inserts between the layer within the graphite structure

Lithium ion insertion into graphite proceeds in five distinct stages until at maximum capacity there are six carbon atoms for every lithium

manganese dioxide, and the other of carbon, often graphite. When the cell discharges, lithium ions in solution insert into the manganese dioxide.

$$xLi^+ + MnO_2 + xe \longrightarrow Li_xMnO_2$$

While at the other electrode a lithium ion is extracted from the carbon matrix.

$$Li_xC_6 \longrightarrow xLi^+ + 6C + xe$$

Reversing the two processes recharges the cell. Such a battery can generate almost 4 volts – the voltage of nearly three normal stick batteries connected in series. These type of cells are known as lithium-ion cells.

Researchers have studied the way in which the lithium atoms insert into the carbon electrode, since an understanding of this process is vital for further development. They have observed a sequence of several distinct steps until finally at maximum capacity there is one lithium atom for every six carbon atoms. Figure 3 shows some of the stages in the process. Ideally these insertion and extraction processes should be reversible so that the battery can be recharged many times – much like a lead-acid cell.

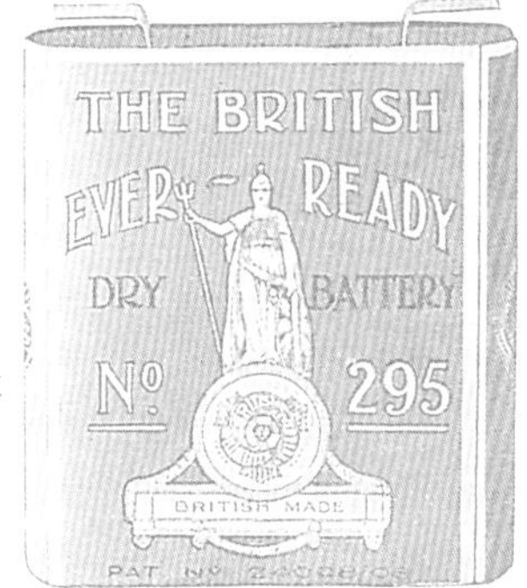

Courtesy of Energizer®

Other types of lithium cells have the same voltage as the average stick battery (1.5 volts) and are available as long-life replacements. They are more environmentally-friendly because they do not contain any heavy metals and are currently employed where low weight is a clear prerequisite.

Batteries powering laptop computers and mobile phones are frequently based on this lithium chemistry. However, a number of scale up problems and safety problems need to be resolved if lithium technology is to be used in electric cars.

For vehicle applications, the main candidates are high-performance secondary batteries based on nickel-metal hydride, sodium-sulfur or zebra batteries. Only nickel-metal hydride batteries are based on an aqueous electrolyte, while zebra batteries are high-temperature batteries working at around 300 °C relying on molten salts (pure salts that are heated above their melting point to become an ionically conducting liquid) for their operation.

Nickel-metal hydride batteries have two distinctly different electrodes. One electrode uses alloys of rare-earth metals that can soak up hydrogen atoms.

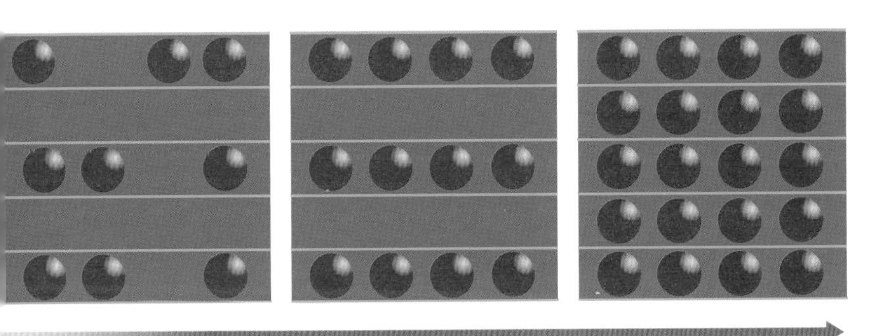

Figure 3. The different stages through which lithium atoms insert into a layered graphite structure as determined by X-ray analysis. Understanding this process is vital in the development of lithium batteries.
(Figure courtesy of Dr A. H. Whitehead.)

Panasonic-Toyota nickel metal-hydride holder
Courtesy of Electric Vehicle Association of Canada.

These alloys can have a general formula AB_5 where A can be lanthanum and B nickel.

Panasonic nickel metal-hydride HEV battery module.
Courtesy of Electric Vehicle Association of Canada.

$$AB_5H + OH^- \longrightarrow AB_5 + H_2O + e$$

The other electrode is based on nickel chemistry.

$$NiOOH + H_2O + e \longrightarrow Ni(OH)_2 + OH^-$$

To complete the cell, hydroxide ions (OH^-) migrate from the cathode where they are generated, to the anode where they are consumed. The cell operates with a potential of 1.35 volts. Unfortunately the high cost of the nickel and the rare-earth alloy electrodes make the cell relatively expensive initially but these cells can be cycled many times.

Both sodium-sulfur and zebra cells have a molten sodium metal electrode contained within a solid ceramic (*beta*-alumina) case through which only sodium ions can pass. The electrochemical reaction is

$$2Na \longrightarrow 2Na^+ + 2e$$

The difference between the zebra battery and the sodium-sulfur battery lies within the materials used for the other electrode. The sodium-sulfur battery employs liquid sulfur (in a separate compartment).

$$xS + 2e \longrightarrow S_x^{2-}$$

(where x=5 to 2.7)

While the zebra battery employs a molten nickel chloride/sodium aluminium chloride system.

$$2e + NiCl_2 \longrightarrow Ni + 2Cl^-$$

Because both these batteries contain very hot, potentially dangerous materials, the safety implications are obvious. For this reason the zebra battery has become the more popular of the two; if the ceramic *beta*-alumina case is broken, then the reaction between the constituents passivates the break and does not cause any serious problems. However, both cells have to be temperature-controlled during operation.

The promise of fuel cells

In batteries, the chemical energy is stored within the electrodes and the solution. There is, however, another method of generating power – by feeding the cell with a continuous supply of chemicals which react at the electrodes to produce electrical energy. This is called a *fuel cell*. Currently, car companies are very interested in them because they could be used to power environmentally friendly vehicles. Fuel and an oxidant are fed into the cell in the same way as petrol and air are fed into an internal combustion engine. However, in a fuel cell the oxidant and the fuel do not combust directly but are 'electrochemically burnt'.

In the simplest case, hydrogen and oxygen gases are converted to water in two processes at different electrodes. These electrodes usually contain platinum or a related catalyst to enhance the rate of the overall process, and are separated by an ion-exchange membrane which allows certain ions through to complete the circuit. This particular type of fuel cell, known as a *proton-exchange membrane fuel cell*, or PEM fuel cell, has a membrane that allows hydrogen ions (protons) to travel between the two electrodes while keeping the gases apart. Figure 4 shows how the system operates.

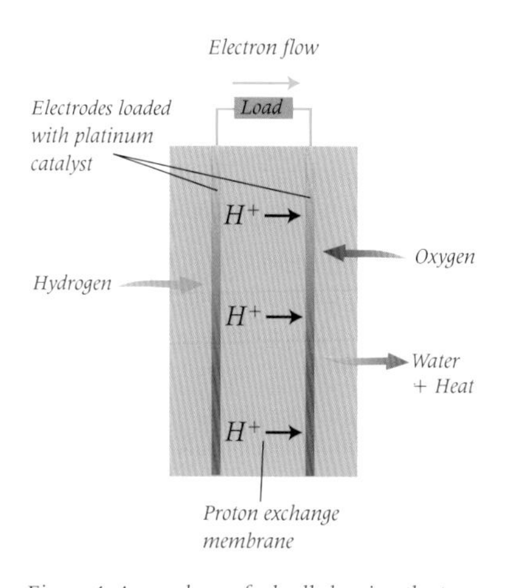

Figure 4. A membrane fuel cell showing the two electrodes. At the hydrogen electrode hydrogen gas is oxidised to form protons. These protons migrate across the proton exchange membrane to the oxygen electrode where they are electrochemically combined with oxygen to form water.

Hydrogen and oxygen gas are fed to the two gas-permeable platinum electrodes. At the hydrogen electrode, protons and electrons are produced, while at the oxygen electrode, oxygen reacts with protons, to produce water and electrons.

$$2H_2 \longrightarrow 4H^+ + 4e$$
$$O_2 + 4H^+ + 4e \longrightarrow 2H_2O$$

The electrons travel around the circuit and do work. To complete the circuit within the cell protons migrate towards the oxygen electrode. The overall reaction produces electrical energy, water and heat.

Fuel cells of this type are ideal for vehicles aiming at zero emissions: water is the only exhaust gas. However, in fairness, the whole energy cycle has to be considered. The source of hydrogen and oxygen may well have been electrolysis of water, using electricity that probably comes from power stations burning fossil fuel. Nevertheless, this clean technology is ideally suited for urban areas, and various companies have developed prototype electric buses running on fuel-cell power. The only other drawback lies in the storage and transportation of liquid hydrogen.

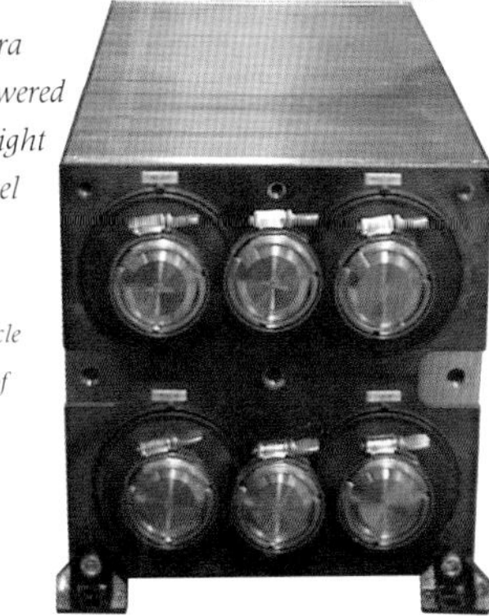

Above Zebra battery-powered EV bus. Right Ballard fuel cell.
Courtesy of Electric Vehicle Association of Canada.

The danger of explosion must be considered although hydrogen is perhaps less dangerous than a liquid hydrocarbon fuel.

This logistics problem has spurred on research into using hydrocarbon fuels for fuel cells so that car-makers can still rely on normal fuel tanks. There are two types: *reforming fuel cells* and *direct fuel systems*. The first uses chemical catalysts that break down the hydrocarbon fuel (for example, diesel) into carbon dioxide and hydrogen gas. This mixture is then fed to a hydrogen-oxygen fuel cell with air as the source of oxygen. Clearly the production of carbon dioxide, a greenhouse gas, is less desirable, and there are other problems as well. First, the need to carry the reforming stage within the vehicle adds to the weight of the unit. Second, the fuel gas produced by the reformer must not contain any measurable amounts of carbon monoxide which would poison the catalyst electrode, and thus rendering the system useless.

An alternative is the *direct methanol fuel cell* (DMFC), which employs methanol as the fuel, and burns it with oxygen to produce carbon dioxide and water. Unfortunately, there are technical problems with its development particularly for powering cars. To be viable, a DMFC needs to produce 200 to 300 milliwatts of energy per square centimetre of catalyst area at a working potential of 0.5 volts. However, both the oxidation of methanol and reduction of oxygen are particularly slow and prone to poisoning by by-products formed in the fuel cell. To overcome this, fuel-cell developers have switched to new catalyst systems based on combinations of platinum and ruthenium. Incorporating the secondary material into the platinum catalyst is thought to reduce poisoning of the platinum surface – particularly by carbon monoxide formed during the oxidation of methanol. High operating temperatures also enhance catalyst performance.

These can be achieved by introducing new polymer materials for the ion exchange membrane. In most cases, a fluorinated polymer called Nafion® is used. Developed by the American chemical company, DuPont, it is related to Teflon and is an extremely efficient proton-exchange membrane able to operate under extreme conditions of temperature and pressure.

The combination of high temperatures, improved catalysts and pressurisation of the whole system has improved the performance of DMFCs. However, there is still one problem, which happens particularly at high operating temperatures and pressures, and has still not been fully solved. The methanol tends to leak from the anode compartment (where it is oxidised) to the cathode (where oxygen is reduced). This 'cross-over' causes the oxygen electrode's potential to be raised closer to that of the methanol electrode, thus decreasing the overall voltage and performance of the cell.

The German high-technology company, Siemens, a group at the University of Newcastle led by Andrew Hamnett, and a team at the Los Alamos National Laboratory in the US are among some of the research teams investigating these new systems. Currently the best performance is around 200 milliwatts per square centimetre at around half a volt. However, a good deal more research is needed to improve on this. These fuel cells still require relatively high catalyst loading (the amount of catalyst per square centimetre of electrode) and still suffer from methanol cross-over. Nevertheless, recent progress is encouraging, and the potential of DMFCs to produce clean and efficient electrical power is high. The American space agency, NASA, is already using large fuel cells in their missions, indicating a possible trend for the future. However, they are unlikely to be used in smaller devices such as watches and laptop computers, which require compact power sources with no exhaust emissions.

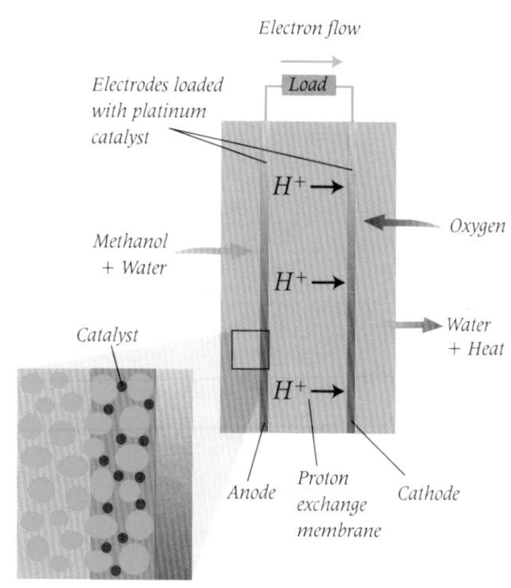

Figure 5. A direct methanol fuel cell (DMFC) employing an acidic solid polymer electrolyte (Nafion®) and finely dispersed platinum catalyst supported on porous carbon electrodes. The insert shows a cross-section of one of the electrodes.

Fuel cells and batteries are clear candidates for electrical vehicles, and perhaps a combination power system may eventually be developed. Such a hybrid would have the instant power of a battery while the fuel cell warms up, and the long range of the fuel cell where large amounts of fuel energy may be stored at a low-weight cost. It is interesting to note that the world's first production-line electric car (of course other electric vehicles such as milk floats and fork-lift trucks have been around for decades), General Motors' EV1 car still

Photo courtesy of NASA.

Courtesy of Electric Vehicle Association of Canada.

employs a lead-acid battery system. Its weight, about 540 kilograms, if replaced with fossil fuel, would drive the car for thousands of miles, while the battery cell has a range of only about 70 miles. Clearly, a major improvement is needed.

Harnessing solar energy

While both fuel cells and batteries produce electrical energy from chemical sources, there is a third type of power source which can also depend on electrochemistry. This is the *photovoltaic cell*, of which some kinds generate power from light via photochemical reactions.

There is an intimate connection between light and electricity. Some molecules – usually electron-rich species – readily absorb light energy of a characteristic wavelength, which then 'excites' certain electrons in the molecules to higher energy levels. These excited electrons may fall back to a lower state, emitting light in the process (see *Following chemical reactions*). Indeed, it is these kinds of processes that give some chemicals their bright colour and result in phenomena such as fluorescence. However, the electrons may absorb enough light energy to be transferred from one part of a molecule to another, or from one molecule to another. There is a great deal of interest in these photochemical processes because they provide a way of converting sunlight into electrochemical energy.

Michael Grätzel and colleagues at the Swiss Federal Institute of Technology, Lausanne have developed such a system. It employs a complex ruthenium-based dye and some clever electrochemistry. The brownish-red compound collects incident light – in the same way as green chlorophyll molecules harvest sunlight in plants. Normally the energy absorbed by the excited dye molecules would be dissipated. However, Grätzel absorbed the dye onto the surface of a semiconducting electrode of titanium dioxide. Once excited by light, the dye molecules lose electrons and are oxidised by the titanium dioxide. They are then reduced by iodide ions present in the solution. The resulting iodine produced is then itself reduced at a second electrode to complete the cycle. The electrons injected into the titanium dioxide travel around the external circuit and do work. Solar cells of this type show relatively high efficiencies (greater than 10%), are cheap in comparison with other solar cells, and are extremely promising as clean energy sources.

Grätzel at work in his laboratory.

A cleaner chemical industry

As seen from the above examples, electrochemical technology does offer a cleaner way of producing energy. The same arguments apply to the production of chemicals. As the demand increases for cleaner, safer industrial processes with zero effluent, the chemical industry is increasingly turning to electrochemistry since in many instances chemicals can be converted directly at an electrode to the product wanted with high efficiency and little waste. Alternatively electrochemical technology can be employed to clean up the waste produced by older chemical processes. An example of this is soil remediation where pollutants are removed from contaminated soil by passing an electric current through it (see Box 2).

Aluminium production. Courtesy of Anglesea Aluminium

Box 2 Cleaning up polluted soil

Surprisingly, electrochemistry can be used to remove pollutants discharged from industrial and agricultural processes into water or soil. It is particularly useful in the reclamation, or more precisely soil remediation, of potentially valuable land polluted by mining heavy metals. The process works like a battery in reverse. Electrical energy, supplied from an external power source, performs work within the interior of the cell which in this case is the contaminated soil. If a current is passed through a soil sample containing, for example, cadmium(II), a heavy metal toxin, then the positive cadmium ions move towards the negative electrode where they can be removed, while negative ions move towards the positive electrode.

In the field, the system operates with several pairs of electrodes placed in the ground covered with appropriate ion exchange membranes. An acid electrolyte is then circulated through the membrane compartment. An electrical potential is applied to the two electrodes which breaks down water into hydrogen and oxygen at the electrode surfaces.

$$2H_2O + 2e \longrightarrow 2OH^- + H_2 \quad \text{(cathode)}$$

$$2H_2O \longrightarrow O_2 + 4H^+ + 4e \quad \text{(anode)}$$

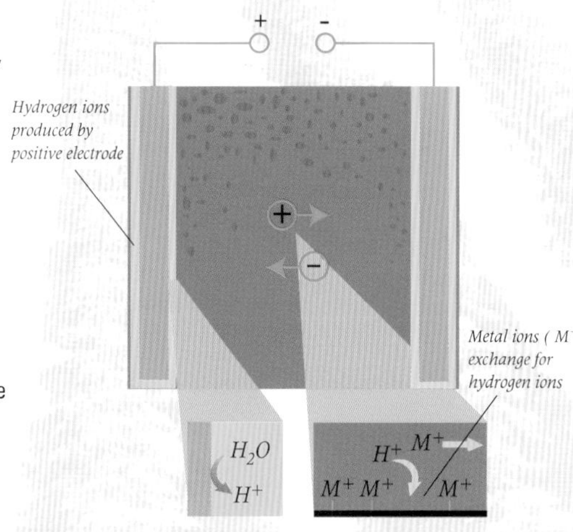

Diagram of the processes involved in soil remediation. Pairs of electrodes are driven into the ground. Each electrode is covered with a membrane so that a solution can be circulated around the electrode to remove ions flowing from the soil to the electrodes as a current is passed through the soil. At the positive electrode water is oxidised to produce protons. These travel through the solution where they can exchange with metal ions absorbed on the soil surface. The metal ion can then travel to the cathode where it is removed by the circulating electrolyte.

As the hydrogen ions (H^+) and hydroxide ions (OH^-) also produced pass through the soil towards opposite electrodes, the hydrogen ions exchange with metal cations within the soil's structure, so releasing the pollutant ions from the soil. These travel to the negative electrode where they can be removed from the circulating electrolyte.

This technique can be used to remove pollutants like heavy metal cations (copper, cadmium, iron, zinc and so on), nitrate from agricultural fertilisers, cyanide and radioactive nuclides. As well as ions, this technique can remove uncharged materials such as toxic organic compounds by a process called *electro-osmosis* whereby water molecules move from the anode to the cathode and take some of the organic toxins with them.

The feasibility of such operations depends strongly on the local cost of electricity. Removing cadmium ions from an area the size of a football pitch would consume around 300 gigajoules (equivalent to the energy used by 60,000 60-watt light bulbs left on for a day). This cost has to be offset against the value of the reclaimed land and the cost of cleaning the land using other techniques. Also significant is that the electrochemical process can, in most cases, be performed on site rather than having to remove the soil for processing.

Electrochemical processes account for a staggering 5 to 10% of an industrialised nation's electricity consumption, most of it being used to extract aluminium from its ore, and for the *chlor-alkali industry* which produces chlorine (and sodium hydroxide).

Aluminium is isolated from aluminium oxide in a molten salt cell at 1000 °C. The cell contains two carbon electrodes, aluminium oxide and the salt, cryolite ($NaAlF_6$). Molten aluminium metal is produced at the cathode while carbon dioxide is released from the anode which is continuously fed into the cell as the carbon is consumed. This process was discovered separately in 1886 by a Frenchman, Paul Heroult, and an American, Charles Hall. The cell has been named after the two inventors as the *Hall-Heroult cell*. Hall founded the Aluminium Corporation of America and became a millionaire as a result of his discovery. Approximately 300,000 tonnes of aluminium are produced electrochemically each year in the UK alone and 20 million tonnes

worldwide. Many other metals are produced electrochemically including copper, lead, nickel and zinc. However, aluminium production is the largest and consumes the most energy.

The other main use of electricity, the electrochemical production of chlorine, generates approximately 42 million tonnes of chlorine per year worldwide. (Vast amounts of chlorine are needed by the chemical industry for making many different chemicals including PVC.) The raw materials are sodium chloride and water only, while the products are sodium hydroxide and hydrogen from the cathode compartment and chlorine gas from the anode compartment. Initially, cells employed mercury and carbon electrodes, but now more efficient devices using safer materials are becoming more commonplace. A particular example is the membrane cell. This cell, shown in schematic cross section in Figure 6, has two electrodes and a Nafion® ion-exchange membrane. The membrane allows only positively charged ions to pass and inhibits the passage of negatively charged ions. At the anode, chloride ions are oxidised to chlorine, while at the cathode hydrogen ions are reduced to form hydrogen gas.

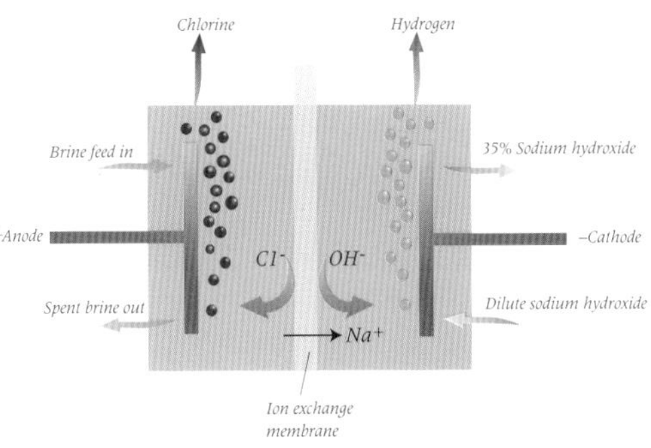

Figure 6. A chlor-alkali cell for making chlorine and sodium hydroxide. The two compartments are separated by an ion-selective membrane that allows only positive ions – in this case, sodium ions – to transfer from one compartment to the other. In the brine-fed compartment, chloride ions oxidise at the electrode to chlorine which is given off as a gas. In the sodium hydroxide compartment, hydrogen gas is evolved as hydrogen ions are reduced. Sodium ions complete the circuit by crossing the membrane and produce concentrated sodium hydroxide as hydrogen ions are removed from the compartment.

Below: a bank of membrane cells in a chlor-alkali cell room. Courtesy of ICI.

$$2Cl^- \longrightarrow Cl_2 + 2e$$
$$2H^+ + 2e \longrightarrow H_2$$

To maintain electrical neutrality, sodium ions move from the anode side to the cathode side through the ion-exchange membrane. The hydroxide ions, produced as the result of hydrogen ion removal at the negative electrode, are prevented from transferring to the positive electrode compartment. The cell as a whole, as its name implies, produces chlorine (in the anode compartment) and sodium hydroxide (in the cathode compartment).

Making organic chemicals

There are also some electrochemical organic processes which are industrially important. The chemicals produced range from speciality compounds to bulk materials used, for example, to make polymers. An example of the latter is the Monsanto process. This is the hydrodimerisation of acrylonitrile which is an important step in the production of one of the precursors to nylon.

$$2CH_2{=}CHCN + 2H_2O + 2e \longrightarrow (CH_2\text{-}CH_2\text{-}CN)_2 + 2OH^-$$

This process produces 270,000 tonnes of adiponitrile every year worldwide.

Another example is the electrochemical oxidation of alkenes to diols (organic compounds with hydroxide groups on adjacent carbon atoms) using two catalysts – the iron complex, ferricyanide and a complex of a rare element, osmium. The osmium complex actually has a 'chiral centre' (which means it can exist in mirror-image forms, see *Make me a molecule*) and is used to introduce chiral centres into the diols – a process extremely important to the pharmaceutical industry. This reaction is a variant of the famous

'Sharpless reaction' – one of the first chemical methods for introducing chiral centres into molecules. The two transition metals act as *redox* catalysts in being sequentially oxidised and reduced, transferring electrons between them in the process. The process is quite neat in that it is carried out in an emulsion of water and an organic solvent such as cyclohexane (because the metal compounds are soluble in water and the organic material soluble in organic solvent).

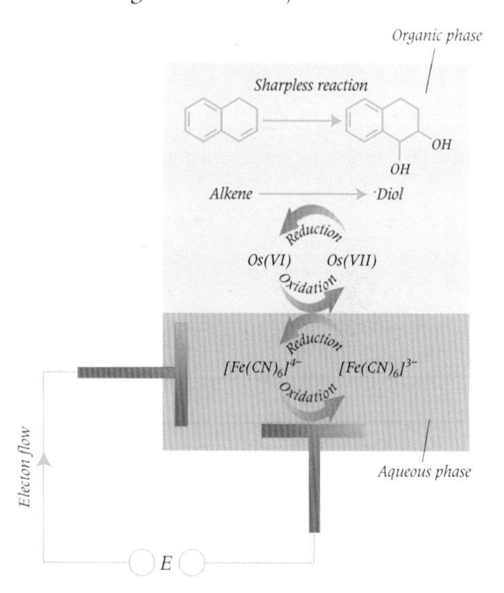

Figure 7. The Sharpless reaction to make diols carried out electrochemically using two transition metal catalysts. The osmium complex is regenerated by ferricyanide ions ($[Fe(CN)_6]^{3-}$) by an interfacial reaction between the organic/water phase. The ferricyanide is regenerated electrochemically.

The ferrocyanide is first oxidised in the aqueous phase to ferricyanide, which then oxidises the osmium complex which in turn oxidises the alkene to the diol in the organic phase. The advantage of such a system is that very little of the toxic osmium complex is required as it is recycled and reused. The overall process is shown in Figure 7.

The electrochemical power of life

Such complex oxidation-reduction cycles power many processes in living matter. Chemists therefore study the role of electron transfer in biology not only to gain insight into biochemical mechanisms but also to investigate possibilities of exploiting in various ways the electrochemistry that Nature uses so efficiently – for making vital chemicals and transferring energy.

One example is the conversion of energy in the mitochondria of cells. Here, electron transfer between specific redox centres (molecular structures that exchange electrons with other centres becoming both oxidised and reduced in the process) is particularly important. A high energy material known as nicotinamide adenine dinucleotide, or NADH, is oxidised via a series of redox proteins and enzymes. The hydrogen ions released are pumped across the mitochondrial membrane to create a trans-membrane potential – an effect similar to the electrical potential generated by a battery. This converts adenosine diphosphate (ADP) to adenosine triphosphate (ATP) and finally, oxygen is reduced to form water. The process is shown schematically in Figure 8.

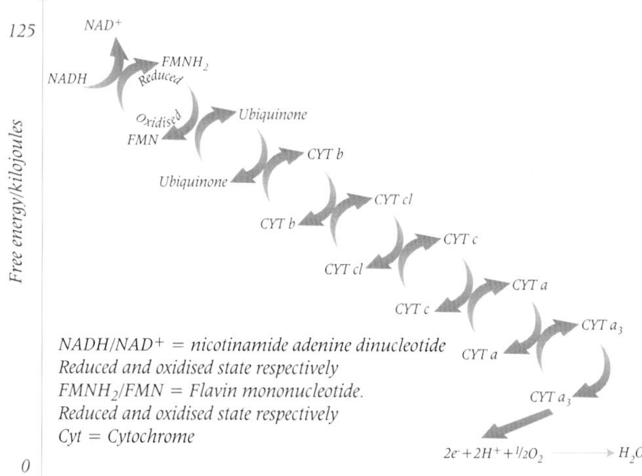

Figure 8. The processes leading to the production of chemical energy in cell's mitochondria. Nicotinamide adenine dinucleotide, NADH, is oxidised by molecular oxygen through a series of electron exchange steps. In each step (there are 7) the biological redox centres are reduced by their higher energy partner and then oxidised by their lower energy partner. In each step the energy released by the electrons is used to transfer a proton across the inner mitochondrial membrane. Finally molecular oxygen is reduced to water.

Biological sensors

One way in which the redox chemistry in living cells can be exploited is in developing sensors to monitor important chemicals in the body, for example, blood glucose in diabetics. Hospitals have routinely done this measurement with an electrochemical analyser incorporating an enzyme, glucose oxidase. This redox protein catalyses the oxidation of glucose (the substrate) and is then regenerated by oxygen, producing hydrogen peroxide in the process. The hydrogen peroxide is

Box 3 Electron transfer in chemistry and biology

The transfer of electrons from one place to another is probably the most important reaction type in chemistry and biology. It pops up in simple reactions such as the well-known 'self-exchange' reaction, learnt at school involving ferrocyanide and ferricyanide complex ions in which the iron is in two oxidation states (II) and (III):

$$[Fe(II)(CN)_6]^{4-} + [Fe(III)(CN)_6]^{3-} \longrightarrow [Fe(III)(CN)_6]^{3-} + [Fe(II)(CN)_6]^{4-}$$

It was the Nobel Prize winner, Henry Taube at Stanford University, California, who first studied such reactions. He showed that the rates of apparently similar reactions could vary by huge amounts. For example, the self-exchange in the two pairs of complex ions – one pair containing the transition metal, ruthenium combined to bipyridyl organic groups, $[Ru(II)(bipyridyl)_3]^{2+}$/$[Ru(III)(bipyridyl)_3]^{3+}$, and the other cobalt linked to ethylenediamine, $[Co(II)(en)_3]^{2+}$/$[Co(III)(en)_3]^{3+}$, differs in rate by 13 orders of magnitude! The physical chemist, Rudolph Marcus won a Nobel Prize for giving a theoretical interpretation of this – and much else in electron transfer behaviour – when he was at the Polytechnic Institute of Brooklyn.

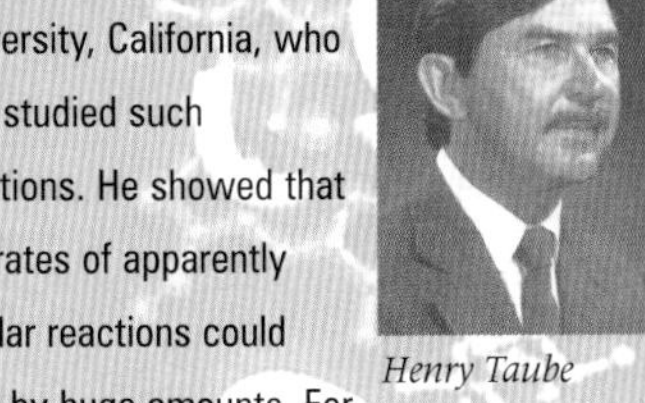

Henry Taube

©The Nobel Foundation

Rudolph Marcus

©The Nobel Foundation

Much more complicated electron transfer reactions happen in biology, some following exposure to light, such as photosynthesis. This complex process, which provides energy to green plants, also involves electron transfer across a highly complex structure consisting of several large biomolecules known as the photosynthetic reaction centre. The best understood process occurs in photosynthetic bacteria. At the reaction centre the captured photon of light excites a 'special pair' of bacteriochlorophyll units, P*, (the asterisk means that the special pair P is electronically excited). The next step, which happens in about 3 picoseconds, or 3×10^{-12} seconds, involves electron transfer to a molecule called bacteriophytin (BP) forming BP^-, and leaves the special pair as P^+. In the following step, which takes about 200 picoseconds, the electron transfers to a quinone (Q). This is followed by a slower step (about 270 nanoseconds, or 270×10^{-9} seconds) in which an electron is transferred from a nearby cytochrome unit (Cy) to P^+. This can all be summarised as:

$$Cy.P.BP.Q. \longrightarrow Cy.P^*.BP.Q \longrightarrow Cy.P^+.BP^-.Q \longrightarrow$$

$$Cy.P^+.BP.Q^- \longrightarrow Cy^+.P.BP.Q^-$$

The overall effect is a charge separation between Cy and Q, and it is this that drives the subsequent chemistry. Analogous, but rather more complex processes, occur in green plants in which overall two photons of light are used to produce useful chemical energy for the plant consuming carbon dioxide and water in the process and producing oxygen.

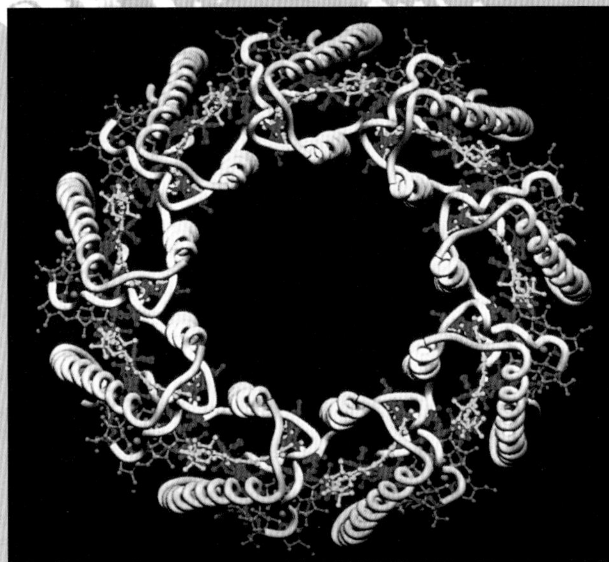

Molecular graphic of Photosystem II. Courtesy of Professor Isaacs, Glasgow University.

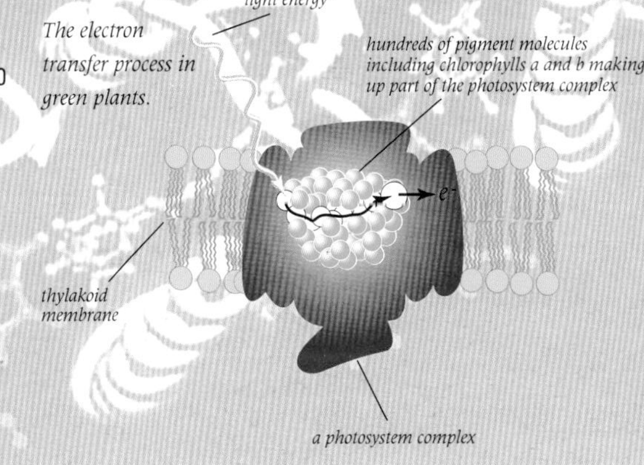

The electron transfer process in green plants.

There has been enormous effort to mimic photosynthesis in the laboratory and therefore harness sunlight – which is free – to do useful chemistry, but success has been limited. One approach has been to use light to excite molecules of transition metal compounds into a so-called charge-transfer state, whereby an electron is pushed across from one portion of the molecule to another ($AB \longrightarrow [A^+B^-]^*$), and to see if these excited states can then react with other molecules to do useful chemistry. Some extremely subtle photochemistry has resulted from these studies but, as yet, no magic solution to the world's energy problems. One obstacle is that the energy contained in the excited state is easily lost in solution. However, Michael Grätzel and colleagues have devised a very ingenious way round this problem as described earlier in the chapter.

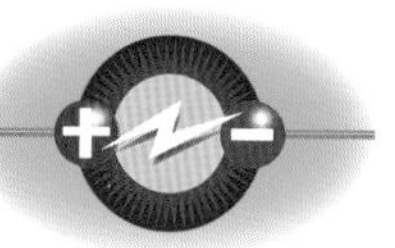

then oxidised at an electrode and the amount of glucose present in the blood calculated from measuring the current flow over a certain time. However, the system used is not portable and is slow. Diabetics need to measure their blood glucose level regularly throughout the day, so they require a cheap, portable device that gives accurate measurements quickly and often.

These limitations have driven researchers to develop faster and more mobile glucose and other *biosensors*, which are based on electrochemical systems involving direct interaction between the electrode and the enzyme. Unfortunately, proteins are very large and the electron transfer to and from the bulky convoluted protein structure to the electrode is very slow. However, Allen Hill and his colleagues at the University of Oxford and at Cranfield University made considerable progress in this area by employing an imaginative approach to the problem. This area of study is called *bioelectrochemistry* (see Box 4).

Allen Hill

Hill added to the enzyme solution small redox-active molecules (in the case of glucose, a derivative of the organometallic compound, ferrocene) that would shuttle electric charge between the active centre of the enzyme and the surface of an electrode – a process he called *mediation*. What happens is that after the redox enzyme has reacted, it is left 'locked' in a reduced or oxidised state. The mediator molecule then converts the enzyme back to its initial redox state ready for another reaction. The mediator itself is electrochemically converted back to its initial redox state by an electrode held at the right potential. The whole process can then be repeated and is thus catalytic (Figure 9). In essence the highly mobile and relatively small mediator species acts as a courier in taking charge to or from

Box 4 The electrochemistry of cytochrome *c*

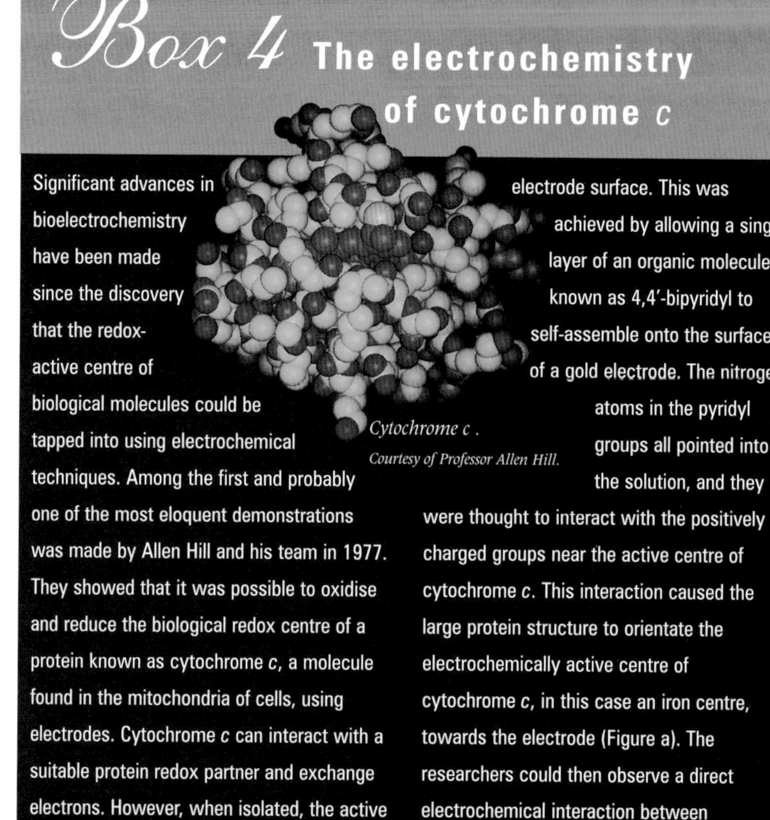

Cytochrome c.
Courtesy of Professor Allen Hill.

Significant advances in bioelectrochemistry have been made since the discovery that the redox-active centre of biological molecules could be tapped into using electrochemical techniques. Among the first and probably one of the most eloquent demonstrations was made by Allen Hill and his team in 1977. They showed that it was possible to oxidise and reduce the biological redox centre of a protein known as cytochrome *c*, a molecule found in the mitochondria of cells, using electrodes. Cytochrome *c* can interact with a suitable protein redox partner and exchange electrons. However, when isolated, the active centre of the molecule is shielded by its large protein structure (with molecular weight of about 12,400) – a property that prevents the molecule from exchanging electrons with an unsuitable partner or, indeed, a bare electrode.

To make the active centre of cytochrome *c* interact with an electrode, the researchers had to modify the electrode surface. This was achieved by allowing a single layer of an organic molecule known as 4,4'-bipyridyl to self-assemble onto the surface of a gold electrode. The nitrogen atoms in the pyridyl groups all pointed into the solution, and they were thought to interact with the positively charged groups near the active centre of cytochrome *c*. This interaction caused the large protein structure to orientate the electrochemically active centre of cytochrome *c*, in this case an iron centre, towards the electrode (Figure a). The researchers could then observe a direct electrochemical interaction between cytochrome *c* and the electrode.

Iron redox centre

4,4'-bipyridyl modified electrode surface. Lone pair on nitrogen sticks out into solution and orientate the active centre of cytochrome *c* to the electrode surface so reducing the distance for electron transfer.

N N

Representation of protein structure cytochrome *c*
Reproduced courtesy of School of Pharmacy, University of London

Figure a. Cytochrome c docking.

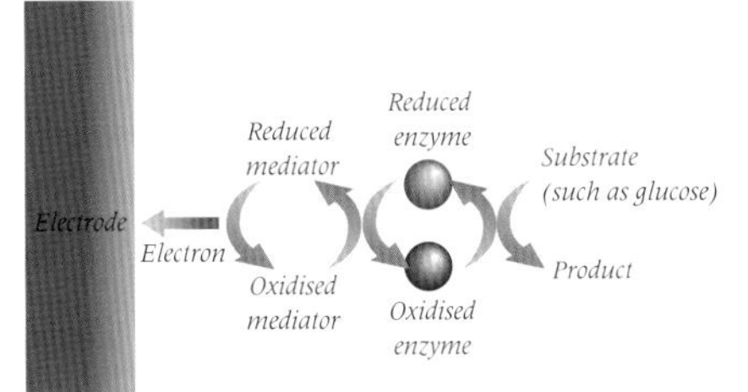

Figure 9. The oxidation-reduction chemistry behind a glucose sensor. A mediator molecule shuttles charge between an electrode and the enzyme which reacts with glucose (the substrate) to form the product.

the electrode surface to the enzyme redox centre.

The Oxford team's work has led to a new blood glucose sensor which is now marketed by a company called MediSense. It was the first company to launch an electrochemical biosensor and is now the leading manufacturer of such devices. This business is incredibly lucrative; after a takeover in 1996 the company was valued at $876 million. Researchers have employed the mediator technique with a huge variety of enzymes and mediator combinations. Lactate, bile acids, neurochemicals, and cholesterol are just a few of the materials that can be measured, and these have led to similar biosensors. Electrochemical sensors that detect more than one chemical are now available. A good example is the i-STAT sensor which can measure the level of six important chemicals in one drop of blood!

A MediSense glucose pen and other products. Reproduced courtesy of MediSense.

Clearly the market in this area is large and this is reflected in the huge amount of research being done in both academia and industry. However, to develop sensor technology further there are problems to overcome. These include: short shelf-life – many biological molecules become inactive after a while because of the natural degradation of proteins; low activity – many enzymes have poor reactivities; electrochemical inefficiency – the reaction between the mediator and the redox enzyme may not be fast enough; interference – blood components, for example, ascorbic acid (vitamin C) can interfere with the electrochemistry of the system. Nevertheless, it is only a matter of time before more electrochemical biosensors come into everyday use.

Many biologically important molecules, for example, antibodies produced by the body's immune response, are present only at very low concentration. It is important, therefore, to be able to measure their low levels accurately. Fortunately today, electrochemistry can offer the technology to measure even a single redox-active molecule.

Detecting a single molecule

Single molecules can be detected with a technique known as *scanning electrochemical microscopy* or SECM (see Box 5). It was first demonstrated by Allen Bard and colleagues at the University of Texas in Austin in 1995.

Two electrodes were immersed in a solution containing a redox-active material so as to trap a small amount (about 1 x 10^{-21} litres) of liquid in the gap between them (Figure 10). The redox-active material was at a concentration low enough to ensure that on average, the electrode gap held only one molecule at any one time. The particular redox-active species was a ferrocene derivative. The potentials of the two electrodes – the large 'substrate' electrode and the smaller microelectrode – were held in such a way that the ferrocene molecule was oxidised at one electrode and reduced at the other. Oxidation and reduction of the ferrocene takes place as the molecule naturally

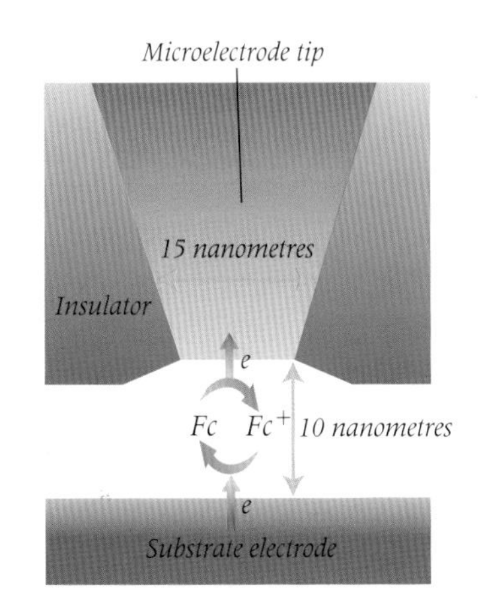

Figure 10. Single molecules can be detected electrochemically in the minute gap between two electrodes. The diagram shows the geometry at the tip of the microelectrode as it is brought down close to a large flat substrate which acts as the second electrode.

Box 5 Scanning electrochemical microscopy

Scanning electrochemical microscopy or SECM is a relatively new technique that can investigate the interface between solids and liquids, liquids and liquids, or liquids and gases. It allows researchers to determine the morphology, or the geometric shape, of the surface as well as the chemical activity of regions on the surface.

The SECM technique relies on a tiny electrode (typically a disc of a conducting material embedded in an insulator less than 50 micrometres across) known as a microelectrode or ultramicroelectrode, which is placed in an electrolyte. The microelectrode is brought close to the surface or the interface and then scanned across the surface in a raster pattern, much like the way a TV picture is generated – only a lot slower, in the order of a few micrometres per second. The microelectrode is held at a certain electrical potential that can oxidise (or reduce) highly mobile electro-active molecules, such as ferrocene or ferro/ferricyanide, dissolved in the solution. As the microelectrode scans the interface, these molecules diffuse to the microelectrode, in effect carrying a current between the interface and the microelectrode. This current can then be measured. If the microelectrode passes over an area that is insulating, the diffusion of molecules to the microelectrode is hindered, and the current recorded is less than that recorded when the microelectrode is far away from the interface. On the other hand, if the surface or interface is electro-active and can, say, regenerate the redox species consumed by the microelectrode, then the microelectrode current will increase compared with that detected when the microelectrode is far away from the interface. These two effects are called negative and positive feedback respectively and are dependent on the distance between the microelectrode and surface (Figure a).

O R
<10a
O R
O=Oxidised chemical species
R=Reduced chemical species
a=Radius of microelectrode
Insulating surface
Conducting surface
a) Negative feedback
b) Positive feedback

Figure (a) The principle of positive and negative feedback in scanning electrochemical microscopy (SECM). In the left-hand figure, the microelectrode is placed close to an insulating surface, and the diffusion of 'O' to the electrode and its reduction to 'R' is hindered, giving a smaller current than expected for the microelectrode held in the bulk of the solution. When the microelectrode is placed over a conducting surface at a potential sufficient to regenerate O from R at the microelectrode, as in the right-hand figure, the microelectrode registers it as increased current. These effects happen at a significant level only when the microelectrode is at a distance from the surface that is less than 10 times the radius of the microelectrode.

The resolution of the technique depends on how small the microelectrode is. However, as the size of the microelectrode decreases, the closer the electrode has to be to interact with the surface. This can cause fabrication problems as the insulating surround of the microelectrode must be flat enough not to crash into the surface as the electrode is moved across it. As an example of how SECM can investigate surfaces, Figure b shows a study on dentine from a tooth. The researchers used a solution containing a redox-active species and applied pressure to the system so that the solution flowed through holes in the dentine. The figure shows the hole, or pore, in the dentine.

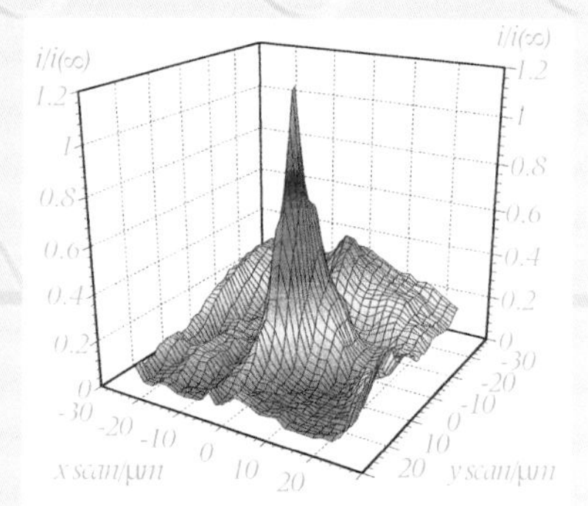

Figure (b) An SECM image over a portion of a tooth. The picture resolves an individual hole in the dentine and is a good example of how this technique can be used to investigate real samples and look at real problems. *Reproduced courtesy of Dr P R Unwin, University of Warwick.*

diffuses between the upper microelectrode and lower substrate electrode. Again, the ferrocene molecule can be thought of as a molecular shuttle (in a similar way to the electrochemical mediator of an enzyme system) carrying electrons between the two electrodes. It is essential that the electrodes are close together to enhance the current recorded at the microelectrode and allow one molecule to be detected at a time. If the single ferrocene molecule diffused out of the space between the two electrodes the current signal was lost or if, by chance, two ferrocene molecules were in the gap at one time the current signal was doubled.

The examples in this chapter have showed how, over the past two centuries, electrochemistry has become a versatile and wide-ranging subject. By the beginning of the 20th century, electrochemistry had progressed

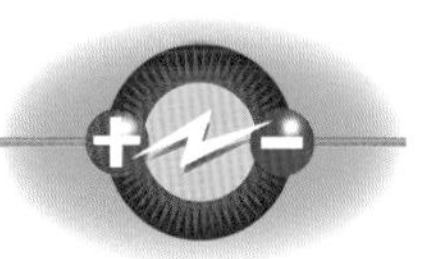

towards investigating the rate and mechanisms by which electrochemical processes happen, and these investigations are still underway. In many instances characterising the structure of the electrode surface is fundamental to the understanding of the processes occurring at the solid-liquid interface of an electrode. It is also important for designing more effective electrode surfaces. Today, the often complex behaviour of chemistry at the interface between an electrode and an electrolyte is being investigated using sophisticated microscopic and spectroscopic techniques coupled to electrochemical systems. Figure 11 shows how a combination of different techniques can be employed to study both the surface of an electrode and the solution immediately in front of the electrode to gain useful information on the processes taking place. Scanning probe microscopies such as scanning tunnelling microscopy (STM) and atomic force microscopy (AFM), as well as the SECM technique described above, are powerful aids to understanding electrochemical processes at the microscopic level. The micrograph below shows iodine atoms being adsorbed onto the surface of a gold, single-crystal electrode as its potential is changed. In the lower half of the picture the potential is such that the gold atoms can be resolved. In the upper half of the image the electrode potential has been switched and iodine atoms are now seen adsorbed onto the surface. Such studies are vital for new developments in electrochemistry.

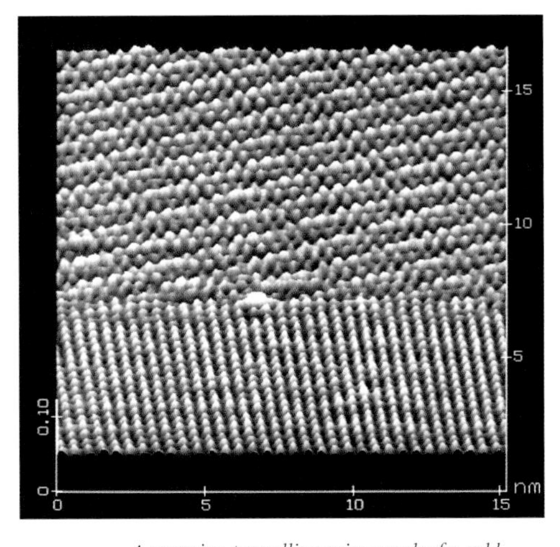

A scanning tunnelling micrograph of a gold electrode surface. The lower half of the image shows single atoms of gold as white bumps. The electrode was stepped to a potential where iodine atoms adsorb on the surface and these can be seen as larger dots in the upper half of the picture.

Reprinted with permission from Michael J Weaver, J Phys Chem, 1996, **100***, 13079-13089.*

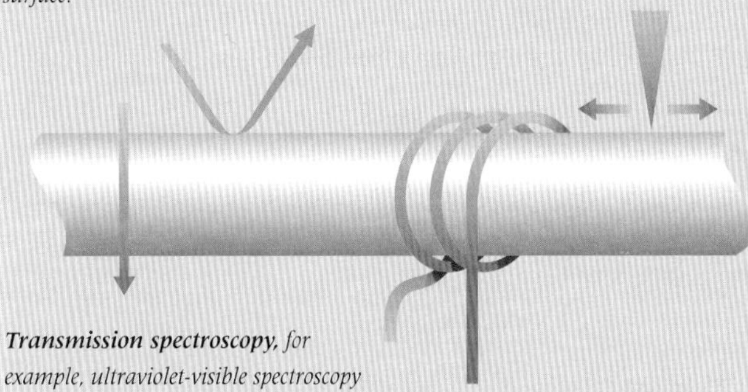

Figure 11. The microscopic and spectroscopic techniques used to study electrochemical reactions at electrodes. To gather spectroscopic information, light is reflected or scattered off the surface or shone through the electrode. The electrochemical quartz crystal microbalance gives data on the mass and viscoelastic properties of the electrode surface by oscillating it at a frequency of 10 megahertz. Resonance techniques (which employ microwaves and magnetic fields) detect unpaired electrons. Scanning probe microscopies (AFM, STM, SECM) probe the surface morphology at a resolution as low as individual atoms.

Electrochemical advances in battery, fuel cell and photovoltaic technology will provide new, clean, efficient and powerful energy storage and generation systems, while new electrochemical industrial processes will be cleaner and more efficient than those used now. Improved sensors and biosensors will allow important substances to be analysed much faster and more accurately. Consequently, electrochemistry has a bright future into the 21st century and beyond.

Further reading

1. *The World of Physical Chemistry,* K. J. Laidler, Oxford: 1993.
2. *Chemical Storylines,* Salters Advanced Chemistry, Heinemann, Oxford: 1994.
3. *Ions, Electrodes and Membranes,* J. Koryta, 2nd edn, Wiley, New York: 1991.
4. *'Harnessing the power of electrochemistry', Chemistry in Industry,* 1996, **18**, 678.

Glossary

Anion Negative ion, which is attracted to an anode.

Anode A positive electrode.

Battery A collection of electrochemical cells used to generate electrical power from chemical energy.

Bioelectrochemistry The electrochemical study of large protein structures.

Biosensor A device designed to monitor and determine biologically important chemicals (for example, glucose) within natural fluids such as blood.

Cathode A negative electrode.

Cation Positive ion, which is attracted to a cathode.

Charge transfer The motion of an electron between two chemical sites.

Chlor-alkali cell An electrochemical process that produces chlorine gas and sodium hydroxide.

Daniell Cell A simple early battery consisting of a zinc metal electrode in a zinc sulfate solution and a copper metal electrode in a copper sulfate solution.

Direct fuel systems A fuel cell that uses fuel directly without any pretreatment (see Fuel cell).

Direct methanol fuel cell A fuel cell that uses methanol directly as a fuel (see Fuel cell).

Electrical potential The driving energy that produces motion of electrons or drives electron transfer.

Electrochemical activity A measure of the ability of an atom or molecule to transfer electrons; related to the redox potential of the material.

Electrochemical cell An assembly of electrodes and electrolytes where electrochemical processes can be observed and manipulated.

Electrochemistry The study of chemical processes driven by electrical potentials

Electrode An electrically conducting or semiconducting material where electrochemical processes take place.

Electrolysis The breakdown of a material driven electrochemically.

Electrolyte A solution containing ions.

Electro-osmosis The removal of, for example, organic pollutants by the motion of water induced by an applied electric field.

Fuel cell An electrochemical cell where chemical energy is converted into electrical energy. The major difference between a fuel cell and a battery is that the reactants are not contained in the cell but fed in as a fuel.

Hall-Heroult cell An electrochemical cell that produces aluminium from a molten salt.

Insertion electrodes Electrodes able to absorb other materials, a good example is graphite into which lithium ions can be inserted.

Ions Positively or negatively charged atoms or molecules.

Lead-acid battery Discovered in 1860 by G. Planté, this is a battery system reliant on the lead(IV) and lead(II) oxidation states. High current densities are possible from this battery; however, it has a poor mass to weight ratio.

Lithium battery A new battery based on lithium chemistry. Lithium is a light metallic element with the lowest redox potential of all elements.

Mediator A small mobile chemical species designed to carry electric charge from the inaccessible centres of large redox proteins or enzymes (see Redox centre).

Nickel-metal hydride battery A promising new secondary battery system.

Oxidation The removal of electrons from an atom or molecule.

Photovoltaic cell A device that produces electrical energy from light.

Primary battery A battery that can be used only once.

Proton-exchange membrane fuel cell A particular type of fuel cell where protons (hydrogen ions) transfer across the membrane to complete the circuit (see Fuel cell).

Redox centre The location, particularly within large proteins, where oxidation and reduction occur.

Redox potential A measure of the ability of an atom or molecule to accept or donate electrons.

Reduction The addition of electrons to an atom or molecule.

Reforming fuel cell A fuel cell system where hydrogen is generated from a hydrocarbon fuel before being fed into hydrogen/oxygen fuel cell (see Fuel cell).

Scanning electrochemical microscope (SECM) A device that electrochemically investigates interfaces.

Secondary battery A battery that can be recharged with energy and thus re-used many times over.

Sodium-sulfur battery A promising new secondary battery system, reliant on high-temperature operation to maintain some of its constituents in liquid form.

Zebra battery A promising new secondary battery system, reliant on high-temperature operation to maintain some of its constituents in liquid form.

Biographical details

Dr Peter Birkin lectures in the chemistry department at the University of Southampton and investigates the chemical and physical effects of cavitation employing microelectrodes as electrochemical prodes.

Acknowledgements

I would like to thank Professor P N Bartlett and Professor G H Kelsall for guidance in the writing of this chapter. I am also grateful to all those people who provided any information, graphics or encouragement.

The age of plastics

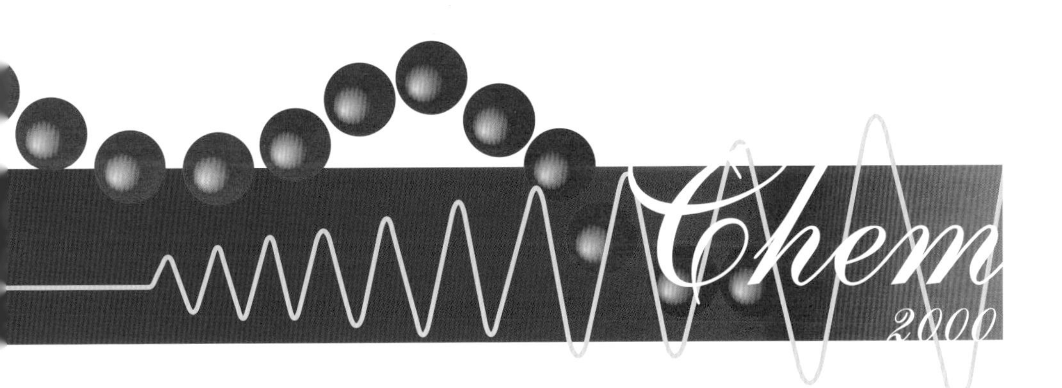

The replacement hip joint, washing-up bowls, paints and video tapes have something in common – plastics. The development and applications of polymers (plastics) have been phenomenal during this century, and are set to continue in the new millennium.

Professor Jim Feast
University of Durham

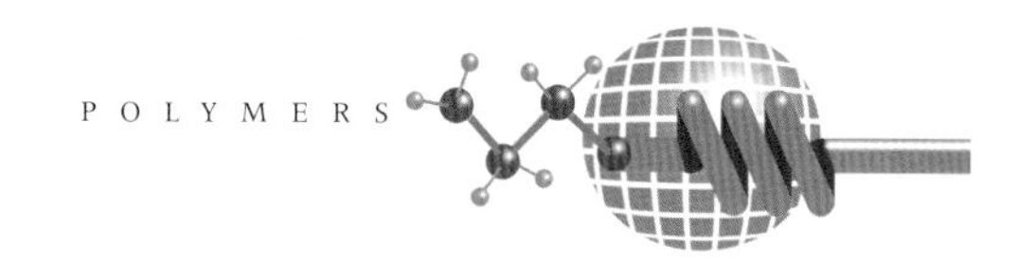

Polymers – chain-like molecules constructed from smaller chemical units – are the basis of many of the materials we use everyday. Indeed, it's hard to imagine living without them. Yet polymers, or plastics as they are more commonly called, are a relatively recent development that resulted from the endeavours of some imaginative and determined chemists.

When you awoke this morning, you were almost certainly lying on a mattress made from several different synthetic *polymers* including, polyester fabric, polyurethane foam, nylon thread and adhesive. Your sheets, pillows, quilt or blankets were probably made with synthetic polymers. These materials are everywhere in our houses and workplaces – in all paints, in PVC window frames, in soft furnishings, toiletries, packaging, electrical insulation and furniture. We rely on polymers for clothing, for transport and even for surgical implants (see Table 1).

We can truly say that the late 20th century has been the 'age of plastics'. Although we also use many other kinds of material, from animal and vegetable fibres, metals and ceramics to the more recent semiconducting materials such as silicon used in electronics, the volume of plastics we use is enormous and growing. In many cases, plastics have replaced wood, stone, glass, leather, natural fabrics and metals from their traditional uses. Plastics have the advantage of being light, strong, hard-wearing, are mostly easy to process and take up colours well. They can be

Table 1.

Applications of polymers
Household furnishings – carpets, curtains and wallpaper
Electrical fittings, wire insulation, casings for electrical goods, printed circuit boards
Household fittings – drain pipes, kitchen fittings
Furniture
Fabrics
Toiletries, cleaning materials
Domestic utensils and containers
Clothing
Surgical prostheses – implants, tooth fillings, contact lenses, absorbants
Transport – bicycles, cars, trains, planes, space modules
Sports materials
Paints and surface coatings

chemically tailored to have specific properties. There are now a huge range of polymers available with a wide variety of chemical structures. Even today, chemists are still developing ever more sophisticated plastics, including novel polymers that possess the electrical properties of metals and semiconductors.

What are polymers?

Polymers are quite different from other compounds in that they are built up from many similar smaller molecules *(monomers)* in long chains. A typical sample is made up from a collection of chains of widely differing lengths. Most common polymers have 'backbones' based on the element carbon which readily forms very stable *covalent* carbon-carbon links (a covalent bond consists of a pair of electrons shared between two atoms). However, other elements also form polymers, such as silicon when combined with oxygen (silicones are also used in many materials), sulfur, nitrogen and some metals (Figure 1).

$(SN)_x$ *An electrical conductor*

Poly (dimethyl siloxane), an inert elastomer (bath seal)

A phosphonitritic polymer

Figure 1. Some inorganic polymers.

The simplest carbon polymer is also one of the most familiar – polythene, or more correctly *polyethene* – and the constituent monomers are ethene molecules. Polyethene consists of a string of linked carbon atoms with each carbon joined to two hydrogen atoms (Figure 2). Some of the molecules in polyethene are incredibly long, having more than 50 000 carbon atoms in one string; they are truly giant molecules.

Figure 2. A fragment of a polyethene molecule shown in three dimensions. The C-H bonds shaped like wedges come out of the paper, whereas the C-H bonds shown as dashes point into the paper.

It is polyethene's string-like structure that gives it its familiar physical properties. In the solid, the molecules can be tidily bundled together in a crystalline array. We can think of the arrangement as being rather like uncooked spaghetti in a jar. When polyethene is heated it first becomes soft and then melts into a sticky, viscous fluid. (Anybody who has left a plastic washing-up bowl on a hot ceramic hob will have seen the messy results of this process.) As the plastic gets hot, the stiff, extended chains soften and begin to wriggle about like cooked spaghetti. In fact, the chain movements are rather livelier than cooked spaghetti – more like a mass of writhing worms. When the molten polyethene cools, it reverts to the highly-ordered crystalline form again and recovers its original properties. This behaviour allows technologists to shape the material into anything we might want, a plastic bag, a washing-up bowl or a hip-socket replacement. Materials with this kind of property are called *thermoplastics*, which simply means they can be shaped when hot.

In fact, the origins of plastics technology can be traced back several centuries to the thermal softening and shaping of horn into utensils or thin translucent sheets used as windows. In the City of London the Worshipful Company of Horners, which was founded in the 12th century, still exists and the fact that many of its members are associated with the plastics industry provides a link with the historic origins of the field. Forms of plastics or rubber technology were also familiar in the pre-Columbian Americas, and lacquers and thin-film coating technologies were well known to the ancient societies of the Far East.

Polymers were used in the music industry to make records.

However, in terms of the quantities of material used, these

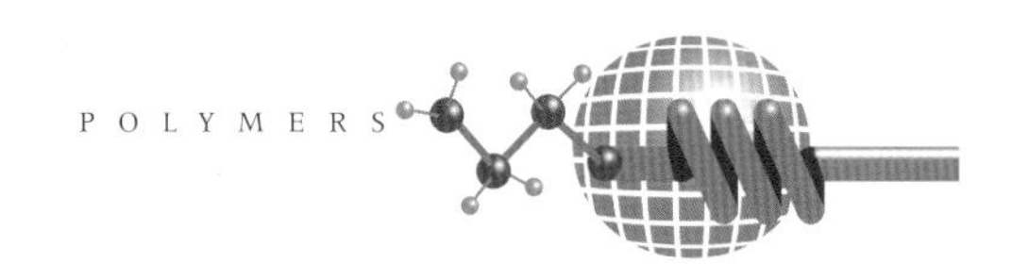

Car tyres are another product of the polymer industry.

were always small-scale and rather specialist activities. A few polymers, like Bakelite and *rubber*, were commercially exploited at the turn of the century, but it was not until the latter half of the 20th century that plastics made a real large-scale impact. Nevertheless the intellectual base for progress in this area of science and technology was laid well before – in the early years of the 20th century.

The rise of modern polymer chemistry

You might be surprised to learn that as recently as the 1920s the notion of very large molecules such as polymers was not accepted. That polymer science became a serious subject for study is owed in no small measure to a determined German chemist called Hermann Staudinger. Staudinger was a successful professor of organic chemistry in one of Europe's premier universities, the Federal Institute of Technology in Zurich (ETH). In 1920 at the age of 39, having already established an excellent reputation based on research on small molecules, he became interested in a dispute about the nature of a group of awkward materials that were then known as *hochmolekulare Verbindung*, or 'high molecular compounds'. This led Staudinger into conflict with the academic chemistry establishment of the day. Science is a human activity and disputes of this kind are quite common. Indeed intense arguments are useful in the development of the subject because if you want to win an argument you really do have to prove that you know what you are talking about, which is a great stimulus to the clear thought and careful experimentation on which progress depends.

Hermann Staudinger.
Courtesy of Exxon Chemical.

The reasons for the conflict and the nature of the problem that attracted Staudinger merit a few moments of our attention. The establishment professors of organic chemistry had achieved enormous success by following procedures developed in the previous century. This classical approach to organic chemistry culminated in an intellectual synthesis that is one of the major achievements of western science. Methods were developed for breaking molecules from natural sources into small, recognisable bits, and for assembling little units unambiguously into more complex structures. These pioneers were trying to find out how Nature works and, if you adopt a mechanistic view, you need to know all about the individual components before you can see how the machine is put together and how it operates.

In the early years of the century, this approach had established structures for thousands of well-defined unique molecules. This part of organic chemistry was and is enormously successful. It is the bedrock on which the pharmaceutical, dyestuffs, perfumery and agrochemical industries are based.

However, there were difficulties in fitting naturally-occurring materials like starch, cellulose, proteins and rubber and the newly invented synthetic organic plastics into this scheme of things. The established theory could not deal with the molecular size of these organic materials. The establishment professors were wedded to the idea that these materials could be accounted for, only if they were treated as bundles of small molecules held together by some unknown force. This was wrong and there was perfectly good evidence that it was wrong. But scientists, being human, tend to have a preference for the comfort of the familiar and it took a brave man like Staudinger to fly in the face of authority.

Staudinger's chosen topic of research – rubbery, apparently high molecular-weight compounds – was treated with undisguised contempt and dismissed as 'grease chemistry' by many of his distinguished contemporaries. For example, Professor Wieland, a colleague and a very successful classical organic chemist, advised him thus: 'Drop the idea of large molecules, organic molecules with a molecular weight higher than 5000 do not exist. Purify your rubber then it will crystallise'.

History has vindicated Staudinger's scepticism and stubborn persistence. Of his many studies, perhaps that on a class of compounds called polyoxymethylenes is one of the most easily appreciated and important milestones in establishing the hypothesis of large molecules. Commercially available paraformaldehyde, familiar to most of us as slug bait, was treated with acetic acid (the main component of vinegar) to obtain a mixture of polyoxymethylene diacetates, a typical example is shown in Figure 3.

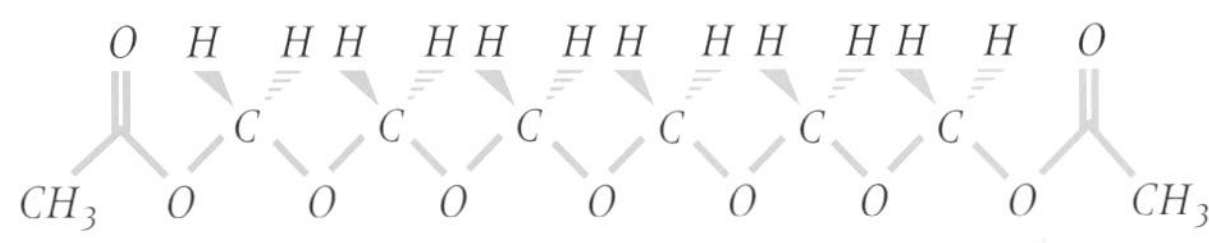

Figure 3. An example of a polyoxymethylene – a family of compounds first studied by polymer pioneer Hermann Staudinger.

Examples of products made by DSM's plastics processing company.
Courtesy of DSM.

Using the techniques of the classical chemistry of the time, Staudinger and his colleagues made a whole series of molecules like the one in Figure 3, each differing from its neighbour in the series by one – CH_2O – unit. Comparing the properties of these molecules in relation to their size established that the small members of this family behaved like any other small molecule, but as the length of the molecule increased their properties gradually changed to those of plastics. In this way, Staudinger and his students established by 1929, that this polymer was not an association of small molecules held together by unknown forces but consisted of chain molecules with normal, covalent bonds of the kind familiar throughout organic chemistry. The chains

terminated in groups of atoms, which if their concentration could be measured, could give a measure of the average chain length of the polymer sample.

Paints and surface coatings are familiar products of the polymer industry.

Acceptance of synthetic polymers as giant molecules was the first intellectual step that ultimately led to the development of polymer chemistry, and Staudinger eventually became polymer chemistry's first Nobel Laureate in 1953. The great polymer chemist Paul Flory, a professor at Stanford University and himself Nobel Laureate in chemistry in 1974, encapsulated Staudinger's achievements in the opening sentence of his 1953 textbook, the most influential in the field of polymer chemistry:

'The hypothesis that high polymers are composed of covalent structures many times greater in extent than those occurring in simple compounds, and that this feature alone accounts for the characteristic properties which set them apart from other forms of matter, is in large measure responsible for the rapid advances in the chemistry and physics of these substances in recent years.'

The second important step in polymer science was the realisation that synthetic polymers are hardly ever single unique molecules but consist of an assembly of molecules with a distribution of sizes and often, but not essentially, a distribution of structures as well. The proportions of chains of different length and thus molecular size has important effects on the polymer's properties (and consequently different applications). Mixtures with very short carbon chains are fluids or soft, waxy solids, whereas mixtures in which very long sequences of carbon atoms predominate can be hard, wear-resistant materials. Although many theoreticians, physical chemists and physicists contributed to establishing these basic points, three figures that stand out are Staudinger, Flory, who established the theoretical basis for understanding the synthesis and properties of polymers, and Wallace Carothers. Carothers, while he is best known as the inventor of nylon, was a major contributor to understanding the synthesis of polymers. Flory was a member of the DuPont research group led by Carothers until Carothers' early death. Flory left industry shortly after Carothers' death to pursue his career in academia.

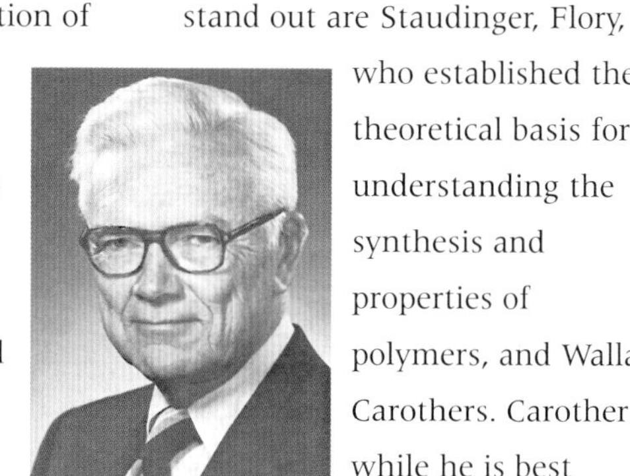

Paul Flory.
©Nobel Foundation

Wallace Carothers.

The story of polyethene

At about the same time as Staudinger was laying the basis of polymer science, industry was getting started on the practical aspects of the subject – that is, making and finding uses for synthetic polymers. A suitable place and time to start telling the strange and exciting story of what was to become the most common synthetic polymer, polyethene, is in the north west of England in March 1933.

At this time a group of experimentalists employed by the Alkali Division of Imperial Chemical Industries at its Wallerscote site in Cheshire were examining the effects of mixing various gases at elevated temperatures and pressures. The reasons for undertaking this programme need not concern us but we should note that 'some 50 experimental high pressure reactions involving different gaseous mixtures had been attempted over the previous 15 months, with disappointing results'. This statement makes it clear that we are not recalling the occupation of dilettanti but the efforts of a band of determined experimentalists.

In the event the key experiment that really starts this story was an attempt to react ethene (or ethylene, C_2H_4) and benzaldehyde (C_6H_5CHO) at a pressure of 1900 atmospheres (about 200 million pascals) and a temperature of 17 °C. Ethene is a gas and benzaldehyde a liquid that smells like almonds. The anticipated result of this experiment is not known, but these open-minded and curious experimentalists found a minute amount of a white waxy solid on the walls of the high-pressure reaction vessel. The result of this reaction created interest among the researchers, who quickly realised that they had for the first time polymerised ethene.

Damage from an explosion, 1942.
Courtesy of Brunner Mond (UK) Limited and ICI Technology.

However, for various reasons the experiment was not repeated for almost two years – in fact not until December 1935, when 8 grams of the material were produced in a small-scale laboratory reactor.

Nobody could then have imagined the outcome of this apparently largely empirical approach to chemistry. Today, in the light of modern health and safety regulations, these early experiments seem crude and even dangerous. The reactors sometimes exploded, some reactions yielded heat, hydrogen and carbon and nothing else of interest. Nevertheless, they were to have enormous and totally unforeseen consequences.

Three years after the 1933 invention, the ICI researchers had progressed to the stage where they believed they might have something saleable. This new material had the advantageous properties of good processing and mechanical behaviour, low water-permeability, good insulating properties, and dielectric properties which made it suitable for use in high-frequency equipment and as an insulator in many kinds of electrical cable. At this stage the researchers needed a name for it and in mid-1936 it was named ALKETH, before it became known as polyethylene or polythene. Its detailed properties depended to a large extent on the conditions of synthesis; for example, the lower the molecular weight the more wax-like the material, and the higher the molecular weight the tougher it was. In 1937 batch syntheses were being carried out in a 9-litre vessel at 900 atmospheres and 200 °C. Material with an average molecular weight of about 10000 was produced under these conditions and soon progress in control of the reaction allowed samples with average molecular weights up to 27000 to be produced at a rate of manufacture of up to 1.6 kilograms an hour.

It is something of a minor miracle that polyethene production ever got off the

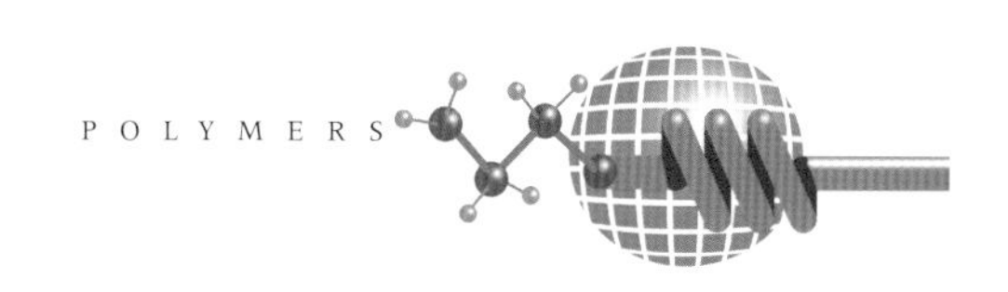

starting blocks. The experimental work was technically demanding and risky, the raw material was expensive – the petrochemical industry hardly existed. The ethene used by the ICI workers was bought from the British Oxygen Company who, in their turn, obtained it by dehydrating ethanol with phosphoric acid. This route is rarely used now. Today, ethene is a commodity chemical produced in enormous quantity from oil by the petrochemical industry. Indeed most of the polymers manufactured are made from petrochemicals.

It is worth noting that if polyethene had been discovered in more recent times it would not have had a chance of reaching commercial production because of the tight budgetary control and accountancy that now dominate research planning. Fortunately our scientific and managerial forebears were made of sterner, more confident and adventurous stuff and clearly not so much in awe of the all-powerful 'bean counters' as we are today.

The chemistry of polymerisation

So how was polyethene formed? The reactant was simply ethene maintained at high temperature and high pressure. Ethene is a flat molecule of six atoms (C_2H_4). The two carbons are linked by a double bond with two hydrogens bonded to each carbon as in Figure 4. These are all covalent bonds formed by pairs of electrons shared between two atoms. The two bonds between the carbon atoms are of different strengths. We can imagine that one of the bond pairs can be decoupled to give a free electron on each carbon atom, which can then combine with free electrons on carbon atoms in other

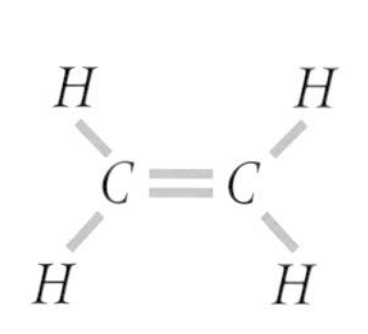

Figure 4. The structure of ethene.

Extruding plastic.

$$X\bullet \quad H_2C{=}CH_2 \quad H_2C{=}CH_2 \quad H_2C{=}CH_2 \cdots\cdots H_2C{=}CH_2 \quad H_2C{=}CH_2 \quad \bullet X$$

$$X\bullet \quad \bullet CH_2{-}CH_2\bullet \quad \bullet CH_2{-}CH_2\bullet \quad \bullet CH_2{-}CH_2\bullet \cdots\cdots \bullet CH_2{-}CH_2\bullet \quad \bullet CH_2{-}CH_2\bullet \quad \bullet X$$

$$X{-}CH_2{-}CH_2{-}CH_2{-}CH_2{-}CH_2 \sim\sim CH_2{-}CH_2{-}CH_2{-}CH_2{-}X$$

Figure 5. The simplest model for the polymerisation of ethene. The dots represent single electrons decoupled from one of the electron pairs involved in the double bond, which then become available for coupling ethene molecules together. X represents an atom that interacts with the spare electrons left over from polymerisation.

molecules to form new electron-pair bonds. Ethene molecules can therefore join in series – in other words, polymerise – forming infinite chains of new carbon-carbon bonds. Figure 5 shows this simple picture. Note, however, that it tells us nothing about the actual reaction mechanism and we shall have to assume that unknown atoms or groups of atoms X are tagged on at the ends of the chains to take up the left-over electrons at the ends of the chains.

This description turns out to be something of an over-simplification in that it suggests that the product polymer will be completely linear. However, a variety of physical measurements indicate, that the structure is more complex than that. Spectroscopic studies showed that there were both long and short-chain branches present, and X-ray diffraction measurements revealed that the material was semicrystalline. What the scientists were now forced to accept was that when the molten polymer cools and solidifies, part of it is organised and crystalline, and part of it is disordered.

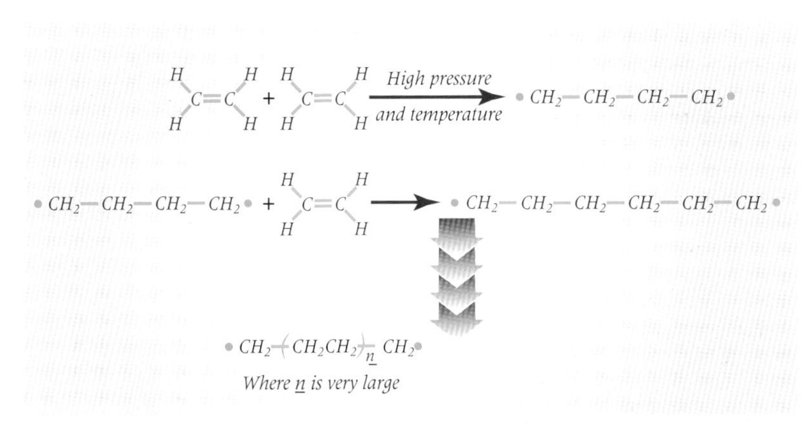

Figure 6. An alternative picture of what happens in ethene polymerisation. At high temperatures and pressures, ethene molecules slam into each other and bond to form new chemical species with unpaired electrons, or radicals. These readily react with more ethene molecules eventually to form polyethene.

Auto fuses

Suppose, however, we consider another approach. Pumping up the pressure and temperature in the reactor forces two ethene molecules to collide so that they bind to form four carbons and eight hydrogens in a linear array of four CH_2 units. As a result of this process there will be unpaired electrons left over sitting at the ends of the chains. These are called *free radicals*. Since electrons usually hate to be unpaired they are extremely reactive, seeking other electrons with which to form a new chemical bond. Indeed they happily react with any ethene molecules that come within bonding distance to build up the molecular chain as in Figure 6.

This free radical description of the polymerisation allows us to explain why branches form in polyethene. The free radical chain ends are so enormously reactive that if an appropriate collision occurs, the chain-end free radical ($-CH_2\cdot$) will simply grab any hydrogen atom within reach, in

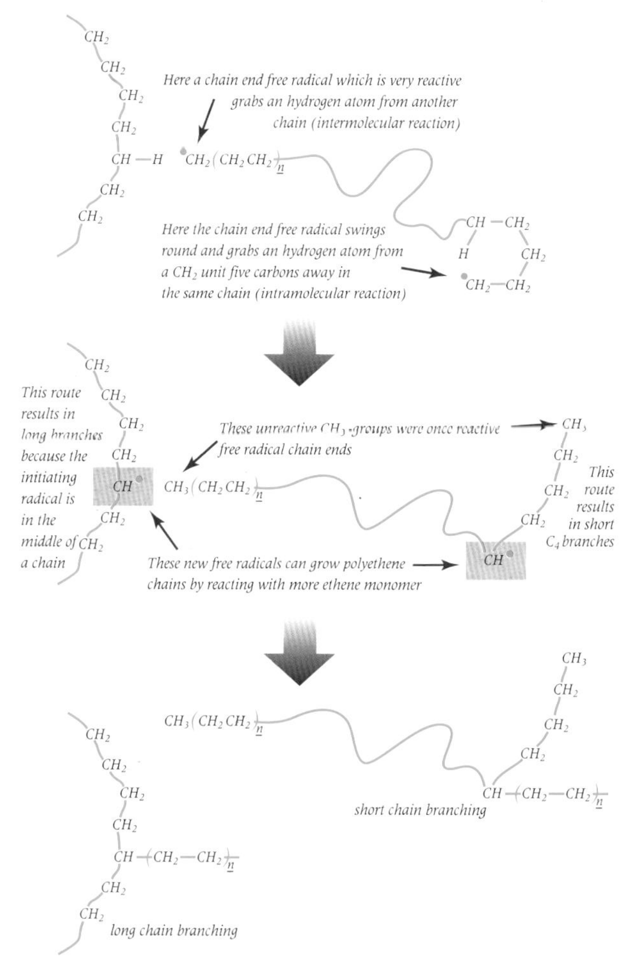

Figure 7. How chain-end free radicals can react to form long and short chain structural defects.

the process severing a carbon-hydrogen bond and recreating a free radical on another carbon atom. This new free radical, which could be in the middle of a long chain or relatively close to the end of a chain, can now grow another sequence of ethene units as before (Figure 7).

It isn't too hard to see that the higher the temperature and pressure of reaction the more frequent the branches in the product polyethene will be. Since the type of branching and its frequency influence the properties of the material, we can begin to see how the chemist can tailor the reaction to make products with specific desired properties for particular applications. For polyethene made in this way the greater the relative proportions of long straight sequences of – CH_2 – units the more crystalline and hard the product. This is because ordered crystalline chain arrangements are harder than disordered amorphous chains and regular straight-chain segments are more easily arranged in a crystal lattice than segments with branches. In practice the crystalline parts of polyethene are associated with the linear chains which are arranged in a zigzag fashion with the carbons in a plane, and with the chains folded backward and forward in a regular way. The chain branches are located in the disordered regions. One molecular chain can pass through a chain-folded crystalline part of the sample into a disordered amorphous part and then into another crystalline part or back into the crystalline part it emerged from. This complexity in the structural organisation is what gives polyethene its variable and valuable properties (Figure 8).

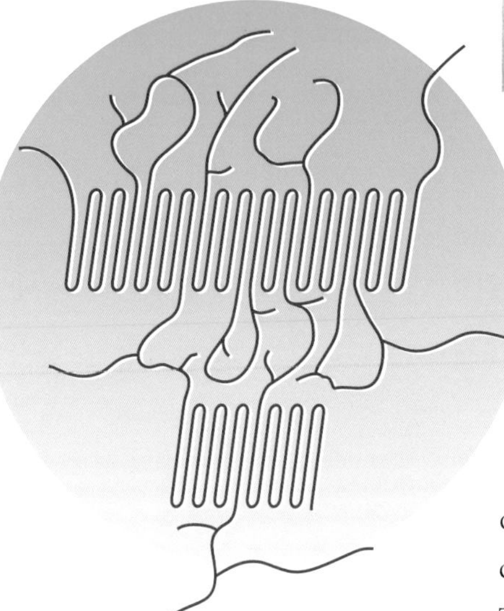

Figure 8. Two-dimensional cartoon showing crystalline and amorphous regions, and long and short-chain branching. The lines represent polyethene molecules.

Shopping bags, a common use of plastics.
Courtesy of British Plastics Federation

immense. That year, three UK cable companies took 17500 pounds (7945 kilograms) of ALKETH with a further 1800 pounds (817 kilograms) going to a manufacturer of candles. With the fall of France, however, the requirement for candles in the trenches was deemed to be of low priority and virtually all production was directed into electrical insulation. The high-frequency cables required for RADAR being the major customer, although no mention of this is to be found in any of the available records of that period. Production spiralled from 8 grams in 1935 to about 8 tonnes in 1938, 557 tonnes in 1942 and 1441 tonnes in 1944. Sadly much of this part of the history of the development and exploitation of polythene was lost to view in the cloak-and-dagger atmosphere that surrounded what became, between 1935 and 1950, a strategic military material.

A secret military material!

Within a relatively short period after its discovery, polyethene was to have an influence on the outcome of the Second World War. By the end of 1939, ICI was selling its new material to a few carefully selected customers. Within a year the selection of customers would be tightly constrained and the need to produce more material became

After a dip in demand following the end of hostilities, new markets for this cheap versatile material began to appear. By the end of 1946 ICI had three production units all operating at a rate equivalent to a total production of more than 4000 tonnes a year. By about 1950 the material had expanded out into packaging (bottles, film, bags) and was beginning to initiate what was to be one of the themes of polymer science during the next 50 years; namely, the process of displacing metals from various market niches.

In the early days of polyethene production, the fairly undemanding targets were things like washing-up bowls, buckets and the like. It required new chemistry before the material could be used to challenge the tensile strength of steel cables, the bursting pressure required for pipes for gas and water mains, and eventually to meet the stringent requirements for surgical implants. How these things came to pass is the next part of our story.

A mature technology?

There is a story, dear to the hearts of research chemists, to the effect that around 1950, when the polyethene production technology stemming from ICI's 1933 invention was widely disseminated and the new material was increasingly widely used, business and finance managers in charge of the various companies now involved formed the opinion that polyethene production and its applications had reached the stage of being a mature technology. In practice this means that there is nothing to learn technically, remaining interest is in driving down production costs and research which brought the product into existence in the first place, ceases.

This assumption, as in so many other cases, was wrong. In 1950 the company decision-

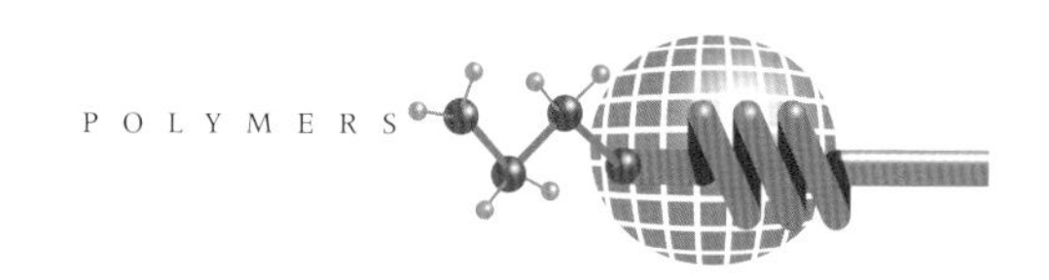

Karl Ziegler.
Courtesy of GDCH.

makers had not heard of the work of Karl Ziegler's group in the Max Planck Institute in Mülheim in the Ruhr. Ziegler was a remarkable, persistent and meticulous chemist. He was fascinated by the formation and reactions of metal-to-carbon bonds. In the early 1950s his research group were considering some new bonding ideas concerning compounds called aluminium alkyls. The key compound they studied was a trialkyl aluminium, R_3Al, in which a single atom of aluminium has three hydrocarbon groups (R) attached to it via aluminium-to-carbon bonds. Ethene reacts with this compound by inserting itself into the aluminium-carbon bond (see right). This process required temperatures of 100 to 120 °C and was carried out in sealed pressure vessels called autoclaves. Just like ICI's earlier work the research had its element of excitement, even danger, because aluminium alkyls are pyrophoric, that is, they catch fire spontaneously when in contact with air.

$$R_3Al + CH_2CH_2$$
$$\downarrow$$
$$R_2Al\text{-}CH_2CH_2\text{-}R$$

Ziegler's group found that from time to time they obtained a white powder which was clearly polyethene, but which was more crystalline and had a higher melting point than the material produced by the ICI route. The aluminium trialkyl was clearly catalysing ethene polymerisation.

In-line skates contain parts made from plastics technology.

Courtesy of DuPont

This discovery promised to extend the range of polyethene's exploitable properties. The researchers also observed that 'traces of colloidal nickel accidentally left in a reactor' and/or 'trace impurities in the steel reactor' made the polyethene-yielding reaction go rather better than was the case in immaculately clean conditions. These astute observations led them to search for better systems to catalyse the polymerisation of ethene, and within a few years they showed that they could produce polyethene using a compound made by reacting trialkyl aluminiums with transition-metal compounds such as zirconium or titanium tetrachloride. What was even more amazing was that this

Household products packaged in plastic containers.
Courtesy of British Plastics Federation.

process produced an essentially linear (with no, or very few branches) high molecular-weight material at room temperature and at about one atmosphere. This polyethene was different from the high pressure-high temperature form. Not only was it more crystalline but it also had a higher molecular weight, higher density, and better machining and wear properties. It was therefore called *high density polyethylene (HDPE)*. Because Ziegler's method allowed the makers to avoid the costly requirements of high temperature and high pressure it offered the potential of a different material – superior in some respects – at lower cost. Cost is partly dependent on market conditions but properties depend on molecular weight, chain structure and the organisation of different segments of the polymer chains. The physicists and engineers fell upon this new material and worked out how to use it in all sorts of exciting new ways. At the everyday level they found that films of the same or higher strength, could be made much thinner than with the conventional material – a fact with obvious economic and environmental benefits. Very high molecular-weight samples have been found to be suitable for making the socket part of artificial hip joints and components of knee and knuckle replacements. When mixed with

Artificial hip joint and components of knee and knuckle replacements.

calcium hydroxyapatite and suitably processed, HDPE makes a satisfactory substitute for bone in restorative surgery. Furthermore, the long, straight, virtually defect-free form of polyethylene obtained using the *Ziegler catalyst* offers the possibility of producing superbly-ordered molecular arrangements of polymer chains which allows the production of fibres and therefore ropes that compete with steel hawsers and have the advantage of being light and rust-free.

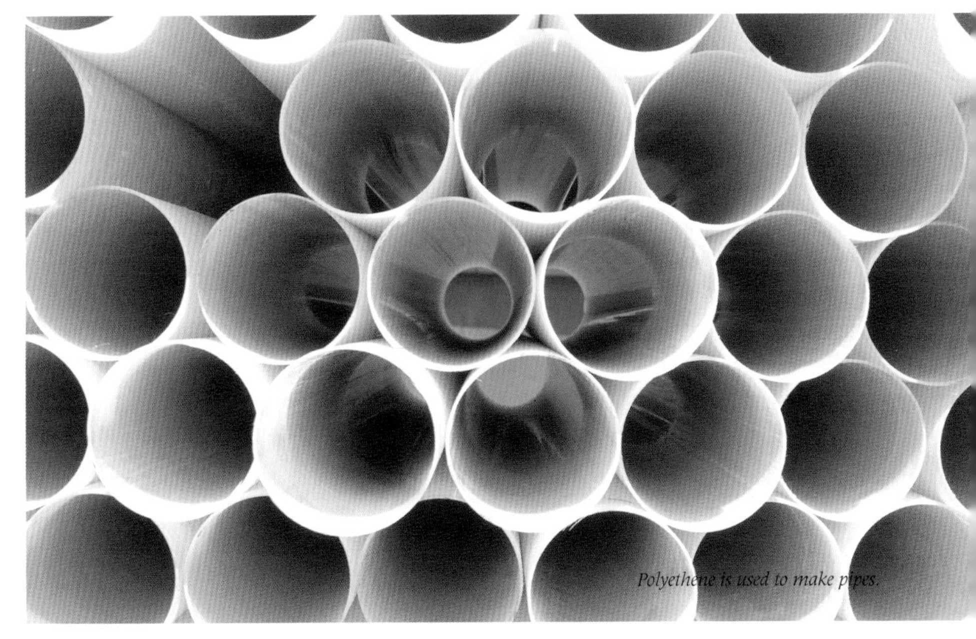

Polyethene is used to make pipes.

This technology is still developing rapidly. It is now possible to extrude polyethene to produce pipes which are replacing metal water and gas mains pipes, and these will be familiar to everyone – pigmented yellow for gas and blue for water. Ian Ward at Leeds University has been a major player in developing the physics and engineering of these 'oriented' polyethenes. A recent development in his laboratory has been to take highly-oriented fibres and process them under pressure at a temperature just below the main melting point of the material. This process is called hot compaction. It gives what is essentially a fibre-reinforced

Ian Ward

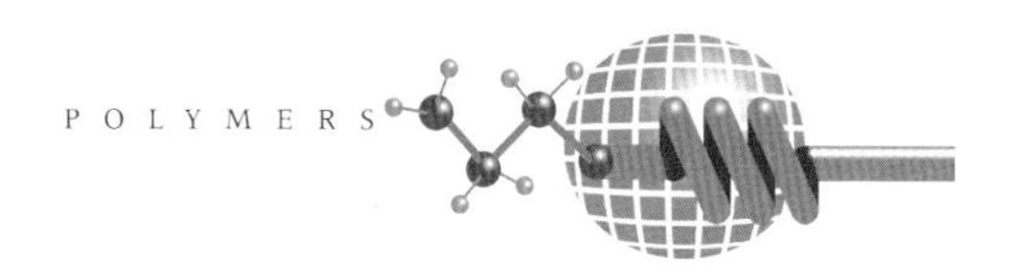

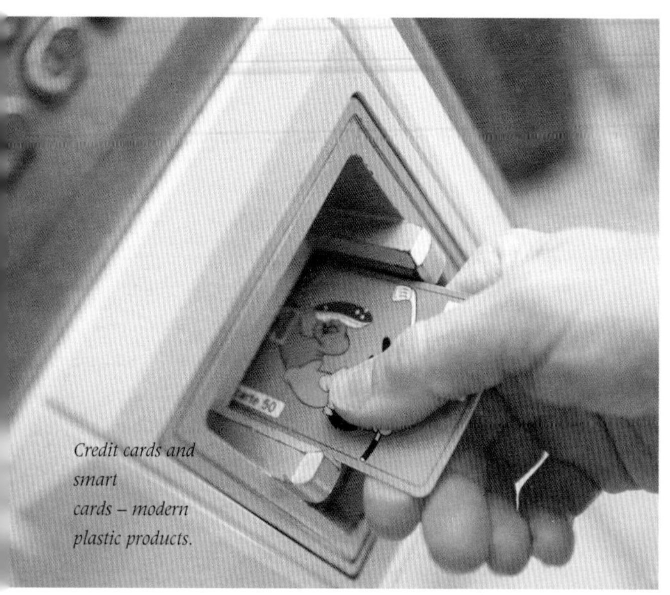

Credit cards and smart cards – modern plastic products.

composite which is composed entirely of polyethene. The 'reinforcing fibres' are highly extended configurations of the polymer chains while the continuous matrix is the conventional semicrystalline thermoplastic with crystalline chain-folded and disordered regions. This form of polyethene is very strong, rigid and resilient and a very far cry from the thermoplastic which was available in the 'mature technology' days around 1950. This new form of the material is just beginning to reach the market-place. It has properties that make it suitable as a protective shield for sensitive electronic components as well as the sensitive parts of people engaged in sports such as American football and cricket. We may well be seeing the beginning of yet another revolution in polyethene science and technology.

Polyethene is used in protective gear worn by American football players.

Car bumpers are made from polypropene.
Courtesy of DSM

Polypropene

Another major development, which was initiated by Ziegler's discovery, took place in the laboratories of the Polytechnic Institute of Milan in a research group directed by Guillio Natta. What Natta's group did was change the monomer from ethene to propene (C_3H_6). It polymerises to produce a material, *polypropene,* with the same basic structure as polyethene except that the hydrogen atoms on alternate carbon atoms in the chain are replaced by methyl, CH_3, groups, as shown in Figure 9.

Guillio Natta

We can see that this introduces complications that don't exist for polyethene in that the methyl groups can be arranged in different ways with respect to the extended carbon-chain backbone. Natta named the polymer in which all the groups are on one side of the extended chain the *isotactic form*; if the groups alternated from side to side he called it the *syndiotactic form*, and the form in which the groups were statistically distributed on either side he called the *atactic form*. Different catalysts gave different forms in a well-controlled way. This was very important from a practical point of view because the isotactic and

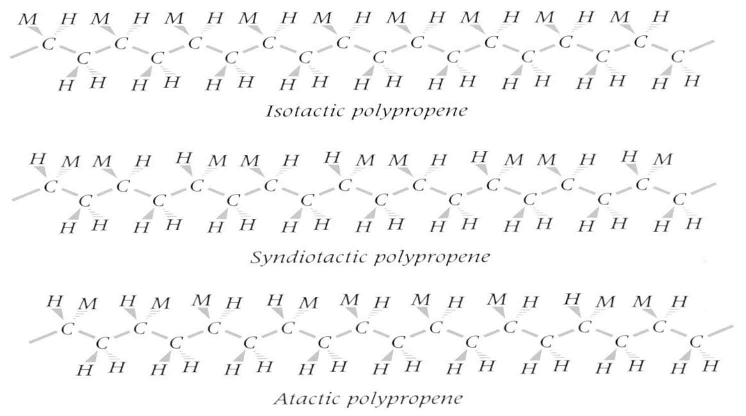

Figure 9. The three types of polypropene (M stands for methyl group). Isotactic and syndiotactic polypropenes are described as stereoregular.

syndiotactic forms, which are called *stereoregular polymers*, have much more useful properties than the atactic form. Stereoregular polypropenes are semicrystalline thermoplastics similar to polyethene. In these polymers the chains in the crystalline parts are not stretched out in a zigzag way as in polyethene but curl up as regular helices. Polypropene is the next most common polymer in use after polyethene, and you find it in all sorts of applications from the fascia of cars to the thermal underwear of skiers and mountaineers. Fittingly, Natta shared the 1963 Nobel Prize for Chemistry with Ziegler for his work using the Ziegler catalysts to stereoregulate the polymerisation of propene.

Ziegler's key patent had a priority date of 18 November, 1953, but the process of invention and innovation did not stop here. Almost 20 years later, Walter Kaminsky and Hansjorg Sinn, working in the University of Hamburg, announced another breakthrough with their invention of new catalysts based on a compound of zirconium activated with methylaluminoxane, which is formed by reacting trimethylaluminium, $(CH_3)_3Al$, with water. These were highly efficient homogeneous catalysts (see *Chemical marriage-brokers*). Another 20 years later, in the 1990s, we are seeing a tremendous explosion of activity in catalyst design which gives improved control of polymer structure and properties. It seems reasonable to assume that the science and technology of making and processing this very variable and valuable material still has a long way to progress, and that its value to society will continue to increase year on year.

In the end, polyethene and polypropene come from oil. A few per cent of the mass of every barrel ends up as useful manufactured goods made of these plastics. A few more per cent end up as goods prepared from other hydrocarbon polymers. Most of the oil is used for energy, heating and transport, the part going to plastics can be recycled and/or recovered by burning safely. Alongside the development of polyethene, several other polymers came into prominence. The *nylons*, or polyamides are probably one of the best known.

Nylons

Polyamides are characterised by possession of an amide link usually as a secondary amide (-CONH-). The secondary amide link has already been used by Nature in the vast and complex range of natural polypeptides found as hormones, enzymes, muscles, hair, tendon and structural materials (see *The chemistry of life*). The best known polyamides are the nylons, and the first of these was invented and developed in DuPont's Experimental Station in Wilmington, Delaware during the late 1930s by the scientist who laid the

foundation of this part of polymer chemistry, Wallace Carothers. Nylon is not a unique material but a name for a whole family of materials, which are identified by numbers at the end of the name.

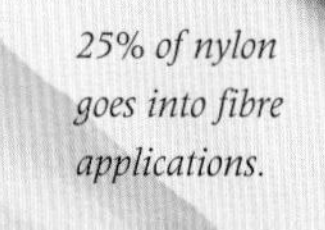

25% of nylon goes into fibre applications.

Tooth brush bristles made from nylon.

Generally, in between the amide links are methylene (-CH_2-) units. The longer the chain of methylene units the lower the melting temperature and water uptake. All these materials are used in fibres and light engineering applications such as pram wheels, cams and gear wheels for light-duty applications and, when reinforced with fibre glass, for more demanding applications including bike wheels. They display low shrinkage in the mould so can be easily manufactured into objects with precise dimensions. All the nylons-X , from nylon-2 to nylon-11, have been made. They have the repeating structural motif shown in Figure 10(a) and include: nylon-3, which is used as the thread in commercial sewing machines because it has a high

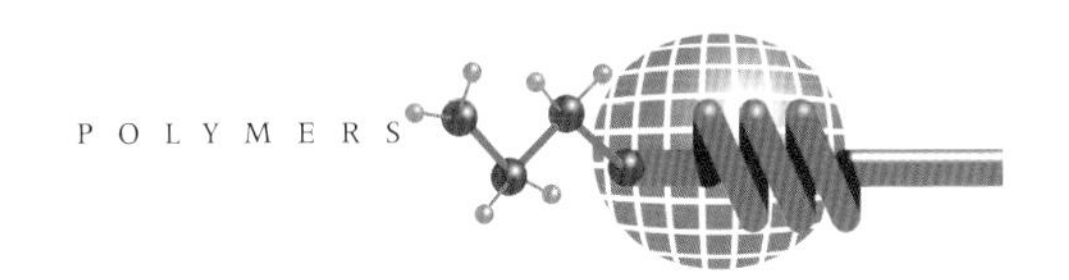

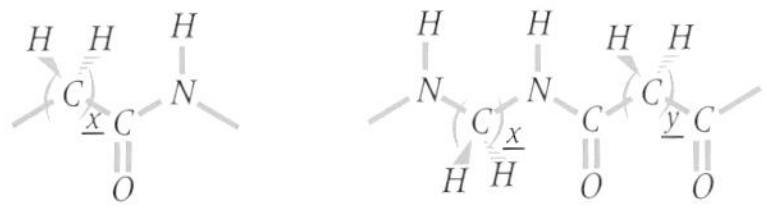

(a) Nylon-X (b) Nylon-X,Y

Figure 10. The general structures for nylons.

melting point and can withstand the frictional heating experienced in machines operating at high speed; nylon-6, also known as polycaprolactam, is found in the walls of car tyres; and nylon-10, poly(11-undecanoamide), which retains its stiffness when wet and is found in applications where that matters, such as some sports goods and tooth brush bristles. The nylons-XY have the structural motif shown in Figure 10(b) (Y=*y*+2) and the 46, 66, 69, 610, 612 versions have been developed for commercial use.

There are many routes to nylons. The so-called salt dehydration method outlined in Figure 11 for nylon-66 was initially developed by Carothers and has found widespread use. In this process, a diamine, $NH_2(CH_2)_6NH_2$ reacts with a dicarboxylic acid, $HO_2C(CH_2)_4CO_2H$ to give a salt by transfer of hydrogen ions, H^+, to the base units, $-NH_2$. This salt, $^+NH_3(CH_2)_6{}^+NH_3{}^-CO_2(CH_2)_4CO_2{}^-$, is formed in a very pure state and when it is heated gives off water with the formation of the amide bonds which hold the polymer together. This polymer is called nylon-66 because there were six carbon atoms in both monomers.

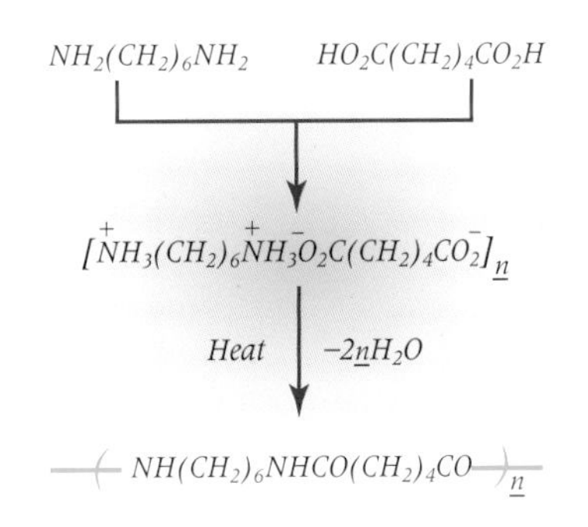

Figure 11. Salt dehydration method for making nylon-66.

There are many other polyamides including the structures shown in Figure 12, the aromatic polyamides. The two structures at the bottom of the Figure, are known as *polyaramids* (they are actually processed as liquid crystals, see *The world of liquid crystals*) and give fibres of exceptional strength, crystallinity, and stability with melting points above 400 °C . Such materials find many specialist applications as fire and impact-resistant fabrics. They are extremely strong so are used as aerospace materials and in bullet-proof vests.

Although such a wide range of materials are available, nylons-6 and 66 account for about 80 per cent of all nylon manufacture at present and 25 per cent of that goes into fibre applications.

Related to the nylons are the *polyurethanes* which are said to have been invented in an attempt to circumvent the nylon patents. This seems plausible when you compare the linking units in Figure 13.

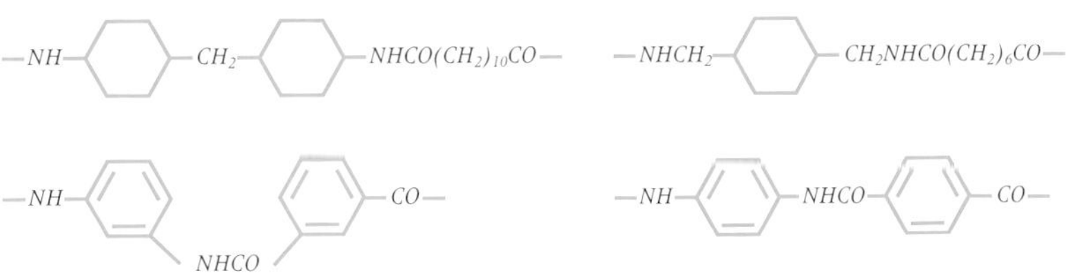

Nomex(DuPont); Conex (Teijin) Kevlar (DuPont), Twaron (Akzo)

Figure 12. Some more examples of polyamides.

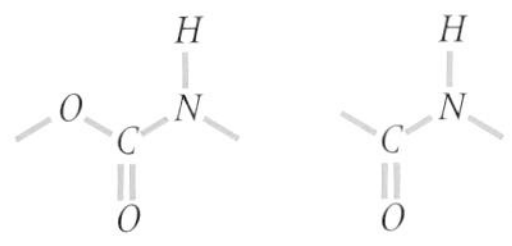

Figure 13. A comparison of the linking units of urethane (left) and nylon.

One method of the many invented for the synthesis of polyurethanes involves the addition of alcohol units $-CH_2OH$, to isocyanates, $-CH_2NCO$. Isocyanates are very reactive and also very toxic so although the chemistry works very well it has to be carried out under very well-controlled conditions.

$$O{=}C{=}N\text{-}(CH_2)_x\text{-}N{=}C{=}O + HO(CH_2)_yOH \downarrow \;\; -\!\!\!\left(OCNH(CH_2)_xNHCOO(CH_2)_yO \right)\!\!\!-_n$$

The products are softer and lower-melting than analogous nylons and find application in encapsulation, coatings and elastomers. A network-forming polymerisation, in which a gas (carbon dioxide or an inert gas) is released or injected at the appropriate time, leads to an expanded honeycomb structure that is relatively stable. The type

of foam – rigid or soft – is determined by the nature of the polymer network and the size of the holes in the foam. Such foams find uses in applications varying from the rigid interiors of doors and surf-boards through to elastic foams for furniture.

Another related group which have more complex structures are the *polyimides*. Probably the best known example of this class of polymer is shown in Figure 14. It is very stable to heat and oxidation, displays little weight loss up to 500 °C, and has a melting point above 600 °C. It is used in electrical insulation, as a printed circuit base film, in composites and as fibres.

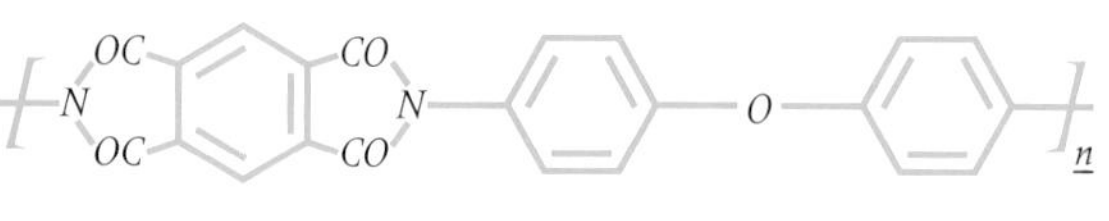

Figure 14. A polyimide.

Water bottles are made from PET. BSIP/Science Photo Library.

Polyesters

Polyesters are another group of polymers that have been developed over a long period of time and are well-known to the public. Although Carothers worked on polyesters in his pioneering research before the Second World War, the systems he chose turned out to be relatively low melting and, so far, have not had any commercial significance. The first practically useful material was *poly(ethylene terephthalate)*, PET, (Figure 15a) which was developed immediately after the Second World War. PET found use as a fibre and film-former at first, and subsequently had a big impact on the blow-moulded drinks bottles market. It remains a major material in the fabric fibres and film markets. When you read 'polyester' on a garment

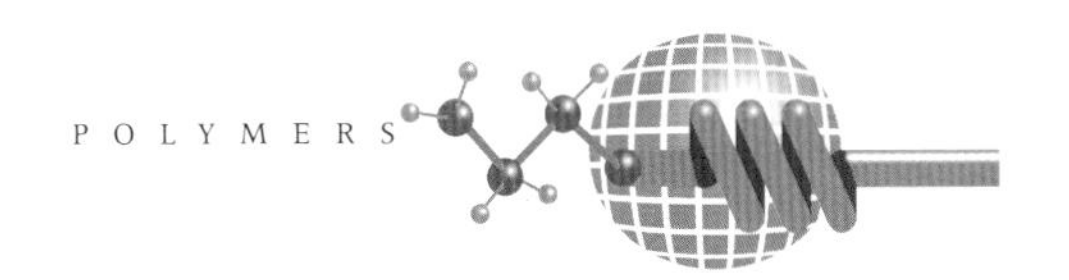

(a) $-[O-C(=O)-C_6H_4-C(=O)-O-CH_2-CH_2]_n-$

(b) $-[O-C(=O)-C_6H_4-C(=O)-O-(CH_2)_4]_n-$

Figure 15. The basic units of polyesters such as polyethene terephthalate (a) and polybutene terephthalate (b).

label it is odds on that it is PET. Similarly when you buy a 'boil in the bag' meal, the bag is probably PET, as is the plastic bottle for fizzy drinks (soft or alcoholic) and the tape for your video recorder. Another important polyester is *poly(butylene terephthalate)*, PBT, (Figure 15b) which is used for making thermoplastic mouldings.

Like polythene, polyester processing continually improves. As a result PET has become the world's most rapidly growing textile fibre. Fibre manufacturers are now able to make very fine, microfibres which are used to make fashionable raincoats and sportswear.

There are lots of ways of forming ester bonds and the majority of ester forming reactions have been used to make polyesters of one kind or another. As was the case with polyamides there has been considerable interest in making higher-melting polymers for use in more demanding engineering applications. Table 2 shows how the structure of a repeating polyester motif influences the melting point of the polymer.

Repeat unit	Melting point
$-CO(CH_2)_4COO(CH_2)_2O-$	50 °C
$-CO-C_6H_4-COO(CH_2)_2O-$	265 °C
$-CO-C_6H_4-(CH_2)_4-C_6H_4-COO(CH_2)_2O-$	170 °C
$-CO-C_6H_4-CH{=}CH-C_6H_4-COO(CH_2)_2O-$	420 °C
$-CO-C_6H_4-COO-C_6H_4-O$ (meta-linked)	209 °C
$-CO-C_6H_4-COO-C_6H_4-O-$	~600 °C
$-CO-C_6H_4-O-$	~610 °C

Table 2. The structure of polyester repeat units and the melting point of the resulting polymers.

Another group, *polycarbonates,* are polyesters of carbonic acid and the structure of the best known example is shown in Figure 16. This semicrystalline polymer gives high clarity mouldings with good electrical properties and high impact strength, which have many applications. This material will be familiar to anybody who has entered a well-run chemistry laboratory since it is the material from which the majority of safety spectacles are produced.

$-[O-C_6H_4-C(CH_3)_2-C_6H_4-O-C(=O)]_n-$

Figure 16. The most common polycarbonate.

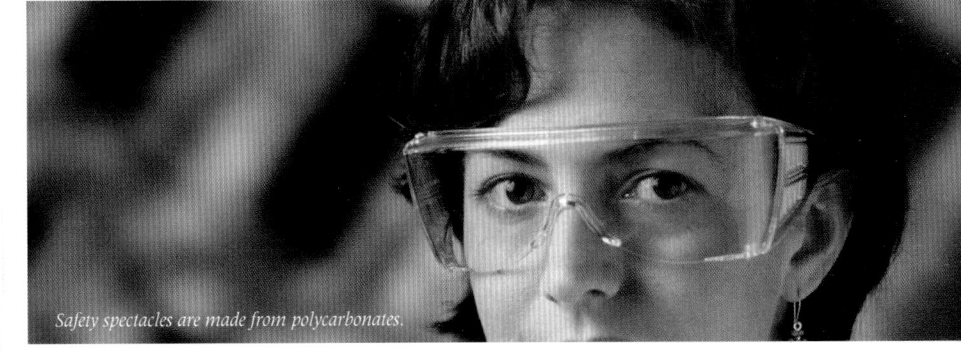

Safety spectacles are made from polycarbonates.

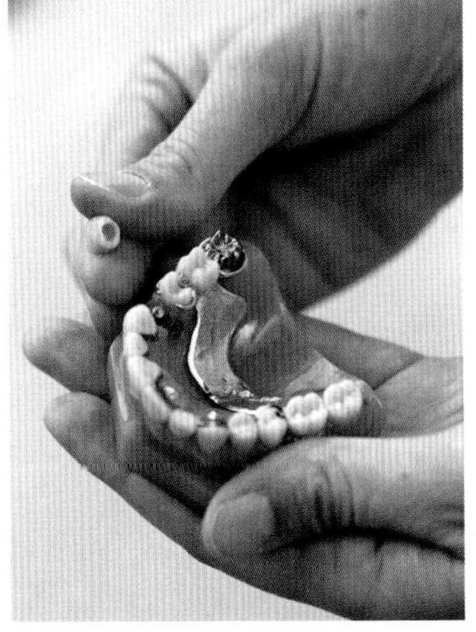

Contact lenses and dentistry are typical applications of polymers.

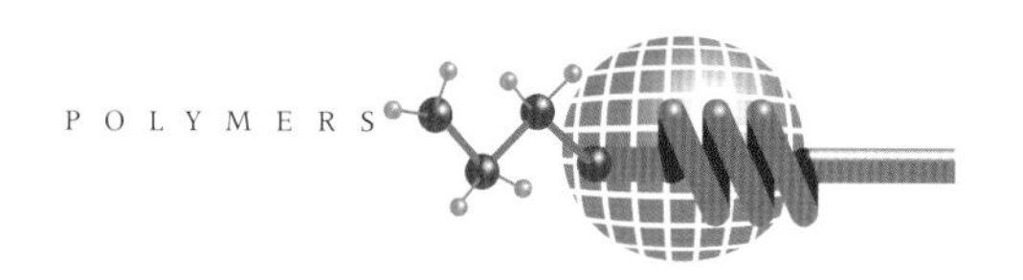

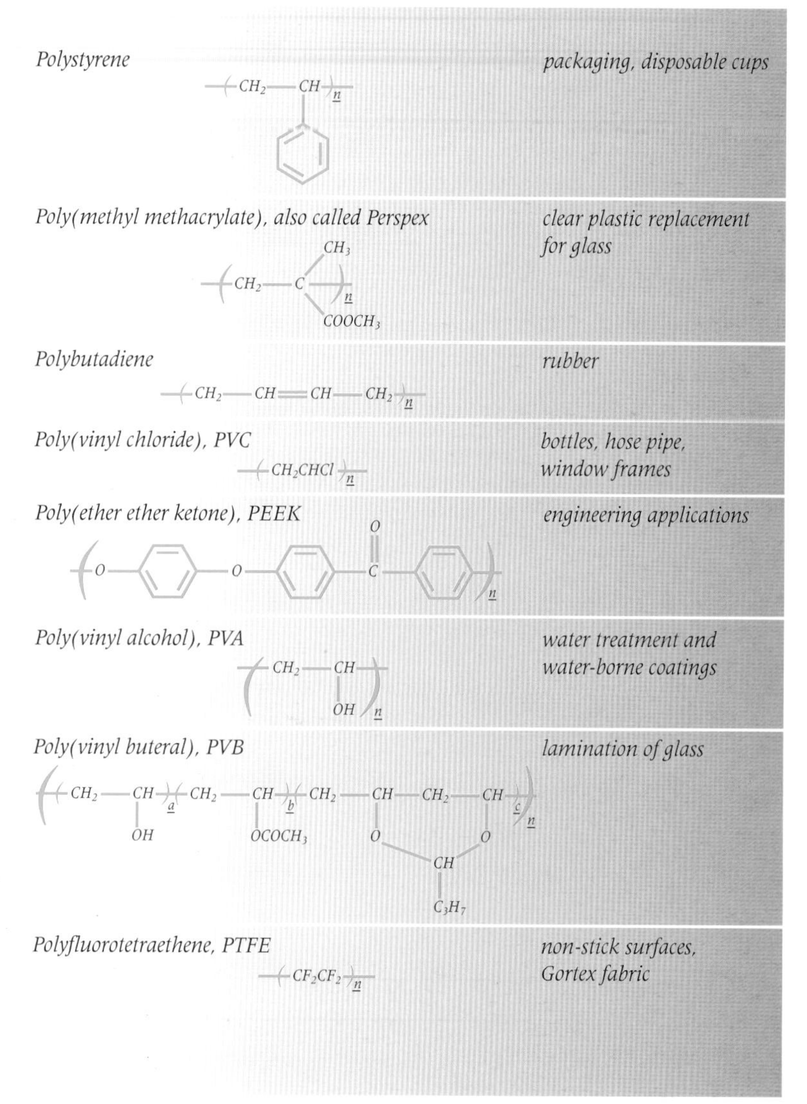

Polymer	Structure	Uses
Polystyrene	$-(CH_2-CH(C_6H_5))_n-$	*packaging, disposable cups*
Poly(methyl methacrylate), also called Perspex	$-(CH_2-C(CH_3)(COOCH_3))_n-$	*clear plastic replacement for glass*
Polybutadiene	$-(CH_2-CH=CH-CH_2)_n-$	*rubber*
Poly(vinyl chloride), PVC	$-(CH_2CHCl)_n-$	*bottles, hose pipe, window frames*
Poly(ether ether ketone), PEEK	$-(O-C_6H_4-O-C_6H_4-C(=O)-C_6H_4)_n-$	*engineering applications*
Poly(vinyl alcohol), PVA	$-(CH_2-CH(OH))_n-$	*water treatment and water-borne coatings*
Poly(vinyl buteral), PVB	$-((CH_2-CH(OH))_a(CH_2-CH(OCOCH_3))_b(CH_2-CH(O)-CH_2-CH(O))_c)_n-$, with the two O atoms joined by $CH(C_3H_7)$	*lamination of glass*
Polyfluorotetraethene, PTFE	$-(CF_2CF_2)_n-$	*non-stick surfaces, Gortex fabric*

Table 3. Some more examples of useful polymers.

The polymer industry is vast and produces a wide array of materials of practical value in our everyday lives. There isn't room here to describe them all – a few prominent ones are given in Table 3 – but there are some new polymers with unexpected properties that deserve special mention.

Plastic is widely used in the electrical and electronics industry.
Courtesy of DSM

Plastics that conduct electricity

All of the materials described so far are electrical insulators. But can a polymer be designed to be electrically conducting? In other words, can it behave as an organic metal? Curiously enough this question was being addressed by theoretical chemists at the same time that ICI's scientists were developing the high pressure-high temperature route to polyethene.

In the 1930s, chemists started developing theoretical descriptions of chemical bonding and reactivity based on the then fairly new quantum theory. They were interested in describing the behaviour of organic molecules like ethene which had 'unsaturated' double bonds with the potential to react

further, as in polymerisation to form polyethene. A German chemist Eric Hückel came up with a theoretical model called the pi molecular orbital theory which was based on the idea that one of the electron pairs of a carbon-carbon double bond was rather loosely bound. This pair is called the 'π' bond. The 'π' electrons were free to move over the whole molecule. This concept nicely explained not only the chemical properties of simple molecules like ethene but also those of other molecules with 'π' bonds, so-called *conjugated structures*, like benzene, a six-membered ring with six 'π' electrons. In benzene, the 'π' electrons can circulate around the whole ring, and are not localised between two atoms.

Taking this idea further you can imagine making a polymeric molecule consisting of long carbon chains of alternating double bonds – a polyene. The 'π' electrons should be able to move along the length of the chain. If so this polymer might be electrically conducting; like a metal. This is what John Lennard-Jones in Cambridge University suggested in 1937 when he used Hückel's theory to predict the 'π'-electron arrangement in an infinite *polyene*. Experimental verification of this interesting prediction had to wait another 20 years, until Natta tried to polymerise acetylene (ethyne) to make *polyacetylene* using a Ziegler catalyst.

$$nHC{\equiv}CH$$
Acetylene
$$\downarrow$$
$$(\text{-}CH{=}CH\text{-}CH{=}CH\text{-}CH{=}CH\text{-})_n$$
Polyacetylene

Sadly, the result of Natta's 1958 experiment was disappointing. The polymerisation worked but the polymer had none of the attractive properties of conventional polymers. It was an air-sensitive, black powder, and couldn't be processed. Natta showed that all the double bonds had a *trans* geometry, (with the hydrogens on opposing sides of each double bond), but then abandoned it (Figure 17).

The story then travelled first to the US where the synthesis of the powder was repeated by two industrial chemists, Donald Berets and Dorian Smith at the Enjoy Chemical Company in New York, who compressed the black powder into a pellet and measured its electrical conductivity. They found that this synthetic organic polymer was a semiconductor rather than a metal, but it certainly wasn't an electrical insulator like polyethene. Furthermore, these chemists showed that by exposing the polymer pellet to an oxidising vapour (iodine) its conductivity could be increased and by exposing it to dry ammonia gas its conductivity could be decreased.

The next step was the serendipitous discovery in 1970, in the laboratory of Hideki Shirakawa in Japan, of the polymerisation of acetylene at an interface between a solution of a Ziegler catalyst and acetylene gas to produce handleable films, and the study of these films by Shirakawa and Alan MacDiarmid and Alan Heeger in the US. There was then an enormous explosion of activity and interest which lasted for about 15 to 20 years.

Figure 17. Fragment of a trans-polyacetylene chain.

Smart cards and security labelling have benefitted from developments in the polymer industry.

Figure 18. Fragment of cis-polyacetylene chain.

It was established that if the material was made at low temperature (at about -70 °C), the films had a golden metallic sheen but were electrical insulators. The most carefully made samples had a conductivity comparable with quartz. These samples made at low temperature had all *cis* double bonds (where the hydrogen atoms are on the same side of the double bond) as shown in Figure 18. If the sample was allowed to warm up to room temperature two things happened spontaneously; first, the molecule switched from a *cis* to a *trans* geometry, and secondly, the number of mobile electrons in the material increased so that the material changed from an insulator to a semiconductor, the exact value of the conductivity depending on the chemical and thermal history of the sample. Exposing this organic semiconductor film to oxidants, either chemical or electrochemical, resulted in an increase in conductivity up to values similar to those of good conventional metallic conductors.

This was amazing and the physicists, engineers and journalists had a very exciting time for several years. In one apparently simple organic polymer we had a material whose properties could be varied at will from an electrical insulator, via a semiconductor into the metallic conductor regime. The conductivity could be varied by 19 powers of 10, that is, by a factor of 10 million million million. No wonder excitement became intense. This tremendous versatility meant that all sorts of devices were made as laboratory demonstrators during the 1980s. Storage batteries, transistors, electro-optic switches and high-speed image processing devices were produced in quick succession using this attractively simple organic polymer. There was a problem in processing this intractable air-sensitive material but this was overcome using a strategy of classical organic chemistry. However, polyacetylene was to remain the paradigm for the field not the practically useful material, at least not yet!

Other interesting conjugated materials have been developed such as polypyrroles and *polythiophenes*. Major

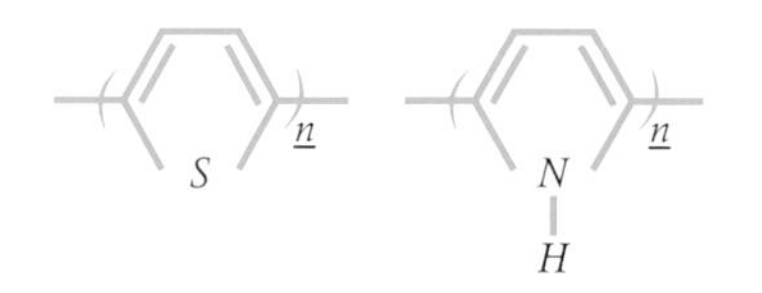

Figure 19. Structure of polythiophene (left) and polypyrrole (right).

electronics companies have now demonstrated that all organic electronic circuitry can be constructed and may, in the fullness of time, find its way into use in low-cost disposable consumer goods such as smart cards and security labelling. However, the work that started with polyacetylene has had many spin-offs and it is one of these which is the basis of our next story.

Courtesy of CDT.

Light-emitting polymers

One evening in a part of the Cavendish Laboratory in Cambridge, a student was switching off the lights after a day's work when he noticed that a part of his experimental set-up was glowing in the dark. Closer investigation revealed that the source of the light was a film of synthetic polymer that was being used as an insulating layer in an electrical device. The polymer was a simple hydrocarbon that had been known for some years and was being used because it could be conveniently processed to give reasonably air-stable, electrically insulating coatings. The material

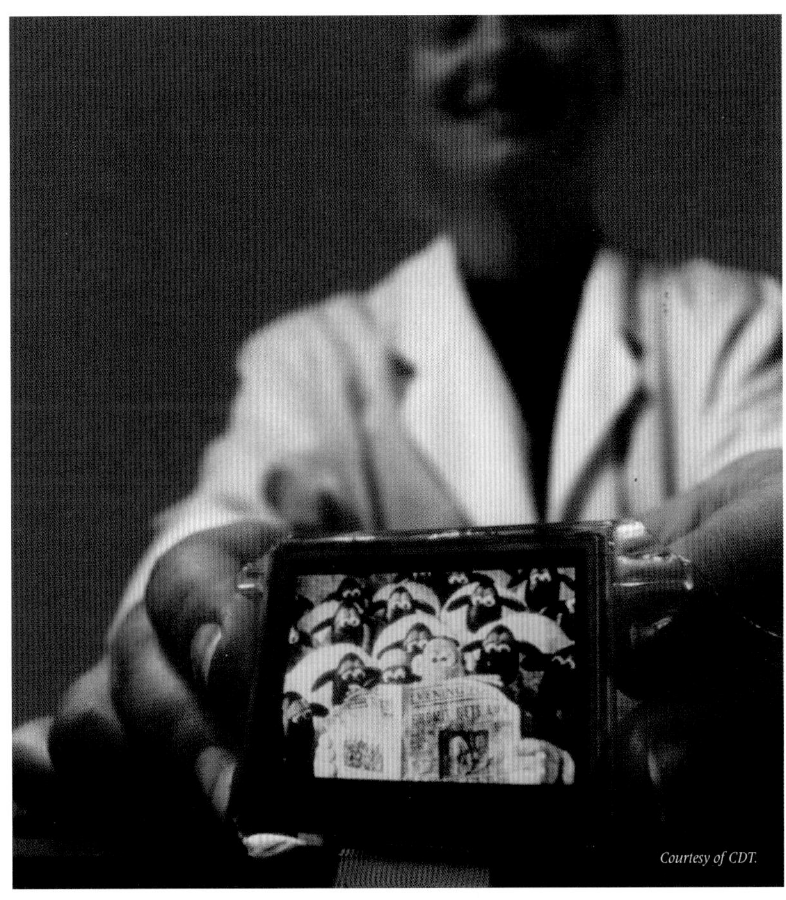

Courtesy of CDT.

was poly(*para*-phenylene vinylene), a segment of which is shown in Figure 20.

Figure 20. The electroluminescent poly(para-phenylene vinylene) is being used to make light emitting diodes.

Richard Friend's research group, to which the student belonged, soon established that thin films of this material showed the interesting phenomenon of *electroluminescence* – that is, it emits light when a voltage is applied across a film of a material. It turns out that many other materials display this phenomenon and by selecting the appropriate chemical structures it is possible to tune the colour of the light emission. A simple sandwich structure of electrically conductive glass, electroluminescent polymer film and metallic electrode constitutes a primitive light emitting diode (LED); these structures work at low voltage (6 to 12 volts) and have long lifetimes.

The implications are enormous. By patterning the electrodes, images can be produced, and fast switching then allows the images to move. In principle, we have a simple hydrocarbon polymer which can act as the active material at the heart of a display device and possibly even as the core of a flat-screen TV. Indeed, a monochrome screen operating at video rate has already been demonstrated.

Richard Friend.
Courtesy of CDT.

The future

How will polymer chemistry affect our lives in the 21st century? We have seen how polymers have developed over the past 50 years from simple polyethene to conducting, metal-like materials. In the future we can expect to see polymers with very novel properties and applications. Chemists and physicists have already demonstrated plastics in the research laboratory which, by applying a voltage, can be made to change shape, or alter optical characteristics such as colour and transparency, or mechanical properties. Plastics used to make semiconductor lasers, to control slow-release drug delivery, to store information like DNA, or behave as self-repairing or 'self-moulding' materials are also real possibilities. These exciting new active materials will surely have just as much an impact on our lives over the new century as the first synthetic polymers have had in the last.

Further reading

1. *Principles of Polymer Chemistry,* Paul J. Flory, Cornell University Press, Ithaca and London: 1953; this book remains in demand despite its age, my copy was from the 16th printing 1995.
2. *Polymers: the origins and growth of a science,* Herbert Morawetz, John Wiley & Sons, New York: 1985.
3. *The development of plastics,* S. T. I. Mossman and P. J. T. Morris (eds), Special Publication No. 141, The Royal Society of Chemistry, 1994.
4. *Comprehensive Polymer Science,* Geoffrey Allen and John C. Bevington (eds), in seven volumes with supplements, Pergamon, Oxford: 1989 onwards.
5. *Polymers: Chemistry and Physics of Modern Materials,* J. M. G. Cowie, Blackie, 2nd edn, Glasgow and London: 1991.

Glossary

Aluminium alkyl A compound of a metal and three alkyl groups (R_3Al); frequently used as a component of Ziegler-Natta catalysts (see Ziegler-Natta catalyst).

Atactic polymer A polymer containing asymmetric carbon atoms in which attached chemical groups are randomly arranged on either side of the carbon chain.

Conjugated structure Compounds in which the bonds are alternating multiple and single such that the π electronic orbitals associated with the double bonds are spread out across all the relevant bonds to give an extended π system.

Covalent bond A bond in which an electron pair is shared between two atoms to form a bond.

Electroluminescence The emission of light from a molecule when placed in an electric field.

Free radical An atom or molecule containing an unpaired electron. Free radicals are usually extremely reactive.

High density polyethene (HDPE) A structurally regular form of polyethene (polythene) with very few branches (less than 7 per 1000 carbon atoms). As a result of this regularity, the polymer chains pack efficiently to afford a highly crystalline material with a higher density than other forms of this polymer.

Isotactic polymer A polymer in which chemical groups attached to carbon atoms of the main polymer chain, when viewed three-dimensionally along the chain, are on the same sides of the chain.

Monomer A small molecule used as the basic building block for making polymers.

Nylon A generic term for the synthetic polyamides.

π-bond A bond formed by sharing a pair of electrons located in a molecular orbital formed by overlapping atomic orbitals with parallel axes.

Polyamide Polymers in which the basic monomer' units are connected through amide links (-CONH-).

Poly(butylene terephthalate) (PBT) A semi-crystalline engineering thermoplastic polyester.

Polycarbonates Polyesters of carbonic acid which are strong, semicrystalline, transparent polymers.

Polyimides A class of polymers which are stable to heat and oxidation. They are used in electrical insulation, in composites and as fibres.

Polymer A large molecule, usually with a chain-like structure (which may be branched), made from many smaller, similar chemical units called monomers. Some polymers may consist of more than one type of monomer.

Polyethene (polythene) A polymer made from ethene (C_2H_4) units.

Polypropene A polymer made from propene units (C_3H_6). Polypropene can exist in three geometrical (stereochemical) forms (see Atactic, Isotactic and Syndiotactic polymer).

Polyurethanes Polymers containing urethane groups, -(-NH-CO-O-)-, in the main polymer chain.

Rubber A generic term for polymers that are highly elastic. Natural rubber is a linear polymer made up of isoprene (C_5H_8) units.

Stereoregular polymer A polymer in which the chemical groups attached to the main chain are arranged in a three-dimensional configuration which is uniform all the way along the chain.

Syndiotactic polymer A polymer in which the chemical groups attached to the main chain, when observed three-dimensionally along the main backbone, are arranged in a regular alternating configuration.

Thermoplastic polymers Polymers that can be melted and shaped.

Ziegler catalyst A group of catalysts used to synthesise high-density polyethene (HDPE) and polypropenes. They are composed of different combinations of organometallic compounds containing metals such as aluminium with a transition metal (such as titanium or vanadium) halide.

Ziegler-Natta synthesis An important industrial route used to prepare polypropene polymers in which the stereochemistry (three-dimensional geometry) is controlled (see Stereoregular polymers).

Biographical details

Professor Jim Feast is Courtaulds Professor of Polymer Chemistry in the University of Durham and Director of the Leeds, Bradford, Durham Interdisciplinary Research Centre in Polymer Science and Technology. His research interests are primarily in the areas of novel polymer synthesis, particularly electroactive polymers and polymers with novel topologies.

Acknowledgements

If some measure of the interest and excitement in polymer science is transmitted to the readers of this chapter they and I owe a debt of gratitude to all the students, teachers and colleagues who have educated me in the field. Any errors or misunderstandings are entirely my responsibility.

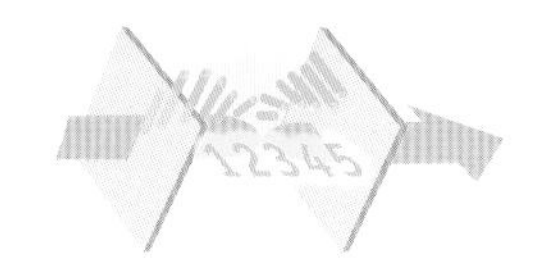

The world of liquid crystals

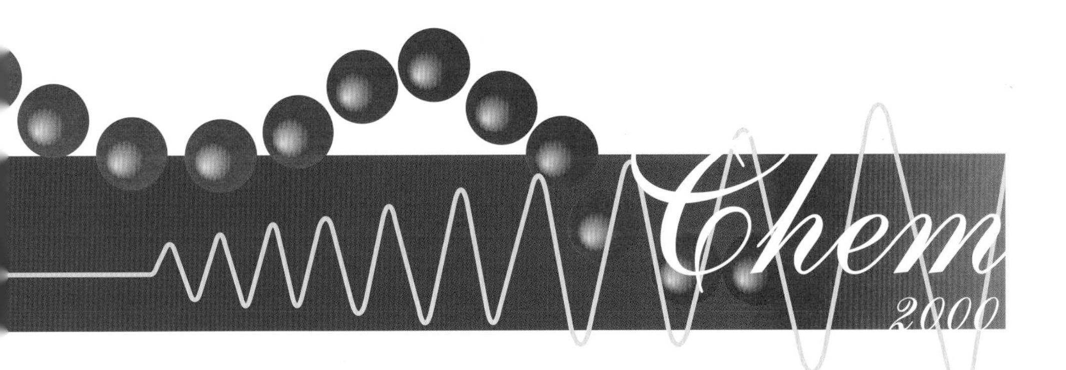

Liquid crystals are part of our everyday lives – in display devices and sensors as well as clothing and sports equipment. Applications of liquid crystals will continue to grow in the future.

Courtesy of DuPont

Dr Corrie Imrie
University of Aberdeen

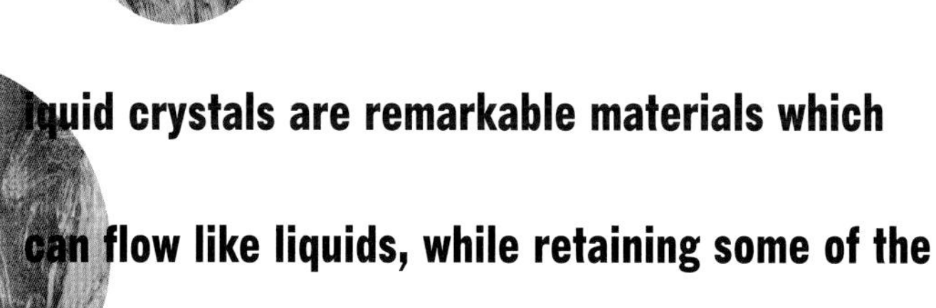

Liquid crystals are remarkable materials which can flow like liquids, while retaining some of the order found in crystals. They have made a huge impact on many areas of science and technology. The electro-optic properties of liquid crystals are widely exploited in display devices and sensors; many of the new high-strength plastics are liquid crystals; soaps and cosmetics rely on liquid crystal behaviour to work. What is even more significant, however, is that life itself could not have evolved without liquid crystals because cell membranes have a liquid crystalline structure.

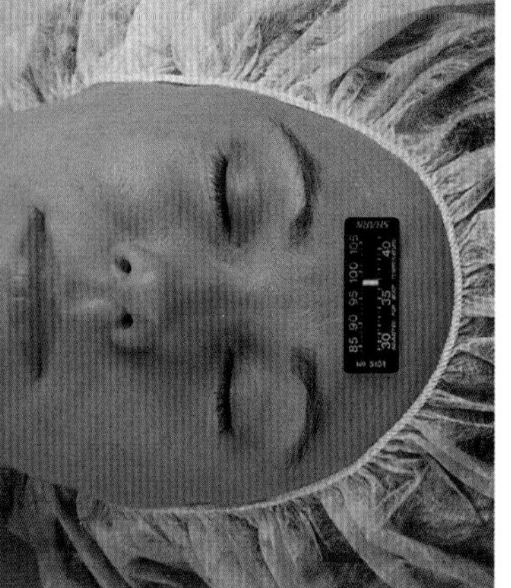

Courtesy of Sharn Inc.

Courtesy of Sharp Electronics (UK) Ltd.

As with so many scientific discoveries, liquid crystals were found by accident. Towards the end of the 19th century, an Austrian botanist, Friedrich Reinitzer, was interested in the role of cholesterol in plants, and so synthesised a large number of derivatives of cholesterol. To characterise each new compound, Reinitzer measured its melting point. However, when he tried to measure the melting point of cholesteryl benzoate Reinitzer was surprised to find that the sample seemed to have two melting points! At 145.5 °C, the crystals coalesced to form a cloudy fluid which suddenly cleared at 178.5 °C (Figure 1).

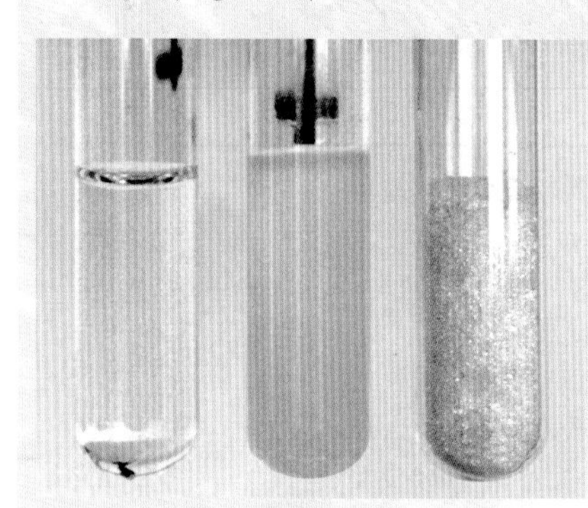

Figure 1. Crystals of cholesteryl benzoate do not melt directly to give a transparent liquid but instead pass through a cloudy 'in-between' or mesophase. The temperature increases from right to left.

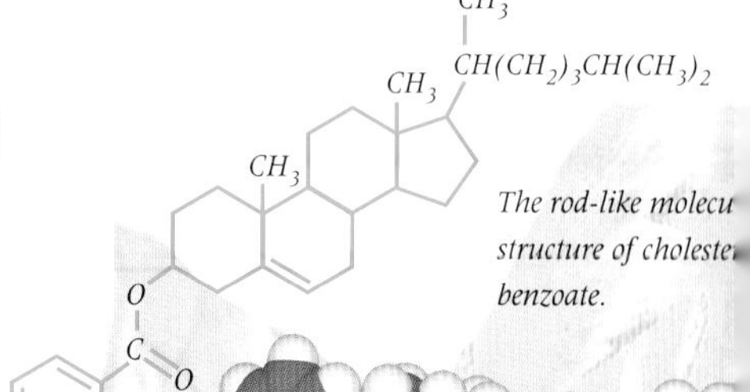

The rod-like molecu structure of choleste benzoate.

What could account for this bizarre behaviour? To find out, Reinitzer turned to a German physicist, Otto Lehmann, for help. Lehmann, whose own research involved studying the processes of melting and crystallisation, examined the cloudy 'in-between phase' – or *mesophase* as it is now called – with a polarised light microscope, a tool he had developed for studying crystals. The microscope incorporates two polarisers (Figure 2) which are devices allowing light waves in one plane only to pass through. In the microscope, the polarisers are 'crossed', in other words, arranged so that plane-polarised light selected by the first polariser cannot then pass through the second, the analyser. Most liquid samples inserted between crossed polarisers would be expected to look black because

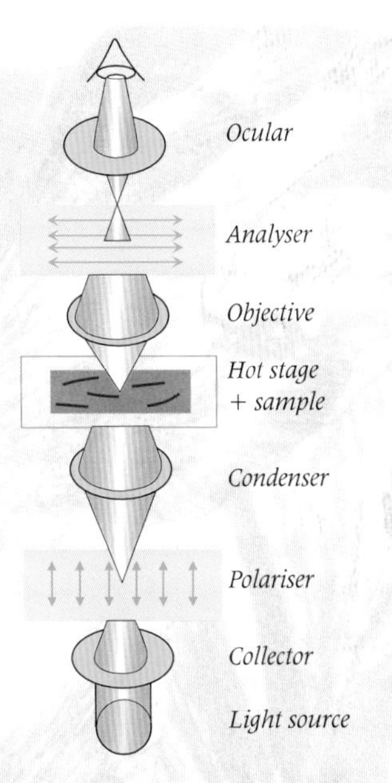

Figure 2. A schematic diagram of a polarised light microscope – the most convenient tool for identifying liquid-crystal phases.

no light would pass through the analyser, but when Lehmann looked through the microscope at Reinitzer's cloudy mesophase, he was astonished to see that it was not black but brightly coloured. The only explanation was that the sample had somehow rotated the plane of polarisation from the first polariser so that light could then squeeze through the analyser.

Lehmann already knew that certain crystals could rotate the plane of polarisation and so appear coloured under the polarising microscope. The reason is that they have different refractive indices in different directions, a property called *birefringence*. The fact that the cloudy phase of cholesteryl benzoate appeared birefringent was, therefore, most intriguing. It seemed to behave like a crystal, yet also flowed like a liquid. Lehmann concluded that "crystals exist with a softness, being so considerable that one could call them nearly liquid". He christened them *fliessende Kristalle* or liquid crystals, a name that has now stuck for more than a 100 years!

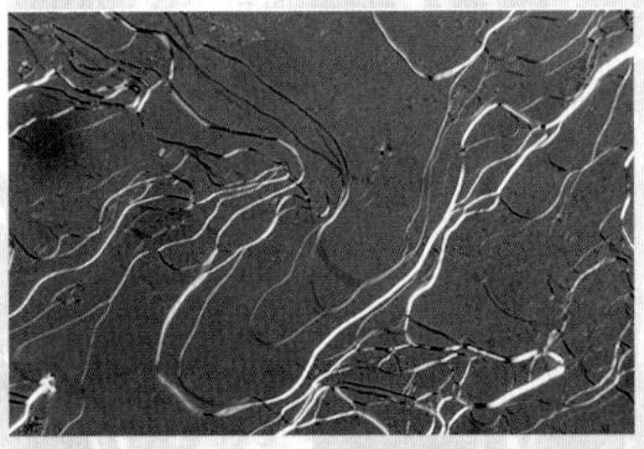

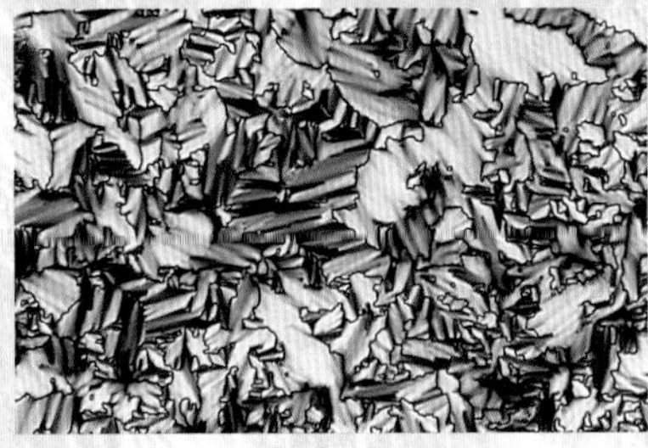

Examples of the cholesteric phase when viewed through a polarised light microscope.

The fourth state of matter

Today, we can explain the observations of Reinitzer and Lehmann at the molecular level. In a crystal, the molecules have an orderly three-dimensional arrangement. Refraction results when light interacts with electrons in the molecules causing them to oscillate. If the molecules have a structure such that they are all oriented in a certain direction, then the interaction with light – and thus the refractive index – varies depending on how easily the electrons can oscillate in a given direction in the molecule. This causes polarised light to rotate, and the crystal is said to be birefringent. When a crystal melts the birefringence is usually lost, as the thermal energy overcomes the electrostatic forces between the molecules holding them in place and the molecules become free to move randomly through the liquid.

It is easy, however, to imagine another melting process in which the molecules can move around, but as a result of residual electrostatic interactions they continue to point in a given direction – thus giving rise to the birefringent liquid with its beautiful colours under the polarising microscope. This is the so-called liquid crystalline phase in which the molecules have lost their positional order but retained their orientational order. Further heating eventually destroys the orientational order as well, giving rise to a second 'melting point' usually called the *clearing temperature*.

Liquid crystals do not, therefore, fit into the classical divisions of matter as solids, liquids and gases that we learnt at school. As a result, the liquid crystal phase can be thought of as the fourth state of matter.

Why liquid crystals form

We now know of many kinds of liquid crystals, but they all have certain chemical characteristics in common. Liquid crystal molecules are usually rod-shaped (although not always as we shall see) with an uneven distribution of electrons leading to

Box 1 Identifying a liquid crystal

The polarising microscope was central to the discovery of liquid crystals. More than a century later, it remains the most widely used technique for characterising liquid-crystal phases. Each phase shows characteristic patterns when viewed under the polarised light microscope. These patterns or *optical textures* consist of a coloured background arising from the birefringent nature of liquid crystals, superimposed on which are dark lines, brushes or points resulting from defects or disinclinations (where the orientation of the director changes abruptly) in the structure of the phase (see Figure). The coloured background indicates that the directors lie at some angle with respect to the axes of the polarisers. If the directors lie parallel or perpendicular to the polarisers' axes, then the field of view becomes black, see Figure (b). All the optical textures shown in the Figure contain adjacent bright and dark regions. At the interface between these, the orientation of the director must change abruptly producing the disinclinations. Point disinclinations are also observed; for example, in the nematic texture (a) the dark areas converge at points where the director is essentially undefined.

It is interesting to note that liquid crystals also have a significant theoretical value. Theorists use the complex patterns of defect structures emerging from liquid-crystal behaviour as physical models for understanding other kinds of phase changes in nature – for example, the phase changes in matter thought to have occurred in the very early Universe.

The optical textures observed for liquid-crystal phases viewed under the polarised light microscope: (a) nematic phase; (b) smectic A phase; (c) smectic C phase. Courtesy of Professor John Goodby

intermolecular forces that cause the molecules to align in a certain direction without being strong enough to hold them firmly in one place. There are two main types of liquid crystal phases. The simplest, but technologically the most important, is called the *nematic phase* in which the molecules are distributed randomly but, on average, all point in the same direction (Figure 3). This common direction is known as the *director*. In the other main type of mesophase, the molecules are arranged into layers and these are called *smectic phases*.

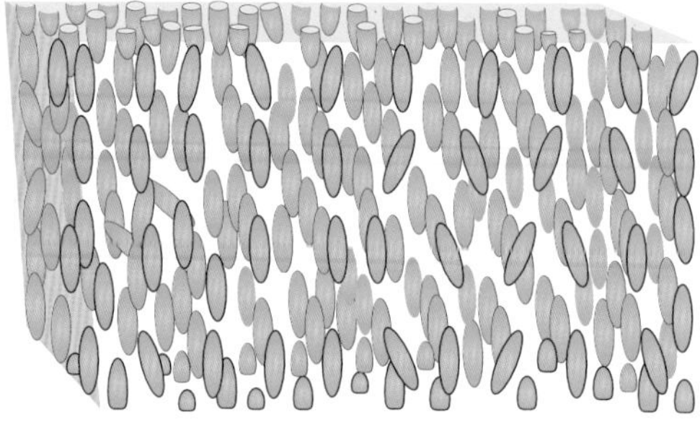

Figure 3. Molecules in the nematic phase tend to point in the same direction, known as the director, but their centres of mass are randomly distributed.

Once liquid crystals had been identified, the key question was how do they arise? Around 1900, a German synthetic organic chemist, Daniel Vorländer at the University of Halle, started a programme of research with the aim of establishing the relationships between liquid crystallinity and molecular structure. He published the fruits of his labours during the 1920s, and these key overviews underpinned much of the research that was to follow. Indeed, it is a great testament to these early studies that many of

Box 2 Smectic phases – layered liquid crystals

In smectic phases, the molecules are arranged into layers. The simplest smectic phase is the smectic A phase in which the molecules are arranged randomly within the layers and there are no positional correlations between the layers, see (a). Smectic phases in which the molecules tilt with respect to the layer planes are also common; the simplest of these is the smectic C phase which is the tilted analogue of the smectic A phase, see (b). There are many smectic phases with higher degrees of order; for example, in the smectic B phase, the molecules within a layer are arranged with hexatic symmetry.

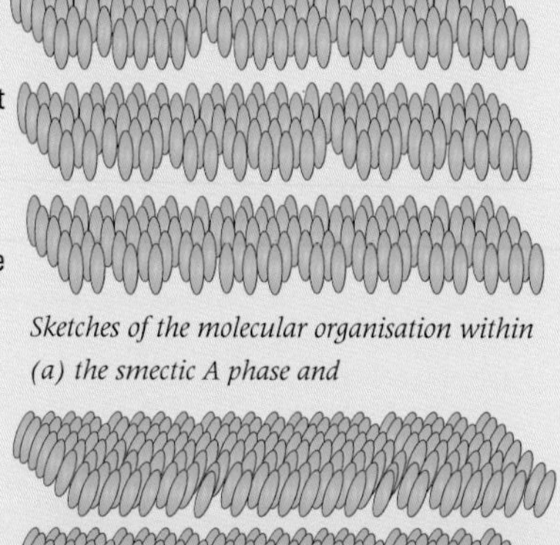

Sketches of the molecular organisation within (a) the smectic A phase and (b) the smectic C phase.

Vorländer's conclusions are still valid today.

Vorländer showed that most liquid crystals are composed of elongated molecules (Figure 4). It is easy to imagine why this is so by filling a box with pencils and shaking it. After shaking, the pencils will all lie in the same direction but their relative positions will be random – just like a liquid crystal nematic phase! The rod-like molecules are usually made by attaching one or two carbon chains to a rigid core (Figure 4). This core normally consists of flat electron-rich, aromatic rings, such as the benzene ring, separated by short linkages containing double bonds. In general, the more rod-like the molecule, then the higher its clearing temperature because the interactions between the molecules and their ability to pack together are enhanced. If we add groups of atoms onto the side of the molecules, then the clearing temperature usually falls because the rod-like shape of the molecules has been reduced.

4(b)

You might suppose that any molecule that is not spherically symmetrical would be a liquid crystal. Most compounds, however, are not. The reason is quite simple: if the temperature at which the compound melts is considerably higher than its clearing temperature then you are unlikely to see liquid crystallinity. In designing a new liquid crystal, therefore, we have to balance the forces between the molecules which give rise to three-dimensional crystals with those responsible for the formation of liquid crystal phases. This is by no means a trivial exercise because it is the strong interactions between the rigid cores that produce both kinds of molecular organisation.

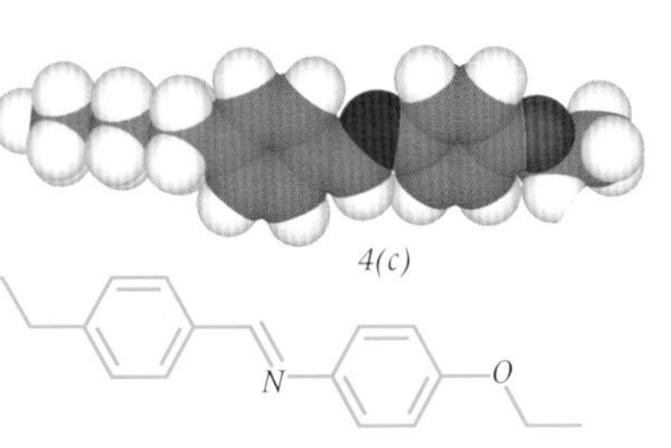

4(c)

We can control the melting points of these compounds to some extent by varying the lengths of the carbon chains attached to the core (Figure 4). Increasing the length of the carbon chains not only tends to reduce the melting point of the compounds by diluting the forces between the rigid cores, but also promotes smectic over nematic behaviour. In other words, molecules containing long carbon chains tend to arrange themselves into layers in which the chains form one region while the aromatic cores form another.

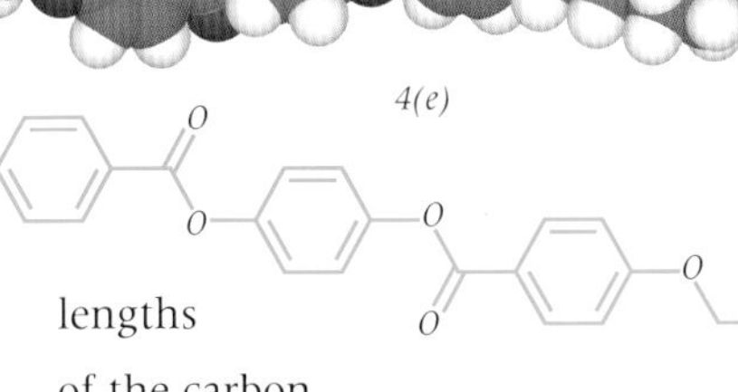

4(e)

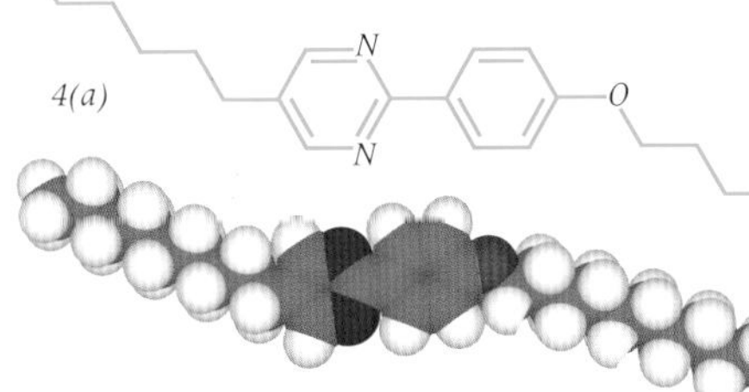

4(a)

Figures 4(a), (b), (c) and (d). Some typical examples of rod-like liquid-crystal molecules containing rigid cores and flexible carbon chains.

The birth of the billion-dollar liquid-crystal display industry

After the flourish of activity at the beginning of the century, the level of interest in liquid crystals declined markedly. By the 1960s, a standard university chemistry textbook was describing liquid crystals as "uncommon and of no practical importance". Yet today,

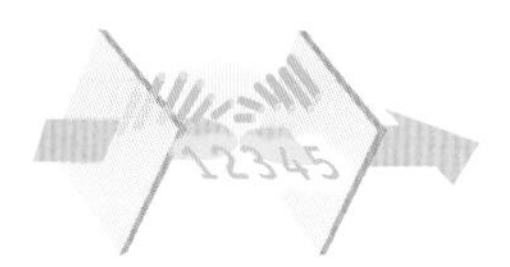

just 30 years later, liquid-crystal displays are a multi-billion-dollar industry. What led to this dramatic change in the fortunes of liquid crystals?

Throughout the 1920s and 1930s, researchers in several countries were studying the effects of electric and magnetic fields on liquid crystals. They already knew that electrostatic forces on a surface could align a liquid crystal and wanted to know whether an external electric field could achieve the same effect. The first patent for a liquid-crystal device was issued as early as 1936 to the Marconi Wireless Telegraph company (UK). These studies themselves did not result in a practical device, but they did lead to an important discovery that was later to underpin the development of liquid-crystal displays.

In 1926, Vsevolod Frédericksz, a Russian scientist, devised an experiment to study how a thin film of a liquid crystal sandwiched between two plates would respond to an applied magnetic field. Liquid crystals tend to lie parallel to glass surfaces because of surface forces. Frédericksz discovered that he could force the molecules to lie perpendicular to the surface by applying a magnetic field. This switch in the molecular alignment has become known as the Frédericksz transition, and the analogous effect in the presence of an electric field is at the very heart of liquid-crystal displays.

However, the fruits of Frédericksz's work did not ripen for 30 years. In the 1960s, the burgeoning technology of electronics was producing ever smaller and cheaper electronic devices, particularly for processing information. There was a snag, however: although these new microprocessors could work off lower voltages, the only device available for displaying images, the cathode ray tube, required high voltages. This bulky device was likened to "a stagecoach in an era of fast cars". The search therefore began in earnest for a new type of flat display that could operate at low voltages. From the field of possible contenders emerged liquid crystals.

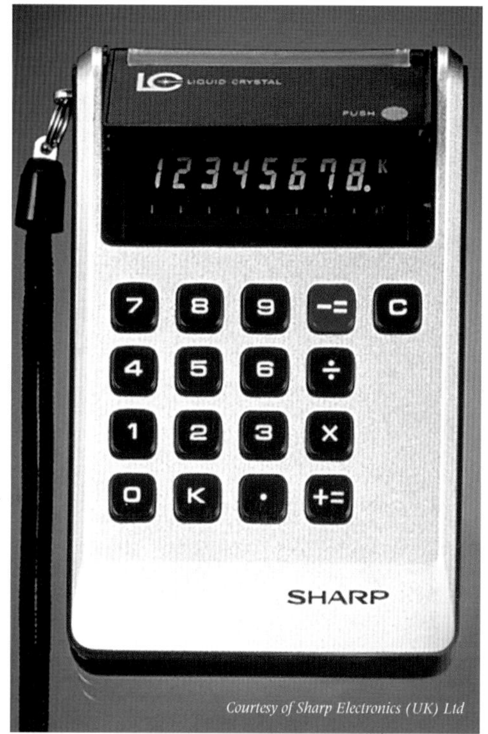

Courtesy of Sharp Electronics (UK) Ltd

The first generation of display device was the *dynamic scattering display*. It was invented in the late 1960s by George Heilmeier and his colleagues at the RCA Corporation in Princeton. In this device, a thin film of the liquid crystal 'doped' with ionic impurities (which occurred naturally in the earliest devices) was sandwiched between two glass slides each coated with indium tin oxide, a transparent conducting layer. Surface forces were used to align the liquid crystal so that in the 'off-state', the cell appeared transparent. When an electric field was applied across the display, the ionic impurities tried to drift across the cell but were opposed by the liquid crystal tending to align perpendicularly to the field. The net effect was that the alignment was destroyed so that the cell appeared opaque. The lifetimes of these devices were very short – normally just a matter of weeks – for the simple but annoying reason that the liquid crystals used were not stable and decomposed in the device. Nevertheless, this type of device was commercialised and in production for three to four years. In fact, the first LCD calculator had a dynamic scattering display, and some of these are still working today.

The dynamic scattering display had other inherent design problems: it was difficult to see

because it relied on a scattering (opaque) texture against a clear background so that in the wrong lighting it was almost invisible. In addition, the electrical current needed for the display to work shortened its operating life. What was needed was a display with higher contrast – that depended on light being transmitted (clear) or absorbed (black) – and which also did not rely on electrical conduction.

In 1970, European and American researchers came up with a better system that is still manufactured today. Martin Schadt and Wolfgang Helfrich at F. Hoffman La Roche in Basel and Jim Fergason working at the International Liquid Crystal Company in Kent, Ohio independently developed the *twisted nematic display* (see Box 3 for how it works). This device, however, required a liquid crystal with three important characteristics: a molecular structure that would allow it to align parallel to an electric field; a nematic phase over a working range of -10 °C to +60 °C; and a very high chemical stability. This combination of properties proved to be a demanding challenge and triggered a dramatic surge of interest in liquid crystals that still has not subsided.

Box 3 The twisted nematic liquid-crystal display

The vast majority of commercial liquid-crystal displays use the so-called twisted nematic effect. In a typical device, a thin layer of the liquid crystal is sandwiched between two plates of indium tin oxide (ITO) coated glass; the ITO layer serves as a transparent conducting layer. The electrodes are, in turn, coated with a polyimide which, if rubbed in a given direction, can align the liquid crystal along that direction.

Polariser

Glass

Liquid crystal molecules

Glass

Polariser

V

(a) Off (b) On

Schematic of the design and operation of the twisted nematic liquid-crystal display device.

The cell is constructed so that the two aligning layers have been rubbed in orthogonal directions inducing surface forces that cause the nematic to twist through 90 degrees across the cell. The cell containing the liquid crystal is then placed between crossed polarisers. Light entering the cell is, therefore, plane polarised, but the twisted nematic phase guides the light through 90 degrees allowing it to pass through the second polariser. This is the off-state of the device and appears bright.

Applying an electric field overcomes the aligning surface forces, and the liquid crystal now aligns parallel to the electric field. In this arrangement the liquid-crystal layer can no longer twist the light through 90 degrees, and it is blocked by the second polariser. This is the on-state and it appears dark. On removing the electric field, the surface forces restore the twist in the liquid-crystal layer.

These displays offer many advantages over competing technologies including low power consumption, compatibility with integrated circuits operating at low voltages and low power, flat design, reliability and fast response-times. They can also be used in displays designed to deal with a lot of information such as those in laptop computers.

George Gray and the cyanobiphenyls

It was research in the UK that eventually solved the problems. At around this time, a Ministry of Defence committee chaired by Cyril Hilsum awarded a contract to a chemist, George Gray, at the University of Hull to work on "Substances exhibiting liquid crystal states at room temperature" with a value not to exceed £2117 per annum. This contract preceded the discovery of the twisted nematic display and, as we shall see, it was this award that led directly to today's multi-billion dollar display industry! It is difficult to imagine that a better investment has ever been made. In fact, George Gray had earlier applied to the then Science Research Council for funding of this work. In the application he had included a description of display devices. The reviewing committee – presumably chemistry – decided

George Gray

he was a physicist and would not support the work unless he teamed up with a chemist!

Gray had been active in liquid crystal science since the early 1950s, and in 1962 had published the first book in English devoted to the subject. Indeed, the availability of this text has been cited as one of the factors that revitalised interest in liquid crystals as candidates for display devices. Until that point, Gray's research had been largely devoted to understanding the relationships between molecular structure and liquid crystallinity – fundamental work that remains highly relevant today. Thus, in 1970, he was in the unique position of being able to apply this knowledge to design the new materials required for display devices.

Although the race for new materials had begun in 1970, little progress was made in the following two years. The commercial liquid-crystal displays were based on the dynamic scattering effect and tended to be unreliable. Understandably, liquid-crystal display designers were depressed. But then came the breakthrough. In the latter half of 1972, Gray and his colleagues designed and synthesised two rod-like organic chemicals based on biphenyls – the 4-*n*-alkyl- and 4-*n*-alkyloxy-4′-cyanobiphenyls (Figure 5a and b). These molecules had three essential features: biphenyl (two benzene rings linked together) units which are chemically very stable, alkyl (hydrocarbon) chains which allowed some control over the melting points, and a cyanide unit which ensured the right response to an electric field.

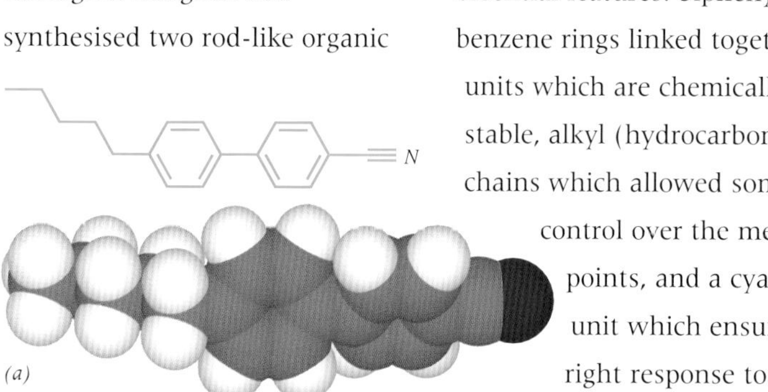

Figure 5. The liquid crystals that revolutionised the display industry: (a) the 4-n-alkyl-4′-cyanobiphenyls, (b) the 4-n-alkyloxy-4′-cyanobiphenyls and (c) 4-n-pentyl-4″-cyanoterphenyl.

These compounds proved to be stable and worked well in test displays, but no single compound provided the sought-after operating temperature range. Mixtures of these compounds showed wider temperature ranges, and one in particular worked from -3 °C to 52 °C but this was not good enough. There was an immediate demand for these new materials, and so a chemical firm, BDH in Poole, was chosen to produce them in larger quantities. In just a matter of months, BDH chemists, Ben Sturgeon and his team, were supplying the new compounds.

Ben Sturgeon, the chemist responsible for the rapid commercialisation of the cyanobiphenyls.

The team who developed these materials were awarded the 1980 Rank Prize for Opto-electronics: (left to right) John Kirton (DERA Malvern), George Gray (now at the Southampton Liquid Crystal Institute), Ken Harrison (now at DERA, Malvern), Cyril Hilsum (now at Unilever), Peter Raynes (now at Sharp Laboratories of Europe).

By 1973, a mixture had still to be found that could operate over the demanding temperature range. The formulation of the mixtures was a bit of a lottery so was very time-consuming. This was to change dramatically when Peter Raynes at the Royal Radar Establishment (later called the Royal Signals and Radar Establishment, and now the

Defence and Evaluation Research Agency) in Malvern, Worcester, developed a thermodynamic method for calculating the transition temperatures of the mixtures. His calculations revealed depressingly that no mixture of the biphenyl compounds existed that could operate over the required temperature range. There was still a missing ingredient. Again Gray's intimate knowledge of liquid-crystal chemistry came to the fore. He reasoned that to widen the temperature range of the mixtures, a component with a high clearing temperature had to be added. With his team, he synthesised 4-*n*-pentyl-4''-cyanoterphenyl (Figure 5c). Gray's intuition proved to be correct: a mixture of this compound with the biphenyl-based materials was found with an operating range -9 °C to 59 °C. This mixture was christened E7 and its availability allowed the twisted nematic display to be commercialised. It is no longer found in these devices but other mixtures containing cyanobiphenyls are still used.

The RRE, Hull and BDH consortium went on to make many significant contributions in the applications of liquid crystals. Gray's personal contribution to liquid crystal science has been acknowledged by numerous scientific prizes but most recently by the award of the Kyoto Prize, Japan's highest scientific honour.

Recently, a new type of nematic display device has been developed by Jim Fergason of Optical Shields, California, and Bill Doane at Kent State University, Ohio – the polymer dispersed liquid crystal (PDLC) display. It works slightly differently and has some potentially exciting applications. The device consists of small droplets of a liquid crystal embedded in a matrix of transparent polymer. This is created by dissolving the liquid crystal in precursor monomer molecules which are then polymerised. As the polymer's molecular weight starts to increase, the liquid crystal separates into droplets. The directors of these droplets are randomly aligned, and as light passes through the PDLC film, the mismatch in refractive index between the droplets and polymer matrix causes the light to be scattered. Thus the film appears opaque. This is the device's off-state (Figure 6a). When an electric field is applied across the film, each droplet's director aligns. Now, if the refractive index of the liquid crystal measured along the director matches that of the polymer matrix, then light is no longer scattered on passing through the film. The electric field, therefore, switches the film from being opaque to transparent.

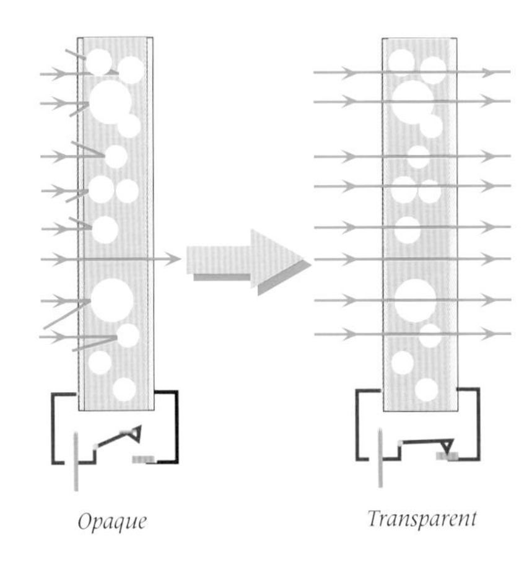

Figure 6. A sketch of a polymer-dispersed liquid crystal (PDLC) display device.

A PDLC-based secrecy window, top transparent, below opaque.

The proposed applications of PDLCs range from projection and large-area displays to secrecy windows which may be switched electrically from opaque to transparent.

Discotic liquid crystal materials and molecular wires

The rod-like nematic and smectic structures are not the only materials to show liquid crystallinity. In the 1920s, Vorländer proposed that 'flake-like' molecules ought to be able to pack face-to-face into piles or columns to give liquid crystal phases. He made many compounds to test this suggestion but failed to find any supporting evidence.

It was not until 1977 that Vorländer's proposal was verified, when Sunil Chandrasekhar and his colleagues at the Raman Research Institute in Bangalore showed that the disc-like molecules of the hexa-alkanoyloxybenzenes (Figure 7a) are liquid crystals. Since then, a great many *discotic liquid crystals* have been discovered. They are usually composed of molecules with planar, aromatic cores to which are attached six or eight carbon chains.

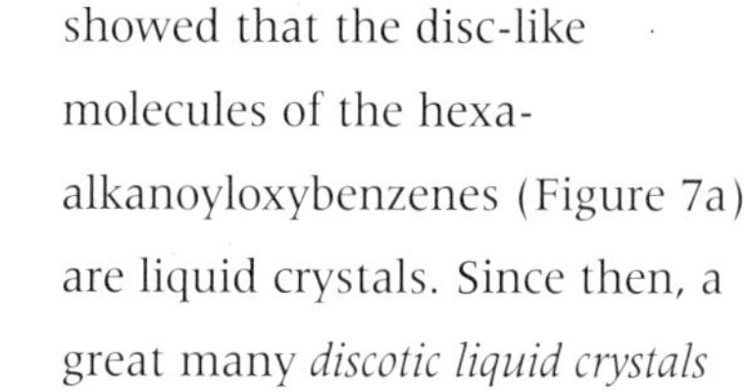

George Gray (left), the 1995 Kyoto Prize Laureate for his outstanding achievements in liquid crystal science, pictured with Sunil Chandrasekhar (right) the discoverer of discotic liquid crystals.

Figure 7(b). A schematic of a nematic phase composed of disc-like molecules.

The simplest discotic phase is the discotic nematic phase in which the short axes of the discs are arranged along the director (Figure 7b). This phase is easy to visualise if you consider throwing a handful of coins into a box. The coins all lie flat in the box – a discotic nematic phase.

Figure 7(c). Sketches of the molecular organisation within columnar liquid-crystal phases.

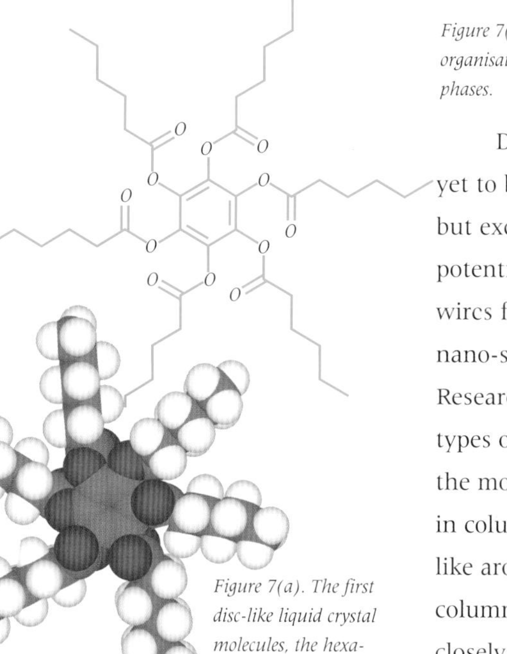

Figure 7(a). The first disc-like liquid crystal molecules, the hexa-alkanoyloxybenzenes.

Discotic liquid crystals have yet to be exploited commercially but excitement surrounds their potential for use as molecular wires for transporting electrons in nano-scale electronic devices. Researchers have found several types of discotic phases in which the molecules are now arranged in columns (Figure 7c). The disc-like aromatic cores within the columnar phases may be so closely stacked together –

separated by only about 0.35 nanometres – that the electron orbitals of adjacent aromatic rings actually overlap. This means that electrons can pass more easily up and down the columns than across them. One possible application for these molecular wires is as environmental gas sensors, or electronic 'noses', because their conductivity depends strongly on the distance between the molecules – which changes when gas molecules are absorbed.

Liquid crystals through the looking-glass

Although the first liquid crystal discovered was cholesteryl benzoate, its mesophase has turned out to be more complex than any of those described so far. Cholesteryl benzoate is, in fact, an optically active, or chiral, compound it can exist in two forms that are mirror images (see *Synthesis* chapter), and is now known to exhibit the cholesteric or *chiral nematic phase*. At the molecular level, the cholesteric phase is indistinguishable from the nematic phase but the molecular chirality induces a helical arrangement of the directors. Thus, as you move through the cholesteric phase, the director traces out a helix with a pitch length typically between 300 and 800 nanometres (Figure 8).

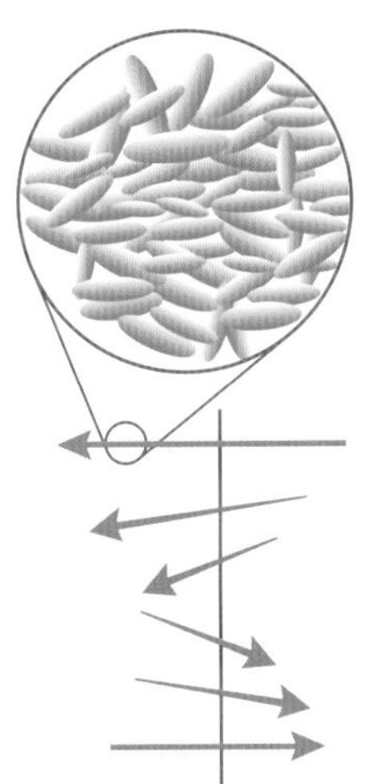

Figure 8. Sketch of the molecular organisation found within the chiral nematic or cholesteric phase. The directors map out a helical structure but on the molecular level the ordering is essentially identical to that found in the nematic phase (see inset).

These materials have found widespread use in temperature sensors (see Box 4).

Recently, Bill Doane and his colleagues at Kent State University in Ohio have invented a new use for cholesteric liquid crystals in displays (Figure 9). The cholesteric phase shows two optical textures (the characteristic patterns seen under a polarising microscope) one of which strongly reflects light while the other does not. The device is essentially the same as the dynamic scattering

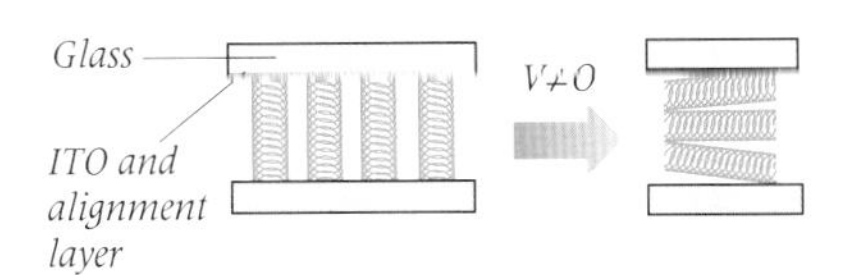

Figure 9. Schematic representation of the cholesteric liquid-crystal display device.

Box 4 Colourful cholesterics in nature and technology

If the pitch of a helical, cholesteric liquid crystal is comparable to the wavelength of visible light, then it reflects just a particular wavelength, or colour. The reflected colour depends on the pitch length, which in turn varies with temperature. On cooling, the pitch length usually increases and longer wavelengths of light are reflected; the accompanying colour change is from blue (hot) to red (cold).

This change is used in temperature-sensing applications. The cholesteric compound is encapsulated in a polymer matrix and sandwiched between a transparent surface and a black support, which absorbs any extraneous light. These heat sensitive materials are used as children's thermometers, as coatings for aircraft to monitor hotspots, and in fabrics and fashion accessories. More recently, it has even been suggested that cat's eyes in roads should be coated in a cholesteric coating to warn drivers of icy conditions. The coating would be chosen so that the cat's eyes would reflect a distinctive colour if the temperature fell below zero.

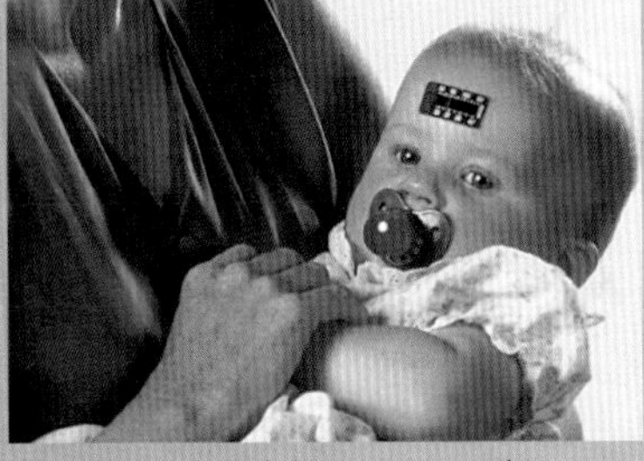

(a) A liquid-crystal thermometer used to monitor the temperature of a patient before, during and after surgery. The thermometer is a non-invasive device and more cost-effective than electronic temperature probes.

Courtesy of Sharn Inc.

Intriguingly, cholesteric phases are also found in nature. For example, the cuticles of Scarabaeid beetles selectively reflect light. The major component of the cuticle is chitin which is a biopolymer aggregate with a rod-like shape. This shows a lyotropic cholesteric phase which is thought to congeal to give the solid cuticle. Thus the structure of the cuticle and its optical properties, including the selective reflection of light, are inherited from those of the cholesteric precursor.

An optically active Scarabaeid (Lomaptera species) beetle. they reflect left-hand circularly polarised light but appear black in right-hand circularly polarised light.

display except that it has a black, light absorbing back. Therefore, regions that reflect light appear bright while those that do not appear dark. The two textures are stable but can be inter-converted using an electric field. This display gives excellent contrast, and because both textures are stable the information can be stored for reading at some later date. These cholesteric liquid-crystal displays will play an important role in the development of erasable electronic newspapers.

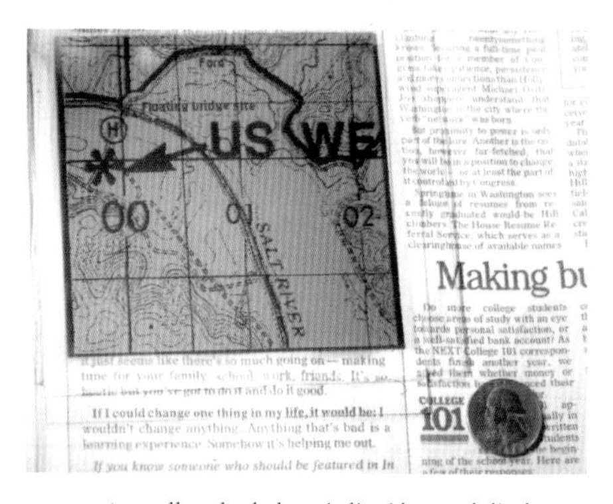

A small-scale cholesteric liquid-crystal display image showing excellent contrast.
Courtesy of J W Doane and B Taheri.

A flat screen TV. Courtesy of Sharp Electronics (UK) Ltd.

Liquid crystals for flat-screen TVs

Chirality does not only result in helical nematic phases but also 'tilted' helical smectic phases (see Box 2). For example, if a *smectic C phase* consists of chiral molecules then the direction in which the molecules tilt traces out a helix on passing in a direction perpendicular to the layers of the phase; this is referred to as the chiral *smectic C* phase.*

In 1974, Bob Meyer and his colleagues at the University of Paris South first predicted and then showed that the chiral smectic C* phase is *ferroelectric* – in other words, it has a permanent dipole moment in the absence of an external electric field. This is a result of the molecular dipoles within a layer tending to line up in the same direction, giving the layer a net dipole moment. As you move through the phase, however, the helical arrangement means that each layer is rotated with respect to its neighbours. The effect is to cancel out the dipole moments, so the ferroelectricity of the phase is reduced to zero. This phase is termed *heli-electric.*

Almost a decade after Meyer's discovery, Noel Clark and Sven Lagerwall at the University of Colorado in Boulder showed that the

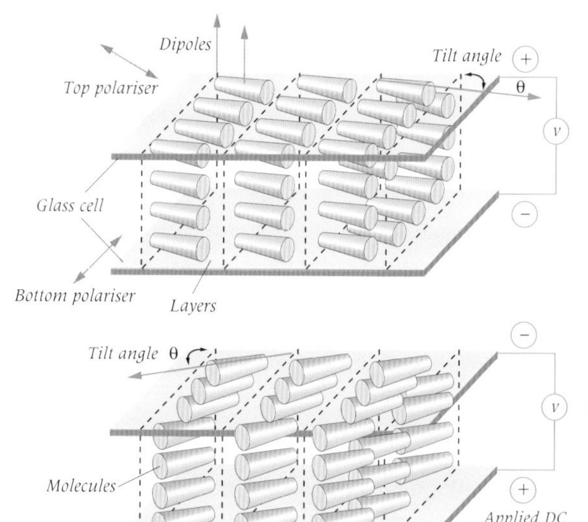

Figure 10. Sketch of the ferroelectric smectic C liquid-crystal display.*

ferroelectric nature of the smectic C* phase could be used to design displays that can switch very quickly – in microseconds as opposed to milliseconds for twisted nematics. The device is similar to the twisted nematic display but the surface forces are now used to unwind the helix (Figure 10). Thus the dipole moments of each layer no longer cancel each other and an overall dipole moment is created across the device. Applying an electric field causes the dipoles to flip, which reverses the average molecular tilt.

More recently, however, Atsuo Fukuda and his colleagues at the Tokyo Institute of Technology discovered a smectic C* phase that is antiferroelectric. In this phase, the tilt direction alternates between layers. Its use in display devices offers several advantages over the conventional ferroelectric smectic C* phase. In particular, the electric fields required for switching are better defined and so the displays are easier to 'multiplex'. Multiplexing is a method of increasing the information content of a display by having rows of electrodes on one side of the display and columns on the other. To select an individual pixel a voltage has to be applied to the appropriate column and row. This greatly reduces the number of electrical connections that have to be made compared with making connections to each and every individual pixel.

Despite these breakthroughs, research into new liquid crystal materials and devices is still a very active area. It has been estimated that in 1997 the liquid crystal display market was worth in the region of $7 billion and account for about one in every five electronic displays. This, however, is a much smaller market share than the 70 per cent held by the cathode ray tubes of conventional televisions and computer monitors. It is these markets that the manufacturers of liquid-crystal displays hope to expand into. Already laptop computers use liquid-crystal displays that are based on twisted nematic technologies, but there are technical difficulties in producing screens larger than 12 inches. Ferroelectric technology also provides a route towards colour liquid-crystal televisions but again there are problems to overcome. Nonetheless, we anticipate that flat colour-television screens suitable for mounting on a wall will become available towards the end of the 1990s – less than 30 years after George Gray's breakthrough.

Liquid-crystal polymers – the new wonder materials

As well as revolutionising electronic displays, liquid crystals have also stimulated another vitally important area of technology – the development of novel materials. Extremely hard polymers such as Kevlar®, used in bullet-proof vests and in aerospace

Courtesy of DuPont.

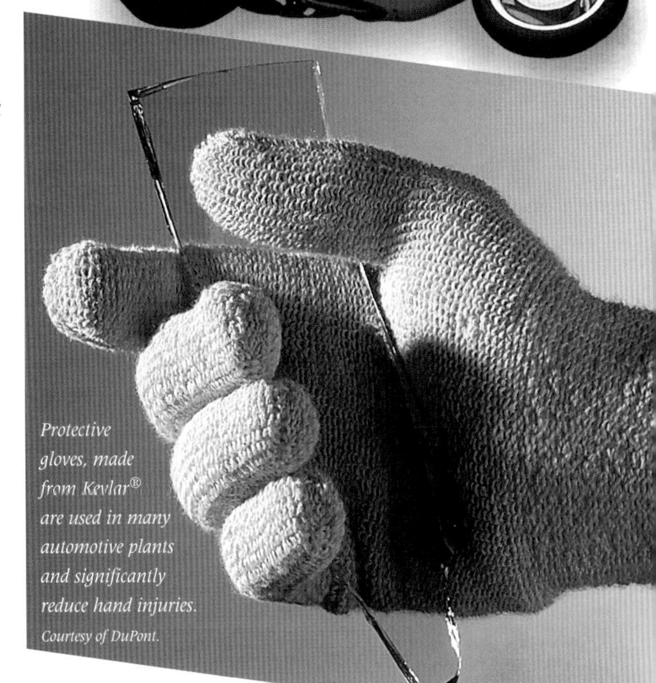

Protective gloves, made from Kevlar® are used in many automotive plants and significantly reduce hand injuries.
Courtesy of DuPont.

components, are in fact liquid crystals. They are now being combined with other materials to make advanced composites with a huge range of applications, from heavy engineering to sports equipment (see *The age of plastics*).

Police vests and military fragmentation vests made from Kevlar®. Courtesy of DuPont.

Again, the concept of liquid crystal polymers goes back several decades to the 1920s, when Vorländer posed the question: "What would happen to the molecules (liquid crystals) when they become longer and longer?" What he found was that short chains composed of repeating segments of hydroxybenzoic acid were indeed liquid crystals. However, as the chains got longer, the resulting polymer soon became intractable, decomposing before melting. Later, in the 1930s, researchers did discover some natural liquid crystal polymers: solutions of the rod-like tobacco mosaic virus particles which were birefringent above a critical concentration; and alkylcelluloses, which had 'softening points' and are now known to be liquid crystalline in nature.

Nothing further happened, however, until the 1950s, when the molecular theories of physical chemists, Lars Onsager and Nobel laureate, Paul Flory, predicted that rod-like molecules would spontaneously order above a certain concentration that depended on the ratio of their lengths to their breadths. Logs floating down a river make a good analogy here they tend to align in the same direction to form the equivalent of a liquid-crystal phase. These materials are referred to as *lyotropic liquid crystals* because their liquid-crystal behaviour depends upon the concentration in solution, as opposed to temperature for so-called *thermotropic liquid crystals*.

Box 5 Soaps and detergents

It turns out that the most common examples of lyotropic liquid crystals are soaps and detergents in water; in fact, the slime you find in your soap dish is a liquid-crystal phase and every time you wash your hands or do the dishes, liquid-crystalline phases form in the dirty water. Indeed, it could be argued that the first liquid crystals to find application were the soaps used thousands of years ago by, amongst others, the Phoenicians.

These materials are composed of amphiphiles – molecules containing two distinct regions, one polar and one non-polar. They are often likened to tadpoles in which the head is strongly water-loving, or hydrophilic, while the tail, normally a carbon chain, hates being in contact with water and so is hydrophobic. Above some critical concentration, these molecules form liquid crystal phases in which only the hydrophilic parts of the molecules are exposed to water so shielding the hydrophobic segments.

The molecular organisation found within some lyotropic liquid crystal phases.

Surfactant
Water
Water
Surfactant

The self-assembly of logs floating down a river: a good analogy for the liquid crystalline behaviour of rigid, rod-like polymers in solution.

The predictions of Onsager and Flory were proved correct almost immediately: researchers at the UK chemical giant, Courtaulds in Maidenhead, discovered that concentrated solutions of the polymers, poly(γ-methylglutamate) (Figure 11a) and poly(γ-benzyl-glutamate), in solvents such as dioxane and methylene chloride were liquid crystalline.

(a)

(b)

(c)

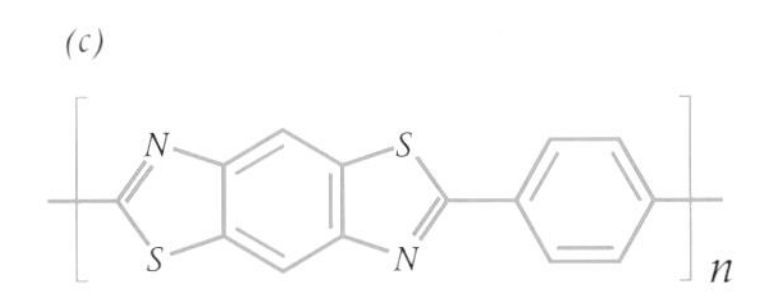

Figure 11. The chemical structures of some polymers which show liquid crystalline behaviour in solution. (a) Poly(γ-benzyl glutamate), the first polymer for which liquid crystallinity was identified. (b) Poly(p-phenylene terephthalimide), better known as either Kevlar® (made by DuPont) or Twaron (made by Akzo). (c) PBZT, a polymer with possible applications in the aerospace industry.

Box 6 Kevlar and spider webs

Kevlar® fibres have exceptional mechanical properties; for example, on a weight-for-weight basis they are 10 times stronger than steel and are used in many diverse areas including anti-ballistic garments, high-strength textiles for sails, rubber reinforcement for radial tyres, body panels for cars and ropes,. The properties of Kevlar® fibres, and more generally polymers processed from liquid-crystalline solutions, fall short of those predicted theoretically. This is partly because defects become trapped in the structure during processing. In order to resolve these difficulties research now focuses on spiders' webs!

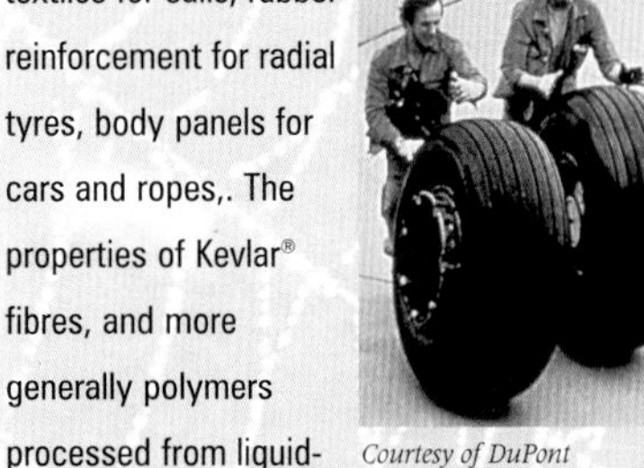

Courtesy of DuPont

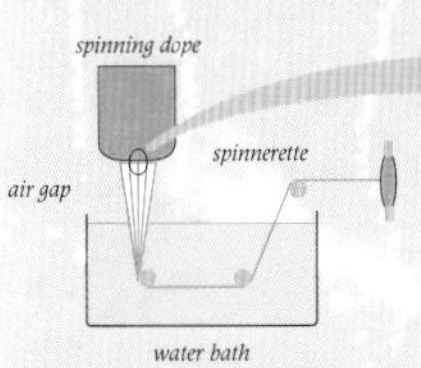

Figure 12 (a)Processing Kevlar® solutions into fibres

Solutions of Kevlar® in concentrated sulphuric acid are liquid crystalline. They consist of domains whose directors are randomly arranged. To obtain a useful fibre from such a solution the liquid-crystal domains must be aligned along the fibre axis. This is achieved using the process shown in Figure (a). As the solution is pushed through a spinneret, the domains are aligned by the shear force and become oriented in the flow direction. On emerging from the spinneret, however, the fibre swells reducing the degree of orientation of the molecules. To recover the orientation and enhance it, the fibre is then stretched. This high degree of orientation is then captured by coagulating the fibre in water.

Spiders produce a dragline silk which they use for the radial threads of webs and also as a safety rope in case of unexpected falls. The combined mechanical properties of this silk outperforms those of any synthetic fibre. For example, Kevlar is slightly more strong but is considerably less resilient. Furthermore, these silks have the highly desirable property of being biodegradable and the spider can recycle them.

People have suggested that liquid crystallinity plays a central role in the processing of the silk (Figure b). The water-soluble silk molecules, fibroin, have a globular structure and are stored by the spider as an aqueous solution in the gland. In the duct leading to the spinneret, these globular molecules aggregate to form rod-like units which assemble into a liquid-crystal phase. Its low viscosity allows the solution to be spun at such a rate that the molecules change into a shape that can pack into a crystalline structure which is actually insoluble in water. It is the combination of interconnected crystalline and amorphous regions in the silk that give the fibres their exceptional properties. Nature has therefore developed a room-temperature process for converting water-soluble single molecules into a high-performance insoluble fibre without the need for chemical change!

(b) A golden orb weaver spider and how it makes its silk.

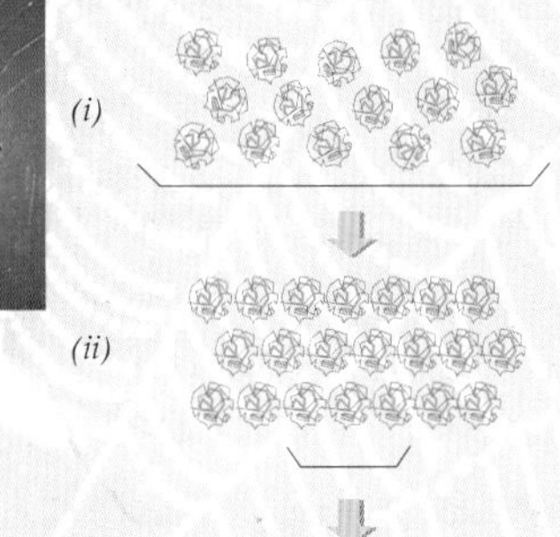

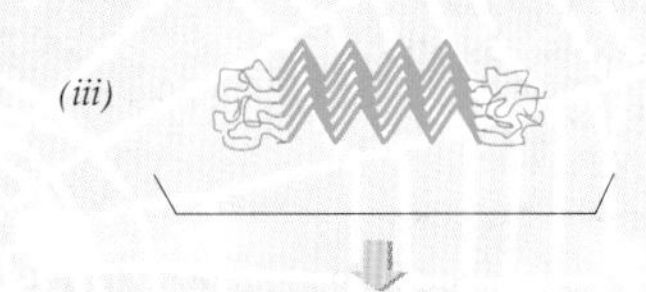

Courtesy of DuPont

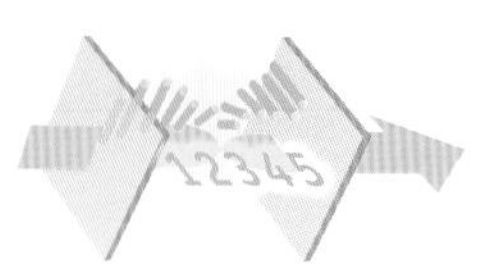

The dissolved polymers adopt an extended helical structure, and so behave like long rigid rods that can pack together efficiently with their long axes parallel.

The real impetus for the development of liquid-crystal polymers started in the early 1970s when Stephanie Kwolek and her colleagues at DuPont in Wilmington, Delaware, found they could make fibres with exceptionally high tensile strength by processing solutions of rod-like polymers identified only later as liquid crystals. In fact, they originally thought that the cloudiness of the solutions was a result of impurities; the spinning unit refused to process the solutions because the impurity particles would plug the holes in the spinneret! These solutions contained relatively rigid aromatic polyamides, or aramids. The first product, Fibre B, was based on poly(benzamide) but this was quickly superseded by poly(*p*-phenylene terephthalamide) (Figure 11b), now known commercially as Kevlar®, which the Dutch company, Akzo Nobel, also markets under the trade name Twaron. Kwolek's central role in the development of liquid crystal polymers was recognised recently when the American Section of the Society of Chemical Industry awarded her the 1997 Perkin Medal. Other industrially relevant aramids include poly(*p*-phenylene benzobis-thiazole), or PBZT (Figure 11c), and its oxazole analogue, PBO, which because of their high strength and excellent thermal stability may be used as structural materials in 'military stealth' aircraft.

Stephanie Kwolek. Courtesy of the Society of Chemical Industry.

Aramids still dominate the industrially important liquid-crystal polymer sector. They are, however, difficult to process and soluble only in strong solvents such as fuming sulphuric acid. These problems have driven research towards thermotropic polymers – materials that could be melt-processed. The first examples were developed by three research groups in the mid-1970s: Antonio Roviello and Augusto Sirigu at the University of Naples, Jerry Jackson and Herbert Kuhfuss at the Tennessee Eastman Co, and a team at DuPont which filed a patent.

Thermotropic main-chain polymers can be divided into two broad classes: rigid and semi-flexible. Industrial interest has tended to focus on the so-called rigid main-chain polymers because of their excellent mechanical properties. This led in the 1980s to several thermoplastic aromatic copolyesters being commercialised, including Vectra made by Hoechst Celanese, and Xydar made by Amoco. Vectra has become the most successful of these polymers. It is made from a basic molecular unit, a 'crankshaft' monomer, 6-hydroxy-2-naphthoic acid, which reduces the ability of the chains to pack efficiently so lowering the melting point (Figure 12). In consequence,

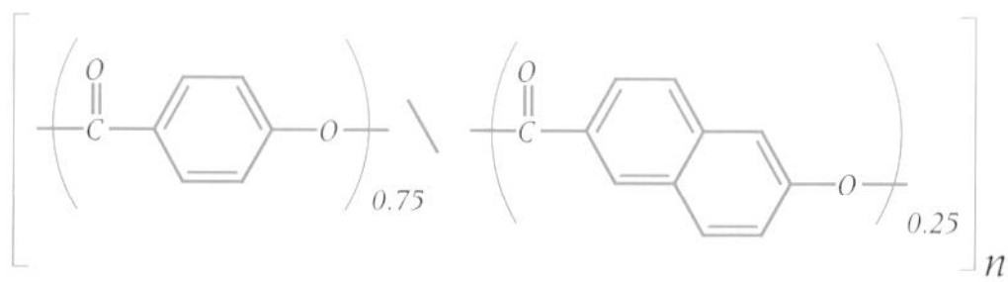

Figure 12. The structure of Vectra (Hoechst-Celanese), a polymer that shows liquid-crystal behaviour on melting.

Vectra is used in drift chamber detectors in particle accelerators such as the Large Electron-Positron collider at the European particle physics laboratory, CERN in Geneva. Courtesy of CERN.

Vectra can be melt-processed from a liquid-crystal phase. An important benefit is that the viscosity of the melt is low, which allows these polymers to be used for injection-moulding parts with complex shapes.

As well as being very strong, these aromatic copolyesters are chemically and thermally stable with low expansion when heated – attributes that have been exploited in several applications. They include mounts for the optical components in compact disc players, fibre optic connectors, utensils for microwave cooking, and filament-wound pressure vessels for fire extinguishers. Another application of Vectra is in drift chamber detectors in particle accelerators such as the Large Electron Positron collider at CERN in Geneva. Their commercial application, however, is severely restricted because the raw materials are expensive, and recycling used polymer has met with only partial success. Thus, the outlook for these materials in structural applications is not encouraging. By comparison, their use as flow aids in the processing of thermoplastic polymers is attracting increasing attention. For example, blends of a polyimide and a liquid crystalline aromatic polyester can be extruded into films which are promising candidates for use in flexible electronic circuitry in the automotive industry.

So far, no-one has found a use for the semi-flexible main-chain polymers, in which flexible molecular links, or 'spacers', and the groups of atoms responsible for liquid crystallinity alternate along the backbone (Figure 13). This increase in molecular flexibility dramatically reduces the transition temperatures resulting in readily accessible liquid-crystal phases ideally suited for fundamental study.

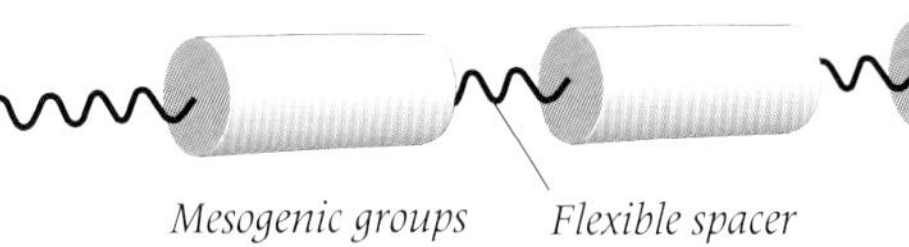

Figure 13. A semi-flexible main-chain liquid-crystal polymer in which the flexible spacers greatly reduce the melting temperatures making them ideally suited for fundamental studies.

Side-group liquid-crystal polymers for optical memories

The other main class of liquid-crystal polymers are the *side-group liquid crystal polymers*. These have exciting potential for application in advanced electro-optic technologies. They were discovered independently by two research teams in the late 1970s: Helmut Ringsdorf and his colleagues at the University of Mainz, and Valery Shibaev and his group at the Moscow State University. In these materials, the liquid-crystal units are attached as pendants onto the polymer backbone via a flexible spacer (Figure 14). The insertion of this spacer between the liquid crystal group and backbone is critical and without it you do not normally see liquid crystallinity. The spacer's role is to decouple the tendency of the liquid-crystal groups to self-assemble, from the opposing tendency of the backbone to become randomly coiled. This decoupling endows a unique duality of properties: polymer characteristics such as glassy behaviour combined with the electro-optic properties such as responding to light or electrical signals – albeit on a much slower timescale.

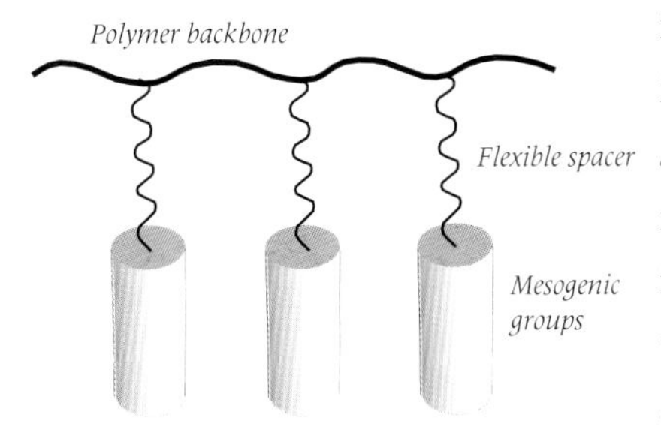

Figure 14. A side-group liquid crystal polymer in which the liquid-crystal group is attached as a pendant to a conventional polymer backbone via a flexible spacer.

This duality was first exploited by Harry Coles, now at the University of Southampton, and his colleague Richie Simon, in a new type of optical information storage device. Later, Joachim Wendorff and his team, now at the University of Marburg in Germany, showed that polymers responding to light can also be used to store information. These photosensitive materials are made by incorporating molecular units that change their shape

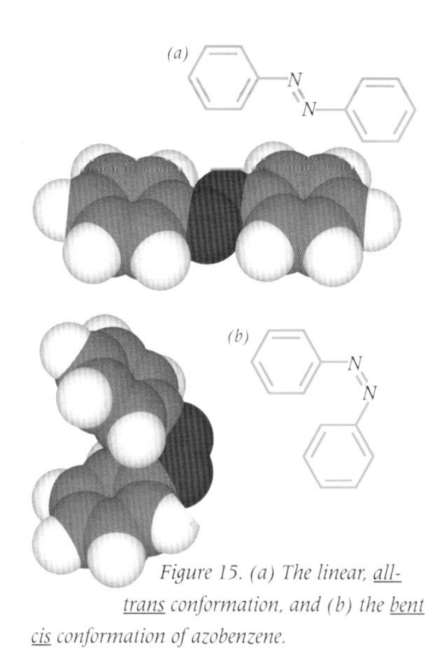

Figure 15. (a) The linear, all-trans conformation, and (b) the bent cis conformation of azobenzene.

when irradiated by light of a certain wavelength. Most commonly, the azobenzene group is used which can exist in two forms, the rod-like *trans* form and the bent *cis* form (Figure 15).

If the liquid crystal group is based around azobenzene, then the rod-like *trans* form promotes liquid crystallinity while the *cis* form destroys the molecular ordering. A nematic phase composed of azobenzene-containing molecules can be converted into the isotropic phase in just 200 microseconds by laser illumination. Tomiki Ikeda at the Tokyo Institute of Technology built a device around this effect in which (Figure 16) an opaque film is obtained by cooling the polymer from the nematic phase. Shining a laser onto the film converts the *trans* form into the *cis* form, producing small areas of the isotropic phase which are optically transparent. Although, the *cis* form changes quickly back into the *trans* form, providing the polymer is below its glass-transition temperature, the side groups are frozen and so are unable to reorganise to form the liquid-crystalline phase. Thus the small transparent regions are stored indefinitely on an otherwise opaque background. To erase this information, the area simply requires heating above the glass-transition temperature and the opaque nematic texture forms very rapidly.

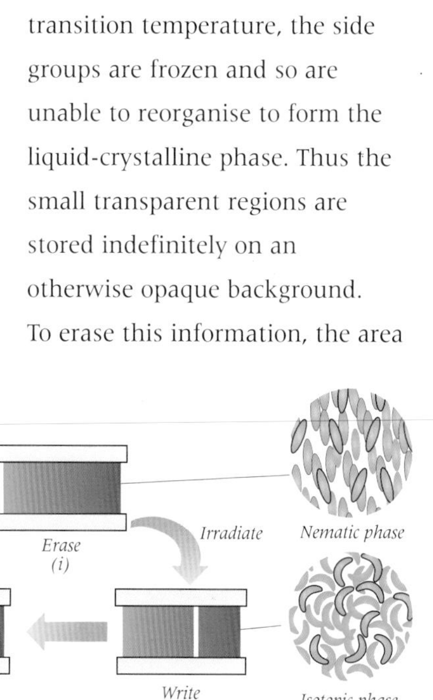

Figure 16. An information storage device based on the light-induced shape change of azobenzene (see Figure 15).

The outlook for liquid crystals

Liquid crystals are now a mature area of science, accounting for about one out of every 100 research papers published on chemistry. There have been many successes for liquid-crystal technology, most notably display devices but also in other areas such as liquid-crystal polymers, thermography applications, as well as soaps and detergents. These successes have in turn generated a considerable amount of fundamental research on how molecules self-assemble, but there remains much to be understood.

As we approach the next century, the new display devices described in this chapter are well-placed to maintain a pre-eminent role for liquid crystals in the display industry. Furthermore, the applications of liquid crystals will become ever more diverse, encompassing areas such as information storage, nonlinear optics, molecular wires and sensors. In addition, as our understanding of the role of liquid crystals in nature increases this, will lead to developments in new materials for application in biological and pharmaceutical areas. Liquid-crystal science will clearly remain an active and exciting area to be involved in during the next century.

Courtesy of DuPont

Box 7 Liquid crystals and life

Although Reinitzer is generally acknowledged with the first observation of liquid crystallinity, some 30 years earlier a German ophthalmologist called Mettenheimer reported that myelin, the material that surrounds nerve fibres, was not only birefringent but also flowed (Figure a). Many years later, people realised that myelin had a liquid-crystalline structure.

Myelin contains regularly stacked layers of membranes. Such membranes are ubiquitous in biology; they surround all living cells and the organelles within cells. They are composed of amphiphiles such as dioleoylphosphatidylcholine, Figure (b), a naturally occurring phospholipid.

In the presence of water, the amphiphiles form double layers, or bilayers,

$^{+}NH_3$
CH_2
CH_2
O
$O{=}P{-}O^{-}$
O
$H_2C - CH - CH_2$
O O
$O{=}C$ $C{=}O$

(b) *A naturally occurring phospholipid.*

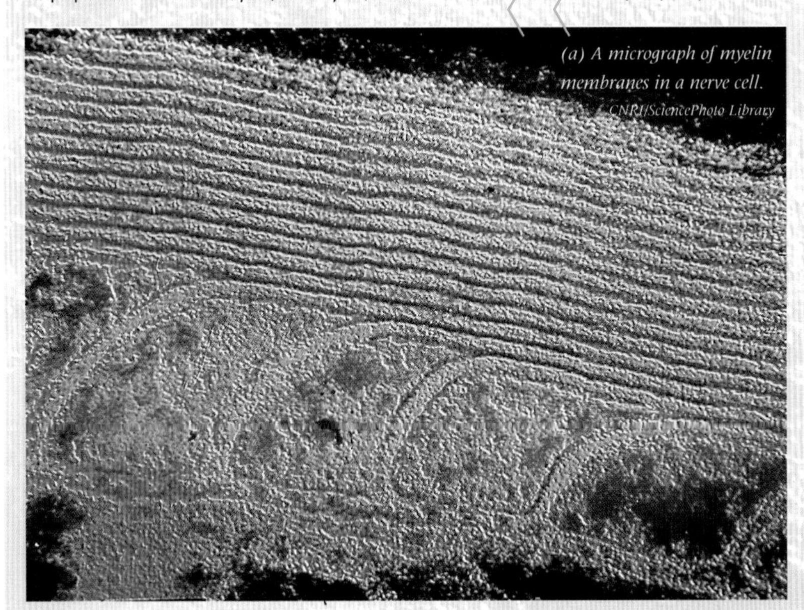

(a) *A micrograph of myelin membranes in a nerve cell.* CNRI/SciencePhoto Library

which stack to give a myelin-like structure. Within the plane of the membrane, the molecules are free to move around, so the layers may be thought of as two-dimensional fluid fabrics. The molecules cannot move between membranes because if a molecule were to quit its position in the membrane it would expose the hydrophobic regions to water. If more water is added to the system, the bilayers break down and globular sacs are formed with single bilayer walls, Figure (c). These are considered to be primitive cells, and it is believed that such structures played a key role in the evolution of life on Earth. It is the liquid-crystalline nature of the membrane that endows the cell with the flexibility to divide and reform, and also permits the proteins to move around the membrane ensuring that the new cells have the same allocation of proteins.

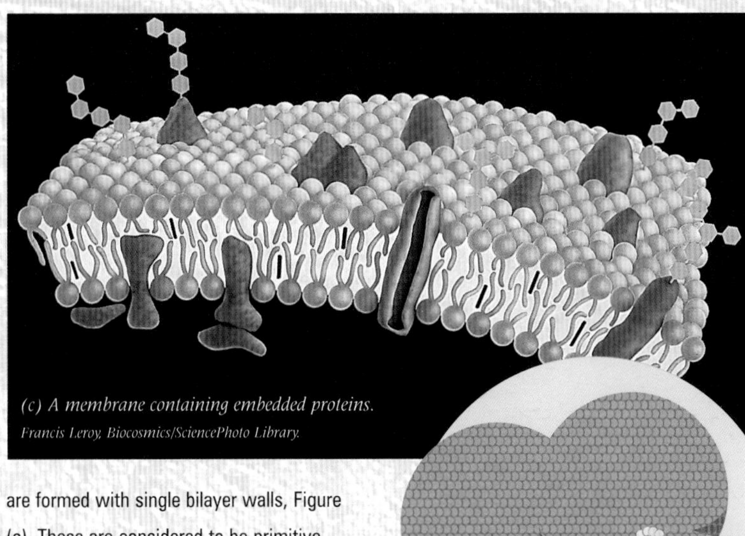

(c) *A membrane containing embedded proteins.* Francis Leroy, Biocosmics/SciencePhoto Library.

Liquid-crystal phases are also thought to play an important role in biological processes. For example, phospholipids extracted from the brain exhibit the hexagonal phase (see Figure in Box 5) although the significance of this has still to be established. Naturally occurring biopolymers such as DNA also show liquid crystallinity although how important this is in biological processes is unknown.

(d) *Vesicles or primitive cells formed from a membrane. Each vesicle contains a small amount of water which is separated from the surrounding water.*

Further reading

1. *Introduction to Liquid Crystals Chemistry and Physics*, P. J. Collings and M. Hird, Taylor and Francis, London: 1997.
2. *Thermotropic Liquid Crystals*, G. W. Gray (ed), Wiley, New York: 1987.
3. *The Molecular Dynamics of Liquid Crystals*, G. R. Luckhurst (ed), Kluwer Academic, Dordrecht: 1994.
4. *Structures of Liquid Crystal Phases*, P .S. Pershan, World Scientific, Singapore: 1988.

Glossary

Birefringence The ability of materials such as liquid crystals to rotate the plane of plane-polarised light, producing coloured fringes under a polarised light microscope.
Chiral smectic C* phase A liquid crystal phase composed of layers of tilted chiral (handed) molecules: the tilt direction traces out a helix on passing through the phase (see Smectic C phase).
Cholesteric, chiral nematic phase A nematic liquid crystal phase composed of chiral molecules: the average direction in which the molecules point traces out a helix on passing through the phase (see Nematic phase).
Clearing temperature The temperature at which a liquid crystal phase, with all the constituent molecules pointing on average in the same direction (the director), becomes isotropic with all the molecules randomly arranged.
Director Preferred direction of orientation in a liquid crystal phase.
Discotic liquid crystals Liquid crystals composed of disc-like molecules.
Dynamic scattering display An early type of liquid crystal display device.
Ferroelectric materials Materials with a net permanent dipole moment.
Lyotropic liquid crystals Materials showing liquid crystalline behaviour when in solution.
Mesophase A liquid crystal phase.
Nematic phase The least-ordered liquid crystalline phase in which molecules on average point in the same direction but the positions of the molecules are random.
Optical textures Characteristic patterns shown by liquid crystal phases when viewed through the polarised light microscope.
Side-group liquid crystal polymers Polymers with liquid-crystal units attached as pendants to the polymer backbone via a flexible spacer.
Smectic phase A liquid crystal phase in which the molecules are arranged in layers.
Smectic C phase A layered liquid crystal phase in which the direction of orientation of the molecules is tilted with respect to a particular layer normal but the tilt direction from one layer to the next is random.
Thermotropic liquid crystals A material that shows liquid crystalline behaviour depending on the temperature.
Thermotropic main-chain polymers Polymers in which liquid crystal groups are incorporated into the polymer backbone and as a result exhibit liquid crystalline behaviour as a function of temperature.
Twisted nematic display A type of liquid crystal display in which the optical effect is achieved by unwinding a twisted nematic phase using an electric field (see Nematic phase).

Biographical details

Dr Corrie Imrie is a senior lecturer in the Department of Chemistry at the University of Aberdeen. He has been involved in liquid crystal research for 14 years and has published more than 50 papers in the area.

Acknowledgements

The author gratefully acknowledges the following for their assistance in preparing this chapter: Duncan Bruce, John Goodby, George Gray, Stephen Picken, Peter Raynes and Christopher Viney.

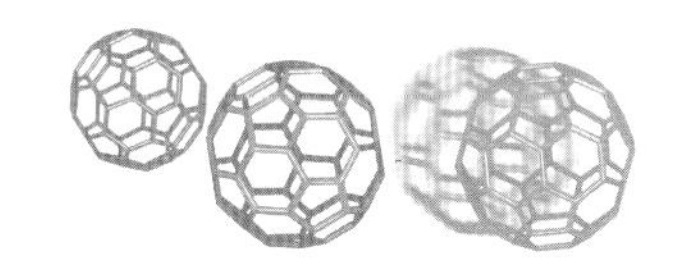

New science from new materials

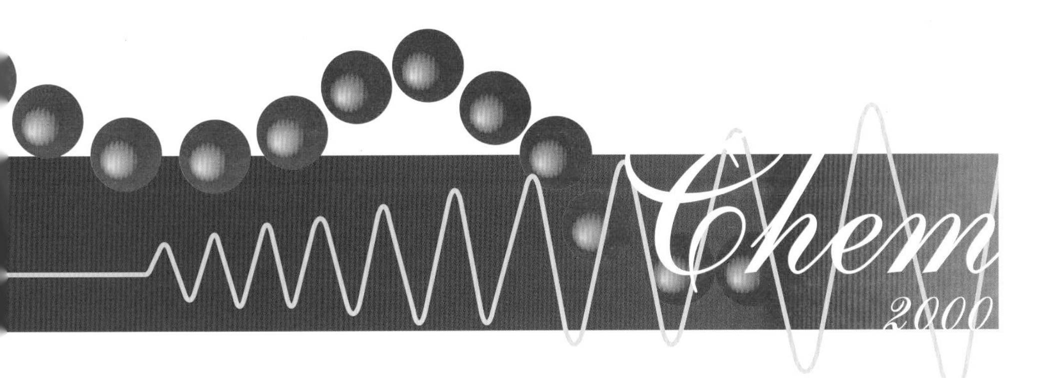

Fascinating new developments in scientific research, such as new materials with surprising electronic and structural properties, may play a vital role in the development of even smaller and faster electronic devices in the future.

Dr Paul Attfield
University of Cambridge
Dr Roy Johnston
University of Birmingham
Professor Sir Harry Kroto
University of Sussex
Professor Kosmas Prassides
University of Sussex

In the past 20 years, researchers have discovered new chemical compounds with exotic physical properties that are broadening our understanding of the physical world at a fundamental level. These structures, which are based on highly inventive synthetic chemistry, look set to provide the next generation of electronic, magnetic and optical devices.

Philippe Plailly /Eurelios/Science Photo Library

Other chapters in this book show how chemical ideas and experimental techniques have developed during the 19th and 20th centuries, leading to a wealth of new molecular structures that have had an enormous influence on society. Chemical research has produced a huge variety of new materials, such as plastics and medicines. As we will see in *The chemistry of life*, chemical behaviour underpins our understanding of living processes, and a large section of modern chemical research is devoted to the life sciences. However, there is another area of scientific endeavour where chemistry is hugely important, and that is in the design of materials with specific electronic properties. Some of these have already been mentioned in the chapters on liquid crystals and polymers.

In the 21st century chemistry is likely to play an increasing role in the development of ever smaller and faster electronic devices. It is now possible to design unusual chemical compounds that could be exploited, for example, as electronic switches. In recent years, novel materials with breathtaking properties have been synthesised which are forcing theorists to re-work their fundamental ideas on condensed matter. Meanwhile, new electronic phenomena based on exotic quantum behaviour (particularly at very low temperatures) have been uncovered which researchers are now trying to optimise through carefully devised structures. Today, chemists and physicists are very much working together to explore and develop these exciting areas for future commercial exploitation.

In this chapter, we give three examples of new materials synthesised in the past few decades which have turned out to have surprising electronic and structural properties. The first are mixed metal oxides that exhibit highly unusual electronic and magnetic behaviour, the second

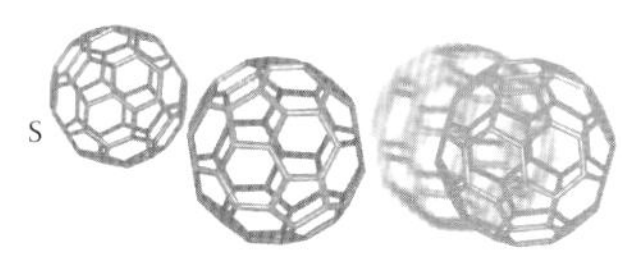

are tiny clusters of metal atoms or inorganic compounds which could be used to create minute devices that depend on quantum effects, and the third are new types of carbon molecules which not only have interesting electronic properties but have also opened up a new area of carbon chemistry.

Amazing oxides

Computers and other modern technologies create demands for entirely new kinds of electrical conductors and magnets. Many of these materials are turning out to be solid oxides of common metals such as iron, manganese and copper, with simple compositions that belie their complex chemistries and remarkable properties.

Much of human progress has followed the extraction of metals from their mineral ores. The terms – Bronze Age and Iron Age – reflect the significant impact that the mechanical properties of these metals made in their times. During the 20th century, the electrical and magnetic properties of metals have become just as important. Copper wires carry electricity in almost every household, and iron-containing magnets are in everyday use. However, many of the new materials with the best electrical and magnetic properties are no longer pure metals or their alloys, but metal oxides similar to the natural ores from which the metals themselves were extracted! The atoms in these compounds are not arranged into molecules, but instead form infinite, ordered arrangements of positively charged metal cations and negatively charged oxide anions reminiscent of simple ionic salts such as sodium chloride.

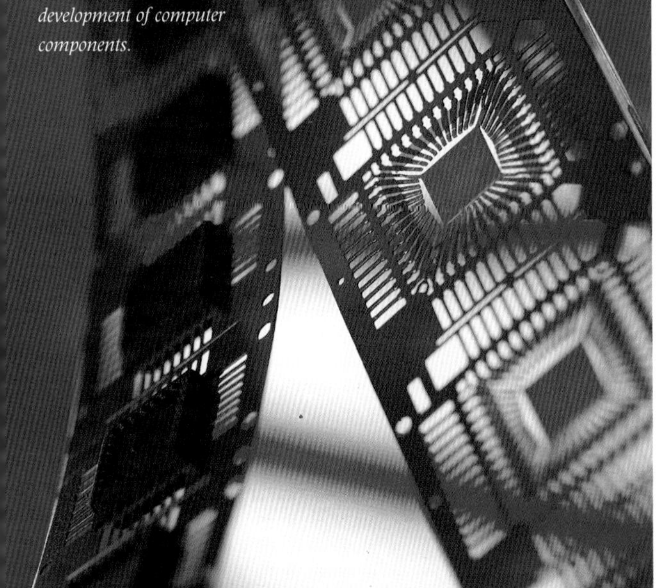

New materials play a vital role in the development of computer components.

Magnetite – the first magnet

Unusual metal oxides are not all new or synthetic. The mineral *magnetite* has been known for many centuries and was widely used as a compass from the Renaissance onwards. The magnetic properties of magnetite are not only useful to human beings but also to some species of bacteria which contain minute particles of magnetite for navigation. By sensing the vertical part of the Earth's magnetic field they are able to move up or down.

Magnetite is an oxide of iron with composition Fe_3O_4. It illustrates two of the key properties of the group of metals known as transition elements, which includes iron, manganese and copper in having unpaired electrons and the possibility of differently charged metal ions such as Fe^{2+} and Fe^{3+}. Every electron in an atom or molecule or ion behaves like a tiny magnet, but in most substances electrons are paired up such that their magnetic fields cancel out. This is called *diamagnetic* (nonmagnetic) behaviour. Transition metal cations are unusual in having some electrons that remain unpaired, and in magnetite the individual electron magnets are all lined up in the same direction so that the entire solid is magnetic or to be more accurate *ferromagnetic*. (In fact, the term

A sample of magnetite.
Courtesy of P P Edwards.

'magnetic' derives from the name of this mineral, which was found in the Magnesia district of Greece.)

The chemical formula of magnetite is formally written as $Fe^{2+}(Fe^{3+})_2(O^{2-})_4$ to show the charges on the ions. An Fe^{2+} ion has one more electron than Fe^{3+}, and when the two ions are close together, then the extra electron can hop from Fe^{2+} (which now becomes an Fe^{3+} ion) to the original Fe^{3+} (which now becomes Fe^{2+}). This is a symmetrical redox (reduction-oxidation) reaction in which the reactants are identical to the products. In solid magnetite there are almost infinite numbers of Fe^{2+} and Fe^{3+} ions close to each other so this self-redox process takes place countless times. In the infinite solid, the transferred electrons hop through the material making it electrically conducting. However, the conductivity is generally lower than in conventional metals so that magnetite is not a useful conductor. In fact this conduction is a disadvantage which prevents magnets made of magnetite being used in electrical transformers and motors, as eddy currents in the magnetite would lead to energy losses.

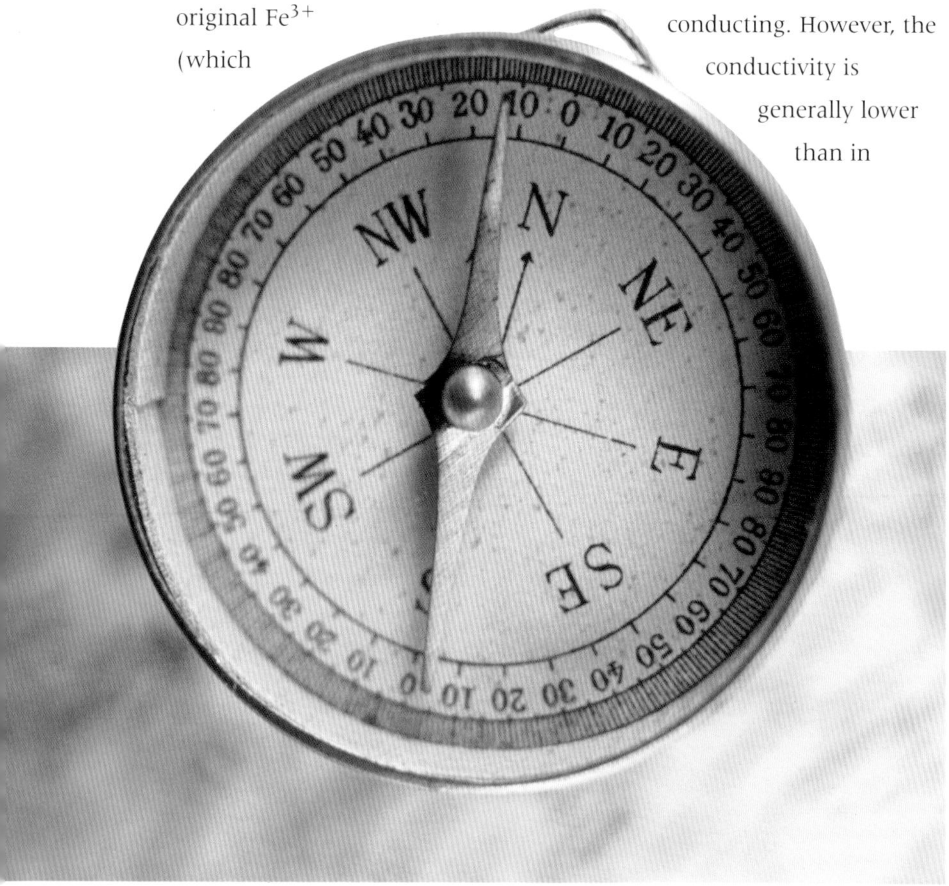

To preserve the useful magnetic properties of magnetite while suppressing the conductivity due to redox transfer of electrons, the Fe^{2+} ions are removed. This can be done by carefully oxidising all the Fe^{2+} to Fe^{3+} without disturbing the arrangement of the iron ions, to produce γ-iron(III) oxide which is widely used in magnetic recording tapes. (The mineral haematite, α-iron(III) oxide, has a different atomic arrangement and is not magnetic.) Alternatively, Fe^{2+} can be replaced by other ions such as manganese (Mn^{2+}), cobalt (Co^{2+}), nickel (Ni^{2+}) or zinc (Zn^{2+}) giving a family of materials known as *spinel ferrites* that are widely used in high-frequency transformers and inductors.

Colossal magnetoresistance

Some oxides do, however possess much more exotic magnetic properties which could have a profound effect on the next generation of electronic and magnetic devices. These relate to a phenomenon called *magnetoresistance*. When an electrical conductor such as copper is placed in a magnetic field, its resistance changes slightly. This can be a very useful property for sensing magnetic fields because it enables magnetically stored information to be read. Information is written onto 'hard' magnetic computer disks by changing the direction in which small particles of a magnetic material such as γ-iron(III) oxide are magnetised. When the disk is moved past a magnetoresistive sensor, the variations in magnetic field due to the magnetised particles on the disk give rise to corresponding changes in the electrical resistance of the sensor.

The magnetoresistance changes of most metals are between 1 and 2 per cent which are too small to be of practical use. However, a breakthrough came in 1988 when Albert Fert and colleagues at the University of Paris-Sud discovered that metal multilayers can produce giant *magnetoresistances* of up to 50 per cent. These multilayers consist of alternating thin layers of iron and chromium metals stacked upon one another. The thickness of each layer is only around 1 nanometre or about seven atoms thick! Other minute constructions of two different metals such as nanometre-sized cobalt particles embedded in solid silver have also been found to give large magnetoresistances. However, the most spectacular effects have been found not in such bimetallic systems, but in a family of manganese oxides.

The ions in lanthanum manganese oxide, $La^{3+}Mn^{3+}(O^{2-})_3$, form a simple infinite arrangement known as the *perovskite* structure (Figure 1). (Perovskite is the mineral name of calcium titanium oxide ($CaTiO_3$) which typifies this arrangement.) $LaMnO_3$ is made by heating a mixture of lanthanum oxide and manganese oxide to temperatures above 1000 °C. If some of the lanthanum oxide in the initial mixture is replaced by calcium oxide, then some of the positions in the infinite structure that would normally be occupied by lanthanum cations are instead occupied by calcium so the product has the formula $(La_{1-x}Ca_x)MnO_3$. (Although molecules always have integral numbers of atoms in their chemical formulae, the essentially infinite number of atoms in repeating ionic structures, such as that of lanthanum calcium manganese oxide, means that the proportion of calcium substituted for lanthanum, x, can take any value between 0 and 1.)

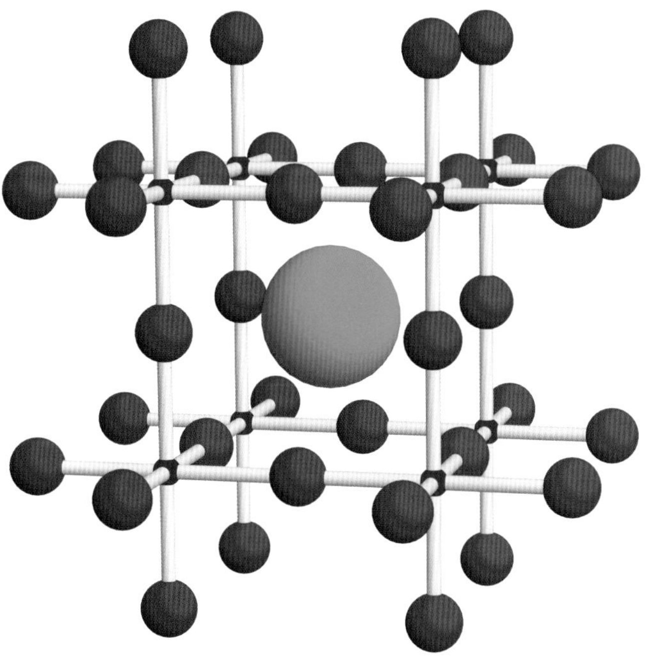

Figure 1. The colossal magnetoresistances of lanthanum calcium manganese oxides $(La_{1-x}Ca_x)MnO_3$ arise from the 'perovskite type' arrangement of their atoms. Electrons move between manganese ions (purple) via the interconnecting oxygens (red) through the network of bonds, while the lanthanum or calcium cations (green) fill the spaces in between.

Lanthanum and calcium are not transition metals and their ions La^{3+} and Ca^{2+} have fixed ionic charges which are compensated for by altering the charge on some of the manganese cations (a transition element for which differently charged ions are possible), giving the formula $(La^{3+}_{1-x}Ca^{2+}_x)(Mn^{3+}_{1-x}Mn^{4+}_x)(O^{2-})_3$. For x values other than 0 and 1, a mixture of two differently charged manganese cations is present. This situation is like that in magnetite, in which a mixture of Fe^{2+} and Fe^{3+} is naturally present, and similar conducting and magnetic properties are found. Electrons hopping from Mn^{3+} to Mn^{4+} cations make lanthanum calcium manganese oxide a good conductor of electricity, and magnetism arises from a parallel alignment of all the unpaired electrons. The importance of having the two different manganese cations present can be seen by looking at the extremes of x. Both $LaMnO_3$ ($x = 0$) and $CaMnO_3$ ($x = 1$) have only one kind of manganese cation, consequently they are both insulating and nonmagnetic.

The $(La_{1-x}Ca_x)MnO_3$ perovskites are better electrical conductors than magnetite but their magnetism is less strong and so they are not used as magnets. However, in applied magnetic fields their electrical resistances decrease markedly – this is a very large magnetoresistance effect. Many different manganese oxide perovskites can be made, some of which give magnetoresistance values in excess of 99 per cent. The superlative 'colossal' is used to distinguish them from the merely giant magnetoresistance bimetallic systems described above. However, these colossal *magnetoresistances* require extremely strong magnetic fields of around 5 tesla, whereas the fields generated by aligned magnetic particles on a typical computer hard disk are only about 0.001 tesla.

Chemists around the world are currently trying to improve the properties of manganese oxide perovskites so that these large magnetoresistances can be induced by small magnetic fields. If this is achieved then ultra-small computer disks holding much more information than conventional media could follow. The magnetoresistive effect may also be used to design new magnetic transistors with greater efficiencies than conventional *semiconductor* transistors that make up the microprocessors of current computers.

High temperature superconductivity

The most spectacular property of transition metal oxides discovered so far is undoubtedly that of high temperature *superconductivity*. In 1911, Heike Kamerlingh Onnes found that when he cooled mercury down to 4 K using liquid helium, there was a remarkable change – at 4.2 K the electrical resistance of mercury sharply dropped to zero. Onnes named this new low temperature behaviour superconductivity – the ability to conduct electricity without any resistance, unlike normal metals which always have some resistance to the flow of an electrical current no matter how pure they are. Materials become superconducting only below a characteristic temperature known as the *critical temperature*, T_c, (4.2 K for mercury) above which some become ordinary metals.

The possibility of superconducting wires that would carry electricity without any energy losses was immediately apparent, but the cost of cooling superconductors with liquid helium outweighed the energy saving, except in small devices such as coils for producing high magnetic fields (as in superconducting magnets for particle accelerators and magnetic

A block of yttrium barium copper oxide, having been cooled into the superconducting state by liquid nitrogen, levitates above a magnet.
Photograph taken by B K Papworth, Cavendish Laboratory, University of Cambridge.

Heike Kamerlingh Onnes.
©The Nobel Foundation.

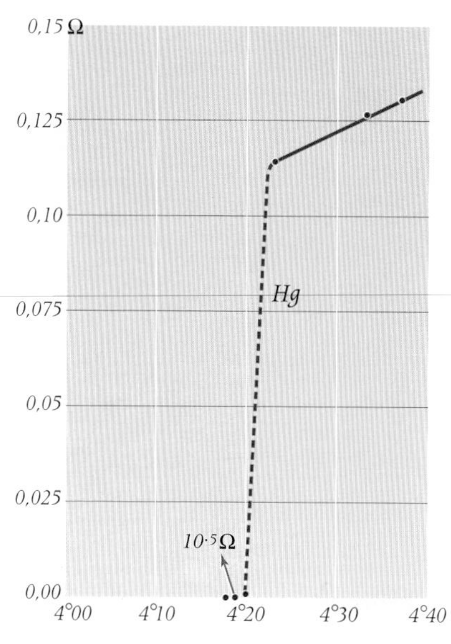

A graph of mercury cooling from Onnes' notebook.

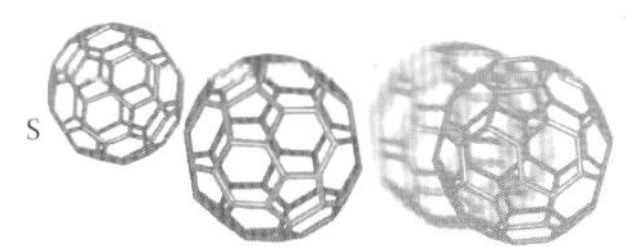

resonance imaging). Higher T_c values would be required to improve the practicality of superconductors. However, in the next 75 years the highest known critical temperature rose only slowly up to 23 K for an alloy of the metals niobium and germanium, which still required liquid helium cooling. Other useful properties of superconductors were uncovered during this period. In 1933, a German physicist Walther Meissner showed that superconductors are perfect diamagnets – they are so strongly nonmagnetic that they will move away from a magnetic field, and so a superconductor can be used to levitate a magnet or vice versa. This *Meissner effect* can be used to levitate trains or rotating parts of heavy machinery so that their motion is almost frictionless.

A further property of superconductors was predicted in 1962 by Brian Josephson, then a Cambridge PhD student. He proposed that a small current could flow between two superconductors separated by a narrow insulating barrier even when no voltage is applied across the barrier. The experimental verification of this *Josephson effect* led to the invention of *superconducting quantum interference devices*, or SQUIDs. These can detect minute magnetic fields such as the biomagnetism from electrical impulses in living organisms. The uses of SQUIDs range from measuring foetal heartbeats to locating the electrical activity caused by thoughts in the brain!

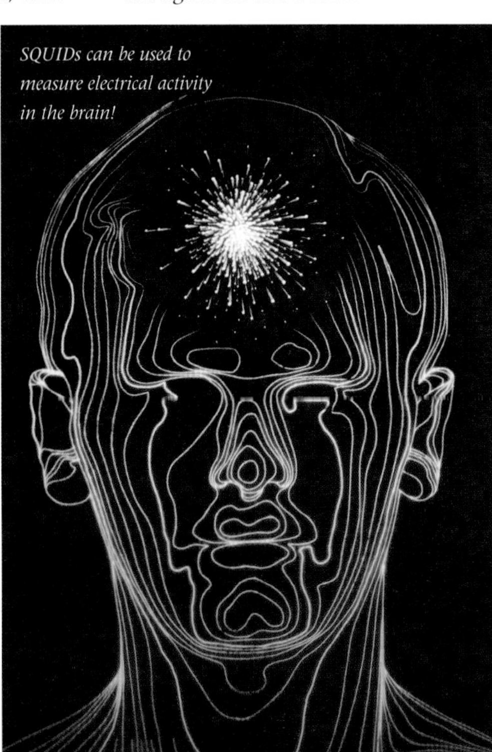

SQUIDs can be used to measure electrical activity in the brain!

Courtesy of William Bardeen.

©The Nobel Foundation

Courtesy of the National High Magnetic Field Laboratory, Florida State University.

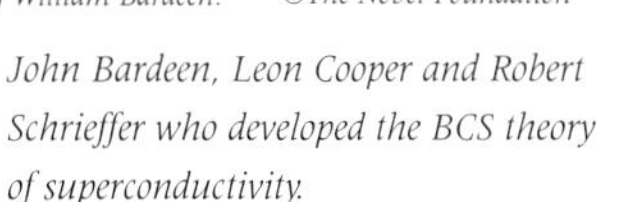

John Bardeen, Leon Cooper and Robert Schrieffer who developed the BCS theory of superconductivity.

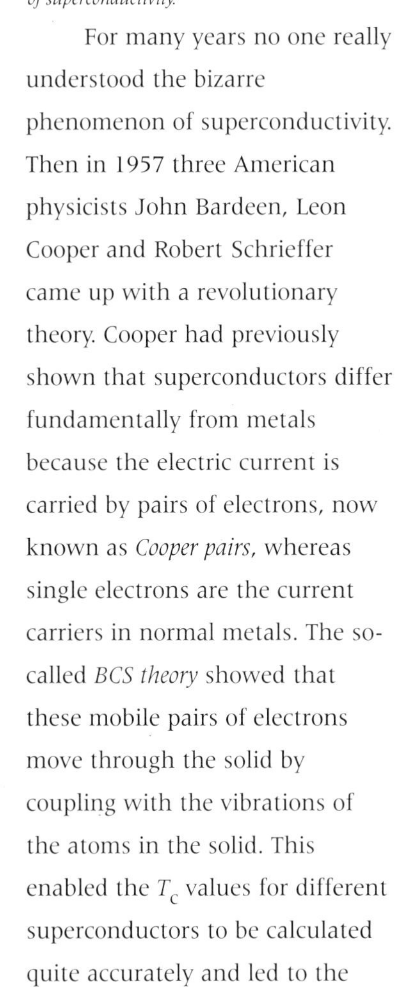

For many years no one really understood the bizarre phenomenon of superconductivity. Then in 1957 three American physicists John Bardeen, Leon Cooper and Robert Schrieffer came up with a revolutionary theory. Cooper had previously shown that superconductors differ fundamentally from metals because the electric current is carried by pairs of electrons, now known as *Cooper pairs*, whereas single electrons are the current carriers in normal metals. The so-called *BCS theory* showed that these mobile pairs of electrons move through the solid by coupling with the vibrations of the atoms in the solid. This enabled the T_c values for different superconductors to be calculated quite accurately and led to the prediction that the highest possible T_c would be around 35 K.

Then in the 1980s things suddenly changed. George Bednorz and Alex Müller, who were working at the research laboratories of the computer company IBM in Switzerland decided to search for new superconductors among oxides of the transition metals. Their hunch was that the strong influence of oxide anion vibrations upon the mobile electrons in some of these conducting oxides might lead to the formation of Cooper pairs and thus superconductivity at high temperatures. After studying a number of oxides they were proved correct, and in 1986 they reported superconductivity in lanthanum barium copper oxide with a T_c of 30 K.

This remarkable discovery hit the headlines early in 1987, starting a gold rush in laboratories around the world to find other copper oxide

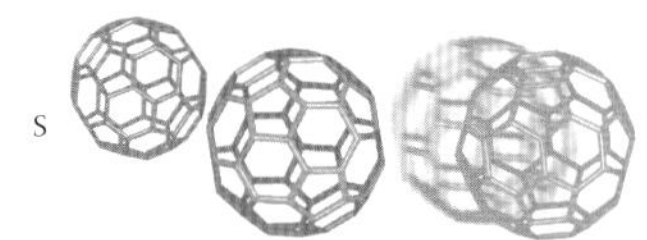

George Bednorz and Alex Müller who won the Nobel Prize for their discovery of the oxide high-temperature superconductors.
Courtesy of IBM.

superconductors. Within a year a new yttrium barium copper oxide had been discovered with an unprecedented T_c of 93 K. An important milestone was passed – that of the boiling point of liquid nitrogen at 77 K. Nitrogen is an abundant gas making up 80 per cent of the atmosphere and is relatively cheap to liquefy (a pint of liquid nitrogen costs less than a pint of milk!). The new high-T_c *superconductors* could thus be cooled cheaply unlike the previous lower T_c materials requiring expensive liquid helium (which costs around the same price as table wine, volume for volume). In the autumn of 1987, Bednorz and Müller received a Nobel Prize, in record time, for their work on superconductors – following the Nobel awards given to Onnes (1913), Bardeen, Cooper and Schrieffer (1972), and Josephson (1973).

The chemistry of the superconducting copper oxides is related to that of the conducting iron and manganese oxides. The lanthanum barium copper oxide discovered by Bednorz and Müller has chemical formula $(La_{2-x}Ba_x)CuO_4$. This is similar to the $(La_{1-x}Ca_x)MnO_3$ perovskites as the replacement of La^{3+} ions by Ba^{2+} is compensated for by oxidising some Cu^{2+} ions to Cu^{3+} according to the formula $(La^{3+}_{2-x}Ba^{2+}_{x})(Cu^{2+}_{2-x}Cu^{3+}_{x})(O^{2-})_4$. The presence of two differently charged copper ions changes the material from being insulating for $x = 0$ to conducting, however, the barium-doped compound is not magnetic but instead becomes a superconductor at low temperatures.

Why are these copper oxides not magnetic like magnetite and the manganese oxide perovskites? An important difference is that the Cu^{3+} ions have no unpaired electrons and so are nonmagnetic. This presence of nonmagnetic

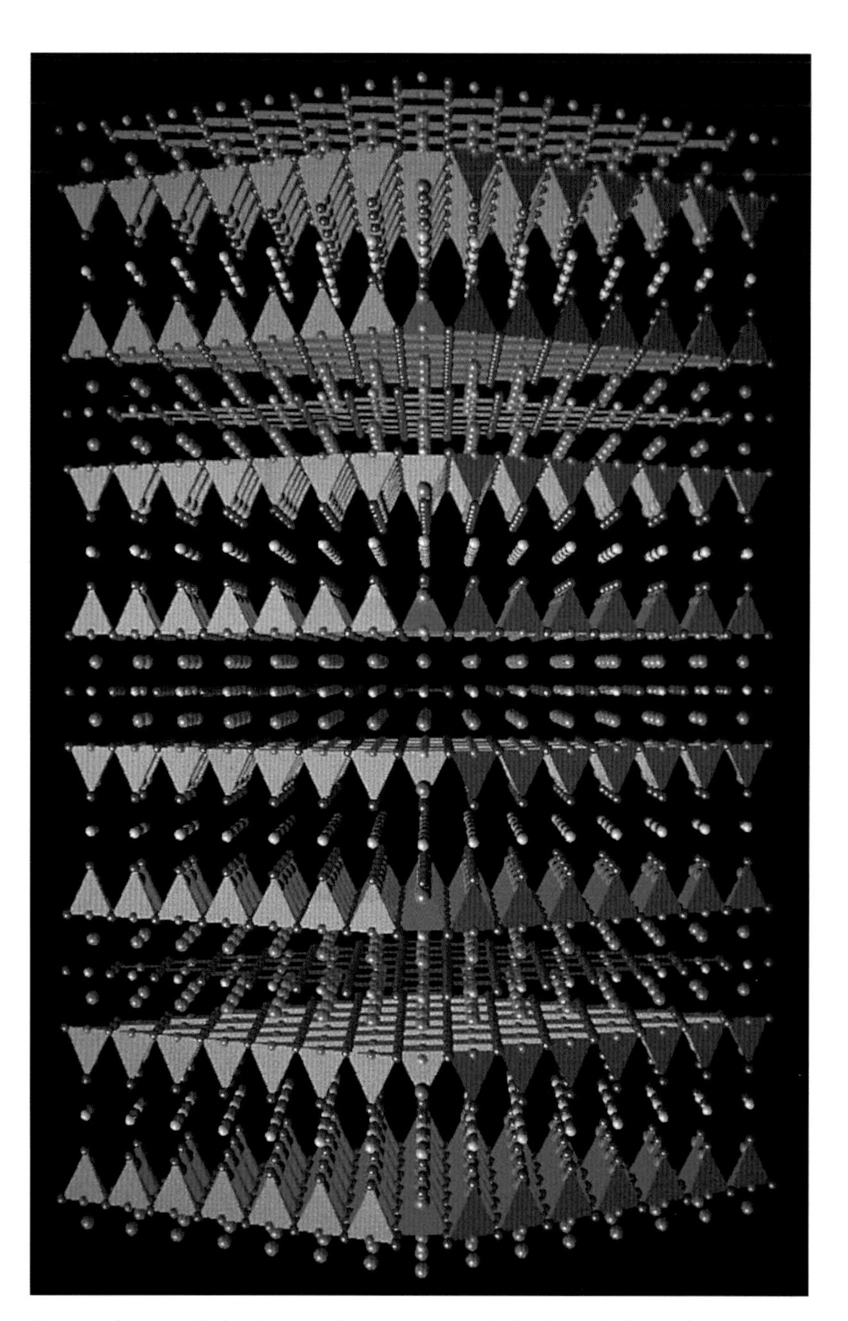

Figure 2. The current highest-T_c material superconducts at 135 K and contains layers of mercury (purple), barium (green), calcium (blue), copper (yellow) and oxide (red) ions stacked in a regular sequence.
Courtesy of P P Edwards and G B Peacock.

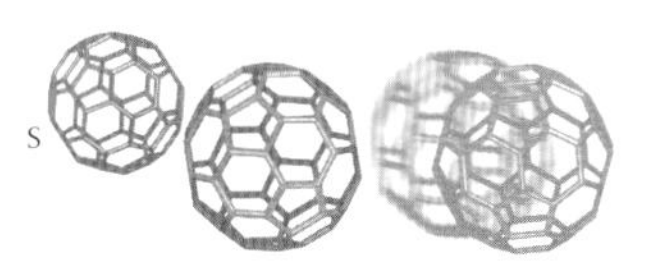

transition metal ions prevents the material becoming magnetic, although why superconductivity occurs instead and with such high T_cs is not yet understood. Many physicists believe that there is a deep theoretical connection between magnetism and superconductivity.

The highest-T_c superconductor now known contains mercury, barium and calcium cations in addition to copper and oxygen. The atomic arrangement is shown in Figure 2. As in all these superconductors, the metal cations are arranged in a repeating sequence of layers. The T_c increases with the number of closely spaced copper-oxide layers and the presence of three successive layers in this compound leads to a T_c of 135 K (-138 °C). This is still a very low temperature in everyday terms (the lowest recorded temperature on Earth was 183 K, or -90 °C, in 1983 in Antarctica) and requires liquid nitrogen cooling, but until recently it would have been considered an impossibly high T_c for any superconductor.

The technological development of high-T_c superconductors has proved difficult because they are brittle ceramic materials, but after a decade of effort, kilometre-long sections of superconducting wire are now being produced and tested. High-T_c SQUIDs are also now available. The discovery of the high-T_c materials has led to an intense search by chemists for other new superconductors. Several types have been discovered, although none with T_cs above 40 K. Most notable among these are the fullerides which are described later on in this chapter.

This chip is a prototype for a new kind of temperature sensor, the quantum roulette noise thermometer. The central component is a SQUID made from the high-T_c superconductor yttrium barium copper oxide.
Photograph provided by Dr J. Gallop, National Physical Laboratory Teddington.

These examples show that apparently simple oxides of the transition metals can display extraordinary conducting and magnetic properties, some of which, in particular high-T_c superconductivity, are not yet understood. They pose problems across a range of scientific disciplines – requiring chemists to synthesise them and tune their compositions, physicists to measure and explain their unusual properties, and materials scientists and engineers to use them in new technological applications. This area of chemistry clearly holds many challenges and opportunities for the new millennium.

Nanoclusters – small but perfectly formed

A second example of materials with novel properties is the class of particles and structures with dimensions of around a nanometre (one-billionth of a metre). In 1960, the famous American physicist, Richard Feynman wrote a visionary article called "There's plenty of room at the bottom", in which he challenged scientists to develop a new field of study where devices and machines could be constructed from components consisting of a small number (tens or hundreds) of atoms. This article inspired generations of physicists and chemists to try and make Feynman's vision a reality. Thus, the field of *nanoscience* was born. Today, there are hundreds of nanoscience laboratories in

The great theoretical physicist and visionary Richard Feynman.
©The Nobel Foundation.

universities and institutes throughout the world.

From nanoscience has sprung the applied field of *nanotechnology* – in which the aim is to build minute devices for electronic, optical, mechanical and even medical applications. Two areas of chemistry are currently contributing to this rapidly growing field. One area is organic/bioorganic chemistry which is generating complex molecular structures with precise physical, chemical and mechanical properties. Often based on the weak interactions between groups of atoms or molecules associated with biological behaviour (supramolecular chemistry, see *Make me a molecule*) these nano-architectures are allowing researchers to design molecular-scale machines and computers. This fascinating subject can in part be traced back to the ideas of Eric Drexler at the Massachusetts Institute of Technology who, a decade ago, wrote the first book *Engines of Creation* speculating on the future use of nanomachines.

> *Neither is it impoſſible that of theſe minute Particles divers of the ſmalleſt and neighbouring ones were here and there aſſociated into minute Maſſes or Cluſters, and did by their Coalitions conſtitute great ſtore of ſuch little primary concretions or Maſſes as were not easily diſſipable into ſuch Particles as compos'd them.*

Robert Boyle was the first scientist to refer to minute clusters.
Reproduced courtesy of the Library and Information Centre, Royal Society of Chemistry.

The other important area of nanotechnology exploits particles consisting of between tens and millions of atoms as the basic building blocks. These *nanoclusters* are formed by most of the elements in the Periodic Table – even the noble gases! The earliest reference to clusters may have been made by Robert Boyle as long ago as 1661. In his book *The Sceptical Chymist* he speaks of: "... minute masses or clusters.... as were not easily dissipable into such particles as compos'd them." Nanoclusters of the coinage metals (copper, silver and gold) are to be found in stained glass windows and in photography. Neutral and ionic molecular clusters, for example clusters of water molecules, are even found in the atmosphere. Carbon nanoclusters are now well known, including the famous soccer ball-shaped carbon-60 (C_{60}) and related structures. These clusters are described in more detail later on in the chapter.

We will focus on two types of nanocluster – those made up of atoms of metallic elements (which conduct electricity) and those formed from atoms of semiconducting elements. These are elements such as silicon and germanium which conduct electricity only under certain conditions such as at high temperature or when small amounts of elements such as phosphorus or arsenic are included in their structures (doping). Both kinds of nanoclusters have tremendous potential in developing novel nanoscale electronic, magnetic and optical devices. As we start the new century, computers and other electronic equipment are developing along three directions – miniaturisation, increased memory capacity, and increased processor and communication speed. Nanocluster technology has the potential to impact substantially on all of these areas.

Such nanoclusters are also of great theoretical interest to both physicists and chemists because they represent a state of matter that lies in between individual atoms and molecules, and crystalline solids where the atoms or molecules are fixed in infinite regular arrays. The electronic structure of single atoms is characterised by discrete quantum energy states (see *Following chemical reactions*). In the bulk material, however, interactions between the outer atomic electrons throughout the solid cause these electronic states to be 'smeared out' into *bands*, each containing many states of slightly different energies. Nanoclusters appear to have novel electronic properties somewhere in between, depending on the cluster size.

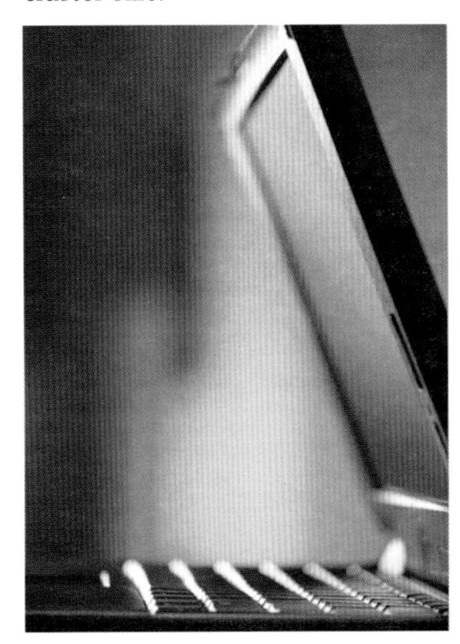

Nanoclusters have huge potential to impact on computer technology.

The theoretical and experimental study of the effect of size on the properties of clusters is a major area of modern research. Because a high percentage of atoms in the cluster are on its surface, clusters and surfaces have much in common. Nevertheless, nanoclusters can be said to constitute a new type of material with some properties that are fundamentally different from those of discrete atoms or molecules, or the infinite solid.

Size does matter

As long ago as 1871, Lord Kelvin (William Thomson) posed the question: "Does the melting temperature of a small particle depend on its size?" Nearly 40 years later, it was predicted that the melting temperature of a metallic particle should decrease as the particle gets smaller, though the unambiguous verification of the prediction had to wait until 1976, when Philippe Buffat and Jean Borel at the University of Lausanne measured the melting temperatures of gold clusters under a transmission electron microscope. Melting temperatures for the smallest nanoclusters turned out to be as low as 300 K – significantly lower than the melting point (1338 K) of bulk elemental gold! Lord Kelvin's specific enquiry can be generalised as the following problem: "How large must a cluster of atoms be before its properties resemble those of the bulk element?" The answer to this 'simple' question is exceedingly complex, and will depend on the type of atoms constituting the cluster and the type of physical property being measured.

For large clusters (comprised of thousands or more atoms) many intrinsic physical properties, such as ionisation energy or electron affinity (the energy needed to remove an electron from a cluster or add an electron to a cluster respectively), melting temperature, and cohesive energy (the energy which binds the atoms in the cluster) show a regular variation with cluster size. Many properties of clusters depend on the fraction of atoms that lie on the surface of the nanocluster; we can easily show that clusters that are roughly spherical in shape containing as many as 10 000 atoms still have nearly 20 per cent of their atoms on the surface, with the percentage of surface atoms only dropping below 1% for clusters of more than 64 million atoms! It is this high surface-to-bulk ratio that has made metal nanoclusters so important as a source of finely dispersed metal for application in heterogeneous catalysis (see *Chemical marriage-brokers*).

The size-dependent behaviour of cluster properties suggests the very exciting prospect of using nanoclusters as building blocks to construct electronic, magnetic or optical devices with characteristics that can be fine-tuned by carefully controlling the size of the component nanoclusters.

Metal clusters – a universal model

In a series of seminal experiments in the early 1980s, Walter Knight and colleagues at the University of California, Berkeley prepared beams of clusters of sodium atoms by heating the metals to give a vapour, which when mixed with a cold inert gas condensed into clusters of various sizes (Figure 3). By ionising the clusters and passing them through a mass spectrometer, where they were deflected by electric and magnetic fields and detected electronically, the researchers obtained a mass spectrum of peaks which could be related to the spread of cluster size. The Berkeley team noticed a periodic pattern of relatively intense peaks in the mass spectra, which corresponded to cluster sizes 2, 8, 20, 40, 58 and so on. The researchers realised that the numbers were extremely significant. These are the so-called *magic numbers*, originally seen in nuclear structure studies in which atomic nuclei with a combination of protons and neutrons adding up to one of these magic numbers are particularly stable.

The magic numbers were the key to devising a theoretical description of cluster structure and behaviour. This was the so-called *jellium model* and it is a quantum mechanical description. In the jellium model, the valence electrons (those involved in cluster bonding) are constrained to move within the cluster sphere under the influence of an attractive averaged electrostatic potential due to the positive nuclei (whose actual positions are not important). Using an appropriate version of the Schrödinger equation, the cluster configuration and energies can be calculated. This is a similar approach to that used by theoretical chemists to describe the electronic structure of atoms except that in the case of clusters there is no central nucleus (see *Computational chemistry and the virtual laboratory*); indeed, the jellium model is much closer to the description used to calculate the motions and energies of protons and neutrons in a nucleus (the so-called shell model). Just as electrons fill up orbitals and shells in atoms, the cluster atoms fill up the jellium orbitals and shells. Magic numbers arise due to the complete filling of a shell in an analogous way to the filling of atomic shells to give an inert gas configuration. Both physicists and chemists have become very excited by the idea of a single theoretical model that can be applied over a wide range of length scales – from the inside of a nucleus via the electronic structure of an atom to that of a cluster of hundreds or even thousands of atoms. Fundamental science is very much about

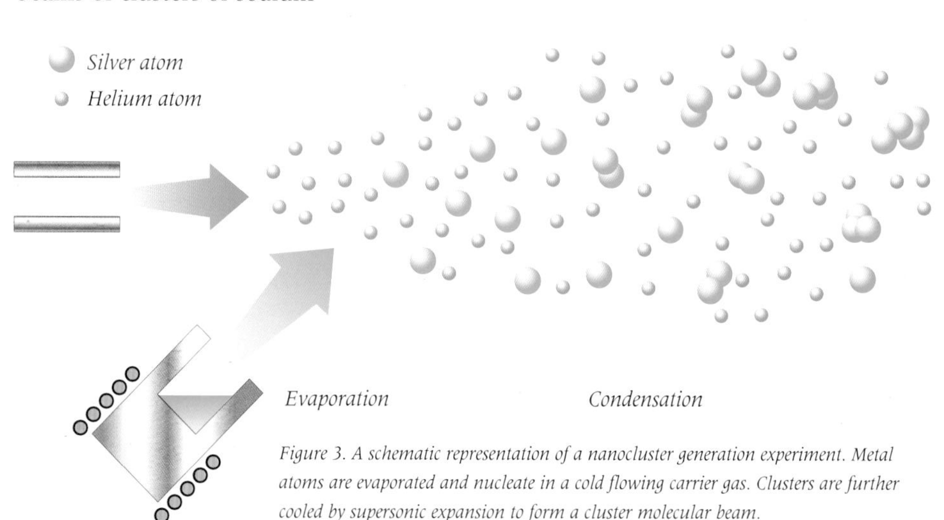

Figure 3. A schematic representation of a nanocluster generation experiment. Metal atoms are evaporated and nucleate in a cold flowing carrier gas. Clusters are further cooled by supersonic expansion to form a cluster molecular beam.
Courtesy of R E Palmer.

discovering major unifying concepts in the laws of Nature.

In the early 1990s, Patrick Martin and colleagues at the Max Planck Institute in Stuttgart carried out similar mass spectrometry experiments on larger sodium clusters of up to 2000 atoms. They observed regular variations in mass spectral intensity as a function of cluster size, which they attributed to the filling of bunches of jellium levels that were very close in energy. Eventually these bunches coalesce to form dense bands as in metals and semiconductors.

For even larger clusters of up to 25 000 atoms, the Stuttgart team saw longer period oscillations in intensity. These were due to the filling of concentric polyhedral, *geometric shells* of atoms rather than electronic shells due to quantum mechanical effects. Filled geometric shells, impart stability to the cluster by reducing its surface energy. They were first detected in the mass spectra of noble gas clusters, by Klaus Sattler at the University of Konstanz in 1981. For alkali metal clusters, the magic numbers correspond to full geometric shells shaped like icosahedra. Recent experiments have shown that magnesium, calcium and strontium clusters are also icosahedral, while aluminium clusters are octahedral, with local face-centred cubic packing, as in solid aluminium. Some typical cluster shell geometries are shown in Figure 4.

The size at which cluster stability becomes governed by geometric, rather than electronic effects depends on a number of factors, such as the density of electronic states in the bands, atomic electron configuration, cluster melting temperature and the temperature of the cluster. Thus, clusters of the transition metals (such as nickel and cobalt), where valence electrons give rise to narrow bands with a high density of states, have mass spectra characteristic of geometric shell structure, even for clusters as small as tens of atoms. Above the cluster melting temperature, the cluster resembles a spherical liquid drop and all geometric shell structure disappears.

icosahedron

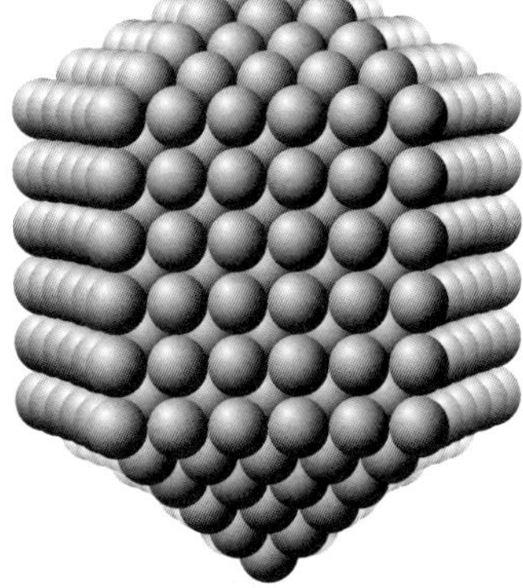

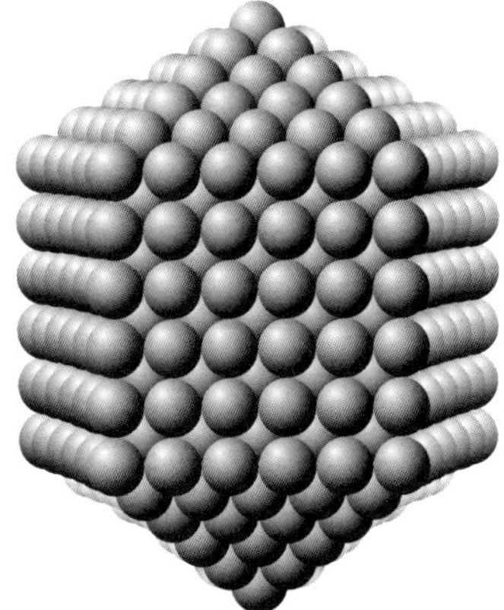

truncated decahedron

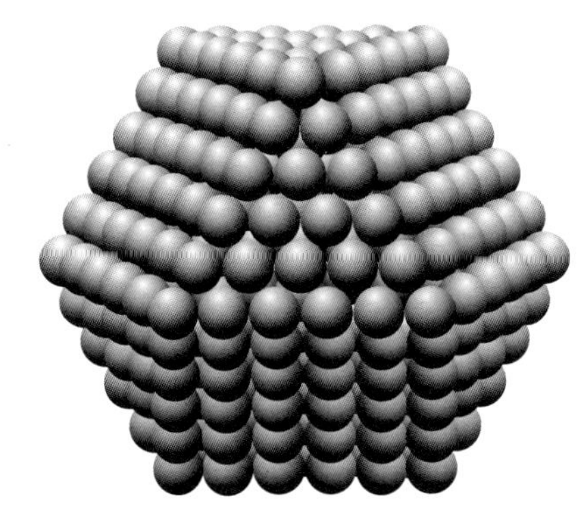

cuboctahedron

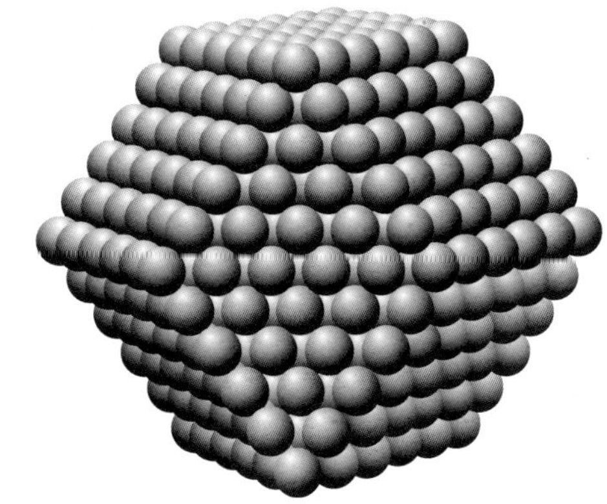

rhombic dodecahedron

Figure 4. Examples of cluster shell geometries found for rare-gas and metal nanoclusters.

Nanomagnets

Magnetism in clusters also shows interesting behaviour that is half way between that of a single atom and the bulk metal. The magnetism of clusters of iron, cobalt and nickel have been studied in molecular beams by measuring their deflection in a non-uniform magnetic field – an experiment which is a

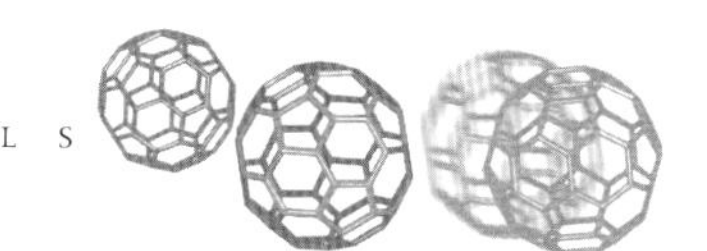

modification of that used by German physicists Otto Stern and Walther Gerlach more than 70 years ago to measure the spin of the electron. As described earlier, an electron behaves like a tiny bar magnet, giving rise to a magnetic moment in atoms and molecules possessing one or more unpaired electrons. Individual atoms or molecules show *paramagnetism* (the unpaired electron aligns with an external magnetic field) while in bulk metals the unpaired electrons show much stronger magnetic behaviour, coupling and aligning to produce the phenomenon of ferromagnetism. Clusters show behaviour that is different again – they are *superparamagnetic*. This means that the individual atomic spins within each cluster are strongly coupled by so-called exchange interactions, with the result that they have high total magnetic moments and thus resemble giant paramagnetic atoms. The average magnetic moment per atom is also higher than in the ferromagnetic solid and decreases as the cluster gets larger, reaching the bulk limit at around 500 atoms. Interestingly, rhodium, which is not magnetic in the solid state, has actually been shown to form magnetic clusters!

Recently, people have become very interested in using such clusters for new types of magnetic devices. The idea is to embed the clusters in a solid substrate. When embedded in metals, or even insulators, magnetic clusters (for example, of chromium, iron, cobalt, nickel or mixtures of these metals) are known to exhibit the phenomenon of giant magnetoresistance, as described earlier. For cobalt clusters embedded in silver, the change in resistance can be as high as 20%. Such magnetoresistive materials are already being used for magnetic recording and storage of data and show considerable promise for other applications such as magnetic sensors.

One problem with studying naked metal clusters such as those created in the molecular beams just described is that they can't be isolated and handled on a preparative scale like conventional molecules; at high densities the clusters just coalesce. To investigate clusters that are roughly uniform in size, we must protect them with a 'shell' of other atoms. Fortunately many metals happily coordinate with groups of atoms (ligands) forming complex structures consisting of clusters of metal atoms bound to various ligands. For clusters of a few atoms or tens of atoms this has led to a wealth of inorganic metal cluster compounds, which are a subject of extensive study for chemists (see Box 1).

Metallic colloids

In the larger size-regime of clusters ranging from nanometre to micrometre dimensions, lie the old and venerable class of *metal colloids* – suspensions of metal particles in another material. Some of these have been known for many centuries. Stained glass windows, dating back as far as medieval times, provide beautiful examples of colours caused by the suspension of small colloidal particles of copper, silver and gold in the glass. In fact, the use of colloidal metals to colour glass probably dates back as far as the ancient Egyptians. Cleopatra may also have used colloidal gold as eyeshadow!

The original colloidal sol produced by Michael Faraday.
Courtesy of the Royal Institution

The scientific study of colloids is much more recent. In 1857 the polymath Michael Faraday made the remarkable observation that in gold colloids "... the gold is reduced in exceedingly fine particles which becoming diffused, produce a beautiful fluid ... the various preparations of gold whether ruby, green, violet or blue ... consist of that substance in a metallic divided state". We now understand the

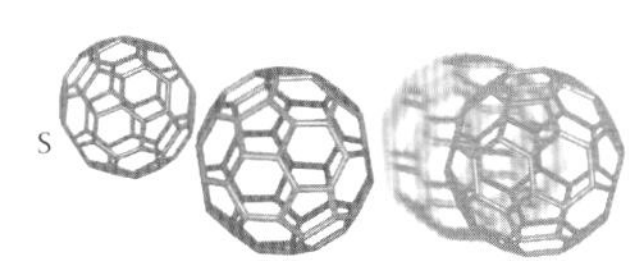

Box 1 Metal carbonyl clusters

Metals not only form clusters on their own but also in combination with other groups of atoms, in particular the carbonyl group. Transition metals can bind directly to carbon, and the first metal-carbonyl compound identified was nickel carbonyl, $Ni(CO)_4$ (by the German chemist and one of the founders of ICI, Ludwig Mond in 1890). Many years later, chemists discovered a range of very exotic metal carbonyl compounds which are composed of more than one metal atom and containing direct metal-metal bonds. These metal carbonyl clusters have presented chemists with many fascinating chemical and structural quandaries.

One of the first clusters made was triiron-dodecacarbonyl, $[Fe_3(CO)_{12}]$ which is still attracting interest today. The X-ray structure of this molecule reveals the three iron atoms in a triangle (Figure a). Ten of the twelve carbonyls coordinate to the metals in a terminal fashion, in other words, they bond to just one metal atom, whereas the remaining two carbonyls coordinate across an edge of the iron triangle. Moreover, at room temperature the carbonyls bridge this edge asymmetrically while on cooling they become more symmetrical. Researchers have observed several other bonding modes for the carbonyl group; it can even bond to a cluster through both the carbon atom and the oxygen atom.

(a) Triirondodecacarbonyl $[Fe_3(CO)_{12}]$, one of the first clusters made.
(b) The ruthenium cluster $[Ru_3(CO)_{12}]$ catalyses the coupling of heteroaromatic compounds such as pyridine.

Clusters have found unique uses in organic synthesis and catalysis. When organic compounds bond to clusters their reactivity may be modified, and they are activated in ways that cannot be mirrored using other methodologies. An example of this is the use of $[Ru_3(CO)_{12}]$ as a catalyst in the selective coupling of heteroaromatic compounds (organic compounds with rings containing atoms other than carbon) with carbon monoxide and alkenes at the *ortho* position of the ring. A heteroaromatic compound such as pyridine, which has a five-membered ring containing a nitrogen atom, bonds to $[Ru_3(CO)_{12}]$ through the nitrogen, while the adjacent carbon-hydrogen bond in the pyridine ring is activated with the formation of a carbon-ruthenium bond. It is this two-site activation that allows the coupling reaction to take place effectively.

Much larger clusters can be made – usually from smaller ones. This work has been carried out mostly by Paolo Chini, Larry Dahl, Brian Johnson and Jack Lewis. An example of this is the vacuum pyrolysis (heating in the solid state) of $[Os_3(CO)_{12}]$ or $[Os_3(CO)_{10}(NCCH_3)_2]$, which produce a number of giant clusters depending upon the precise temperature and times allowed. Two clusters which can be isolated from this type of process include $[Os_6(CO)_{18}]$ and $[Os_{20}(CO)_{40}]^{2-}$.
The metal core of the first cluster looks like three fused tetrahedra and is commonly referred to as a bicapped tetrahedron.
The cluster with 20 osmium atoms resembles a pyramid of osmium atoms following a cubic close-packing growth sequence and is shown in Figure c.

(c) The cluster, $[Os_{20}(CO)_{40}]^{2-}$, which has 20 metal atoms, starts to behave more like a metal particle.

One type of reactivity peculiar to clusters is illustrated by the reduction of $[Os_6(CO)_{18}]$ to $[Os_6(CO)_{18}]^{2-}$. It results in a structural change involving the rearrangement of the osmium core from a bicapped tetrahedron to an octahedron. This seemingly complicated transformation happens because large clusters do not contain the usual simple chemical bonds involving a pair of electrons, one donated from each bonding atom. Instead the overall bonding in the metal cluster core has to be considered holistically – with the bonding electrons spread throughout the structure. This bonding model readily explains the transformation and a number of rules have been devised (mainly by Ken Wade and Mike Mingos) especially to help understand cluster bonding.

The 20-metal atom cluster, $[Os_{20}(CO)_{40}]^{2-}$, has properties that begin to approach those of the small metal particles mentioned in the main text. One major difference between $[Os_{20}(CO)_{40}]^{2-}$ and a metal particle is that the metal-carbonyl cluster is soluble (in polar organic solvents). The cluster can therefore be studied in considerable detail using the wide range of solution techniques available to chemists. Evidence for actual metallic-like properties stems from an electrochemical study of the cluster which has been found to have five different oxidation states. This broad range of oxidation states can be likened to the broad range of energy states in the conduction band present in pure metal clusters.

Dr Paul Dyson
University of York

Stained glass window. Parish of St Mary the Virgin, Kent. The colours are due to collodial metal particles.
©Paul Burgess

colours of colloidal suspensions in terms of strong absorption bands in the visible region of the spectrum caused by collective oscillations of the cluster

The colours of colloidal metallic suspensions depend critically on the size of the metallic particles. Powdered gold is shown on the left.
Courtesy of P P Edwards.

electrons. The absorption maxima of such 'plasmon modes' move to lower energy as the colloidal particles get smaller.

Studies of colloidal metal particles using electron microscopy and scanning tunnelling microscopy have shown that platinum and palladium colloids grow as face-centred cubic single crystals, even for particles with diameters as small as 0.40 nanometres. Gold and silver colloids, by contrast, are often found to consist of icosahedral and decahedral particles. Figure 5 shows an electron micrograph of a colloidal metal particle of silver, together with a derived crystal packing diagram.

Exciting advances have been made by a number of research groups around the world in producing colloidal suspensions with narrow distributions of size, and in making thin films of gold and silver nanoclusters with surfaces protected by organic molecules. An electron micrograph of a hexagonal film of silver clusters coated with organic

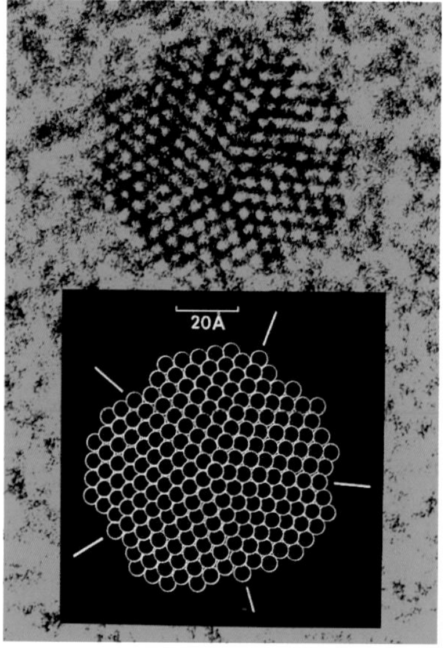

Figure 5. An electron micrograph of a decahedral particle of silver (with a diameter of 4 nanometres). The cluster contains just over 1000 atoms.
Courtesy of P P Edwards, D A Jefferson and B F G Johnson.

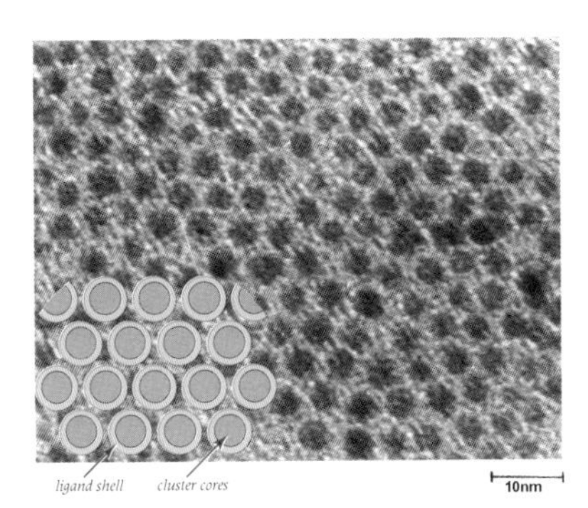

Figure 6. An electron micrograph of a hexagonal film of gold nanoclusters coated with organic thiol ligands. (Inset – schematic representation of the film).
Courtesy of C N R Rao.

thiols is shown in Figure 6. These thin films show great promise for making new types of electronic and optical devices. For example, recent experiments by Jim Heath and colleagues at the University of California in Los Angeles, have shown that a thin film of 4 nanometres thickness of thiol-coated silver clusters becomes metallic (that is electrons can tunnel from one silver cluster to another – through the organic ligands) when the film is compressed so that the inter-particle spacing is less than 0.5 nanometres. This process, which is reversible, may allow researchers to fabricate electronic switches for nanoscale devices.

Trapped nanoclusters and nanowires

For any useful applications we will need a lot of clusters and some method of storing and transporting them. As mentioned earlier, naked metal clusters tend to fuse spontaneously to form bulk metal so they will have to be stabilised against coalescence. This can be done either by coating them with protective ligands (as in the colloids mentioned earlier), or through immobilisation by tethering them to, or implanting them in, a support – which may be either an inert surface or the inside of a porous material such as a zeolite. Peter Edwards, Paul Anderson and colleagues at the University of Birmingham have managed to synthesise and trap a number of highly reactive charged alkali metal clusters inside zeolite cavities. They have demonstrated how, as more and more metal atoms are added, the white zeolitic powder turns blue and finally black, and that the zeolites become semi-conducting as electrons can hop easily from cluster to cluster. The ultimate goal of this work is to form chains of metal atoms in the zeolites and thus assemble dense bundles of around 10^{19} one-dimensional conductors or *nanowires*, as shown in Figure 7. Such nanowires could be used in the assembly of electronic devices on the nanoscale.

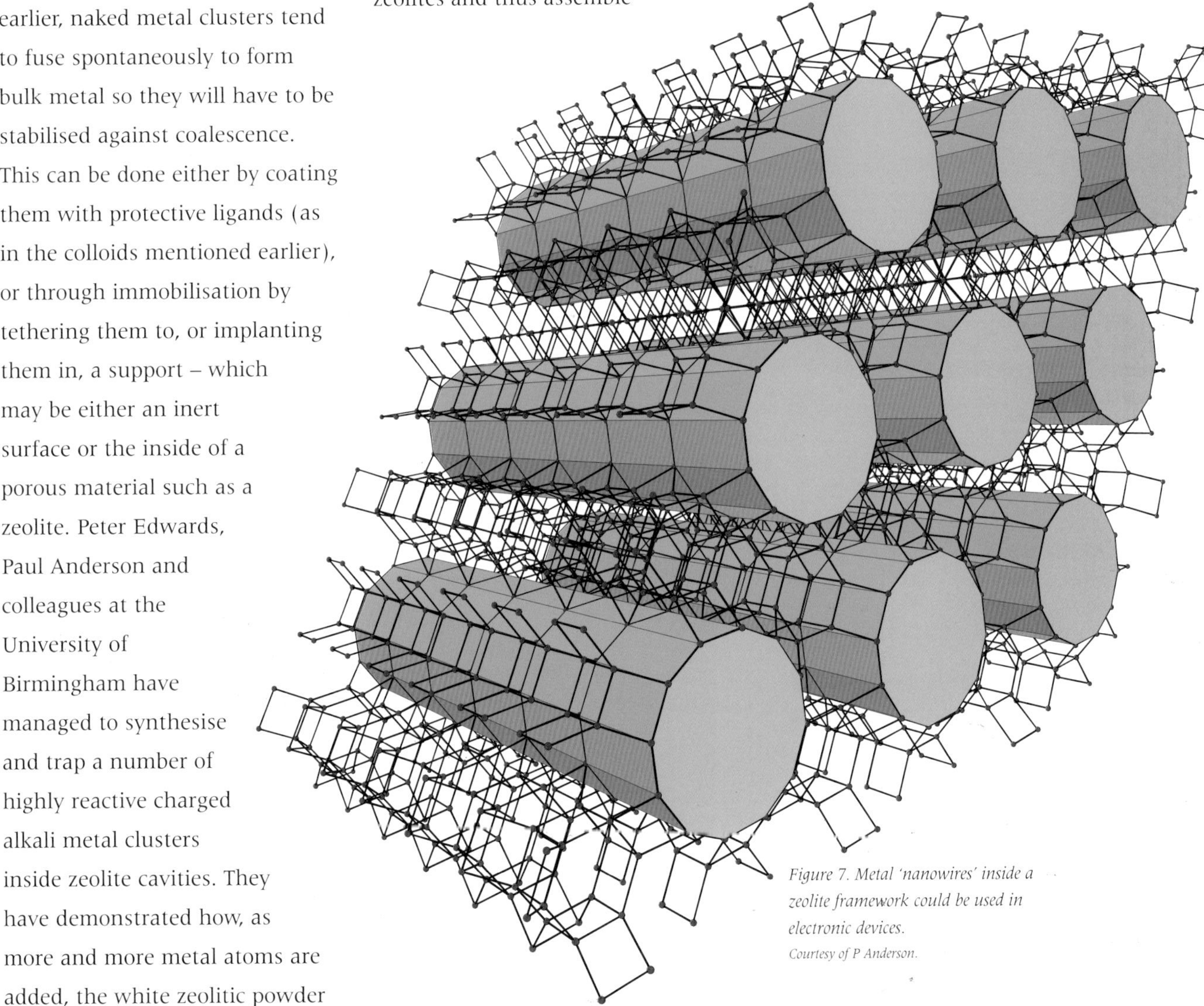

Figure 7. Metal 'nanowires' inside a zeolite framework could be used in electronic devices.
Courtesy of P Anderson.

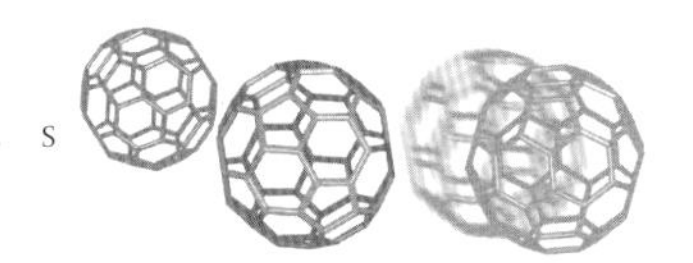

Semiconductor clusters and quantum dots

Electronic devices such as transistors depend on the imaginative application of semiconducting materials, and semiconductor nanoclusters have great potential here. To understand how needs a little more explanation of band theory. As explained before, in solids, the electrons are located in energy bands rather than single levels. The energy bands fully occupied by electrons are referred to as *valence bands*. Electrons that escape into unoccupied higher energy bands are free to move around the solid (electrical conduction) so these bands are referred to as *conduction bands*. In insulators and semiconductors, however, there is an energy gap between the highest energy occupied valence band and the lowest unoccupied conduction band. This *band gap* is extremely important in semiconductors where it may be narrow enough for electrons to traverse if enough energy is supplied (heat or light) or if the gap is modified in some way. The gap may conveniently lie in the visible or ultraviolet regions of the spectrum. If atoms which can donate or accept electrons are allowed to diffuse into the semiconductor (a process called 'doping'), they introduce 'impurity' states in the band gap which let through electrons or holes (a hole is a positive region resulting from a missing electron) into the conduction band when the semiconductor is heated.

Manipulating the band gap in semiconductors is one of the main tasks in designing new electronic and optical devices, and there has been considerable interest in generating semiconductor nanoclusters and studying how the cluster size affects the band gap in relation to that of the bulk. The clusters may be based on single elements such as silicon or germanium or on semiconducting compounds such as cadmium sulfide and gallium arsenide (these are referred to as II-VI and III-V semiconductors respectively after the Groups in the Periodic Table that the component elements belong to).

Semiconductor nanoclusters may be useful in optical computers where information is transmitted along optical fibres.

Studies of the electron energy levels in semiconductor nanoclusters have shown they are more widely spaced in small clusters so that the band gap increases as the clusters get smaller. For cadmium sulfide, for example, the band gap in small clusters is practically double that in the bulk material.

Semiconductor clusters could have one very exciting application – as *quantum dots*. These are minute regions of electrical charge (containing from one electron to several thousand) confined in what may be termed 'zero-dimensional' particles. The multitude of electrons trapped in these regions behave like electrons in isolated atoms. They sit in a series of energy levels and can be excited by absorbing energy (as light), just like atomic electrons. The electrons in nanoclusters embedded in an insulating matrix behave in just such a way. Quantum dots, which promise great sensitivity for measuring currents and local fields, may eventually be used as switching devices or transistors involving just a single electron. Their small size (nanometre dimensions) means that a large number of quantum dots can be packed onto a computer chip.

Semiconductor nanoclusters may also be of use in optical computers, where digital information is transmitted as light pulses travelling along optical fibres, as they open up the possibility of manufacturing very small, highly efficient *quantum dot lasers* which operate at low power, but which can produce laser light of high frequencies. Such lasers should be tunable, because the light emitted will depend on the band gap, and therefore the size of the component nanoclusters. Large metal oxide nanoclusters (or nanocrystals), derived from zinc oxide (ZnO), tin oxide (SnO_2) and other compounds have also found use as gas sensors, since the absorption of small amounts of gas molecules are known to affect their conduction properties.

The development of devices based on quantum dots depends on being able to fabricate semiconductor nanoclusters connected in one, two and three-dimensional arrays (so that information can pass between them in the form of electrons) while at the same time being protected from coalescence. A chemical solution to the problem is to precipitate out colloidal quantum dots (which are protected by a layer of organic surfactant molecules) as three-dimensional 'crystalline superlattices', analogous to the colloidal metals described above. Researchers have fabricated such quantum dot superlattices with colloidal cadmium sulfide or selenide nanoclusters carefully selected for size. These superlattices (either of semiconducting or of metallic nanoclusters) are fascinating examples of materials that are ordered on two different length scales: there is a regular arrangement of ions (or atoms) within the nanoclusters, with spacings of the order of 0.1 nanometres; and the nanoclusters themselves form a regular three-dimensional arrangement with nanometre spacings.

As we move into the 21st century, nanoclusters look set to play a pivotal role as components in novel electrical, magnetic and optical devices. Chemists will continue to contribute to this 'nanorevolution', as they develop improved methods of synthesis, stabilisation and assembly of single nanoclusters and arrays of nanoclusters.

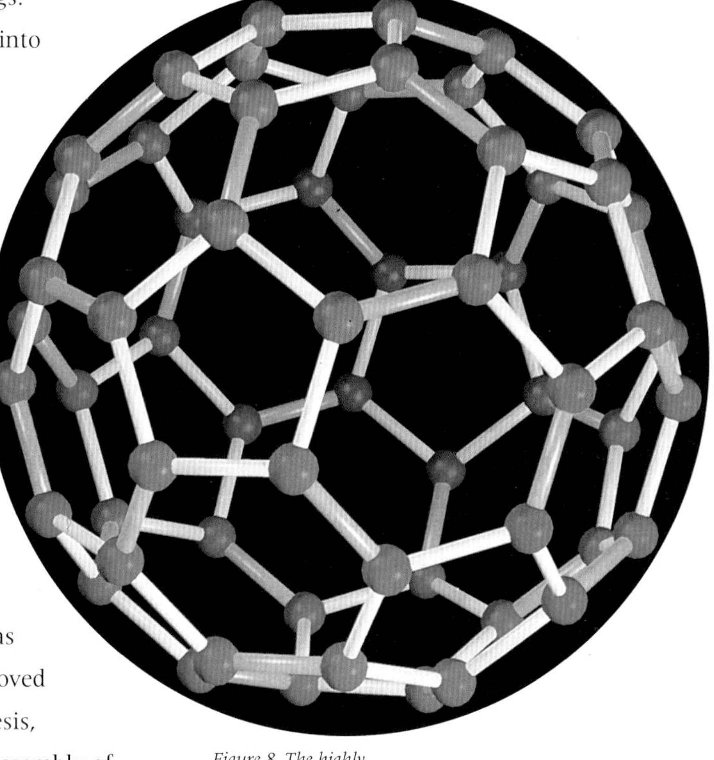

Figure 8. The highly symmetrical structure of C_{60}
J Bernholc et al, North Carolina State University/Science Photo Library.

C_{60} – the most symmetrical cluster

Perhaps the most famous nanocluster of all is the wondrously symmetrical football-shaped carbon molecule, C_{60} (Figure 8). Identified for the first time by the team of British and American scientists in 1985, it represents a third form of carbon (the other two are graphite and diamond). Since its discovery, C_{60} has caused tremendous excitement among scientists. It has opened the door onto a new world of related cage and tube-shaped carbon molecules with new chemistry (and physics) to explore. Like other clusters, C_{60} and relations have not only provided insights into our understanding of

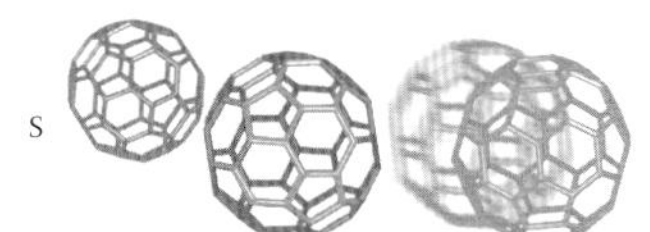

chemical bonding and structure but also offer the tantalising promise of novel applications in electronics, materials science and even medicine.

Although C_{60} became a subject of serious study only in the mid-1980s, people had speculated earlier on what would happen if you rolled up the hexagonal 'chicken-wire' structure of graphite into a ball. Indeed, in 1966, David Jones, writing his whimsical 'Daedelus' column in the popular science weekly, *New Scientist*, suggested that graphite might be persuaded to close up into a ball-shape consisting of hexagons and pentagons. The earliest serious research paper which describes C_{60} was by Eiji Osawa now of Toyahashi University in Japan published in 1970. It suggested that C_{60} should be a stable molecule, and not long after, Osawa and others published theoretical studies on what the properties of such a molecule might be. However, no-one managed to synthesise it, and very few appear to have tried – indeed few noticed these early studies.

The story of a discovery

The discovery of C_{60} is a story worth telling here because it beautifully illustrates how modern chemical advances are made: an idea leading to an experiment; an unexpected result which eventually brings together several strands of research involving international teams of scientists. It also shows how progress in chemistry depends on imaginative experiments combined with theory and a variety of complementary analytical techniques (see *Analysis and structure of molecules* for more information on these techniques).

The story starts in outer space with Harry Kroto (now Sir Harry) who was a microwave spectroscopist at the University of Sussex. He was interested in very long-chain carbon molecules called cyanopolyynes that might form in the atmospheres of carbon-rich stars. In 1984 another microwave spectroscopist Robert Curl at Rice University in Houston, Texas suggested that Kroto visit the laboratory of a colleague at Rice, Richard Smalley who was doing interesting work on clusters. He was using a laser to vaporise atoms from a solid target. The beam of atoms was then allowed to cool and condense into clusters which could be studied by mass spectrometry (in a similar manner to the study of sodium clusters mentioned earlier). When Kroto saw the apparatus Smalley was using, he concluded that it might shed light on his ideas on the creation of carbon chains in stars. Kroto suggested that he, Curl and Smalley do an experiment with carbon atoms. At the beginning of September 1985, the three researchers, together with postgraduates Jim Heath, Sean O'Brien and Yuan Liu set to work using a graphite disc

The British-American team of scientists who identified C_{60} for the first time in 1985.

Courtesy of H Kroto, University of Sussex.

to create the sooty clusters. The researchers already knew that clusters with certain magic numbers were favoured but were surprised to find that one number, 60, was much more 'magic' than the others; the mass spectrometer peak for C_{60} was bigger. Another peak that stood out was that for C_{70} (Figure 9). This was a great

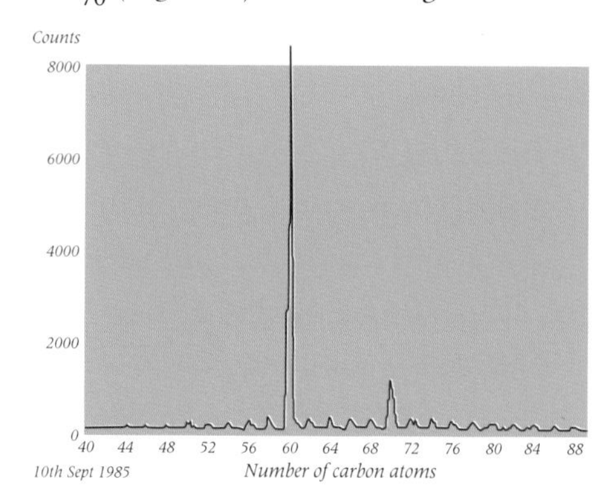

Figure 9. The mass spectrum obtained by vaporising carbon with a laser and allowing the atoms to condense. The peaks for C_{60} and C_{70} can be clearly seen.

surprise because a team at Exxon's laboratories in New Jersey had carried out similar experiments but had not noticed the larger C_{60} and C_{70} peaks.

What was the reason for the special stability of C_{60}? Clearly, there was something unusual about its structure. (The researchers had not then heard of the earlier speculations on C_{60}). They came to the conclusion that C_{60} was a closed spherical cage of carbon atoms similar in shape to the geodesic domes designed by the American engineer and philosopher R. Buckminster Fuller. It consisted of a hollow spherically shaped polyhedron composed of 20 hexagons and 12 pentagons with carbon atoms sitting neatly at the 60 vertices. When Smalley asked the head of Rice's mathematics department what this shape was called the answer was: "I could explain this to you in a number of ways, but what you've got there, boys, is a soccer ball!"

In fact, the 18th-century Swiss mathematician Leonard Euler had calculated that such a polyhedron must have 12 pentagons to close up into a spherical shape but can include a range of hexagons (any number greater than one). C_{70}, which is more elongated, has 25 hexagons. The researchers found that all the even-numbered carbon clusters of greater than 24 atoms were metastable (and less stable so than C_{60} and C_{70}) which inferred that they all had similar geodesic structure. C_{60} was dubbed *buckminsterfullerene* – later popularly shortened to 'buckyballs' in the popular press – in honour of the imaginative architect, while the new class of molecules were given the generic name of *fullerenes* (Figure 10).

A new type of carbon structure was thus born, and with it the potential for new chemistry. The evidence was still indirect although well supported by further experiments and theoretical calculations. The problem was that the researchers could produce only a few tens of thousands of molecules, which was not enough to characterise the molecules spectroscopically (which would establish their structures beyond any doubt), let alone do any real chemistry. Nevertheless, some informative experiments were carried out. The Rice-Sussex team managed to enclose a metal atom in the C_{60} cage by soaking a graphite sheet in a solution of a metal salt (such as those of potassium, lanthanum or uranium) and repeating the laser vaporisation-condensation experiments. Mass spectrometry results revealed that the C_{60} molecules had metal ions attached which could not be removed by irradiating them with an intense laser beam. This result reinforced the idea that the metal atoms were trapped inside the C_{60} structure (Figure 11). What is more, further laser-blasting of the C_{60}-metal compounds ruptured the carbon cage and released two carbon atoms before closing up again. The researchers found that they could gradually reduce the size of the carbon cage so as to 'shrink-wrap' the metal ions, the final cage dimensions depending on the ionic radii of the enclosed species.

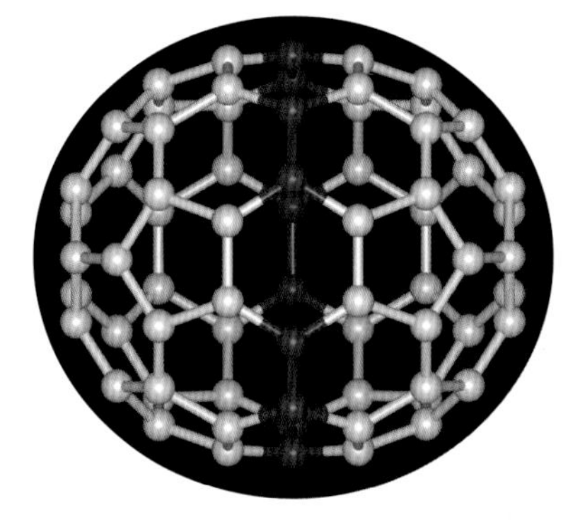

Figure 10. Carbon forms a range of closed cage-like structures – the fullerenes. C_{70} is shown here.
Courtesy of R Smalley, Rice University.

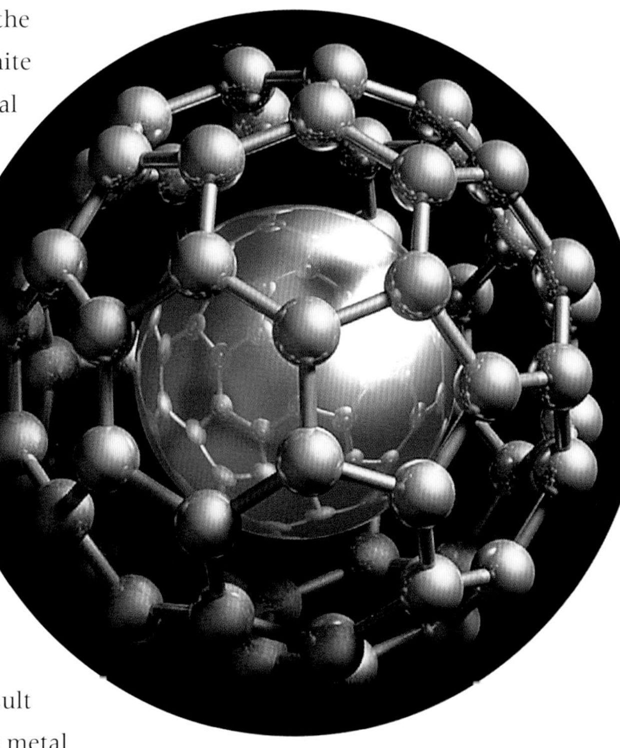

Figure 11. Metal ions can be trapped inside a C_{60} structure.
Ken Edward/Science Photo Library.

A breakthrough at last

Kroto, Smalley, Curl and their teams spent five frustrating years looking for a way to make larger amounts of C_{60} that they could analyse. Here's where another strand of the story links in, also starting in space. Two physicists, Donald Huffman at the University of Arizona in Tucson and Wolfgang Krätschmer at the Max Planck Institute for Nuclear Physics in Heidelberg, were interested in interstellar dust which they thought was probably mainly soot-like particles.

Huffman and Krätschmer

In the early 1980s, they started experiments to model the formation of soot in space by evaporating a graphite rod in a bell-jar of low-pressure helium. Huffman and Krätschmer collected the soot and measured its absorption in the far ultraviolet part of the spectrum. They noted that there were two strange camel-like humps in the spectrum which they thought must be due to contamination but did not examine the anomaly any further. However, later, when Kroto and his colleagues announced their putative C_{60} molecule, Huffman and Krätschmer decided to have another look at the artificially-produced soot, ingeniously speculating that the camel humps in its spectrum might be due to C_{60} (Figure 12). The Krätschmer – Huffman research teams measured the mass spectrum, and the ultraviolet and superb infrared spectra. They found that the results fitted the theoretical predictions for C_{60}, as did further infrared experiments using samples prepared from carbon-13 isotope instead of normal carbon-12. These results provided the crucial evidence for the presence of C_{60}.

By 1989 the physicists could make milligram quantities of what they now believed was C_{60}. The soot sample dissolved in benzene to form a red solution, which when evaporated the solution left behind red crystals. These turned out to be a mixture of 75 per cent C_{60} and 23 per cent of C_{70} and a few per cent of higher fullerenes, which Krätschmer and Huffman called 'fullerite'. Their research teams then went on to measure the crystal structure of fullerite

Polarised light micrograph of a thin film containing crystals of C_{60}
Michael W Davidson/Science Photo Library

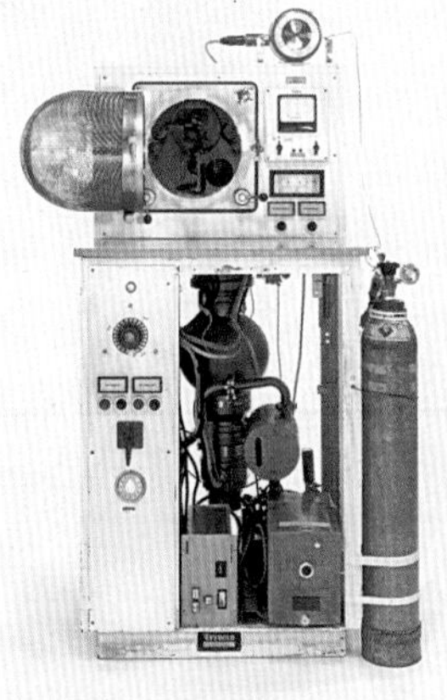

The Heidelberg carbon evaporator. With this apparatus bulk amounts of fullerenes could be produced for the first time.
Courtesy of Wolfgang Krätschmer, Max Plank Institute.

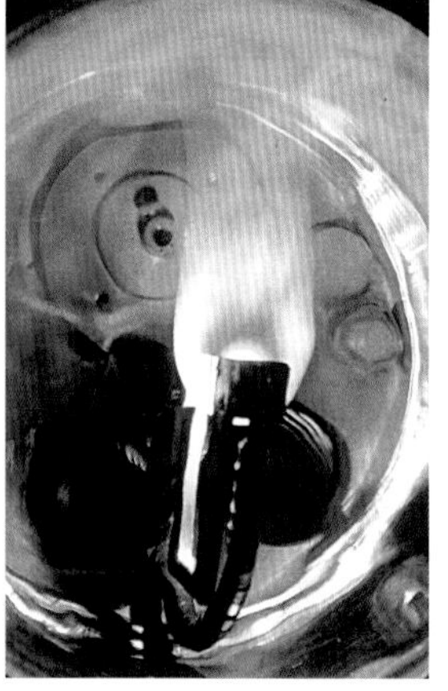

Carbon dust particles floating upwards in the current of helium gas, looking like a cloud of cigarette smoke.
Courtesy of Wolfgang Krätschmer, Max Plank Institute.

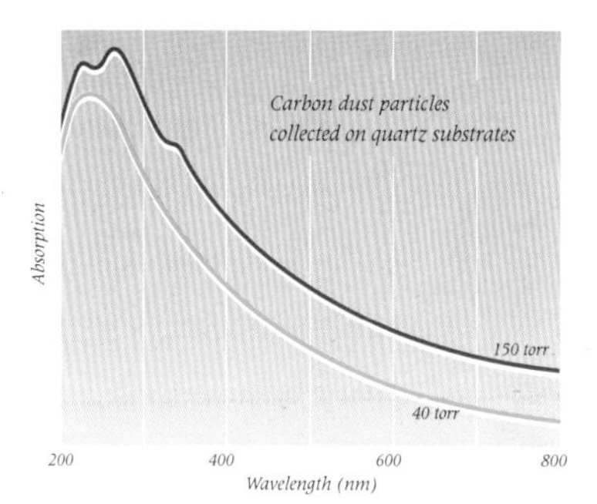

Figure 12. The UV spectrum of Huffman and Krätschmer's samples of 'soot' showed two camel-like humps which are due to C_{60}.

using X-ray and electron diffraction, confirming that the structure of C_{60} was indeed that of a spherical cage.

When the physicists announced their results in 1990, the researchers at Rice and Sussex had mixed feelings. They were delighted that their early ideas on fullerenes had been vindicated (there had been a lot of controversy in the preceding five years) but disappointed to be pipped at the post. Although dogged by lack of funding, Kroto's team had carried out similar experiments and were only just beaten to publication of results. Nevertheless, Kroto and colleagues at Sussex extracted C_{60} independently and subsequently were able to separate C_{60} and C_{70} chromatographically and take their nuclear magnetic resonance spectra. The carbon-13 NMR spectrum of C_{60} revealed the expected single peak (in this highly symmetrical molecule all the carbon atoms are in equivalent positions). The structure of C_{60} was confirmed beyond doubt. In 1996 the discovery of a new form of carbon was recognised through the award of a Nobel prize to the original team of Kroto, Smalley and Curl.

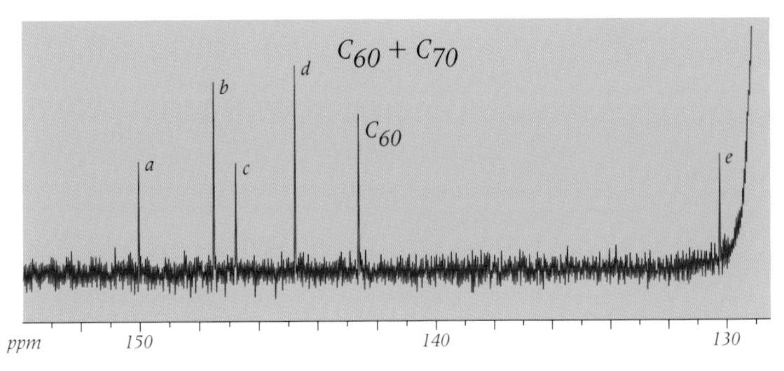

The ^{13}C NMR spectrum for C_{60} and C_{70}, where a-e denotes peaks for C_{70}.

©The Nobel Foundation.

Courtesy of Leah Bernard-Boggs, Rice University.

©The Nobel Foundation.

Sir Harry Kroto, Richard Smalley and Robert Curl who won the Nobel Prize for the discovery of C_{60}.

And now some chemistry

Once it was possible to make C_{60} in reasonable amounts, research groups around the world set about exploring its chemistry. Because of its unusual structure, C_{60} would be expected to have some fascinating electronic properties. The electron density across its surface is uneven with somewhat higher electron density in the six-membered carbon rings and lower electron density in the five-membered rings. In the solid state, C_{60} molecules pack together in a face-centred cubic lattice but are free to rotate at random, which is quite surprising (most molecules in the solid state don't rotate, see *Analysis and structure of molecules*). Such properties mean that C_{60} like silicon is electronically versatile. In 1991, researchers at AT&T Bell Laboratories in New Jersey showed that when crystalline C_{60} is 'doped' with metals such as potassium which can donate electrons, C_{60} forms negative fulleride, or 'buckide' ions and becomes electrically conducting. The highest conductivity is achieved with three potassium ions for every C_{60} molecule. However, the addition of further metal ions turns C_{60} into an insulator. The Bell Labs team further discovered that K_3C_{60} becomes a superconductor below about 18 K. This was improved on

Computer graphics representation of K_3C_{60}

Biosym Technologies, Inc/Science Photo Library.

by researchers at the NEC Fundamental Research Laboratories in Japan, who substituted potassium with rubidium and caesium to obtain a material with a T_c of 33 K – the highest T_c for a molecular material apart from the metal oxide superconductors mentioned earlier.

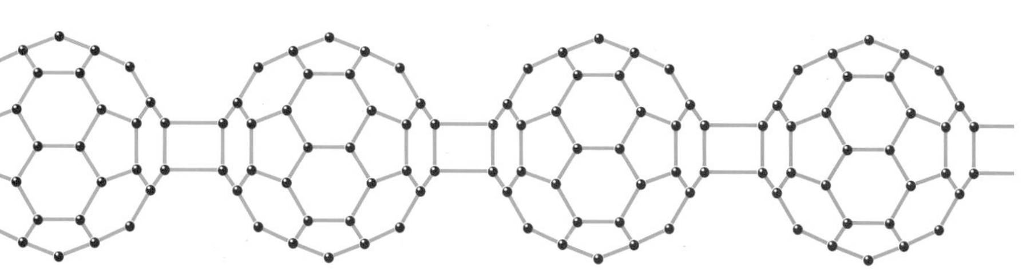

Figure 13. Fullerene polymers.

C_{60} also reacts with organic electron donors. For example, it reacts with tetrakis-(dimethyl-amino)ethylene (TDAE) to give (TDAE)C_{60} which is ferro-magnetic below 16 K – the highest T_c for a purely organic ferromagnet. Indeed, C_{60} seems to behave more like an electron-deficient 'super-alkene' (an alkene is an organic molecule with a carbon-carbon double bond which can add on various chemical species) rather than an aromatic molecule like benzene (C_6H_6) even though C_{60} contains benzene-type rings. C_{60} can add on osmium tetroxide, amines, hydrogen and various alkyl groups. It also undergoes addition reactions with halogens (fluorine, chlorine and bromine) to form a variety of C_{60}-halogen compounds. UK chemists, including the team at Sussex, with colleagues at Leicester University have prepared C_{60} derivatives in which each carbon atom has attached fluorine atoms sticking outwards. This represents a flourishing new area of organic chemistry.

Of particular interest is the formation of exotic C_{60} polymers. Visible or ultraviolet laser light causes C_{60} molecules to 'fuse' into a linear chain linked by four-membered carbon rings between each C_{60} molecule (Figure 13). The mechanism is thought to be similar to a very famous organic addition reaction called the *Diels-Alder cycloaddition* in which a molecule with a single double bond reacts with a molecule containing two adjacent double bonds to form a ring. In addition, C_{60} polymers can form under high hydrostatic pressure (5 gigapascals) at temperatures between 500 and 800 °C. There is now a rich and growing family of double and single-bridged polymers; some are made from alkali fullerides (such as RbC_{60} and Na_2RbC_{60}) and display metallic properties.

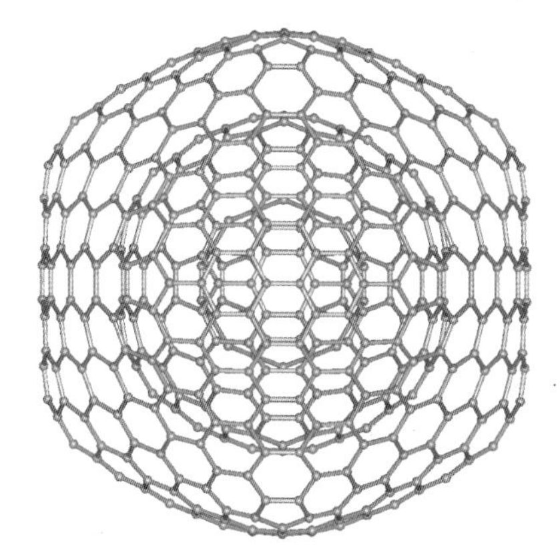

Figure 14. A typical very large fullerene.

A whole series of variants on the fullerene theme has now been synthesised including giant molecules (Figure 14) such as C_{240}, C_{540} and C_{960}, and nested molecules where the cages are arranged like a set of Russian dolls. It is most probable that soot particles formed in flames contain these kinds of species. Fairly early after the discovery of C_{60}, Japanese scientists prepared tube-shaped carbon molecules – *carbon nanotubes* which promise a revolution in materials science and electronics. Bundles of nanotubes should have the tensile strength between 50 and 100 times that of steel, conduct as well as copper and have electronic properties that could be exploited for quantum devices (see Box 2).

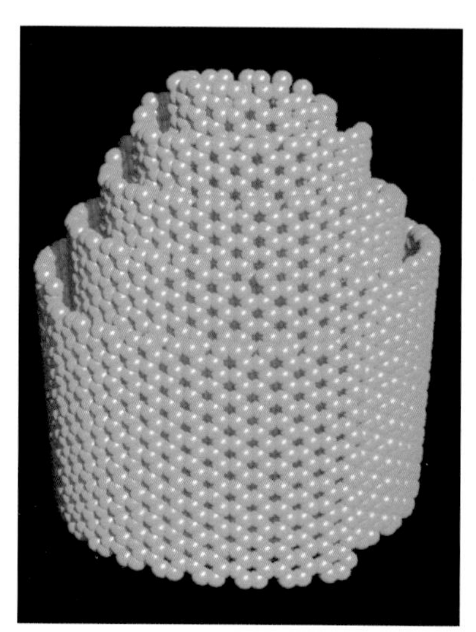

Molecular graphic of a nanotube.
Biosym Technologies, Inc./Science Photo Library.

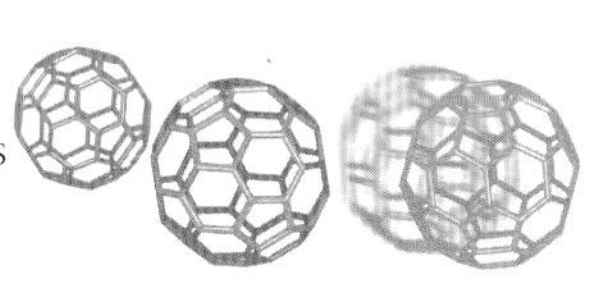

Box 2 Nanotubes

In 1991, Sumio Iijima at NEC Corporation in Japan discovered a new carbon variant consisting of multiple sheets of graphene (the chicken-wire graphite structure described earlier) rolled-up along an axis to give concentric *nanotubes* which are hollow in the middle. The nanotubes vary in length from 100 nanometres to several micrometres and the concentric graphene cylinders (between two and twenty arranged like Russian dolls) have an internal diameter of between 1 and 20 nanometres. They are 'capped' at each end by fullerene-type structures containing five and six-membered carbon rings (Figure a).

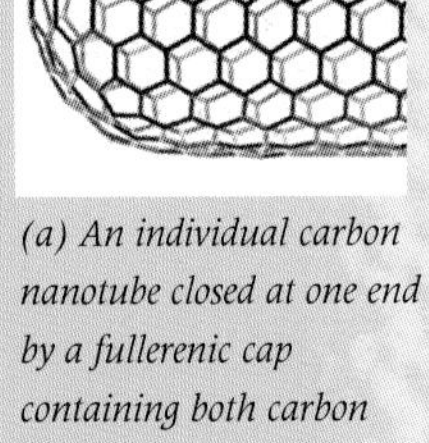

(a) An individual carbon nanotube closed at one end by a fullerenic cap containing both carbon pentagons and hexagons. The main cylindrical body consists solely of carbon (graphene) hexagons.

Iijima made the nanotubes using the principle of arc vaporisation (as for fullerenes) – in this case by passing an electrical discharge between two carbon electrodes in a chamber under dynamic vacuum with a partial pressure of helium. The nanotubes formed on the cathode. Later, several research groups produced nanotubes with just a single wall of carbon, using hollow carbon rods filled with various metals. Interestingly, the single-walled nanotubes (Figure b) are found on the walls of the arc chamber rather than on the cathode. These structures are truly molecular capillaries consisting of individual graphene cylinders with internal diameters only 1 to 5 nanometres across, as shown in the micrograph below.

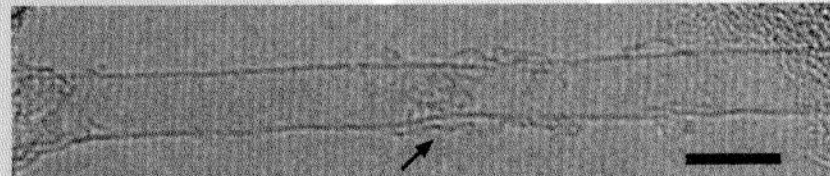

A high resolution transmission electron microscope (HRTEM) image of a single-walled carbon nanotube (scale bar = 4 nanometres). The 'kink' in the nanotube possibly arises from defects consisting of carbon heptagons either side of the arrow.

The potential applications of both kinds of nanotubes may be even more exciting than those of fullerenes. Shortly after Iijima identified his nanotubes, he and his colleagues showed that heating them to 400 °C with molten lead in air opened up the tubule ends and caused the liquid metal to be sucked into hollow cavities by capillary action – as with a drinking straw. The fullerene caps are more prone to chemical attack (because they contain carbon pentagons as well as hexagons) than the tubule sides (which contain only hexagons) so the nanotubes can readily be filled. There are two methods of filling: one is the drinking-straw method just described, which works well for liquid phases with a low surface tension at temperatures below 600 °C; the other is the solution approach which might, for example, involve filling an opened nanotube with a solution containing a metal salt, after which the material in solution is decomposed to give either a metal oxide or the base metal.

$$Ni(NO_3)_2\text{/MWNTs (in solution)} + \text{MWNTs} \xrightarrow{\text{(heat)}} NiO\text{/MWNTs} \xrightarrow[\text{(reduce)}]{} \text{Ni metal}$$

(MWNTs stands for 'multiple-walled nanotubes')

Both multiple-walled and single-walled carbon nanotubes have now been filled with a wide variety of different materials, including: metal oxides, base metals, mixed metal oxides, metal salts, metal sulfides, metal carbides, main group elements, gases and proteins. It is easy to guess what the uses might be. Because the internal dimensions correspond to only a few atomic layer distances of the encapsulated materials, surface effects predominate over the bulk properties. There exists, for example, the potential for performing size and site-selective catalysis on particles (see *Chemical marriage-brokers*) incorporated within nanotubes.

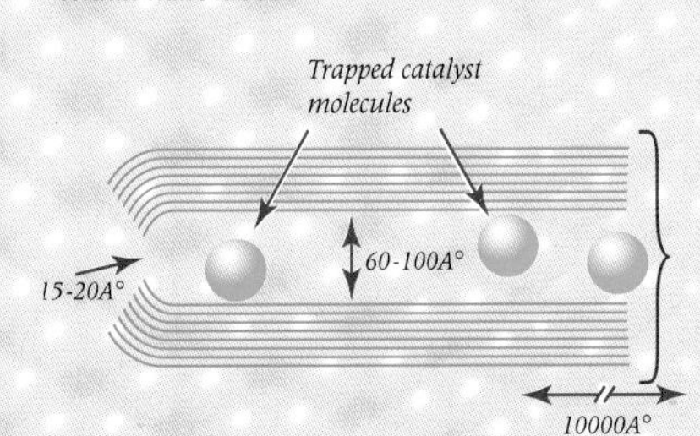

(b) A catalyst molecule trapped inside a multiple-walled nanotube. Size and shape-selective catalysis may be performed in such confined environments.

The properties of materials seen only at a microscopic level also become important, for example, electronic and optical quantum effects. Researchers thus hope to use carbon nanotubes to create new types of nanocomposite materials that could be incorporated into nanoscale electronic devices. Most recently, mixtures and individual metal halides, such as potassium and uranium chlorides ($KCl\text{-}UCl_4$), and silver chloride and silver bromide (AgCl-AgBr), have been inserted into single-walled nanotubes by capillarity. The silver halides can be decomposed to give 'metallic nanowires' of pure silver metal. These are the thinnest encapsulated wires in existence and can vary from 20 up to 100 nanometres in length.

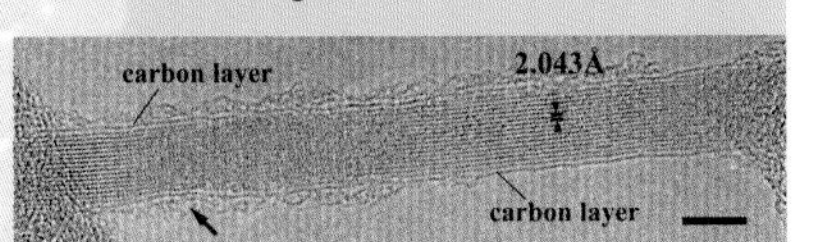

A micrograph showing a large diameter single-walled carbon nanotube containing silver metal after filling with silver chloride using the capillary method, followed by photolytic reduction (scale bar = 3 nanometres). The indicated d-spacing is the characteristic d(200) plane of silver metal. Note that the 'kink' in the nanotube wall (arrowed) causes a corresponding distortion in the silver wire.

Even large biological molecules such as cytochrome *c* and β-lactamase have been inserted into nanotubes, so there also exists the potential for developing biomedical applications – either as supports or biosensors. The potential for the development of new types of materials and devices based on nanotubes is vast.

Dr Jeremy Sloan and
Professor Malcolm Green
Oxford University

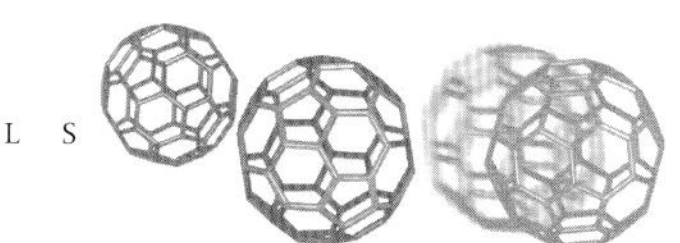

Finally, the simplest variant of C_{60} structure itself results from substituting one of the carbon atoms with nitrogen (an element with an appropriate number of available bonds – *ie* three) to give azafullerene ($C_{59}N$). The introduction of the nitrogen atom into the fullerene cluster strongly perturbs the electronic and geometric character of the parent C_{60} molecule, resulting in a very reactive $C_{59}N$ radical (a radical is a molecule with an unpaired electron). This radical rapidly joins with itself to form a $(C_{59}N)_2$ dimer (Figure 15) which readily reacts with alkali metals to afford an azafulleride ion $(C_{59}N)^{6-}$. The ion has been isolated and structurally characterised as a potassium salt and turns out to be electrically conducting. This kind of chemistry opens the way to yet another exciting set of materials with electronic properties potentially as rich as those of their fullerene antecedents.

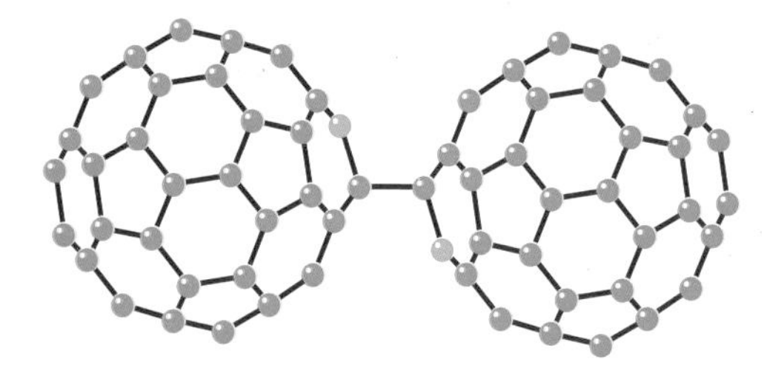

Figure 15. Azafullerene, $(C_{59}N)_2$.

Applications

What about applications? As you might expect, there are plenty of ideas for exploiting C_{60} and other fullerenes. C_{60} could provide a new kind of conducting or semiconducting material for batteries, transistors and sensors. Scientists have grown highly ordered thin films of C_{60} on other semiconductor substrates such as gallium arsenide so C_{60} could lend itself to the fabrication of the new generation of advanced electronic devices which are based on thin films. Thin films of the potassium buckide (K_3C_{60}) superconductor can also be made. C_{60} also has unique optical properties in that it is transparent to low intensity light but nearly opaque above a critical intensity. These 'nonlinear' optical properties mean that C_{60} could be used as an 'optical limiter' in optical digital processors and for protecting optical sensors from intense light.

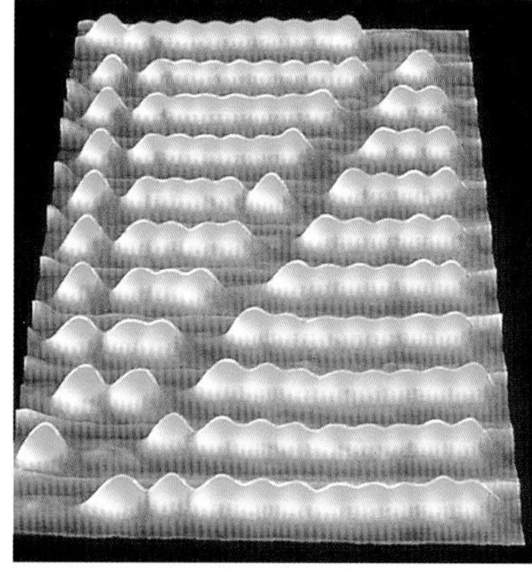

STM of C_{60} on a surface.

Another set of applications arises from the fact that various chemical species can be attached to the surface of fullerenes or even trapped inside. Finely divided carbon is a common component of industrial heterogeneous catalysts, often acting as a large surface substrate (see *Chemical marriage-brokers*). C_{60} balls could be coated with various catalytic materials, affording a huge surface area for catalytic activity. Chemists have managed to squeeze various molecules such as helium and other inert gases inside the fullerene cages. If the radioactive gas radon was incorporated into C_{60} molecules, they could be used to 'label' potential pollutants such as crude oil and toxic waste, or as an anticancer drug. Radon preferentially kills cancer cells, so radon-laden fullerenes could be conveyed to tumours by attaching tumour-targeting antibodies to the outside of the fullerene. In medical imaging, the deployment of radioactive tracers trapped in fullerenes would result in lower doses to patients since the fullerene would prevent the tracer from interacting with body tissues other than the one being targeted.

Conclusion

It has been said, in the recent past, that chemistry is a mature subject in which no new fundamental discoveries are likely to be made. The discovery and development of the fullerenes and other nanoclusters, and the high T_c superconductors has shown that chemistry can reveal new types of structures and phenomena in matter which not only extend our understanding of Nature at a very basic level but will also generate new technologies in the future.

Further reading

1. *Superconductors: the Breakthrough*, Robert M. Hazen, Unwin Hyman, London: 1988.
2. *The Engines of Creation,* K. Eric Drexler, Fourth Estate, London: 1990.
3. *"Microclusters",* M. A. Duncan and D. H. Rouvray, Scientific American, December 1989, p60.
4. *"Metal clusters and magic numbers",* Scientific American, December 1997, p30.
5. *"The colour of metal clusters and of atomic nuclei",* R. A. Broglia, Contemporary Physics, 1994, **35**, 95.
6. *Clusters (special issue), Science*, 1996, 271.
7. *"Welcome to clusterworld",* Richard Palmer, New Scientist, 22 February 1997, p38.
8. *"Fullerenes",* Robert F. Curl and Richard E. Smalley, Scientific American, October 1991.
9. *Perfect symmetry*, Jim Baggott, OUP, Oxford: 1994.
10. *The most beautiful molecule in chemistry,* Hugh Aldersey-Williams, Aurum Press, London: 1995.
11. *The chemistry of the fullerenes*, A. Hirsch, Thieme, New York: 1994.
12. *The fullerenes*, H. W. Kroto and D. R. M. Walton (eds), CUP, Cambridge: 1993.
13. *The fullerenes*, H. W. Kroto, D. E. Cox and J. E. Fischer (eds), Pergamon, Oxford: 1993.
14. *The chemistry of the fullerenes*, R. Taylor (ed), World Science, Singapore, 1995.
15. *Buckyworks,* J. Baldwin, Wiley, New York: 1996.
16. *"Carbon nanotube chemistry",* Jessica Cook, Jeremy Sloan & Malcolm H. Green, Chemistry & Industry, 19 August, 1996.
17. *"Through the nanotube",* Philip Ball, New Scientist, 6 July 1996.
18. *"Giants in their field",* Peter Rodgers, New Scientist, 10 February 1996, p34.

Glossary

Band theory The quantum theory that explains the behaviour of electrons in condensed matter, particularly metals and semiconductors. In molecules, the electrons occupy states with well-defined energies; in solids, interactions between neighbouring atoms in the lattice cause the electrons to occupy a range of states close in energy called bands.

Band gap The separation in energy between electronic bands in a solid, in particular the valence band (highest filled energy band) and the conduction band (lowest unfilled energy band).

BCS theory The theory devised by John Bardeen, Leon Cooper and Robert Schrieffer to explain the phenomenon of (low-temperature) superconductivity in metals and alloys. It postulates that pairs of electrons are able to 'surf' though the metal crystal lattice without resistance by coupling with the lattice vibrations. BCS theory does not explain the behaviour of high temperature superconductors, however (see *Cooper pair, Superconductor and High temperature superconductor*).

Buckminsterfullerene The name given to the spheroidal C_{60} molecule after the engineer R. Buckminster Fuller who created geodesic domes which C_{60} resembles.

Carbon nanotube A variant of carbon consisting of multiple graphite sheets which are rolled-up into hollow tube-like structures. They are 100 nanometres to several micrometres long, have an internal diameter of between 1 and 20 nanometres, and are 'capped' at each end by fullerene-type structures containing five and six-membered carbon rings (see *Fullerene*).

Colloid A finely divided suspension of one phase in another.

Colossal magnetoresistance A newly discovered effect shown by some manganese oxide perovskites whereby their electrical resistance changes hugely on application of a strong magnetic field (see *Perovskite, Magnetoresistance and Giant magnetoresistance*).

Conduction band A higher energy unoccupied band into which electrons can escape and move around in a solid (electrical conduction) (see *Band theory*).

Cooper pair The electronic entity responsible for superconductivity. Pairs of electrons whose spins are coupled travel through a crystal lattice without any resistance (see *Superconductor*).

Critical temperature (T_c) A temperature at which some kind of phase change occurs; it is used in particular to refer to the temperature below which some materials become superconducting (see *Superconductor*).

Diamagnetism A property by which the spins of all the electrons in a substance are paired up so that their magnetic fields are cancelled out. When placed in a magnetic field, the substance is weakly repelled by the field.

Diels-Alder cycloaddition A well-known organic reaction in which a molecule with a single double bond reacts with a molecule containing two adjacent double bonds to create a ring-like structure.

Ferromagnetism A special case of magnetism in which all the unpaired electrons in a substance are aligned so that a large magnetic field is produced (see *Magnetism and Paramagnetism*).

Fullerene A generic name for a form of carbon consisting of spheroidal molecules containing carbon atoms (see *Buckminsterfullerene*).

Geometric shell Concentric polyhedral shells of atoms found in clusters with hundreds of atoms or more.

Giant magnetoresistance An effect shown by stacked thin layers of iron and chromium whereby the electrical resistance of the layers changes dramatically when a magnetic field is applied (see *Magnetoresistance*).

High temperature (high-T_c) superconductor One of a group of mixed metal oxides which become superconducting at relatively high temperatures – above the temperature of liquid nitrogen (77 K) (see *Superconductor*).

Jellium model A quantum theoretical model that describes the electronic structure and behaviour of metal clusters. The electrons involved in cluster bonding move within the cluster under the influence of an attractive averaged electrostatic potential due to the positive nuclei. It bears similarities to the 'shell model' which accounts for the structures and stabilities of light and medium-sized nuclei (see *Nanoclusters*).

Josephson effect A quantum effect whereby a small current flows between two superconductors separated by a narrow insulating barrier (see *Superconductor*).

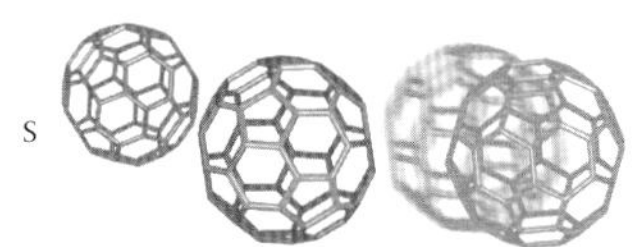

Magic number A number associated with exceptional stability in some groups of interacting particles obeying the laws of quantum mechanics. Magic numbers (2, 8, 20, 28, 50, 82) were first observed in the case of protons and neutrons in atomic nuclei but they are also seen in atomic clusters (see *Nanoclusters*).

Magnetism A fundamental physical property associated with charged particles which have spin, such as the electron. The spin, or magnetic moment, can be regarded as being either up or down. Usually, the electron spins in materials are paired in a spin-up spin-down configuration so that their magnetic moments cancel. In materials containing atoms with unpaired electrons, however, the electron spins may order in such a way as to give rise to various macroscopic magnetic effects (see *Ferromagnetism*).

Magnetite A common ore of iron with composition Fe_3O_4 which behaves as a natural magnet.

Magnetoresistance A phenomenon whereby the resistance of a material changes when a magnetic field is applied.

Meissner effect The expulsion of a magnetic field by some types of superconducting materials.

Metal carbonyl cluster A discrete molecular compound containing three or more metal atoms held together with metal-metal bonds and carbon monoxide molecules as ligands.

Nanoclusters Tiny particles that contain between tens and millions of atoms with properties lying between those of single atoms and bulk materials.

Nanoscience The study of particles and structures with dimensions of typically one-billionth of a metre (10^{-9} m).

Nanowire An electrical conductor only a few nanometres in diameter, but hundreds of micrometres (and ultimately metres) in length.

Nanotechnology A new technology developed on the nanometer scale in which functional structures (mechanical or electronic) are constructed, often on a molecule-by-molecule basis.

Organic metal Organic molecules or polymers which possess electrical (and often optical) properties similar to that of a metal, for example, electrical conduction.

Paramagnetism A property arising in substances whose atoms or molecules contain one or more unpaired electrons. Paramagnetic substances are weakly attracted towards a magnetic field.

Perovskite A type of inorganic structure which is a mixed metal oxide, whose name derives from the natural mineral calcium titanium oxide ($CaTiO_3$). Some perovskites show interesting electronic properties such as colossal magnetoresistance and superconductivity (see *Colossal magnetoresistance*).

Quantum dot A minute region of confined electrical charge in a solid (for example, a metal cluster embedded in an insulator) which has distinct quantum properties, behaving in effect like an isolated atom (see *Nanoclusters*).

Quantum dot laser An ultra-small highly efficient semiconductor laser based on quantum dot technology (see *Quantum dot*).

Semiconductor A material that becomes electrically conducting only above a certain temperature.

Spinel ferrites A family of technologically important mixed metal oxides containing iron, and other metals such as manganese, cobalt, nickel and zinc, with the general formula MFe_2O_4.

SQUID (Superconducting quantum interference device) A small electronic device that exploits a quantum property of superconductors called the Josephson effect to detect minute magnetic fields (see *Superconductor and Josephson effect*).

Superconductor A substance that has no electrical resistance below a certain characteristic temperature (called the critical temperature, T_c).

Superparamagnetism A property shown by clusters of metal atoms whereby individual atomic spins are strongly coupled by exchange interactions to give a high total magnetic moment. The cluster behaves like a giant paramagnetic atom.

Transition element An element in which the valence *d* orbitals are partially filled. They are all metals, conducting heat and electricity well. In general their chemistry is characterised by the formation of highly coloured salts and ability to exist in more than one oxidation state.

Valence band The energy band fully occupied by electrons bound in atoms in solids (see *Band theory*).

Biographical details

Dr J Paul Attfield is a lecturer in chemistry at the University of Cambridge and a member of the Cambridge IRC (Interdisciplinary Research Centre) in Superconductivity. His research interests are centred on the chemistry, crystal structures, and physical properties of transition metal oxides.

Dr Roy L Johnston is a lecturer in inorganic chemistry in the School of Chemistry, University of Birmingham. His research includes theoretical cluster science; structures and dynamics of clusters; size-induced metal-insulator transitions; electronic structures of molecules, clusters and solids, and the application of genetic algorithms to optimisation problems in chemical physics.

Sir Harry Kroto (see Epilogue)

Professor Kosmas Prassides is Professor of Solid State Chemistry in the School of Chemistry, Physics and Environmental Science, University of Sussex. His research interests encompass the chemistry and physics of materials with novel conducting and magnetic properties, including fullerenes and transition metal oxides.

Acknowledgements

The authors would like to thank Professor Peter Edwards for overseeing this chapter.

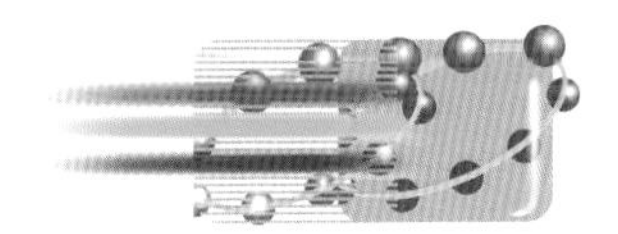

Computational chemistry and the virtual laboratory

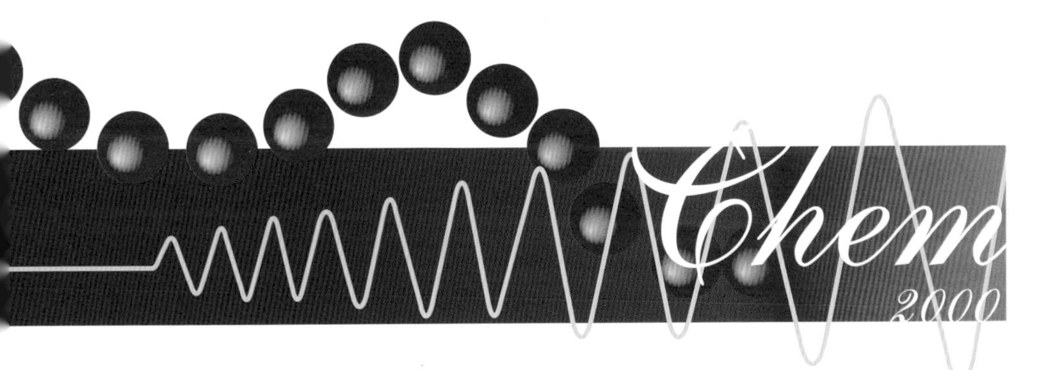

Dr Andrew Leach
GlaxoWellcome

Computational chemistry is an indispensable tool in scientific research and is extensively used throughout industry and universities. Computer graphics images are also widely used in popular scientific journalism and in TV science programmes.

Created by T. Larsen, The Scripps Institute, La Jolla, California.

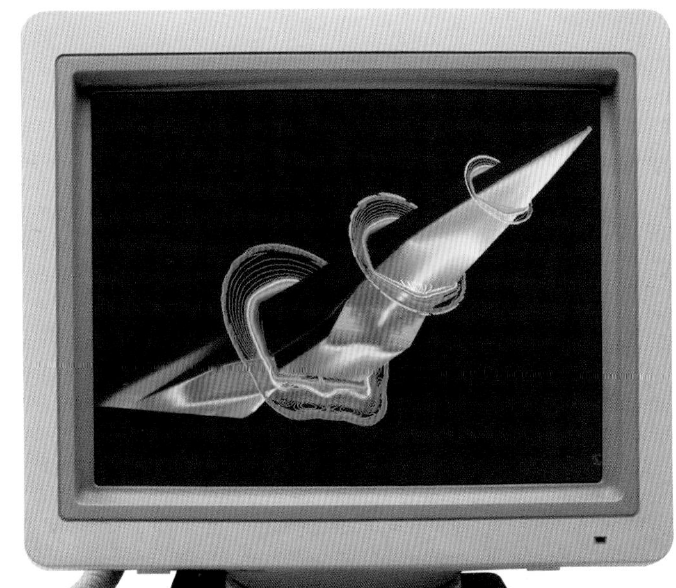

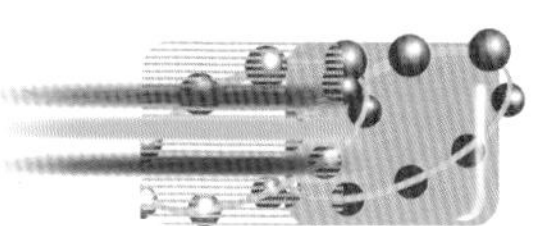

The rapid development of powerful computers over the past 20 years has allowed chemists to investigate a wide range of chemical behaviour without dirtying a single test-tube. Highly usable software packages based on a wide variety of ingenious computational methods are now available. These can be used to build computer models that probe the intimate details of simple reactions, predict the characteristics of materials such as catalysts and polymers, and visualise the interactions of biomolecules and drugs. Computational chemistry is now an indispensable tool in chemical research.

GlaxoWellcome

Isis facility, Rutherford Appleton Laboratory.

Pick any chemistry journal from the library shelf and there is a good chance that in it you will find at least one piece of research that involves some form of computational chemistry. This branch of the subject (also referred to as molecular modelling or theoretical chemistry) used to be restricted to a few specialists with access to expensive computer equipment and expert programming skills. The situation is greatly different today; computational chemistry is widely available and used throughout academia and industry. At Glaxo Wellcome, for example, many of the medicinal chemists (whose main role is to synthesise potential drug molecules) are proficient users of molecular graphics software, visualising the three-dimensional structure of their target protein and considering what modifications might enhance the potency of their molecules.

The most visible face of molecular modelling is undoubtedly computer graphics. With current software packages it is possible to conjure wondrous images of all kinds of complex

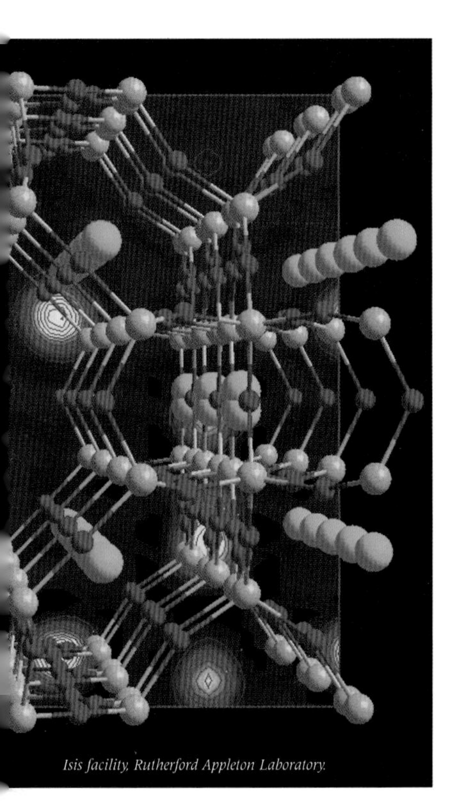

Isis facility, Rutherford Appleton Laboratory.

molecules. Computer graphics images pervade not only the chemical and biochemical scientific literature but are widely used in popular scientific journalism and in TV science programmes. Nor has the power of such images been lost on Hollywood, which has made regular calls on the modelling community to provide suitable pictures to enhance the realism of their movies.

Nevertheless, molecular modelling is much more than just pretty pictures. Today's modeller has available an extensive armoury of computational methods to build up a description – usually presented in the form of these easily understood visual graphics – that reveals details of chemical behaviour: for example, what happens during the transitional stages as one molecule reacts with another, how simple organic molecules 'adsorb' onto metal surfaces (which may be important in catalysis) and the way a protein embedded in a cell membrane can change its shape (important in designing drug molecules to interact with cellular receptors).

There are three main reasons why computational chemistry is so widely used. First, it can provide insights into the underlying molecular behaviour of a system – information that may be difficult, if not impossible, to obtain by any experimental technique. Secondly, it enables researchers to evaluate many possible choices, in terms of reactants, reaction conditions and so on, before undertaking any experimental work, thereby enabling them to make more effective use of available resources. Thirdly, computational methods can be used to investigate systems under extreme conditions such as very high pressures and/or temperatures – which cannot be reproduced in the laboratory but which exist in the Earth's core or in stars.

The key to the widespread use and acceptance of molecular modelling is undoubtedly the phenomenal rate of growth in the power and availability of computers. In recent times, computational power has doubled every couple of years. Even more astonishing is how the amount of computational power obtained for a given sum of money has increased (Figure 1). The personal computers that are so widespread both at work and in the home are

Box 1 Computers used today for computational chemistry

Twenty or so years ago, computational chemistry was the province of a small and select group of researchers who had access to the large and cumbersome mainframe computers. As computers have become cheaper and faster, and thus more accessible, so the methods of computational chemistry have become available to a wider audience. This is also due to the availability of the software packages that can be obtained from one of the many commercial software vendors, or from academic research groups.

The mainstay of computational chemistry is the graphics workstation. Usually running the Unix operating system, these computers can not only be used for quite detailed calculations but also offer the capability to view the results using computer graphics. However, as the graphics performance of such machines can be seriously affected if other programs are running simultaneously, it is usual to perform long calculations on a server machine. Such machines, which may be located many miles away, sometimes even on a different continent and accessed via the internet, will often have gigabytes of memory and terabytes of disk storage available. These servers may contain many processors running in parallel or just a few superfast processors. The performance of such machines is often measured in FLOPS (which stands for floating-point operations per second). The current generation of machines is able to perform 10^{12} such operations and so are known as Teraflop machines.

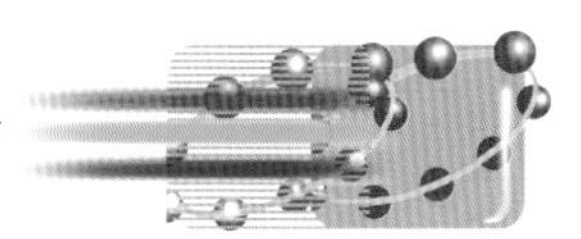

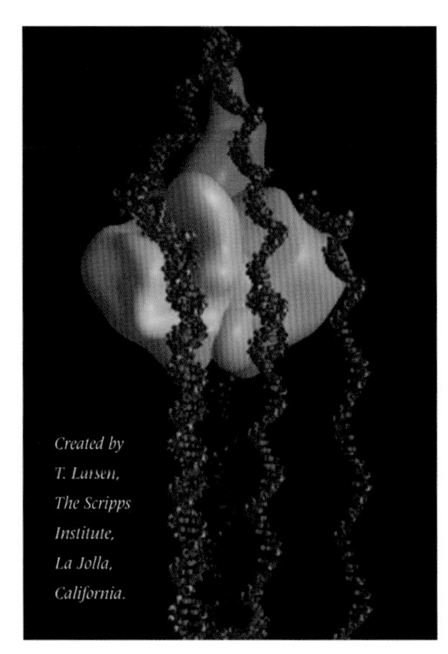

as powerful as the expensive mainframe computers of only a few years ago. A vivid illustration of this is provided by the personal experience of Graham Richards, a leading computational chemist. The Ferranti Mercury mainframe computer he used when working in Oxford as a graduate student in the early 1960s, filled rooms of an old house. Yet it was only a 32K machine which had to be programmed in machine language or autocode!

The other reason for the spread of molecular modelling is the availability of off-the-shelf software packages that are easy to use without being a computer expert. Many programs can now be freely obtained from the Internet, so reducing the cost even more. The increases in computer power and new developments in the methodology mean that we are now able to apply computational chemistry methods to problems in a way previously only dreamed of.

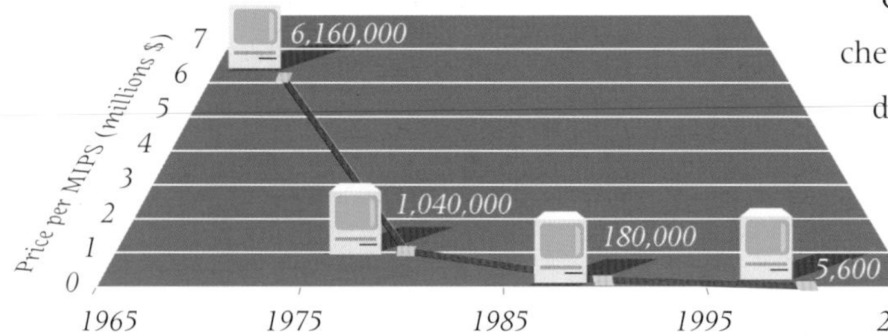

Figure 1. Graph of computer power vs cost per year. Price per MIPS (Millions of Instructions Per Second) of an entry level IBM manufactured over the last three decades. Data courtesy of ComputerWire plc

Computational chemists have at their disposal a variety of methods for studying a particular system. Deciding which is the appropriate method to use depends upon the kind of questions that they want to ask and on the available computational resources. Modellers study systems ranging from single isolated molecules to proteins containing thousands of atoms immersed in a sea of solvent molecules, yet at the heart of all these calculations is some procedure for calculating the energy of the system. Energies can be calculated using two basic methods known as *quantum mechanics* and *molecular mechanics.* Quantum mechanics offers the most fundamental approach but is restricted to relatively small systems. Molecular mechanics is particularly useful for modelling large molecules and assemblies of molecules.

Quantum mechanical methods

A quantum mechanical calculation starts with the famous *Schrödinger wave equation,* which describes the behaviour of a system of particles such as an atom or molecule in terms of their wavelike, quantum nature. It then tries to solve the equation and calculate the energy levels of the system. In most quantum mechanical calculations, the equation is first simplified by separating the motions of the nuclei and electrons and then ignoring the nuclear motion – the so-called *Born-Oppenheimer approximation.* This is possible because the nuclei are much heavier than the electrons, which can therefore react almost instantaneously to any changes in nuclear positions. If the nuclei do alter position then the equation can be solved for each position.

The Schrödinger equation can be written concisely, almost cryptically, as:

$$H\Psi = E\Psi$$

Here, Ψ is the wavefunction, and H is the so-called hamiltonian operator. The wavefunction has no physical meaning but its square is a probability, often an electron density. The hamiltonian is a mathematical device used to calculate the energy of the system and other physical properties. When written out in full, the equation consists of a very large

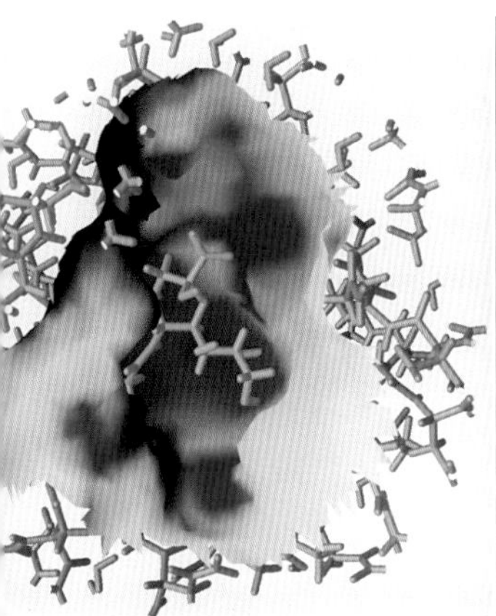

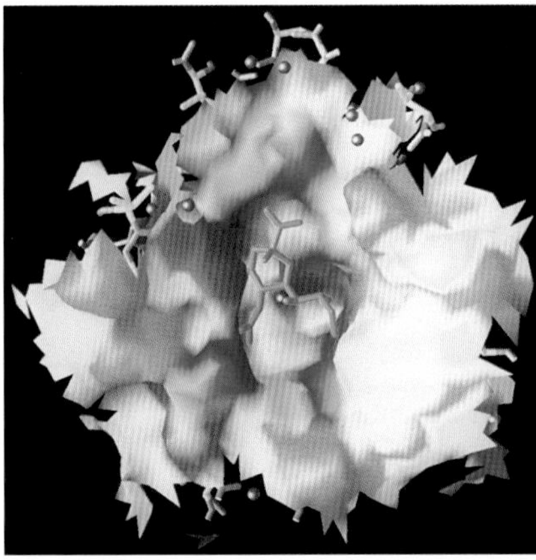

number of terms involving differentials and summations which take account of the attractions and repulsions between all possible pairings of particles and their kinetic energies. Evaluating these terms is out of the question without computers, and progress to bigger molecules has been coupled to computational advances. The resulting energy, E, can tell us about energies as a function of molecular shape or composition, and from the wavefunction, Ψ we can compute physical properties such as dipole moments (the distribution of electronic charge in the atom or molecule).

The Schrödinger equation can be solved exactly only for a limited set of simple problems. Indeed, the only chemical problem for which an exact solution can be obtained is the simplest species, the hydrogen atom. For all other systems we are forced to accept that we can only ever find an approximate solution (though that approximate solution may well be extremely close to the 'real' solution). This is mainly because systems with more than one electron constitute a 'many-body problem' which is notoriously mathematically intractable. There are no exact solutions for even a three-body problem such as the next simplest atom, helium, with two electrons and one nucleus.

Theoretical chemists therefore need to think up ingenious ways of finding the approximate solutions. A key component of most of these approaches is the *variation principle*, as this provides a simple way to compare and evaluate one approximate solution with another. This principle states that the 'better' the wavefunction (in other words, the closer it is to the 'true' wavefunction) the lower the energy. Hence the 'best' wave-function is associated with the most stable or lowest energy requirement.

The self-consistent field approach

One of the first, and still probably the most widely used ways to find solutions to the Schrödinger equation is the *Hartree-Fock self-consistent field* method. It works like this. Suppose we were to 'freeze' all of the electrons bar one, then we could then find a solution by considering an approximate wavefunction for the single, free electron moving in the field generated by all the others. We could then consider another electron (freezing all others, as before), and so on, until all the electrons had been treated. However, the drawback with this approach is that any solution we might find for the first electron will influence the solution for the remainder (because of interactions between the electrons). In the actual strategy used, we obtain an initial set of trial solutions for all the electrons simultaneously. These solutions are then used in the second iteration to provide better solutions (of lower energy). The results from the second iteration are applied to a third, and so on until the results for all the electrons are unchanged at which point they are said to be self-consistent. William Hartree did all these tedious calculations at Cambridge in the 1930s and 40s

on mechanical calculators aided by other members of his family; among his students was the current President of The Royal Society, Nobel Prizewinner Aaron Klug.

Aaron Klug.
Courtesy of The Royal Society.

There are, in fact, many types of calculation to choose from. In general, the 'better' the calculation the more correct the results, though this is not always the case. However, more accurate calculations require bigger and faster computers – as might be expected! Two important classes are *ab initio methods* and *semi-empirical methods*. The first type of calculation, as the name suggests, works from first principles, using as input only certain physical constants. Semi-empirical methods, on the other hand, replace some of the complex integrals required for *ab initio* calculations with information derived from experiment. This enables rather accurate calculations to be performed in a fraction of the time that a comparable *ab initio* calculation would take.

Cray© T94™.
Courtesy of Cray Research, Inc.

As quantum mechanics includes the electrons in the calculation as well as the nuclei, (and the number of electrons typically greatly exceeds the number of nuclei) quantum mechanical calculations, especially *ab initio* calculations, are often very demanding of computer resources. Fortunately, the pace of improvement in computer hardware, together with software developments have allowed quantum mechanical calculations to be performed on fairly large molecules including even small proteins and DNA, for example, by Ian Hillier at the University of Manchester employing a CRAY computer.

Walter Kohn won the Nobel Prize for Chemistry in 1998 partly for pioneering work on density function theory.
Associated Press

One methodological development in recent years has been the *density functional theory* (DFT) which enables accurate calculations to be performed much more efficiently than traditional quantum mechanical methods, at least for certain types of systems. DFT still uses the self-consistent field approach but does not rely on the wavefunction. Instead it calculates all the electron interactions in terms of the electron densities within the system. This approach has the potential to give the system's exact energy and is now being used to perform accurate calculations on crystalline solids (especially those with defects), molecules on surfaces and large transition-metal complexes. Nevertheless, for practical purposes, quantum mechanical calculations are restricted to single molecules containing a few tens of heavy (non-hydrogen) atoms and the most detailed calculations are restricted to systems with just a few heavy atoms.

Comparing theory and experiment

The true test of any theoretical technique is how well its predictions agree with the values obtained from experiment. The systems chemists are interested in are often so complex that the experimental data are usually regarded as 'definitive'. Any discrepancies are usually put down to deficiencies in the theory! One celebrated case where theory proved to be in advance of the experiment is that of the methylene (CH_2) radical. This is an extremely simple species, with just two hydrogen atoms bonded

Methylene radical
H C H — Linear; H C H θ — Bent

Figure 2. Linear and bent versions of the methylene radical.

to a central carbon atom (Figure 2). The carbon atom carries an additional pair of electrons. The geometry of this radical had been investigated using both theoretical and experimental techniques. The early debate concentrated on the lowest energy or ground state of the molecule and whether its geometry is linear or bent.

The first theoretical studies were performed by Foster and Frank Boys at the University of Cambridge in 1960. They concluded that the equilibrium geometry had an H-C-H angle of 129°. There was, then, no experimental data for CH_2, but in the following year one of the leading spectroscopists of the time, Gerhard Herzberg at the University of Ottawa, concluded from his experiments that the molecule was linear. Unfortunately for Boys, his head of department at Cambridge – Christopher Longuet-Higgins – was not at that time convinced by computational techniques but favoured an empirical approach which predicted a linear geometry.

Events came to a head when Charles Bender and Fritz Schaefer at the University of California at Berkeley also calculated a bent geometry (with an H-C-H angle of 135.1°) and concluded that the energy barrier between the bent and linear structures was so large that no further improvements to the theoretical model would remove it. Soon after Bender and Schaefer's paper, several experimental research groups reported new data on the molecule showing that the molecule was indeed bent. In the light of these findings, Herzberg re-examined his data and concluded that they were in fact consistent with a bent structure; he had originally chosen the linear structure because it seemed more reasonable.

Frank Boys is now considered to be one of the founding fathers of computational quantum mechanics. For those of us who have grown up with modern-day computers it is perhaps difficult to imagine why there was such scepticism from his peers as to the usefulness of his work. It is, however, important to realise that in those days it was a far-sighted individual indeed who could have predicted the dramatic advances in computer hardware and software.

Two other key individuals who have helped to bring quantum mechanics to the non-expert user are John Pople and Michael Dewar – both British-educated scientists who subsequently emigrated to the US. Pople was instrumental in developing the theory of both *ab initio* and semi-empirical methods. His 'Gaussian' series of programs rapidly became the standard for *ab initio* calculations. Pople was awarded the Nobel Prize for Chemistry in 1998, for his pioneering work in computational chemistry. Dewar developed the semi-empirical theories and incorporated them into a series of packages which are also widely used around the world. These two pioneers had for a time contrasting philosophies, with Pople aiming to make his semi-empirical calculations produce answers that aspired to the more exact *ab initio* calculations, while Dewar aimed at direct agreement with experiment.

John Pople.
Associated Press

Quantum mechanics really scores for problems that involve the making or breaking of chemical bonds and so has been extensively used to investigate the detailed mechanisms of reactions – the nature of the intermediate or the transition state that forms when molecules react. Indeed, it is only now, with the advent of lasers producing ultra-short pulses that are we able to observe directly the course of a chemical reaction experimentally, and even then only simple reactions are currently amenable to the technique (see *Following chemical reactions*). For most of the reactions of interest to organic chemists, only indirect experimental evidence is available from which we can try to infer features about the actual mechanism.

Professor K. Seddon and Dr T. Evans Queen's University Belfast/Science Photo Library

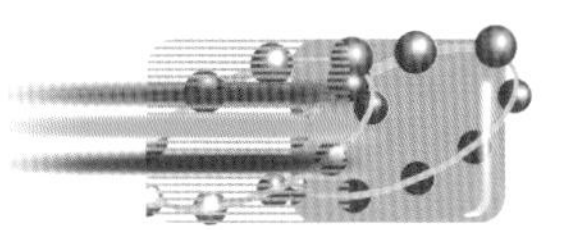

Using quantum mechanics it is possible to 'observe' the reaction all the way from the starting materials through any transition state or reaction intermediate to the products. The famous Diels-Alder reaction for introducing cyclic structures into organic molecules is a classic example that has been studied theoretically and experimentally in some detail (see Box 2).

The Diels-Alder reaction is one of a set of reactions (pericyclic reactions) whose outcome can be predicted by the Woodward-Hoffmann rules. Robert Woodward at Harvard was perhaps the world's most famous chemist in his heyday during the 1960s and one of the first to embrace quantum theory (see *Make me a molecule*). Roald Hoffmann who is based at Cornell University in the US is well-known for his many talents (which include poetry). In the area of computational chemistry he developed a particular type of semi-empirical theory, also in the 1960s. Whilst being simple in nature, it had the key advantage that it can be applied to most systems including those containing metals which are difficult for many of the alternative approaches to deal with. This is one reason why Hoffmann's methods have proved to be so popular with bench chemists working on organometallic compounds. In a complicated iron complex such as $(CO)_4Fe(C_2H_4)$, the Hoffmann approach considers the overall interaction as that between fragments – in this case the three fragments consisting of four carbonyl (CO) groups, ethylene (C_2H_4) and a central iron atom (Figure 3). Although this does not provide very accurate energies it

Roald Hoffmann.

Courtesy of Charles Harrington, Cornell University Photography.

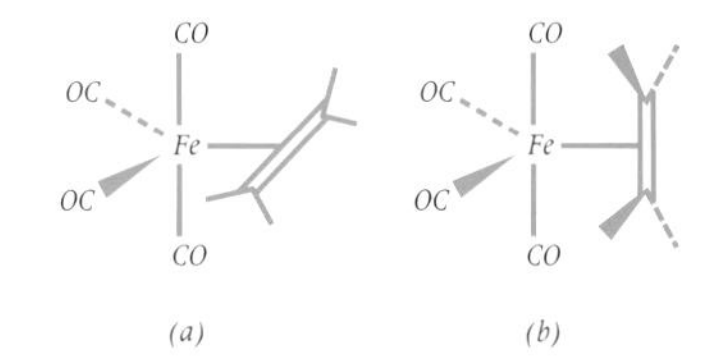

Figure 3. The structure of iron complex $(CO)_4Fe(C_2H_4)$. There are two possible orientations of the ethylene molecule, (a) and (b), giving two different structures of the complex. Simple calculations show that (a) is more stable, in agreement with experiment.

Box 2 The Diels-Alder reaction – one step or two?

The Diels-Alder reaction in which a diene (two double bonds separated a by single bond) adds on to a 'dienophile' (a molecule with an isolated carbon-carbon double bond) to form a ring-shaped structure is an extremely important route to synthesising new organic molecules. Chemists have, therefore, been interested to know the intimate details of the mechanism of addition. A useful way to consider these different mechanisms is to think about how the energy and geometry of the system changes as the reactant molecules progress through to the products. The Diels-Alder reaction involves the formation of two new bonds. In the *stepwise* mechanism these bonds are formed one at a time. The system passes through a reaction intermediate which is the high energy minimum. In the *concerted* mechanism the two new bonds are formed simultaneously. The energy passes through a single maximum at the transition structure. Two different types of concerted mechanism are possible; in the synchronous mechanism the bonds are formed to equal degrees, whereas in the asynchronous mechanism there is an imbalance in the degree of formation.

The big question is: "Does this pass through a one-step, or concerted transition state or is it a stepwise mechanism? If concerted, is the transition state symmetrical or non-symmetrical?" Computations indicate that a concerted symmetrical transition state is indeed involved, as is also suggested by certain key experimental data known as kinetic isotope effects.

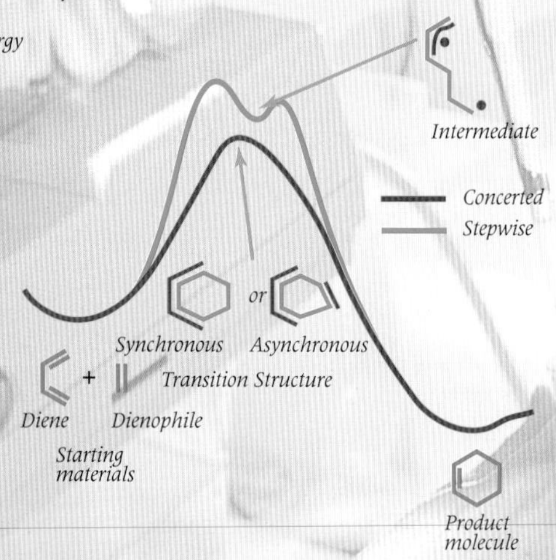

The Diels-Alder reaction: are the two new bonds formed stepwise or simultaneously? The stepwise mechanism passes through a distinct intermediate species as indicated by the well in the energy profile. The concerted mechanism passes through a transition structure – there is no well in the energy profile. It could be either synchronous (involving a symmetrical transition structure) or asynchronous (involving an asymmetrical transition structure).

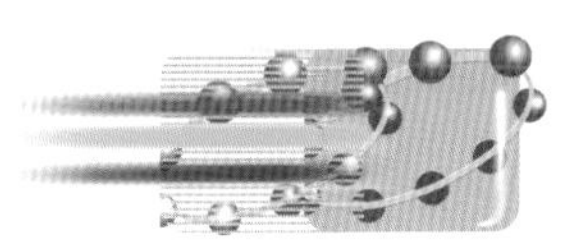

does offer a good picture of the shape of the molecule – that the ethylene molecule is in the plane of a trigonal bipyramidal structure – and how the energy should vary as the ethylene is rotated. This so-called *extended Hückel molecular orbital method* has proved to be powerful for understanding organometallic reactions, including those important in catalysis.

The modelling power of molecular mechanics

When we wish to model larger molecules, or assemblies of molecules, quantum mechanics is usually not feasible. Such systems are the realm of the molecular mechanics, or *force field method*. The molecular mechanics approach considers the energy of any arrangement of atoms – for example, a particular geometrical shape governed by rotation of bonds, the *conformation*, of a molecule – to be broken into several distinct parts, as shown in Figure 4.

First, there is a contribution from the stretching or compressing of bonds. Each of the bonds in the molecule has an 'ideal' value – the equilibrium value of the bond length. For a carbon-carbon single bond, the ideal bond length is around 0.154 nanometres; for a C=O bond it is about 0.122 nanometres and for a nitrogen-nitrogen triple bond it is approximately 0.110 nanometres. Energy must be expended to force a bond to deviate from its ideal value. To a reasonable first approximation we can model the variation in the energy with the degree of stretching or compression using a Hooke's law relationship, similar to that used to explain the behaviour of a real-life physical spring (in other words, the energy depends on the square of the deviation from the ideal value). The force constant, which determines by how much the energy changes for a given deviation, varies from one bond to another, reflecting the underlying strength of each bond. Thus the force constant for the nitrogen-nitrogen triple bond is greater (so the bond is stiffer) than that for a carbon-carbon single bond. The second contribution to the molecular mechanics energy is that due to 'angle bending'. As with bond stretching, there is an 'ideal' value for each angle, and deviations from this ideal value require the expenditure of energy. A Hooke's law relationship is again most commonly employed, with ideal values for atoms having a tetrahedral arrangement of bonds (such as carbon single bonds) being around 109.5°, and for carbon atoms with double bonds being 120°.

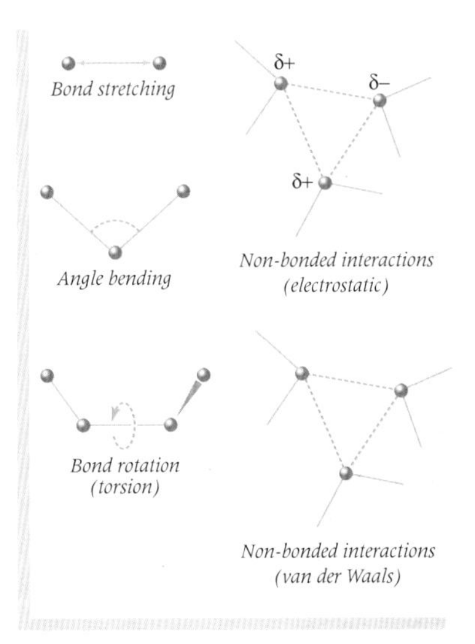

Figure 4. A schematic illustration of the four most important contributions to a molecular mechanics force field, due to the stretching of bonds, the bending of angles, the rotation about bonds (torsion) and van der Waals and electrostatic interactions between atoms.

Most of the variation in the structure of a molecule comes not from bond-stretching or angle-bending, but from rotation about bonds. It is well known that the energy of a molecule varies with its conformation; in other words, the particular three-dimensional configuration arising from bond rotation. Perhaps the simplest example of this is ethane (C_2H_6). As the carbon-carbon bond rotates, the interaction between the hydrogen atoms changes, creating a three-fold periodicity in the energy of the molecule. To enable molecular mechanics to model this

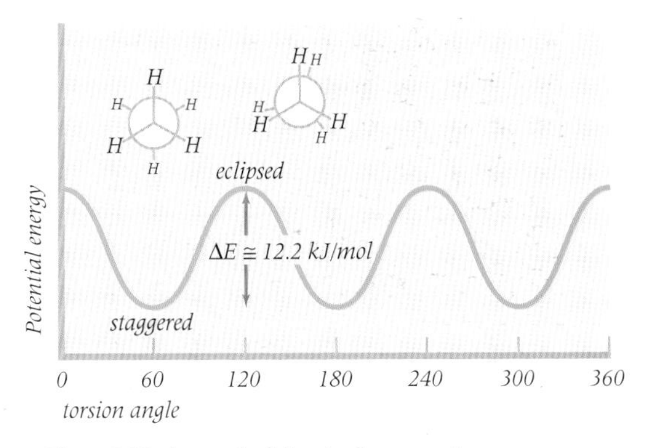

Figure 5. Maxima and minima in the energy of ethane. As the central carbon-carbon bond in ethane rotates the energy of the molecule changes in a periodic fashion, as shown. The minimum energy structures occur when the hydrogen atoms are in the staggered conformation; the maximum energy structures arise when the hydrogens are eclipsed. These two conformations are schematically represented in the figure.

behaviour, a so-called torsional potential is employed which enables the locations and relative magnitudes of the potential energy minima and maxima to be reproduced (Figure 5).

Motions of this kind can be detected by infrared spectroscopy (see *Analysis and structure of molecules*). The fourth contribution to the molecular mechanics energy arises from non-bonded forces. The non-bonded forces are usually modelled by a combination of electrostatic interactions and so-called van der Waals interactions. The electrostatic force arises from the interaction between permanent charges within the molecule due to uneven distributions of electron density in the atoms. A common way to model this is to associate a fractional point charge with each atom in the molecule. The interactions between these charges are determined using Coulomb's law, which states that the energy between two charges is proportional to their product divided by their separation, with charges of opposite sign attracting each other and charges of like sign repelling.

Electrostatic forces cannot be the whole picture, however, for it is well-known that non-bonded interactions are present in systems such as the rare gases, where there cannot be any permanent charge distribution. These interactions are due to van der Waals forces. Their origin is rather more subtle than the straightforward electrostatic interaction. Although the electron distribution around an atom is spherically symmetrical on average, the distribution at any one instant is non-uniform, leading to an instantaneous electric dipole. This instantaneous dipole can then induce a dipole in a neighbouring atom. The two dipoles now attract each other. As the atoms approach closer and closer they attract more and more until an energy minimum is reached, whereupon they start to repel each other (Figure 6).

To calculate the energy of even a simple molecule using molecular mechanics requires a large number of parameters for

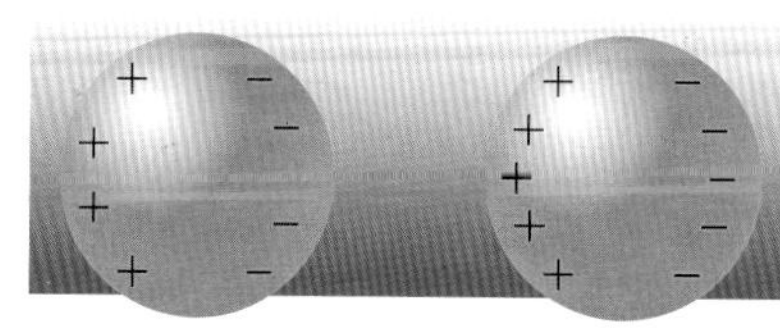

Figure 6. An instantaneous dipole in one atom (for example, in the atom on the left) can induce a dipole in a neighbouring atom, causing the two atoms to be attracted to each other.

the different terms in the energy expression. Determining these parameters is a time-consuming process and is one of the main drawbacks with the approach. A key difficulty is that the parameters are all interrelated to a greater or lesser degree, and so changing, say, one of the van der Waals parameters may require the electrostatic and torsional parameters to be modified as well. In contrast, all that is required for an *ab initio* quantum mechanics calculation are the atomic numbers and the coordinates of the atoms. Some of the published force fields are intended for use on a very limited set of molecules (such as proteins or nucleic acids), whereas others are designed to be much more general insofar as they are able to deal with a much wider range of molecules. The ability to transfer parameters from one system to another is a key attribute of a force field, for it enables a set of parameters, developed using a relatively small set of molecules, to be used much more widely. In particular, parameters developed for small systems can be applied to much larger molecules such as polymers.

A frequent problem with molecular mechanics, particularly when applying it to many different types of molecules (as happens in the pharmaceutical industry) is that some of the parameters needed will be missing, thus necessitating what is often a time-consuming process of 'parameterisation', producing new coefficients for the various terms in the energy expression. This parameterisation can be done by using available experimental data or quantum mechanical calculations.

Conformational analysis

One of the most common applications of molecular mechanics is in *conformational analysis*. This is the task of

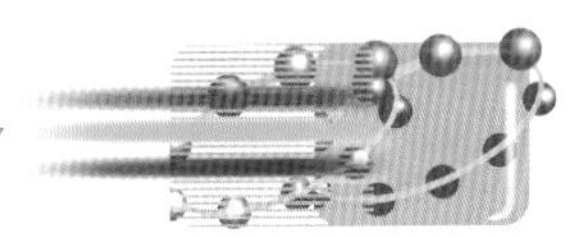

exploring all the ways in which the various parts of a molecule rotate in relation to each other – their so-called *conformational space* – in order to identify the conformations with the lowest energy. These are obviously the conformations the molecule is most likely to have. The importance of the conformation in affecting the behaviour of molecules dates from the pioneering work of Derek Barton, then at Imperial College, London and Odd Hassel at the University of Oslo (who were jointly awarded the Chemistry Nobel prize for their work) (see *Make me a molecule*). Even simple molecules can show some conformational variation, one of the most celebrated being butane (C_4H_{10}). As the central carbon-carbon bond in butane rotates, the energy of the molecules varies in the manner shown in Figure 7a. As we can see, there are three minimum points which correspond to three stable arrangements. Moreover, one of these minima is more stable (lower in energy) than the others. The conformations corresponding to these minima are also shown in Figure 7b. A more typical example of the type of molecule for which conformational analysis is required is Enalapril (Figure 8), used to treat hypertension.

Derek Barton. Reproduced with permission from Chem. Eng. News, April 3 1995, 73 (14) p28. ©1995 American Chemical Society.

(a) Potential energy

14 kJ/mol

3.8 kJ/mol

ca.20 kJ/mol

torsion angle

0 60 120 180 240 300 360

(b)

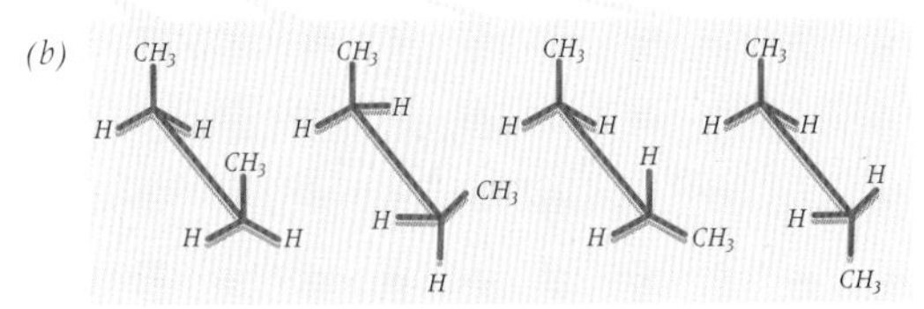

Figures 7. The conformational energy surface of butane contains three minimum energy structures. The lowest energy or anti conformation occurs when the torsion angle about the central carbon-carbon bond is 180 degrees. The two so-called gauche conformations correspond to torsion angles of 60/240 degrees. These gauche conformations are equal in energy but higher than the anti structure.

As a molecule varies its conformation, the changes in energy can be likened to the variations in altitude on the surface of the Earth. The altitude varies with the latitude and longitude. In some locations, such as the English Lake District, it can alter dramatically with location.

In the case of a molecule, the energy (which is equivalent to the altitude) varies with the torsion angle values of the rotatable bonds. There are three special types of points in both the geographical and the molecular systems. Points of maximum altitude correspond to the tops of fells; these may be either local or global. In the Lake District the highest peak is Scafell Pike; this is the global maximum. Helvellyn,

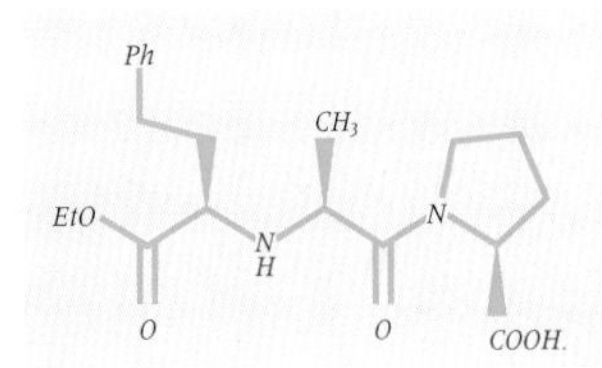

Figure 8. The structure of Enalapril.

Nigel Morrison

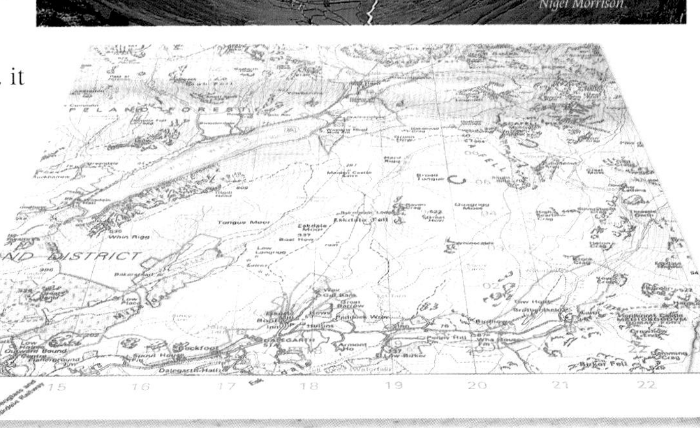

The variation in the energy of a molecule as it changes its conformation can be likened to the way in which the height above sea-level on the Earth varies with the geographical location. The height is a function of just two variables (the longitude and the latitude), whereas a molecule's energy can be a function of many variables. The altitude is commonly represented using a contour map or relief diagram. The English Lake District provides a dramatic illustration of the way in which the altitude can vary with position. Three key features on both energy surfaces and contour maps are the maxima (conformations higher in energy than their neighbours, or the tops of mountains), minima (conformations lower in energy than their neighbours, or valley floors) and saddle points (conformations at the highest point between two minima, or mountain passes).

Map courtesy of Ordnance Survey, the National Mapping Agency of Great Britain.

on the other hand, is only a local maximum as it is the highest peak in its locality but is lower than the global maximum. Points of minimum altitude are found in the valleys (or, more precisely, at the bottom of the lakes).

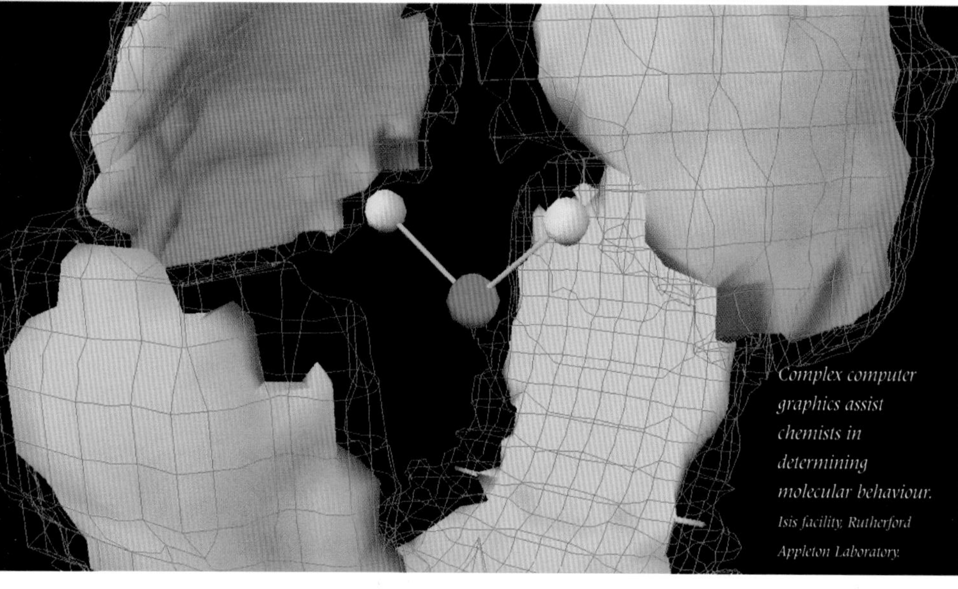

Complex computer graphics assist chemists in determining molecular behaviour. *Isis facility, Rutherford Appleton Laboratory.*

Again, the concept of local and global minima are important here; the global minimum point of altitude is located somewhere at the bottom of the deepest lake (Wastwater). The third important type of feature are the saddle points; these correspond to mountain passes and are the easiest path from one minimum to another. On the molecular energy surface we are interested primarily in conformations corresponding to energy minima because these are the most stable structures of a molecule. It may also be important to take account of the actual energy value; only those conformations that are within a few kilojoules of the global energy minimum are likely to be accessible to any significant extent.

There are several different ways in which to carry out a conformational search. Two of the simplest methods are known as the systematic search and the random search. The purpose of methods such as these is to locate conformations of molecules in the region of conformational minima. These three-dimensional structures are then passed to a *minimisation program* which actually locates the true position of the nearest minimum energy structure.

Viagra. *Professor K. Seddon and Dr. T. Evans, Queen's University Belfast /Science Photo Library*

A systematic search algorithm assigns to each rotatable bond in the molecule (typically every single bond) a range of permitted torsion angles. The easiest way to do this is to start at 0° and add some chosen constant increment (say, 30°). This is repeated for every rotatable bond. All possible combinations of the torsion angle values are then considered by the program. A systematic search can be thought of as equivalent in our geographical analogy to drawing a rectangular grid over a contour map of the region and systematically examining the altitude at each point in order to locate the minima.

The random search is completely different in philosophy to the systematic search. Whereas the systematic search moves over the energy surface in a predictable fashion, the random search jumps all over it in an unpredictable manner. To generate a new conformation, each rotatable bond is assigned a random value. The resulting conformation is then minimised. Another conformation is then randomly generated and minimised, and so on.

In a systematic search, it is obvious when the end of the search has been reached. In a random search, there is no defined end-point. To be sure that all of the conformational space has been explored, it is necessary to run the search until we are confident that no new conformations can be found. This invariably means that each minimum energy conformation will be produced several times during the calculation.

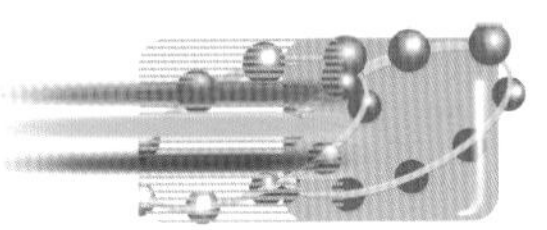

Ensembles of molecules

So far, we have been concerned only with calculations performed on single molecules. Usually, the molecule of interest will be in a solution, a solid, a glass or a liquid crystal. Such systems contain many molecules all interacting with each other. (Single isolated molecules are typically observed only ever in the gas phase.) So what are the important differences between single molecules and ensembles of molecules? An obvious practical difference is that the systems are much larger. Another key difference is that ensembles can exist in many different degrees of order, and the *entropy* plays a much more significant role. The molecular mechanics and quantum mechanics methods that we have met so far provide energies equivalent to the internal energy, or *enthalpy,* of a molecule. What really matters, of course, is the *free energy*. This is related to the enthalpy and the entropy by the well-known thermodynamic relationship $\Delta G = \Delta H - T\Delta S$.

The entropy is a measure of the number of different arrangements (or states) accessible to the system. A single molecule has contributions to its entropy from translational, rotational and vibrational motion as well as some contribution from the different conformational states available to it. It is possible to calculate these contributions quite accurately for small systems. For a macroscopic system, such as a solution, there are vastly more states available. The dilemma we face is that even with the fast computers of today we cannot model the behaviour of systems containing macroscopic numbers of molecules. Recall that Avogadro's number is 6 x 10^{23}, whereas the largest computers can deal with only terabytes of information ('tera' stands for one million million, or 10^{12}). Fortunately, *statistical mechanics* provides us with a theoretical framework for calculating 'bulk' properties using much smaller systems – containing a manageable number of molecules – and still obtain meaningful results that can be compared with experiment. This is the realm of *computer simulation.*

Molecular simulation

The two major techniques used in molecular simulation are *molecular dynamics* and the *Monte Carlo method*. The molecular dynamics method solves Newton's equations of motion for the atoms in the system. The very first molecular dynamics simulations, performed in the late 1950s by Bernie Alder and T. E. Wainwright at the University of California at Berkeley, used a 'hard sphere' model for the atoms in the system. This is analogous to a set of snooker balls moving about on a snooker table. Simple application of Newton's equations of motion enables us to predict the locations and velocities of each ball in an imaginary box at any point in time. In this model there are no interactions between the particles except when they collide. The collisions are perfectly elastic.

Even with such a simple model it was possible to show the existence of both solid and fluid phases with a defined phase transition between the two. This is beautifully shown in some of the earliest examples of the use of molecular graphics in computational chemistry

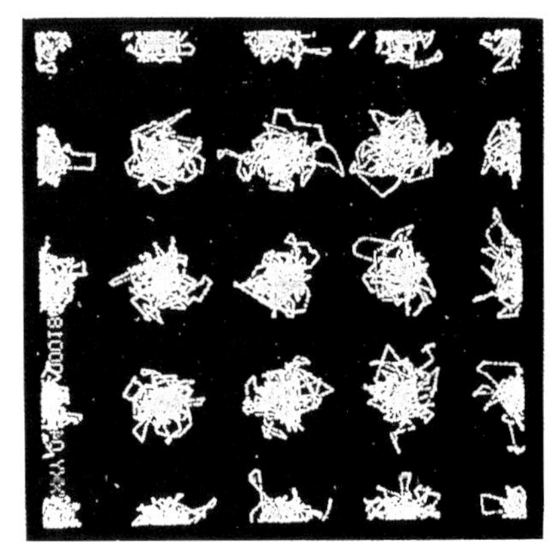

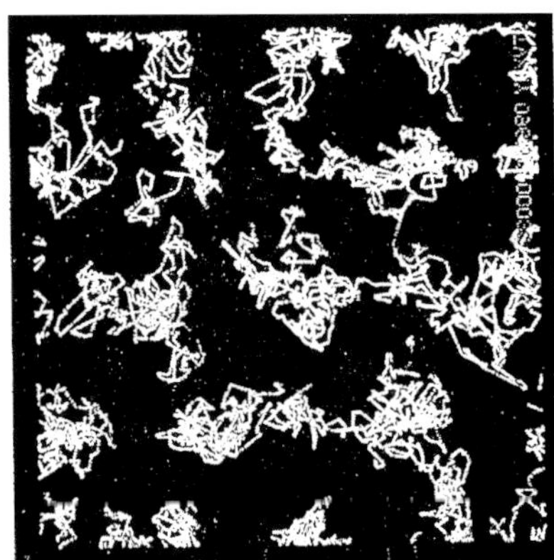

Figure 9. This early molecular graphics image shows the paths generated by 32 hard spherical particles (atoms) from molecular dynamics calculations. The picture on the left corresponds to the solid phase material and that on the right to the fluid phase. As can be seen the particles move much further away from the original positions in the fluid phase.

Reprinted with permission from B. J. Alder and T. E. Wainwright, Journal of Chemical Physics, 31, 459-466, (1959).©1999 American Institute of Physics.

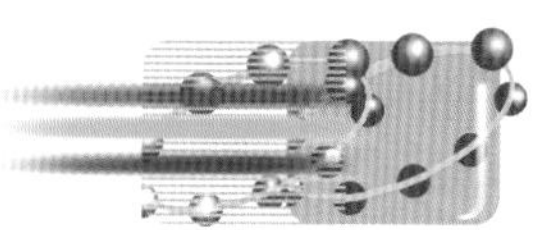

(Figure 9). In reality there are both attractive and repulsive forces between the molecules due to the non-bonded forces. Within a molecule, of course, there are even stronger forces due to the bonding between the atoms. The first simulation using 'realistic potentials' was of liquid argon by Rahman in 1964, who also performed the first molecular dynamics calculations of liquid water with Stilinger at Bell Laboratories in 1971. The simulations done today frequently use the most sophisticated molecular mechanics force-field models with their realistic description of the inter- and intra-molecular interactions.

In current molecular dynamics simulations, the calculation is broken down into many, very small steps, of about 1 femtosecond (a million-billionth of a second). At each time-step, the forces on the atoms are determined, from which their positions at the next small time-step ahead is deduced, and so on. The restriction to small time-steps means that the length of a typical simulation is limited to picosecond (thousand-billionth of a second) or nanosecond (billionth of a second) timescales. Although these appear to be very short times in our macroscopic world, at the molecular level they are often sufficient to enable many interesting phenomena to be studied.

Monte Carlo methods

Unlike molecular dynamics, a Monte Carlo program does not generate a trajectory that shows how the positions of each of the atoms vary with time. Rather, it generates a series of arrangements of the atoms in the system (called configurations) from which it is possible to apply statistical mechanical formulae to derive thermodynamic properties, basically by totting up the energy from the interactions of all pairs of molecules. At the heart of nearly all Monte Carlo methods is a random number generator (hence the name!). Monte Carlo methods were introduced by John von Neumann who needed to be able to predict just how much fissile material was needed to make an atomic bomb during the Manhattan project at Los Alamos. Calculations starting with different numbers of atoms could show how many were needed for the number of neutrons to grow exponentially as a result of fission. The most common type of procedure is known as the Metropolis Monte Carlo algorithm (see Box 3).

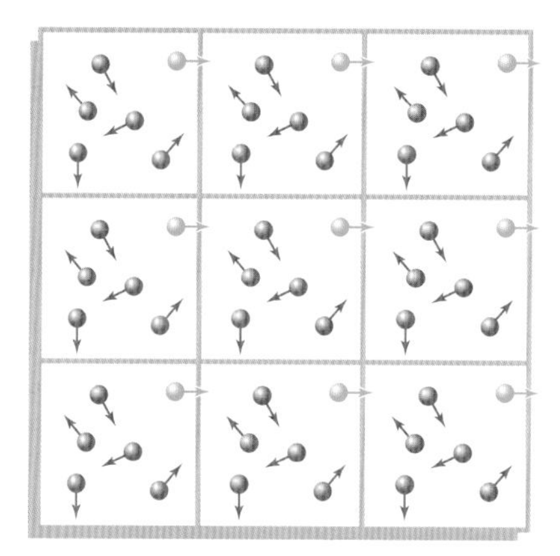

Figure 10. One of the problems with using small numbers of molecules in a computer simulation is that a large proportion of them would be near the edge of the sample. This gives rise to unwanted 'edge effects'. In a bulk sample a much smaller proportion of the molecules are at the surface. When periodic boundary conditions are used, the simulation box is replicated in all other directions. This is illustrated here for the two-dimensional case, where there are eight boxes surrounding the central one. The atoms or molecules in the central box 'see' not only the other particles in the central box but also the images of atoms or molecules in the neighbouring boxes. This has the effect of removing the edge effects from the calculation.

The other key development in molecular simulations was the use of *periodic boundary conditions*. This enables us to treat a relatively small system as if it were right in the middle of a much larger one, thus limiting the size of our system. The basic idea is best illustrated in two dimensions (Figure 10). By replicating the box in all directions, the atoms or molecules at the edge of the box 'see' the atoms not only from their own box but also the images of atoms from neighbouring boxes. This means that all atoms in the central simulation box behave as if they were located in the middle of a very large system, so that 'edge effects' are eliminated.

A variety of simple physical properties can be calculated from both molecular dynamics and Monte Carlo simulations. These include the temperature, pressure and heat capacity. In addition to these 'bulk' properties, molecular information can also be extracted. One commonly calculated quantity is the *radial distribution function*. This shows how, on

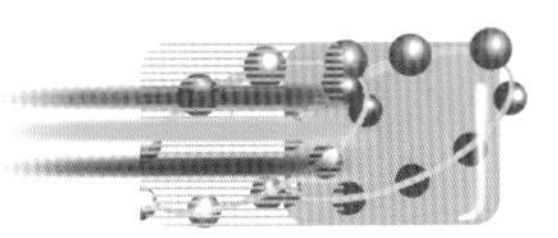

Box 3 The Metropolis algorithm*

The Metropolis algorithm consists of five steps as follows:

1. Randomly generate a new configuration of the system.
2. Calculate the energy of the new configuration.
3. Compare the energy of the new system (E_{new}) with that of its predecessor (E_{old}).
4. If the new system has a lower energy, accept it.
5. If the system has a higher energy, do the following:
   a. generate a random number between 0 and 1;
   b. compare this random number with the Boltzmann factor, $\exp[-(E_{new}-E_{old})/kT]$;
   c. If the random number is lower, accept the new configuration. If the random number is higher, reject the new configuration.

The crucial part of the Metropolis algorithm is embodied in steps 4 and 5. These enable the system to move to states of higher energy with a probability that is dependent upon their energy (step 5c so very high-energy states are unlikely to be accepted). The new configurations can be generated in several ways, depending upon the type of system being considered. The simplest way is to add a random amount to the *x*, *y* and *z* coordinates of one of the atoms. It is in steps 5a to 5c that the Metropolis algorithm permits the system to change its energy to a higher (less favourable) value. Such uphill jumps are needed to enable the system to explore other regions of the energy surface. One of the key results from statistical mechanics is that the probability of the system existing in a particular state with an energy E is proportional to the exponential, $\exp(-E/kT)$. The Boltzmann factor, $\exp[-(E_{new}-E_{old})/kT]$ has a value close to one when the difference between E_{new} and E_{old} is small. There is a good chance that the Boltzmann factor will be larger than the random number and so a relatively high probability that the new state will be accepted. By contrast, when the energy difference is large and the Boltzmann factor has a value close to zero the probabilty of accepting the new state is much lower – but because a random number is being used it is still possible that such a change will be permitted, as in the real system.

**So-called because N. Metropolis was the first-named author on the paper.*

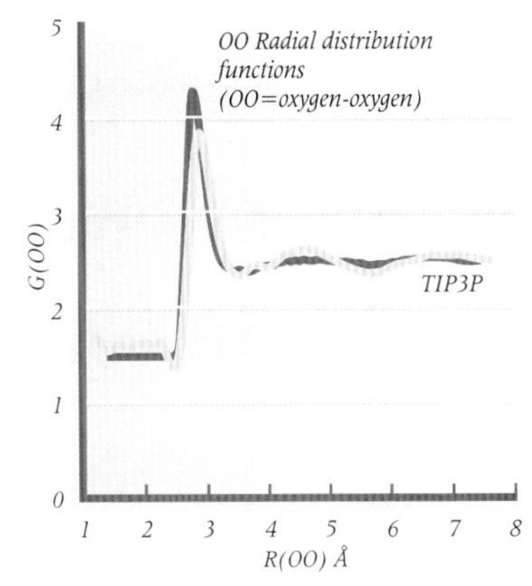

Figure 11. The radial distribution function of water (how the molecules are distributed) calculated using a popular molecular mechanics model called the TIP3P.
Adapted from W. L. Jorgensen, J. Chandrasekhar, J. D. Madura, R. W. Madura and M. L. Klein, Journal of Chemical Physics, 79, 926-935, (1983). ©1999 American Institute of Physics.

average, the molecules are distributed relative to each other. A typical radial distribution function (in this case for liquid water) is shown in Figure 11. The peaks and troughs indicate the probability of finding another molecule at the specified distance. In a crystal, the radial distribution function would have very sharp spikes, reflecting the precise nature of a lattice structure. In an ideal gas, the distribution function would be a horizontal line, reflecting an even distribution. The liquid is intermediate between the two with peaks and troughs, as shown. This means that at certain distances another molecule is more likely to be found than at other distances. These theoretical liquid distribution functions can be compared with those obtained experimentally using X-ray scattering.

Computational chemists routinely apply molecular dynamics and Monte Carlo methods to systems containing thousands, even tens of thousands of atoms. Such simulations enable them to tackle the behaviour of some very large and industrially relevant systems in a realistic manner. One example is a biological membrane, which consists of two layers of lipids – molecules with a hydrophilic 'head' and a long-chain hydrocarbon 'tail' – with additional cholesterol molecules and protein components embedded in the layers (Figure 12). In contrast to the traditional 'textbook' picture of a lipid bilayer, in which all of the hydrocarbon tails are aligned neatly parallel to each other, the

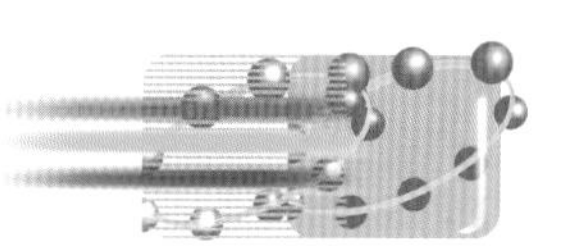

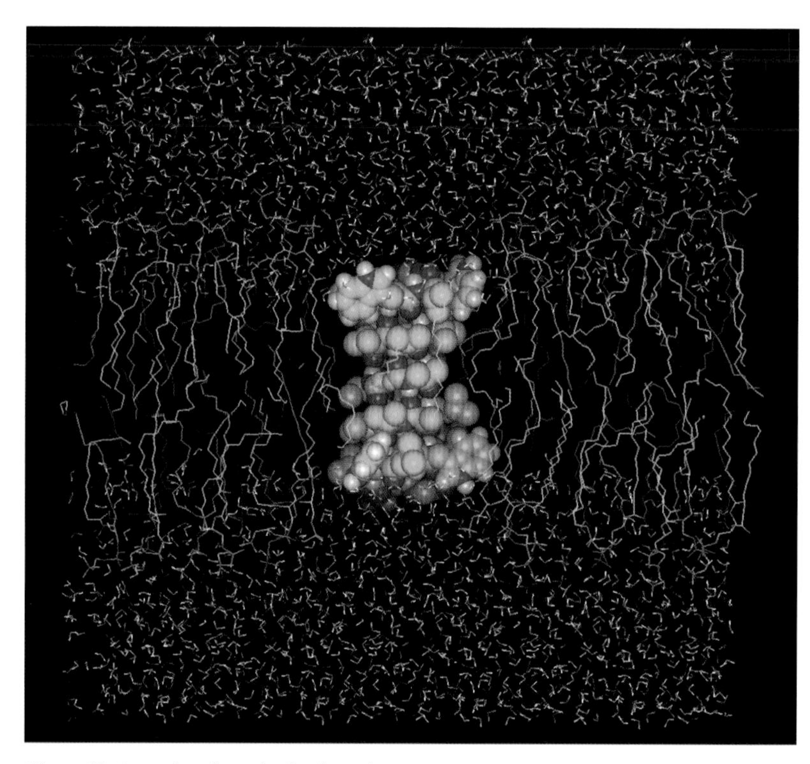

Figures 12. A snapshot of a molecular dynamics simulation of part of the peptide antibiotic Gramicidin A in a biological membrane with water. The disordered nature of the lipid molecules is clear. The whole system contains 34372 atoms. *Courtesy of Alan Robinson and Graham Richards*

simulation suggests a much more chaotic and disordered picture.

Another example, from the petrochemical industry, concerns the adsorption of alkanes in zeolites (highly porous aluminosilicate minerals used as catalysts in the petrochemical industry, (see *Chemical marriage-brokers*) and the calculation of how the amount of material adsorbed varies with the pressure at a given temperature (adsorption isotherms). In this case, the short-chain (between one and five carbon atoms) and long-chain (greater than 10 carbons) hydrocarbons showed simple adsorption isotherms, whereas alkanes of intermediate length (hexane and heptane) had kinked adsorption isotherms. This anomaly was traced to the fact that hexane and heptane are about the same length as 'zig-zag' channels in the zeolite, Figure 13. Once these channels are full, a special 'entropic penalty' has to be paid before the straight channels can be occupied. This penalty does not have to be paid by either the short – or long-chain alkanes.

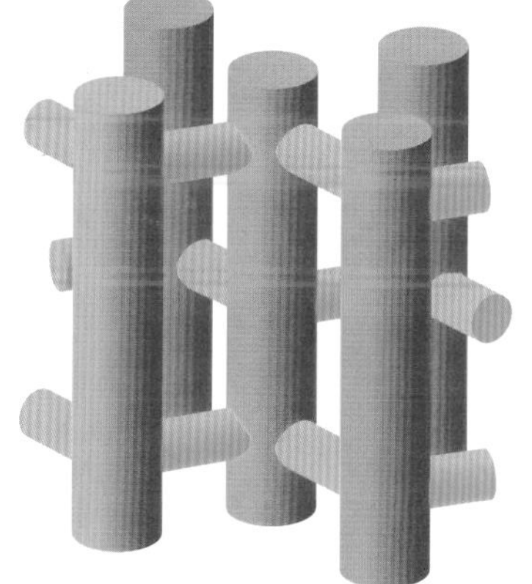

Figure 13. The zeolite silicalite contains both straight and zig-zag channels which can be filled with alkane molecules during adsorption. The ease of adsorption relative to the chain length of the alkanes can be simulated on a computer.

Back to quantum mechanics

A major development in recent years has been the use of quantum mechanics, rather than molecular mechanics, to determine the energies and forces in the system. It is only the combination of the new quantum mechanical methods such as density functional theory with the fastest computers, that such simulations have become possible. The computers employed for such simulations often use parallel processing. A traditional computer operates serially; it performs a sequence of operations one after another. There is a limit to how fast such computers will be able to operate. A parallel processor works by dividing the problem into many sub-problems, each of which is solved simultaneously by a processor and the results then combined at the end.

The research groups of Mike Gillan at Keele and Chris Payne at the University of Cambridge have used such approaches to study the dynamics of the molecular adsorption process. One system they studied was the adsorption of molecular chlorine (Cl_2) on a silicon surface. They were particularly interested in determining whether the chlorine molecule dissociated into chlorine atoms and whether this depended upon how the molecule hit the surface. The researchers carried out a variety of simulations in which chlorine molecules were 'fired' towards

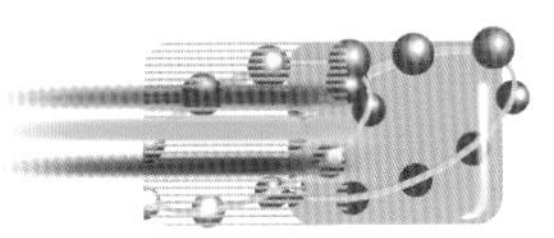

the silicon and the resulting reaction followed. The simulations suggested that the chlorine molecule did, in fact, dissociate for all trajectories, with the chlorine atoms forming new bonds with the surface silicon atoms.

Molecular modelling and the pharmaceutical industry

The pharmaceutical industry has long been a heavy user of, and investor in, molecular modelling techniques. In addition to the array of methods that we have already considered, there are a number of specialised techniques designed to address problems specific to the industry. Many drugs work by interacting with a biological macromolecule, often as not a protein such as an enzyme or receptor (but sometimes DNA). The drug molecule locks into and blocks the so-called active site on the protein structure so inhibiting its normal biological action. The binding power and thus activity of the drug depends on its conformation and the nature of its constituent functional groups (groups of atoms that react as a unit). In an increasing number of cases, the three-dimensional structure of the protein is known from X-ray crystallography or from nuclear magnetic resonance measurements (see *Analysis and structure of molecules*). If so, then using the protein structure as a guide, we can design ligands – small molecular structures that bind to the protein's active site – to help us identify inhibitor molecules. One of the most popular programs for performing this *structure-based ligand design* was developed by Peter Goodford at the University of Oxford. His GRID program constructs a regular grid around the protein. At each of the vertices of the grid, a number of small molecular fragments are positioned which act as 'probes'. Typical probes are hydroxyl (OH), carboxylic acid (COOH), and protonated amine (NH_3^+); these are chosen to represent the range of functional groups commonly found in drug-like molecules. The interaction energy between each fragment and the protein is calculated using molecular mechanics. These energy grids, along with the protein, can then be visualised using molecular graphics. In drug design we are particularly interested in regions corresponding to low energy as these represent regions where that particular functional group could bind. A good example of this approach is in the design of inhibitors of the enzyme, neuraminidase. This enzyme is involved in the mechanism by which a flu virus passes from one host cell to another. Inhibitors of this enzyme are therefore potential anti-flu drugs. The X-ray structure of the enzyme with a previously known (yet not very potent) inhibitor was determined by Peter Colman and colleagues in Australia. When the binding site was analysed (Figure 14), it was suggested that a relatively simple modification would lead to a much more potent compound, as turned out to be the case when the synthesis was undertaken.

In many cases we are not in the fortunate position of having a protein structure available. Nevertheless, it is still possible to deduce some useful information about the nature of the binding site on the protein. The process called *pharmacophore mapping* takes a series of molecules that

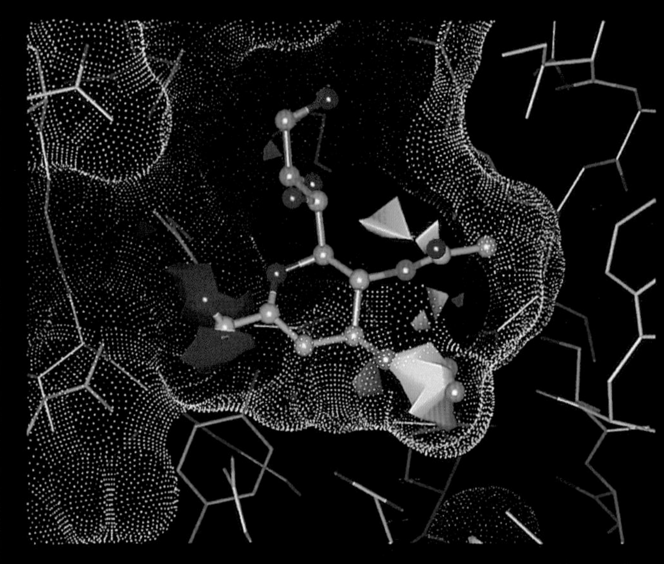

Figure 14. Analysis of the binding site of the flu virus enzyme, neuraminidase, using the GRID program. Molecular fragments such as carboxylate (red) and amidine (blue) are used to probe the part of the enzyme that might bind to a potential drug molecule inhibiting its action. Andrew Leach, Molecular Modelling: Principles and Applications. Reprinted with permissiom of Addison Wesley Longman Ltd.

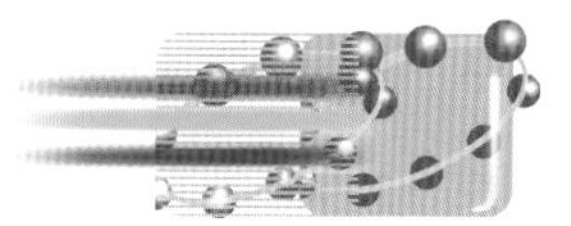

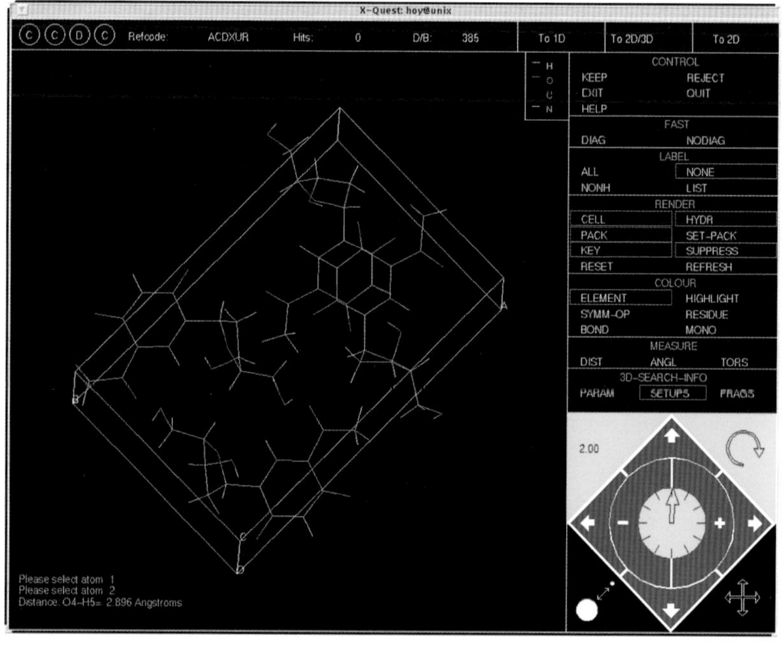

Courtesy of Cambridge Crystallographic Data Centre

are active at the same site and attempts to derive an abstract model called a *3D pharmacophore*. A simple example of a 3D pharmacophore is shown in Figure 15. Such a pharmacophore includes the functional groups that must be present and their spatial relationship necessary to make the molecule pharmacologically active. 3D pharmacophores are often defined in terms of general features, such as a basic group, a hydrogen-bond donor or the centroid of an aromatic ring. To generate meaningful 3D pharmacophores is a difficult problem because the conformational space of all the molecules together with the different ways in which the functional groups could be matched must be considered.

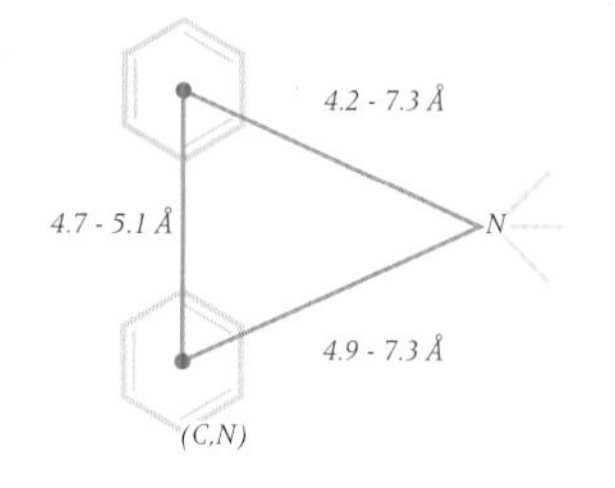

Figure 15. A commonly used 3D pharmacophore (a theoretical construct containing the chemical features essential for drug activity) for compounds with antihistamine properties.

There are several uses that can be made of a 3D pharmacophore once it has been derived. A major application is in searching a database of three-dimensional structures for molecules with similar biological properties. Such molecules may represent lead structures for a new drug development programme. An approach favoured by many pharmaceutical companies is to derive a 3D pharmacophore from known molecules (from the literature or from competitor compounds) and then search the company 3D database of compounds to identify new molecules unique to the company. The critical feature of a 3D database is that it takes the three-dimensional, or conformational, properties of the molecules into account. This enables us to identify a series of molecules with the same three-dimensional properties but not necessarily similar chemical structures. One of the earliest examples of a 3D database system is the Cambridge Structural Database (CSD), established in 1965 by Dr Olga Kennard, which has data on more than 190,000 small molecule crystal structures. The CSD contains a wealth of structural information which can be searched to find specific molecules as well as more general information about molecular interactions.

A key development for 3D database systems was the introduction of programs that could take the computer equivalent of the two-dimensional chemical drawing and generate a sensible three-dimensional structure for the molecule. This was important because crystal structures are not routinely determined for all compounds. Moreover, such programs can process tens of thousands of structures per day, enabling a large database to be converted rapidly and automatically.

A 3D database system matches the three-dimensional structures of the molecules it contains to the query 3D pharmacophore. Most systems

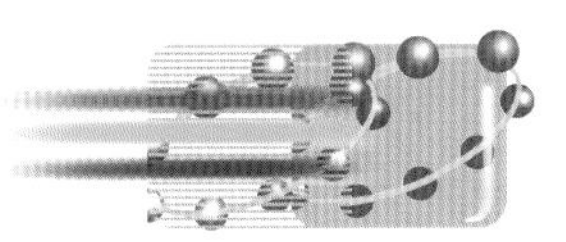

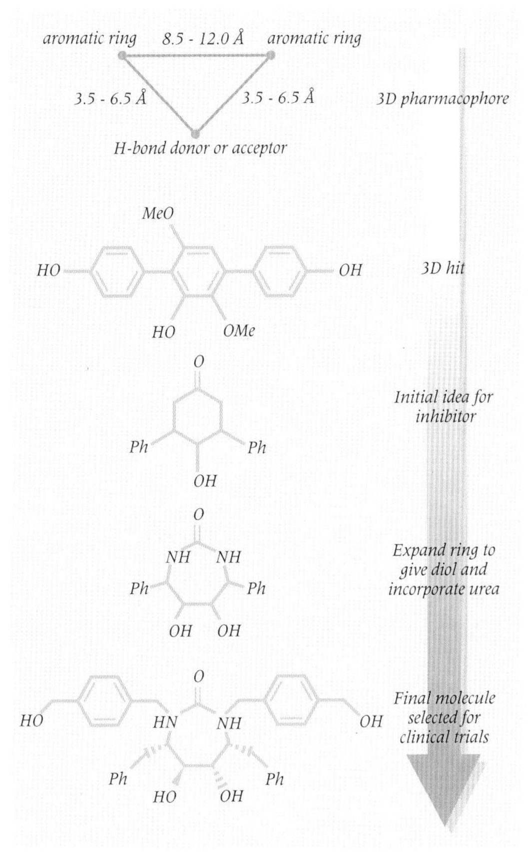

Figure 16. Flow chart to illustrate the key steps involved in the development of an HIV-protease inhibitor (an AIDS drug). A 3D pharmacophore was constructed based on a knowledge of the structure of the protease and existing inhibitors. This pharmacophore was then used to search a database, leading to the 'hit' shown and the initial idea for a new inhibitor. After several further stages of chemical synthesis and molecular modelling, the seven-membered ring structure shown was identified as a particularly potent scaffold upon which appropriate substituents could be placed to interact with the protease.

used today permit the molecules to explore their conformational space and are not limited to the single pre-stored conformation. Pharmaceutical companies now consider 3D searching based on pharmacophores as a valuable tool which enables them to identify molecules entirely different from those already known.

One topical example of the way in which 3D database searching has directly contributed to a drug design programme is the work of the computational chemists at the pharmaceutical company, DuPont Merck in Wilmington, Delaware in the US. They searched the Cambridge Structural Database for potential inhibitors of the HIV protease, the enzyme involved in creating the proteins and enzymes of the AIDS virus from their polypeptide precursors. The 3D pharmacophore consisted of two hydrophobic (water-hating) functional groups and a group that interacts with aspartate amino acids in the active site of the enzyme. One of the structures found not only contained these features, but also other chemically useful ones. After several rounds of chemical synthesis and molecular modelling, a series of cyclic seven-membered compounds was identified that were yet more potent. These compounds also had the key property of remaining active when taken orally. Subsequent X-ray crystallographic analysis showed that the final molecule (Figure 16) does indeed bind in the manner predicted. This story illustrates how a close partnership between experimental and theoretical approaches led to a very successful outcome, far in excess of what either could achieve in isolation.

The future

Computational chemistry has come a long way since the days when programs had to be input to the computer using punched cards and a machine with 1Kb of memory was considered to be state-of-the-art. Certainly, increases in computer performance have enabled major strides to be made in our understanding, at the molecular level, of molecular systems. However, there still remain some important problems for which solutions are still sought, such as the famous 'protein folding problem', in which attempts are made to predict the three-dimensional structure of a protein solely from its 'one-dimensional' amino acid sequence. It may be that ever-faster computers will enable us to solve these problems using current approaches, but history has shown us that it is often the conjunction of technology with a fertile mind that provides the breeding ground for the most significant advances.

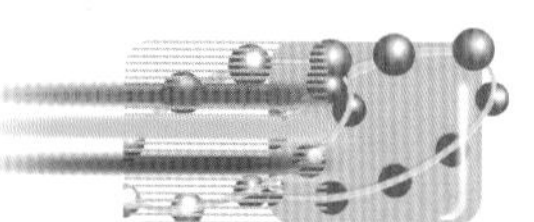

Further reading

1. *Computational Chemistry,* G. H. Grant and G. W. Richards, Oxford Chemistry Series, Oxford University Press, Oxford: 1996.
2. *Molecular Modelling – Principles and Applications,* A. R. Leach, Addison-Wesley-Longman, Harlow: 1996.

Glossary

3D pharmacophore A term used in pharmaceutical chemistry to describe a set of features that are common to a group of biologically active molecules and the three-dimensional relationships between the features.

***Ab initio* methods** A type of quantum mechanical theory which attempts to calculate, from first principles, solutions to the Schrödinger equation without incorporating empirical data.

Born-Oppenheimer approximation An approach to calculating the electronic structures of molecules in which the nuclei are treated as being static.

Computational chemistry The use of computational methods to understand and predict the nature and behaviour of chemical systems.

Computer simulation An approach most generally defined as the mimicking of a molecule's behaviour using a computer model. It is often used to refer to the molecular dynamics and Monte Carlo methods.

Conformation The conformations of a molecule are the three-dimensional shapes that it can adopt, often considered to be those structures which can be obtained by rotation about single bonds.

Conformational analysis The study of the conformations of a molecule and their influence on its behaviour.

Conformational space The range of conformations available to a molecule.

Density functional theory An approach to calculating electronic structure in which the properties of the lowest energy-state of a system are functions of the electronic charge density.

Enthalpy change The heat generated or absorbed by a system at constant pressure.

Extended Hückel molecular orbital methods A simple semi-empirical theory, much developed by Roald Hoffmann, which is able to deal with a particularly wide range of systems, especially those containing metals.

Force field methods Methods that use molecular mechanics to calculate the energy of a system (see Molecular mechanics).

Free energy Thermodynamic expression which determines whether a process will proceed spontaneously.

Hamiltonian operator A term in the Schrödinger wave equation of quantum mechanics denoting the sum of the kinetic and potential energies of a system.

Hartree-Fock self-consistent field An approach to solving the Schrödinger equation in which the equation is solved for a single electron moving in an 'averaged' potential of the other electrons. The solution provided by such an approach is then used in the next iteration until convergence is achieved (see Schrödinger wave equation).

Minimisation program Minimisation is a computational technique frequently used to identify low energy chemical structures or conformations (see Conformation).

Molecular dynamics A computer simulation technique in which the particles in the system are able to move according to Newton's laws of motion.

Molecular graphics The use of computer graphics to create and manipulate images of molecular systems and their behaviour.

Molecular mechanics A description of a molecular system in which the energy depends on the nuclear positions alone.

Molecular modelling A term synonymous with computational chemistry often used when some form of molecular graphics is involved.

Monte Carlo method A computer simulation method which involves random changes to a system from which thermodynamic properties can be calculated using statistical mechanics.

Periodic boundary conditions A theoretical approach in which a small system is infinitely replicated in all directions thus enabling 'bulk' properties to be determined even from relatively small numbers of atoms.

Pharmacophore mapping Computational methods used in drug design for generating 3D pharmacophores from a series of active molecules (see 3D pharmacophore).

Quantum mechanics A mathematical description of a molecular system which includes both nuclei and electrons and describes them statistically in terms of their wave-like quantum nature.

Radial distribution function The probability of finding another atom or molecule at a particular distance.

Schrödinger wave equation A differential equation which describes a system, such as an atom, obeying the laws of quantum mechanics.

Semi-empirical methods A type of quantum mechanical theory that incorporates parameters derived from experimental data.

Structure-based ligand design The use of a protein structure (typically obtained from X-ray diffraction) to design small molecules that bind tightly to it.

Variation principle A mathematical approximation technique commonly employed in quantum chemistry in which solutions to the Schrödinger wave equation of ever lower energy are derived iteratively until a consistent and thus correct result is achieved.

Wavefunction A solution of the Schrödinger wave equation for a quantum system from which all properties of the system may be derived (see Schrödinger wave equation).

Biographical details

Dr Andrew Leach is currently a member of the computational chemistry group at GlaxoWellcome Research and Development in Stevenage, Hertfordshire. His interests include conformational analysis, molecular docking, chemical databases, search and optimisation algorithms, structure-based ligand design and combinatorial libraries.

Acknowledgements

I should like to thank Professor Graham Richards and Dr Mike Hann for their help in preparing this chapter.

The chemistry of life

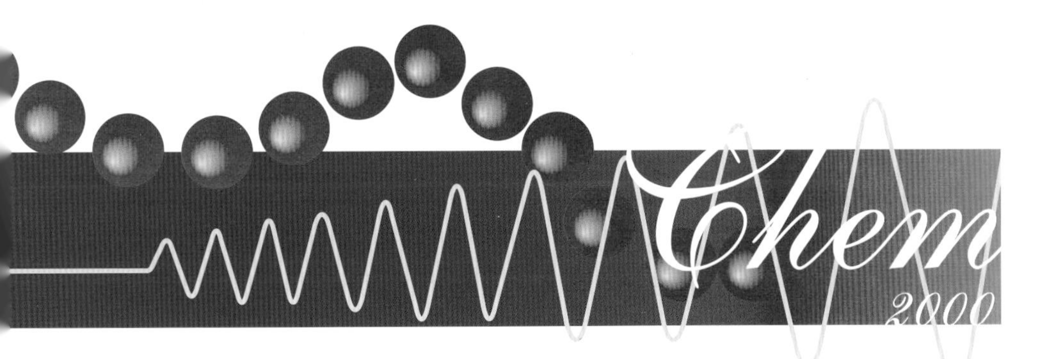

Life depends on chemistry in living things. From the synthesis of penicillin to designing the perfect pesticide, unravelling the structure of DNA and a cure for AIDS, chemistry is fundamental to the quality of our lives.

Professor John Mann
University of Reading
Dr Neil Thomas
University of Nottingham

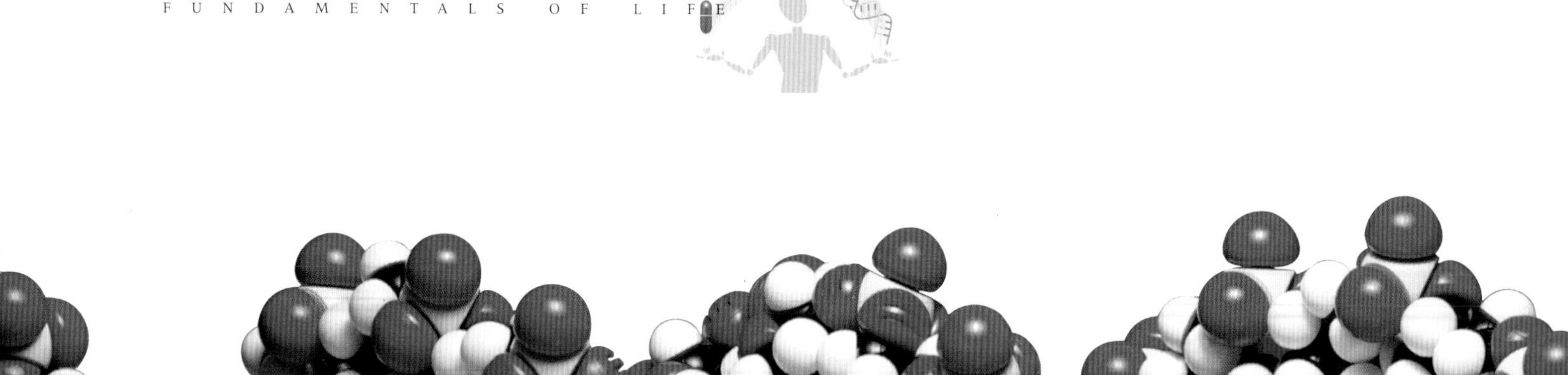

Life is chemistry at its most complex. Over the past 30 years, chemists and biologists have been uncovering the extraordinary molecular processes by which living organisms function. This new area of knowledge is helping to improve many aspects of our lives – from radically improved healthcare to cheaper food and other everyday essentials.

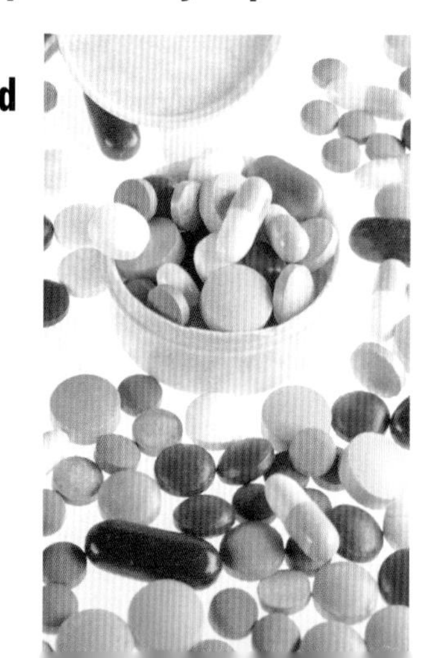

When the 17th-century Dutch scientist, Antonie van Leeuwenhoek, first observed bacteria, or 'animalcules' as he called them, through his primitive microscope, he could never have imagined the complexity and beauty of the chemical structures that lay before him. These humble bacterial cells (from the plaque on his teeth!) comprised a complex cell wall made from carbohydrates, lipids and proteins, which enclosed a watery brew of enzymes and genes. Yet these lowly, single-celled organisms have been among the most successful inhabitants of our planet for the past three and a half billion years. Their genes provide the blueprint for the production of enzymes that are responsible for organising the assembly of the cell wall, for energy production, reproduction, and all the other processes associated with life. We humans are, of course, made up of billions of cells, yet they share these same basic processes of life that are all underpinned by chemistry.

One of the triumphs of recent decades has been the unravelling of the complexities of

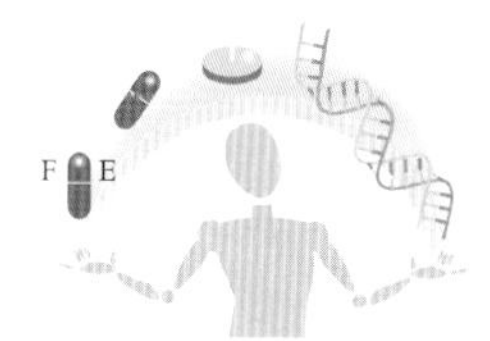

the chemistry of life at the molecular level, and this has spawned a brand new science called *molecular biology*. This new knowledge is not only fascinating, but can also be used to help in designing new drugs, agrochemicals, and biodegradable plastics, that will further enhance the quality of our lives.

Friederich Wöhler showed there was no vital force associated with chemicals of natural origin.
Reproduced courtesy of the Library and Information Centre, Royal Society of Chemistry.

We have to go back rather more than 40 years, however, if we want to find how chemistry first connected with biology. In 1828, the German chemist, Friedrich Wöhler, heated the 'inorganic' chemical ammonium cyanate and converted it into the 'organic' chemical urea, which was identical with the material isolated from urine, thus refuting the concept of a 'vital force' that many people at the time thought was associated with living things and their 'organic' products. Wöhler wrote to a fellow chemist Berzelius: "I must tell you that I can prepare urea without requiring a kidney or an animal, either man or dog".

$$(NH_4)CNO \longrightarrow CO(NH_2)_2$$

Meanwhile, Wöhler's contemporaries were busy isolating and characterising biologically interesting molecules produced by plants, like morphine, atropine and quinine. These materials had been used for centuries in folk medicine, witchcraft and so on, but of course nothing was known about how they worked at the molecular level. Nevertheless, these early adventures in natural product chemistry, which spanned the period 1800 to 1930, set the scene for all the amazing current developments in the design of new medicines and therapies.

It is worth recalling that, at the time of Queen Victoria's death in 1901, the leading causes of death in Britain were pneumonia, tuberculosis and bacteria-induced diarrhoea – all infectious diseases. To combat them, a doctor had available only aspirin and a number of plant and animal extracts, many of which had been in use since before the time of Christ. For example, the *Ebers* papyrus, written in Egypt around 1500 BC, detailed more than 800 remedies involving animal organs, plant extracts and minerals. The collected works of Celsus (*De Medicina*) and Disocorides (*De Materia Medica*) from the Graceo-Roman world, were also widely known, and recommended similar medicines.

New knowledge emerged only slowly, and the *London Pharmacopoeia* of 1809 still listed plant extracts (aconite, belladonna, cinchona, colchicum, hemlock, henbane and opium), and inorganic salts such as *regulus antimonii* (metallic antimony) and *mercurius sublimatus corrosivus* (mercuric chloride), and formed the basis of prescribing by doctors and pharmacists at this time.

A variety of plants used in herbal medicine .
Shelia Terry/SPL

The first synthetic drugs

At the start of the 20th century, the chemical industry was primarily concerned with explosives, fertilisers and dyestuffs, not medicines. Paul Ehrlich in Berlin was the first person to use a totally synthetic chemical compound as a drug, when his investigations into the use of dyes to stain bacteria led to the use of the dye methylene blue

for the treatment of malaria in 1891. Other dyes were active against the trypanosomes that cause African sleeping sickness, and his studies with arsenic-containing analogues of so-called azo-dyes provided the compound (arsphenamin) in 1909. This had excellent activity against the protozoa *Treponema pallidum*, which caused syphilis, a chronic venereal disease affecting millions of Europeans including Randolph Churchill and Isak Dinesen (Karen Blixen of *Out of Africa* fame), and whose long-term effects were dementia and death.

Ehrlich introduced for the first time the concept of a *'magic bullet'* – a chemical designed to target a particular disease: "Here we may speak of magic bullets which aim exclusively at the dangerous intruding parasites, strangers to the organism, but do not touch the organism itself and its cells". There then followed numerous triumphs of drug discovery – the development of the *sulfonamides* and *penicillins* surely rank as two of the supreme achievements of the 20th century. Penicillins were the first *antibiotics*.

Paul Ehrlich was the first person to use a synthetic compound as a drug.
Reproduced courtesy of the Library and Information Centre, Royal Society of Chemistry.

It is often forgotten that average life expectancy at the beginning of the 20th century was only about 45 years, and that around 15 per cent of children died before their fifth birthday (Figure 1). Most of these premature deaths were a result of bacterial infections. Leading on from Ehrlich's research, the German chemist Gerhard Domagk began a systematic screening study in 1927 of the effects of gold compounds and various dyestuffs against a strain of *Streptococcus* that he had isolated from a patient who had died from septicaemia. One particular red dye, Prontosil Rubrum (Figure 2a), displayed antibacterial activity, and in 1933, Domagk administered the dye to a baby dying from staphylococcal septicaemia. The baby survived. This was little short of miraculous, and after further tests in Britain and Germany, Prontosil Rubrum was made available for general use.

	1900	1910	1920	1930	1940	1950	1960	1970	1980	1990
Women	48	52	55	62	65	71	73	75	77	79
Men	46	48	54	58	61	66	67	67	70	72

Figure 1. The life expectancy of women and men at birth in the 20th century.

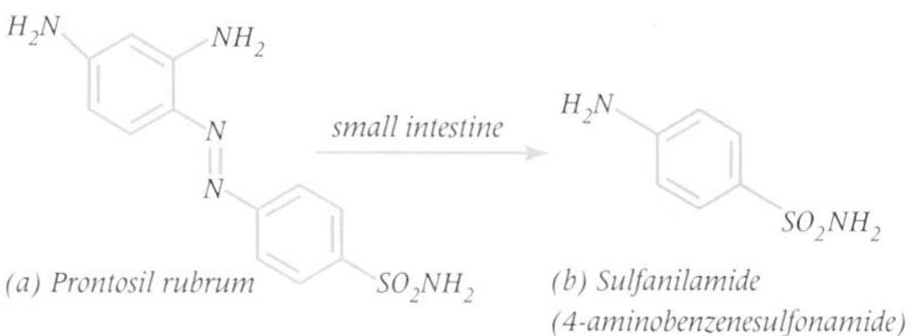

Figure 2. A red dye, Prontosil Rubrum (a), which is converted into (b) 4-aminobenzenesulfonamide – the first antibacterial drug.

However, it was not until 1940 that a biochemist working at Oxford University worked out the drug's mode of action. Donald Woods found that the dye was broken down in the small intestine to release a compound called 4-aminobenzenesulfonamide (Figure 2b), and this prevented the normal functioning, or inhibited one of the key bacterial enzymes. Bacteria need to make two growth factors (cofactors) called dihydrofolate and

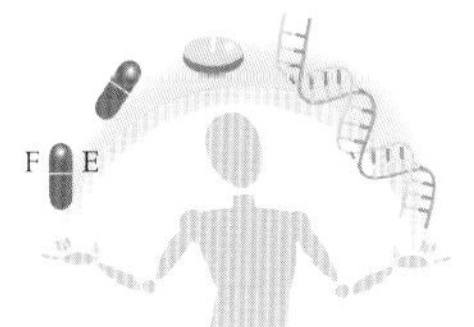

tetrahydrofolate vital for the production of bacterial DNA. In the presence of either 4-aminobenzenesulfonamide or one of the many sulfonamides subsequently prepared, the bacteria cannot produce these cofactors. Humans also need these compounds, but we make them from folic acid, which must be present in our diet. This means that the sulfonamides have selective antibacterial activity and are relatively harmless to humans – an important prerequisite of any drug. One of the sulfonamides, made by the British company May & Baker – so-called M&B 693 – (Figure 3) achieved star status when it was used to cure Winston Churchill of pneumonia in 1943.

H_2N — SO_2NH — N

M&B 693 (sulfapyridine)

Figure 3. The sulfonamide M&B 693 which cured Winston Churchill of pneumonia.

Who knows what the outcome of the Second World War might have been without it!

The wonderful penicillins

The sulfonamides also had a major impact in reducing the number of maternal deaths after childbirth from 50 in 10,000 births in 1900 to less than 5 per 100,000 by the mid-1940s. However, they were soon superseded by the penicillins. Joseph Lister and others had observed that moulds like *Penicillium brevicompactum* had modest antibacterial activity, but it was left to Alexander Fleming to herald the beginning of the antibiotic era with his serendipitous discovery in 1928 of the much more potent activity of the mould *Penicillium notatum*. Fleming was a bacteriologist, and his collaborators were also biologists rather than chemists, so he was not able to isolate or characterise chemically the substances responsible for the antibacterial activity of the mould. This had to await the attentions of the scientists at the Sir William Dunn School of Pathology in Oxford. During the early 1940s, the chemists Edward Abraham and Ernst Chain, under the leadership of the bacteriologist Howard Florey, carried out their seminal studies that culminated in the isolation of penicillins. The first clinical trials were carried out in 1941, at a time when Britain was expecting an imminent invasion, and produced some miraculous results. Several patients who were at death's door through advanced septicaemia made complete recoveries, and the need for large quantities of this new wonder drug to treat battlefield casualties suddenly became of paramount importance.

Alexander Fleming who discovered penicillin, in his laboratory at St. Mary's Hospital in London.
Audio Visual Services, ICSM (St. Mary's).

Fleming's penicillin culture plate showing that there is no bacterial growth near the mould.
Audio Visual Services, ICSM (St. Mary's).

Howard Florey who isolated the first penicillin compound.
©The Nobel Foundation

Because of wartime difficulties, the scale-up of penicillin production was entrusted to the Americans, and with a combination of deep-tank fermentation technology (akin to brewing) and the use of a new strain of mould (*Penicillium chrysogenum*), found on a rotten cantaloupe melon, these aims were achieved. In hindsight, these fantastic advances are all the more extraordinary since the chemical structure of penicillin was then completely unknown. In fact, the structure of penicillin G (the major constituent of this mould) was finally solved by Dorothy Crowfoot Hodgkin in 1945, using X-ray crystallography (Figure 4). The core structure of penicillin is a four-membered ring consisting of a nitrogen atom, three carbon atoms, one of which is bound to oxygen in a carbonyl bond (C=O). This *beta*-lactam is bound to various side-chains and fused with a five-membered sulfur-containing ring (a thiazolidine). Because the *beta*-lactam ring is so unstable to acid or alkali, it was another 12 years before John Sheehan of the Massachusetts Institute of Technology first synthesised a penicillin.

Dorothy Hodgkin who solved the structure of penicillin G, and won the Nobel Prize for Chemistry in 1964 for her determinations by x-ray techniques of the structure of important substances.

Figure 4. Penicillin G – a major natural penicillin.

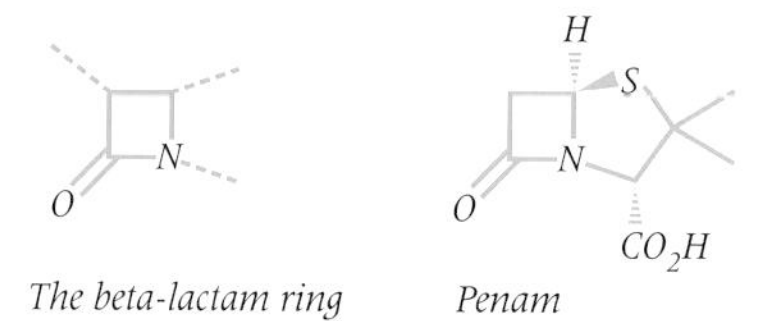

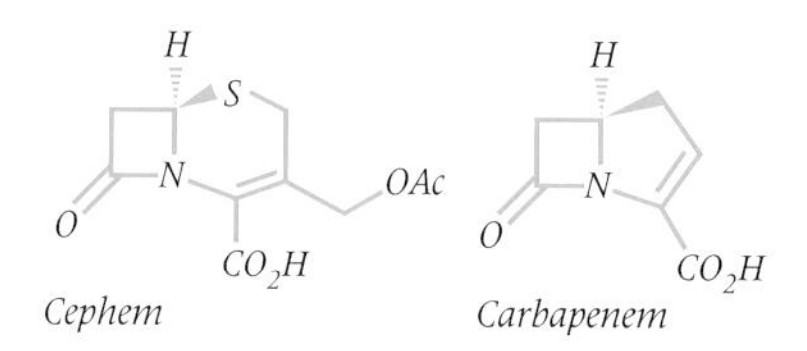

Figure 5. Some beta-lactam structures.

In the interim, researchers at the pharmaceutical company Eli Lilly had found that they could vary the structure of the side-chain of the penicillin, if they added the side-chain of their choice to the fermentation brew. In this way, a range of novel, semi-synthetic penicillins could be produced. Penicillin V (Figure 6a) was probably the most successful drug to arise from this method, because it was reasonably stable to acid and so could be administered by mouth – unlike penicillin G, which had to be injected. Another major breakthrough was the discovery, by a research group at Beechams, that 6-aminopenicillanic acid, which was the most important intermediate for the production of semi-synthetic penicillin, could actually be isolated from the fermentation broth. This allowed them to prepare an almost unlimited array of structures, of which methicillin (Figure 6b) was possibly the most important since it was active against strains of bacteria that had become resistant to the earlier penicillins.

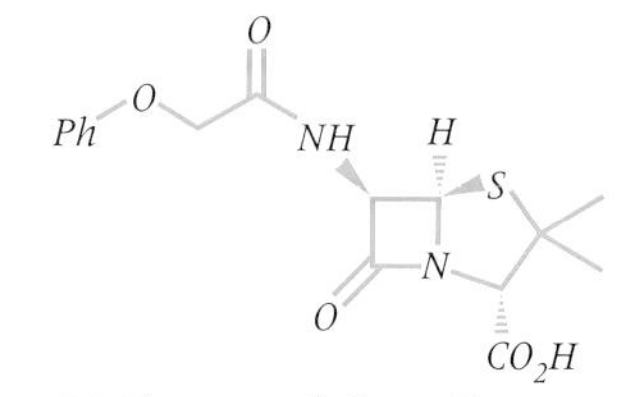

(a) Phenoxymethylpenicillin - penicillin V

(b) Methicillin

Figure 6. The semi-synthetic penicillins, penicillin-V (a) and methicillin (b).

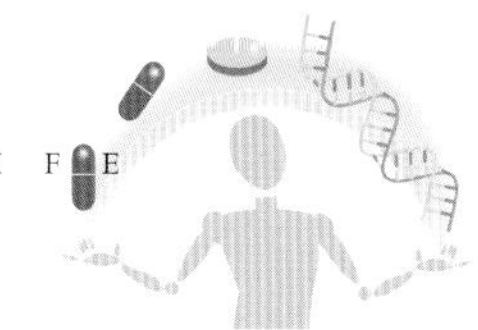

The cephalosporins, tetracyclines, aminoglycosides, and macrolides (Figure 7) are four other major classes of antibiotics that have been discovered and used during the past 50 years; and there are also a host of totally synthetic antibacterial compounds that have been invented during the same period. Indeed, the pharmaceutical industry has mounted a nonstop campaign to isolate novel antibacterial substances, in order to provide drugs with better clinical efficacy and to stay one step ahead of the bacteria. As mentioned earlier, these unicellular organisms have been evolving for billions of years, and have now had a further 50 years to evolve new means of combatting these new foreign chemicals, becoming resistant to their antibacterial effects.

But despite the marvellous properties of these drugs, in every case, they were introduced into clinical use without any real understanding of their modes of action. This was also true for other drugs, like aspirin, morphine, and ephedrine, that arose from the study of natural products; and those like chloroquine, mepacrine and other antimalarials, which were spin-offs from the dyestuffs industry. Realistic insights into how these drugs worked was impossible without a knowledge of the chemistry of *enzymes* (proteins) and the genetic blueprint *deoxyribonucleic acid (DNA)* the molecules that control every aspect of our lives.

(a)

(b)

(c) *Streptomycin (from Streptomyces griseus)*

(d)

Figure 7. Examples from four major classes of antibiotics: (a) cephalosporins, (b) tetracyclines, (c) aminoglycosides and (d) macrolides.

Proteins – the workhorses of life

Until the 1930s, proteins were thought to be random aggregates consisting of chains of amino acids linked together by amide (CON-H) bonds, rather as in the polymer nylon (Figure 8) (see *The age of plastics*). It was only with the advent of X-ray crystallography that a particular kind of geometrical structure – a motif – common to many proteins, the *alpha-helix,* was first identified by Linus Pauling at the California Institute of Technology and Robert Corey in 1950.

The development of a method of determining the amino acid sequence of proteins, by Frederick Sanger at Cambridge University in 1945, and the subsequent improvements and additions provided by Pehr Edman of the University of Lund in 1950, showed that proteins were much more ordered than had

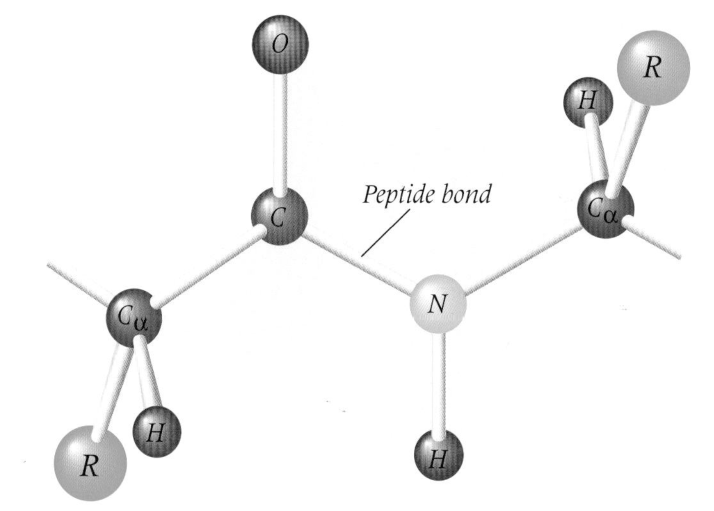

Figure 8. The peptide link between two amino acids.

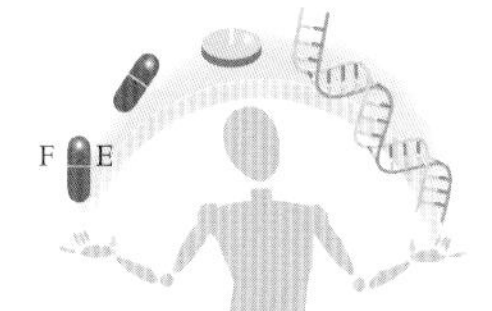

Box 1 How does a protein fold?

Proteins are involved in virtually every biological process. In order to function, after biosynthesis their amino acid chains must fold into a characteristic and unique 'native' three-dimensional structure. This folding process in a cell is assisted by several families of proteins, the best-known of which are called molecular *chaperones*. These, like their human counterparts, prevent 'improper' interactions between polypeptide chains until they are fully developed. Chaperones do not, however, provide any information themselves about the specific structure into which a given protein folds. This is contained fully within the sequence of amino acids. There are, however, nearly 100,000 different protein sequences encoded in the human genome, each one of these has a specific fold, and can be classified into at least 1000 distinct architectures. Establishing how a newly formed polypeptide sequence finds its way to its correct fold rather than the countless alternatives is perhaps the greatest challenge in modern structural chemistry.

Protein folding involves a complex 'molecular recognition' phenomenon that depends on the cooperative action of a vast number of weak, non-covalent interactions (rather than distinct bonds) involving thousands of atoms. The number of possible conformations of a polypeptide chain is astronomically large. For a chain of 100 amino acids (a small protein) there are about 10^{49} possible conformations. Even assuming that the chain can convert from one conformation to another is only one-hundred billionth of a second, a random search of all possible conformations would take about 10^{29} years! The fact that this is incompatible with the observation that natural proteins can often reach their native conformations in less than 1 second has come to be known as the *Levinthal Paradox*, after Cyrus Levinthal who first highlighted this problem in the 1960s.

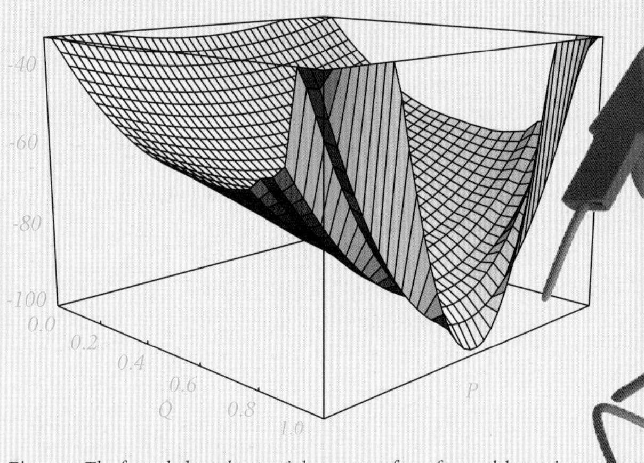

Figure a. The funnel-shaped potential energy surface of a model protein. E is the potential energy, P relates to the number of possible protein conformations and Q indicates the progress of folding.

Recently, real progress has been made towards solving this paradox and understanding the mechanism of folding. This has come about through ingenious experimental strategies, particularly involving nuclear magnetic resonance (NMR) spectroscopy, (see *Analysis and structure of molecules*), for following the folding behaviour of proteins in the laboratory, and by using computers to simulate and analyse folding. The experiments often involve using a chemical 'denaturant' such as urea to 'unfold' the protein, and then diluting the solution rapidly to reduce the concentration of the denaturant such that the protein folds back into its stable native state. To give an idea of the complexities involved, not only do structural changes often happen in mere seconds or less, but there are so many possible conformations that every protein molecule in a sample is likely to have a different one until the final stages of the reaction!

Ironically, it is the abundance of routes to folding that provides the solution to the Levinthal Paradox. An individual protein molecule will pass through very few of the possible conformations on its journey from an unfolded conformation to its native state.

Figure b. A partially folded 'molten globule' state of the protein equine lysozyme. The amino acids (shown in yellow) are key hydrophobic contacts that orchestrate the final folding of the polypeptide sequence into its native state.

To understand how this can happen, we need to look at the relative energies of the possible conformations, which can be represented on a 'free-energy surface' (Figure a) (see *Following chemical reactions*).The surface looks like a funnel in that the number of conformational states accessible to the polypeptide chain becomes smaller as the protein folds. In fact the folding of an ensemble of protein molecules can be thought of as being a bit like a volume of water running down a real funnel. If the funnel has the right shape all the water molecules end up in the same place regardless of how dispersed they were initially. And although the surface as a whole may be covered in water molecules, any individual molecule makes contact with very little of it. All the water can therefore get to the lowest point on the surface far more rapidly than any simple calculation based on the total number of possible locations of the molecules would suggest.

The question remains, however, as to how the sequence of a protein can result in an energy surface of the appropriate shape for rapid folding to a specific structure. This we don't fully understand. For example, despite many attempts no one has yet been able to fold a protein correctly in a computer using a series of potential energies that describe the interactions between the different amino acids. Nevertheless, the outline of the folding process is becoming clear.

The first steps in protein folding appear to be dominated by the way in which different types of amino acid interact with each other relative to water. Polypeptide chains tend to collapse in water to bury hydrophobic (non-polar) amino acids in the interior of a 'molten globule' which has hydrophilic (polar) amino acids on the surface. This simple process turns out to generate the correct overall fold for a given sequence, including the helices and sheets familiar from the many known protein structures (Figure b). At this stage of folding, the amino acid side-chains are not yet properly organised. The search for the unique structure that allows all the amino acids to lock tightly together is the final step in the folding process, and involves the optimisation of many subtle cooperative molecular interactions.

Figure c. An electron micrograph of amyloid fibrils seen in diseases such as Alzheimer's. They form as a result of proteins unfolding and aggregating under pathological conditions.

The value of understanding how proteins fold is becoming more and more evident. It is crucially important in attempts to modify proteins or to design new ones with novel functions. Also, enormous effort is now being invested in the *human genome project* which will result in knowledge of the sequences of all the proteins in our bodies. The vast majority of these proteins do not have known structures, and if we could predict them from their sequences it would help dramatically to identify the likely functions of the proteins encoded by the different genes. This is a crucial aspect of the development of new strategies to combat disease such as gene therapy.

Most remarkable, however, has been the discovery that some diseases, such as cystic fibrosis, are in fact the result of the failure of certain proteins to fold correctly. In addition, other diseases including Alzheimer's and 'mad cow disease' are associated with the fact that some proteins unfold under aberrant conditions in the body and then aggregate to form insoluble and toxic fibrils, Figure c. Understanding the nature of protein folding and misfolding could therefore be crucial for the development of novel therapeutic strategies for treating or preventing many of the most debilitating of human diseases.

Professor Christopher Dobson.
Oxford University.

been originally believed. The X-ray crystal structures of myoglobin from muscle (Figure 9) and haemoglobin from blood, solved by John Kendrew and Max Perutz at Cambridge in 1958 and 1960 respectively, revealed fascinating three-dimensional protein structures made up of portions with particular geometries, or *folds* – for example, the *alpha-helix* and *beta pleated sheet* (Figure 10) (see Box 1).

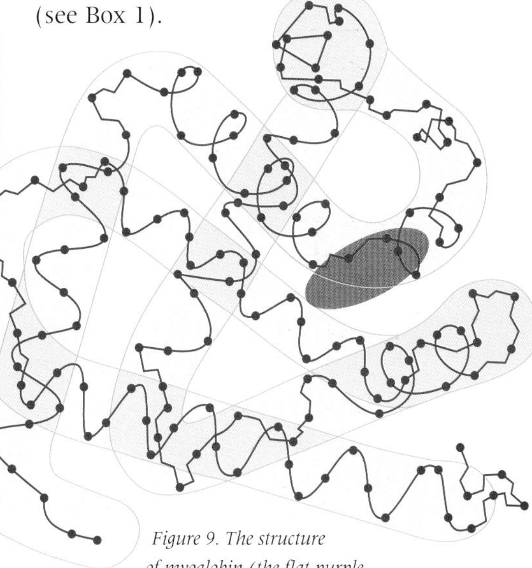

Figure 9. The structure of myoglobin (the flat purple structure is a haem unit).

Ten years later, the X-ray structure of the first enzyme, lysozyme, was worked out by David Phillips at Oxford University, and these structural

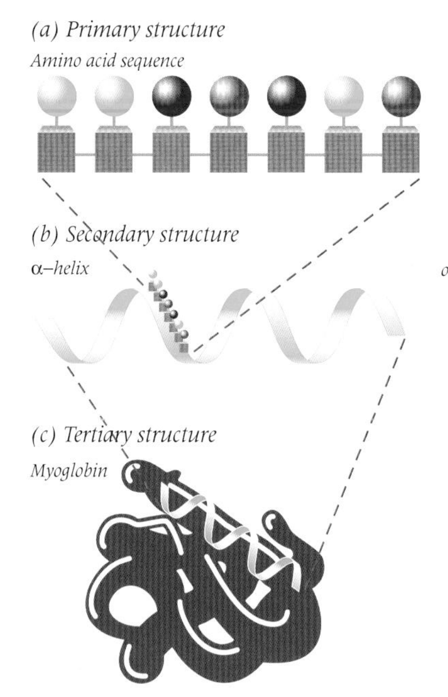

Figure 10. A protein has several layers of structure: its amino acid sequence (a); the types of folds in the structure (b) and the overall folded configuration (c).

revelations allowed investigators to demonstrate how lysozyme destroys the carbohydrate coats of bacteria and thus exerts its antibacterial effect.

Once chemists had determined the amino acid sequence of a number of small *polypeptides* (molecules of less than 50 amino acids strung together), like insulin and human growth hormone, it was quite natural that they would want to synthesise them. To achieve this, Bruce Merrifield working at the Rockefeller University in New York developed the technique of *solid-phase synthesis* which bears his name. The technique involves attaching the terminal amino acid to a polystyrene or similar inert support, and then chemically linking the second and subsequent amino acids. It isn't necessary to purify the growing polypeptide chain; the unreacted reagents and solvents just have to be washed off. Another advantage is that more of the reagents can be used than is actually needed so ensuring maximum yields (100 per cent) of product at each step. At the end of the synthesis, the complete polypeptide is chemically removed from the polymeric support. It then usually folds spontaneously to attain the required natural three-dimensional form of the protein. In 1969, using this technology, Merrifield was able to prepare the enzyme ribonuclease, consisting of 124 amino acids in an overall yield of 17 per cent – a staggering achievement. The method is now routinely used to prepare both natural and unnatural polypeptides and proteins.

Bruce Merrifield who won the Nobel Prize for Chemistry in 1984.
Courtesy of Ingbert Grüttner, The Rockerfeller University.

As the complexities of protein structures were revealed, the importance of the dynamic three-dimensional shapes of these molecules in relation to their biological functions became the focus of attention. Almost every one of life's processes depends upon the interaction of proteins with small organic molecules, be they hormones, neurotransmitters, vitamins, or carcinogens. Proteins may be enzymes catalysing biochemical reactions in the watery cellular environment, or *receptors* sitting in the membranes of cells. In most instances, the protein folds around these small molecules in order to bring about the binding and subsequent catalysis or other chemical event. But how does this translate into a biological response, and how does an understanding of these complex interactions help with drug design? The interactions and responses elicited by adrenaline and the oestrogen hormones, provide excellent examples of these facets of bioorganic chemistry.

The 'adrenaline cascade'

Adrenaline (epinephrine) is the 'fear, flight and fight' hormone, released from the adrenal glands in response to a wide variety of stimulating situations. It has a major effect on the heart and leads to an increase in blood pressure, heart rate and output, and thus prepares the body to react to the stimuli. These physiological effects are the result of a complex chain of chemical events that follow the binding of adrenaline to its protein receptor on the surface of heart-muscle cells. This is the classic *lock and key* type of binding, as first described by Emil Fischer in 1894, where

the shape of a so-called *active site* on the protein fits the shape of the molecule being bound to the protein (Figure 11), in this case adrenaline. The receptor has seven polypeptide strands that traverse the cell membrane.

Enzyme *Possible substrates* *Only one fits the active site*

Figure 11. Lock-and-key binding: receptor proteins bind only to molecules (substrates) with a specific shape (just as a key is specific for one lock).

This structure is, in turn, attached to a so-called *G-protein* on the inner face of the cell membrane. This G-protein comprises three subunits – α,β, and γ-subunits – and the binding event causes the α-subunit to undergo a chemical change and then separate from the G-protein. It migrates along the inner face of the cell membrane until it encounters an enzyme called adenylyl cyclase, to which it binds. This induces a change in this enzyme's three-dimensional structure leading to its activation and the subsequent production of a small molecule called cyclic adenosine monophosphate or cAMP.

Despite the apparent complexity of these initial interactions, they are but the prelude to a cascade of intracellular events (secondary signals). Already one molecule of adrenaline will have triggered the release of several hundreds and perhaps thousands of molecules of cAMP. This is a superb secondary messenger (the first 'message' was provided by adrenaline): it is small and mobile, and is rapidly synthesised and subsequently broken down by another enzyme – phosphodiesterase.

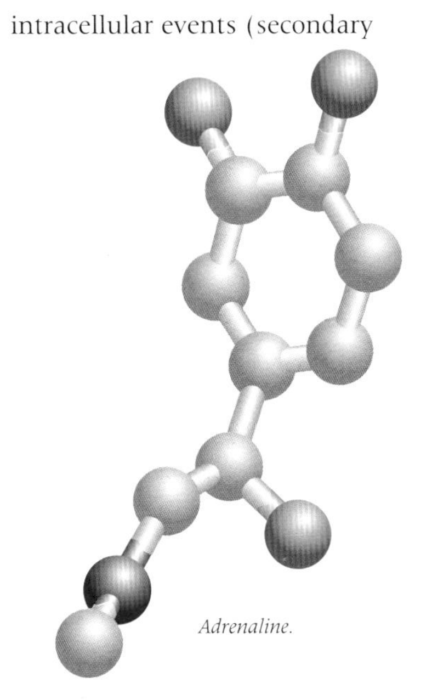

Adrenaline.

The next stage of the signalling cascade usually involves a series of enzymes, cAMP-dependent protein *kinases*, which catalyse the addition of a phosphate group (*phosphorylation*) to other cellular enzymes. This regulation of enzyme activity through phosphorylation, and the reverse process of phosphate-removal catalysed by various phosphatases, is one of the most important means of control available within cells. Given that one molecule of cAMP may activate several kinases, which in turn can switch on or switch off numerous other enzymes, the amplification of the effect of the arrival of one drug molecule at its receptor can be quite staggering.

So how does understanding these complex *signalling pathways* explain how adrenaline acts? One major effect of the cascade of events caused by binding of adrenaline is the release of calcium ions from storage sites within the cell's cytoplasm, and this leads to activation of muscular activity, and thus to increased heart rate. This knowledge also helps us to understand how heart drugs called beta-blockers (for example, propanolol) function (Figure 12).

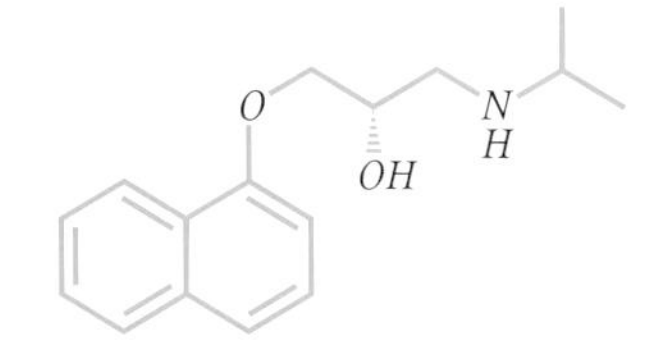

Figure 12. The heart drug, propanolol.

These were first prepared by a research team at ICI led by James Black. The team was looking for molecules that would prevent the cardiac excitation caused by adrenaline and a structural analogue noradrenaline. This was an important therapeutic goal, because these compounds can precipitate an attack of angina in

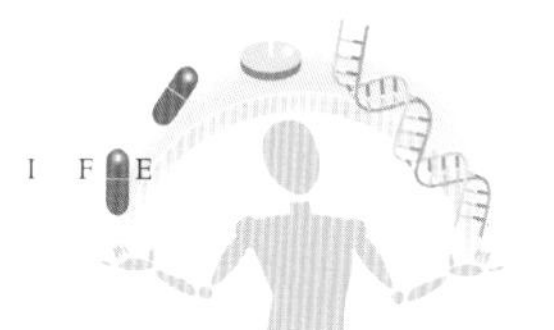

patients suffering from congestive heart disease. The drug's molecules compete with adrenaline and noradrenaline in binding to 'beta-adrenergic' receptors on the surface of heart muscle cells. When bound, however, they do not elicit cAMP production, so inhibiting or blocking the ensuing process – hence the name *beta-blocker.* It is worth noting that the beta-blockers were 'invented' long before this bilogical process was elucidated, but the countless thousands of angina sufferers who have benefited from therapy were none too bothered about that.

This strategy of identifying a key receptor implicated in a disease, and the molecule that normally bonds to it to elicit a biochemical response such as cAMP production – the *agonist* – and then searching for another agent (usually an unnatural molecule) that preferentially binds to the receptor site but does not elicit the response – an inhibitor, or *antagonist* is now an important activity in the pharmaceutical industry. It is the major component of so-called *rational drug design*.

Adrenaline and noradrenaline are just two of the many molecules that exert their effects through receptors coupled to G-proteins. Histamine, which can cause the typical symptoms of hayfever, or elicit release of acid in the stomach, depending on the type of receptor to which it binds; oxytocin, which causes contraction of the uterus at fullterm of pregnancy, and assists in the release of breast milk; and the prostaglandins, which are involved in a plethora of physiological responses, all interact with receptors that are linked to G-proteins. Just as the beta-blockers prevent the activation of heart muscle, so inhibitor compounds called the histamine (H_2)-antagonists such as cimetidine, prevent the activation of cells in the stomach wall, and thus reduce acid secretion. Excellent news for anyone with a stomach ulcer, and another triumph for the pharmaceutical industry, this time the company Smith Kline & French and its research team of Robin Ganellin, Graham Durant and John Emmett. Once again the original research was inspired by James Black, who was a highly deserving winner of the Nobel Prize for Physiology or Medicine in 1988. Cimetidine (Tagamet) (Figure 13) and the subsequently discovered ranitidine (Zantac, developed by Glaxo), were the first true blockbuster drugs, making fortunes for their respective companies, and thus funding research on other drugs.

James Black.

Figure 13. Cimetidine or Tagamet, the antiulcer drug.

The Tagamet team, (left to right) Robin Ganellin, John Emmett and Graham Durant who developed Cimetidine.

Courtesy of SmithKline Beecham plc.

And along came the Pill

Not all drugs, however, cause physiological responses through this type of chemical cascade. The *steroid hormones* like the oestrogens (female sex hormones), progesterone (the hormone of pregnancy), and the androgens (male sex hormones) all pass into cells without interacting with a surface receptor. Once in the cytoplasm, they attach to special protein receptors, and then the

Box 2 The prostaglandins

Another excellent example of a family of compounds that has excited the interest of chemists and biologists for much of the century, is provided by the eicosanoids – most importantly the prostaglandins and leukotrienes. In 1930 two New York biologists, Raphael Kurzrok and Charles Lieb, demonstrated that human semen could affect the contractile state of tissue from the human uterus. A few years later Maurice Goldblatt in the UK and Ulf von Euler in Sweden, showed that seminal fluid could not only induce contractions in smooth muscle but could also affect the blood pressure if injected into animals. Von Euler suggested that this factor (it actually proved to be a family of compounds) should be called prostaglandin because the prostate gland is the source of seminal fluid in human males.

It was to take nearly 30 years before the structures of these intriguing compounds were elucidated, as this had to await the invention of techniques like paper chromatography, gas-liquid chromatography and mass spectrometry (see *Analysis and structure of molecules*). The pioneering structural studies of Sune Bergstrom, Sixten Abrahamsson and Bengt Samuelsson ultimately provided structures for the prostaglandins PGE_1, $PGF_{1\alpha}$ and $PGF_{2\alpha}$ in 1962/3, and those of the PGD_2 and PGA_2 followed soon afterwards.

The highly potent (at the microgram level) biological activity of these molecules was soon revealed. Their ability to reduce secretion of gastric acid (especially the E series), and control fertility – at least in animals – provided the most excitement, though they were also clearly implicated in causing pain and inflammation. This plethora of biological activities provided the incentive for the synthetic chemists and the prostaglandins became the trendy synthetic targets of the 1970s. Literally hundreds of synthetic routes to the primary prostaglandins and their structural analogues have appeared since 1969, and as each new advance in synthetic methodology was unveiled, so it was applied to a prostaglandin synthesis. Further interest was generated in 1970 when John Vane and his colleagues in London demonstrated that aspirin stopped prostaglandin being made in the body. This important discovery provided the basis for an understanding of the mechanism of action of a drug that had been in use since 1899 without anyone having an idea how it exerted its analgesic and anti-inflammatory effects. The Vane research group went on to discover several other prostaglandins, most notably PGI_2 and thromboxane A_2. Both of these are intimately involved in the control of blood vessel tone and contractility.

CO_2H · O · C_5H_{11} · HO · H · OH · PGI_2

O · O · CO_2H · C_5H_{11} · OH · Thromboxane A_2

Two prostaglandins, PGI_2 and thromboxane A_2, play a part in controlling blood vessel tone and contractility.

The even more potent leukotrienes were discovered in the late 1970s thanks to a collaboration between Samuelsson in Sweden and Elias Corey at Harvard. These eicosanoids are principally mediators of inflammation and allergy, and not surprisingly a significant amount of research has been directed towards designing and synthesising leukotriene antagonists for the treatment of asthma and arthritis, and other inflammatory conditions.

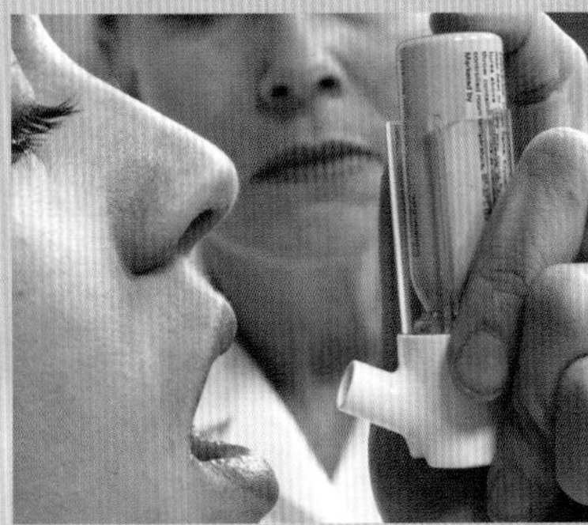

Leukotriene antagonists may be used in the future for the treatment of asthma

The entire family of eicosanoids now numbers well over 100 members, and all are derived from the all-*cis*-tetraunsaturated fatty acid eicosa-5,8,11,14-tetraenoic acid or arachidonic acid. The unravelling of the so-called arachidonic acid cascade, together with the synthesis of the natural products and numerous analogues, has provided chemists and biologists with three decades of highly rewarding research. The excitement is certainly not over, since it has been shown recently that there are two forms of the enzyme cyclooxygenase, which catalyses the conversion of arachidonic acid into the prostaglandins. The normal form COX-1 is responsible for the production of the small, requisite amounts of prostaglandins needed to maintain healthy body functions. A second form, COX-2, is produced in relatively large quantities when a disease state like arteriosclerosis or rheumatoid arthritis has been established. The design of drugs specifically to inhibit this second enzyme thus offers the exciting prospect of controlling such life-threatening or debilitating diseases.

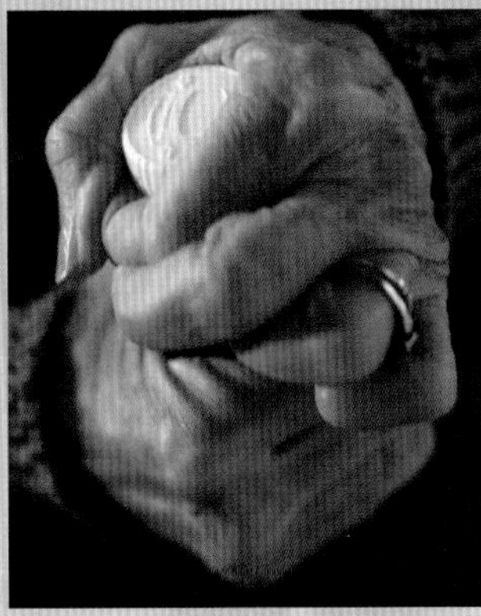

The prostaglandins offer the exciting prospect of controlling life-threatening or debilitating diseases.

whole steroid-receptor complex passes into the nucleus of the cell, where it interacts with DNA. Before we consider the molecular mechanisms of these interactions, it is worth reviewing the fascinating history of the steroid hormones, because they have contributed so much to the improvement in quality of life during the past 50 years.

In the mid-17th century, the herbalist Nicholas Culpeper had reported that plant extracts could be used to improve fertility and as a method of contraception. However, it was not until the 1930s when Adolf Butenandt at Berlin University, Edward Doisy at St Louis University and others isolated and characterised the sex hormones that people appreciated the potential of these substances for the control of fertility. Synthetic work by researchers such as Leopold Ruzicka at the ETH in Zurich, and Carl Djerassi at Syntex in Mexico City provided access to larger quantities of these compounds, and the first oral contraceptive, ethisterone, was in restricted use from around 1937.

However, the major breakthrough, in terms of supply, was made by Russell Marker in the late 1940s. He was a chemist working at Pennsylvania State University, and had spent many years trying to identify a cheap, renewable plant source for the preparation of semi-synthetic steroids. In 1940, he showed that the plant steroid diosgenin, isolated from the Mexican yam *Dioscorea mexicana*, could be efficiently converted into progesterone. The pharmaceutical company Syntex (recently absorbed into the Hoffmann-La Roche empire) was formed to exploit this procedure. This research culminated in the synthesis of the orally active contraceptive norethindrone (Norlutin), a progestin, which was marketed in the early 1960s (Figure 14). Contemporaneously, the American company G. D. Searle marketed a very similar compound, norethynodrel (Enovid), and between them the two companies controlled much of the market for these early forms of the 'Pill'. For the first time in human existence, women had the opportunity to control their own fertility and in many countries the birthrate dropped dramatically.

OH

C≡CH

O

Norethindrone (Norlutin)

Figure 14. Norlutin, the first contraceptive pill.

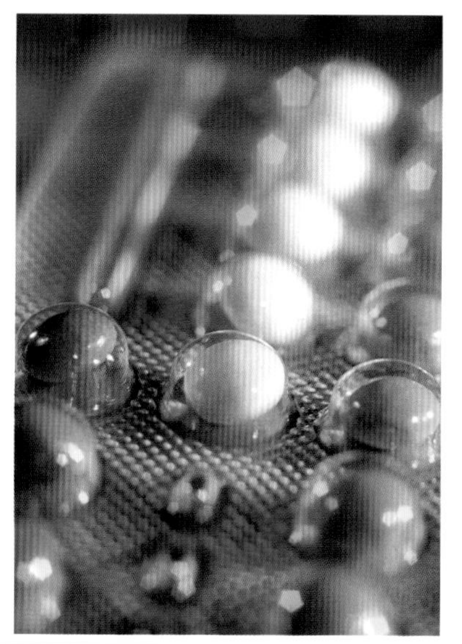

As a result of the contraceptive pill, the birthrate dropped dramatically in many countries.

Tek Image/Science Photo Library

An additional benefit of these synthetic steroids is their use in hormone replacement therapy (HRT) in menopausal women. In this context, the drugs not only help to reduce the symptoms associated with the menopause but also dramatically reduce the incidence of osteoporosis (brittle bone disease) – another major enhancement of the quality of life brought about by synthetic steroid hormones.

As the mode of action of the steroids – both natural and synthetic – became clear, further therapeutic strategies were developed. Researchers noted that the growth of tumours in around one-third of breast cancer patients depended on a supply of oestrogen, so an obvious form of therapy was to deny these steroids to the tumour cells. The drug tamoxifen (Novaldex) developed by ICI (now Zeneca) does just this. It binds to the oestrogen receptors in breast-cancer cells and thus prevents natural oestrogens from attaching. The whole receptor-tamoxifen complex is now translocated into the cell nucleus but does not elicit a burst of DNA activity like the normal complex. Cell division and growth are no longer possible and the tumour regresses.

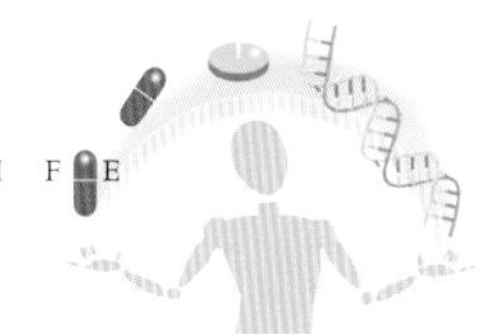

Box 3 How the molecules of life are made

Chemists have for decades been studying the chemical pathways by which biological molecules are made in living organisms. At first elucidating the *biosynthesis* of natural products relied on using radioisotopes carbon-14 and tritium (a radioactive form of hydrogen) to 'label' small molecules. These could be introduced into plants, fungi and microorganisms and traced as they were processed by the organism to yield the final radio-labelled natural products. Using these techniques, John Cornforth and Georg Popják at the MRC laboratories in Mill Hill London worked out completely the route of biosynthesis of cholesterol and similar natural products.

The subsequent availability of the nonradioactive carbon isotope, carbon-13, revolutionised this work because carbon-13 can be used in NMR studies to obtain structural and biosynthetic information. The unravelling of the biosynthetic pathway to the cofactor haemin in blood and vitamin B_{12} by Alan Battersby at Cambridge and Ian Scott at Texas A&M University with their respective groups, was another great triumph made possible by this new technique.

More recently, the isolation of the individual enzymes of biosynthetic pathways, has allowed further insights into their intricacies. Examples are the seminal work of Jack Baldwin at Oxford and Ian Scott on the biosynthesis of penicillin; and the work by Mohammed Ahktar at the University of Southampton on the enzymes that activate oxygen (*cytochrome P450* enzymes). These are of vital importance in the biosynthesis of all classes of natural products, such as steroids, alkaloids, prostaglandins, and also in the metabolism of these compounds and just about every type of drug.

The structure of DNA

It is clearly time to enquire how these hormone-receptor complexes activate DNA. To understand the complexities of this branch of molecular biology, we must return to the early part of the century to chart the progress in understanding the chemistry of deoxyribonucleic acids (DNA) – the other molecules that are so intimately involved in all of life's processes. The structure of the DNA *double helix* (duplex), as first proposed by James Watson and Francis Crick in 1953, was the culmination of a worldwide effort involving many researchers in chemistry and biology for the 75 years leading up to this date. DNA is a linear polymer made up of four different building blocks, known as 2′-deoxynucleotides. Each nucleotide is composed of a 2′-deoxyribose 5-phosphate sugar and a 'base'. This is either a purine (adenine or guanine) or pyrimidine (thymine or cytosine) (Figure 15).

The bases were identified by Albrecht Kossel and his colleagues at Heidelberg University in the early years of this century. Phoebus Levene (a former chemistry student of the Russian composer Borodin) identified D-ribose as the sugar component of the related ribonucleic acids (RNA) in 1912, and later showed that 2′-deoxy-D-ribose provided the sugar component of DNA. He also synthesised the bases. One belief dispelled by Levene, was that there were two types of nucleic acid that depended on the source, these being 'thymus-derived nucleic acid' (DNA), found in animal cells, and pentose nucleic acid (RNA), found in plant cells. His work showed that all cells contained both types of molecules, but there was still no awareness that DNA was important in heredity. Structural confirmation took a little longer,

Adenine Guanine

Purines

Cytosine Thymine

Pyrimidines

2′-deoxyribose 5-phosphate sugar

Figure 15. The building blocks of DNA.

and the total synthesis of the basic RNA and DNA structures was completed only in the 1940s by Alexander (later Lord) Todd and his research group at Manchester, then Cambridge. They deduced the position of the linkage between the base and the sugar, and later that the (deoxy)nucleotide building blocks were linked as their phosphate diesters through hydroxyls of the (deoxy)nucleotide.

In 1944, evidence for the biological significance of DNA was first produced by Oswald Avery at Rockefeller University in New York when he identified DNA as the 'transforming principle' that could turn non-pathogenic bacteria into pathogenic ones, but this was a long way from showing how DNA stored genetic information. The original belief was that DNA consisted of cyclic molecules (molecular weights up to about 1000) comprising equal amounts of the four bases, joined together by the phosphate groups. It was assumed that the DNA was merely a structural scaffold used to support the associated proteins called histones, and it was these that carried the genetic information. Once Todd had deduced the precise structure of the linear nucleic acid polymer in 1951, this hypothesis had to be discarded (Figure 16). William Astbury and Florence Bell at Leeds University conducted some X-ray studies on fibres of DNA, and concluded that the bases were stacked face-to-face like a pile of coins in groups of eight or sixteen. The three-dimensional structure thus resembled parallel ladders standing out from a central axis.

John Gulland at the University of Nottingham carried out pH titration measurements and predicted that the hydroxyl (OH) groups of the bases formed *hydrogen bonds* with one another – the first time that this key form of bonding had been invoked as a stabiliser of DNA structure. Had it not been for Gulland's untimely death in a train derailment in 1947, the solution of the DNA problem may not have had to await the seminal work of Watson, Crick and Wilkins. Gulland's results were also overlooked by Linus Pauling and Robert Corey, who in 1953 proposed that DNA existed in a triplex structure, with externally projecting bases surrounding an internal column of phosphates. However, a careful examination of this structure showed that the negatively charged phosphate groups should repel one another under physiological conditions, thus making this structure unacceptable.

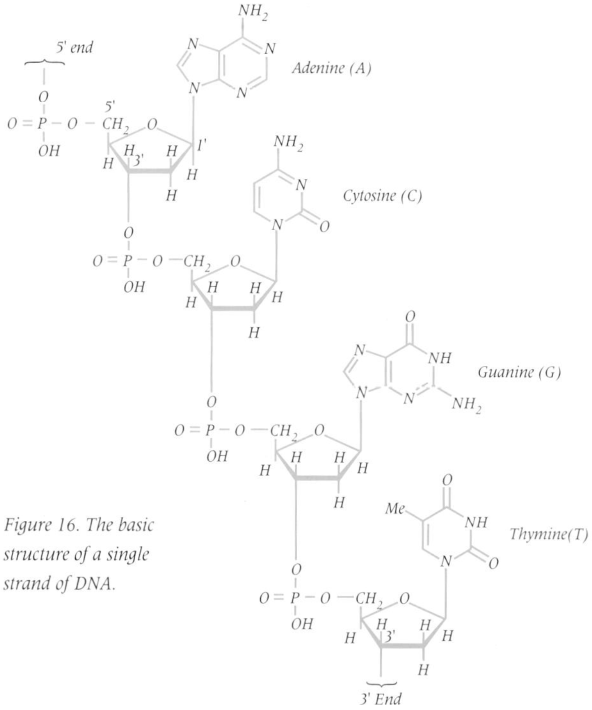

Figure 16. The basic structure of a single strand of DNA.

Meanwhile, Edwin Chargaff of Columbia University in New York, had conducted a number of studies on DNA, and these resulted in what became known as Chargaff's rules (1950). They stated that:

(i) the total of purines (adenine and guanine) always equalled the total of pyrimidines (thymine and cytosine);

(ii) the molecular quantity of adenine equalled that of thymine, and the molecular

quantity of guanine equalled that of cytosine.

Thus: A + G = T + C and A = T and G = C. This information was absolutely crucial for the final solution of the structural problem.

Back in the UK, X-ray diffraction patterns produced by Maurice Wilkins and Rosalind Franklin of Kings College London, suggested that there were two different forms of DNA, but that the dominant form possessed a helical structure with a diameter of 2 nanometres and a pitch of 3.4 nanometres. So it was that James Watson and Francis Crick, armed with all of this information, and bits of Meccano, wire and sundry other ironmongery from the workshop at Cambridge, proceeded to make models during the period 1951 to 1953. Their efforts were ultimately rewarded by their discovery of the DNA double helix, with its antiparallel strands of DNA held together by hydrogen bonding, A to T and G to C – arguably the most famous chemical structure of the 20th century (Figure 17). The structure explained how DNA could replicate itself, so that genetic information could be stored and transmitted. The double helix partially unwinds and then each single strand acts as a blueprint for the production of a new, complementary strand.

James Watson and Francis Crick and their marvellous model of the DNA double helix.
A Barrington Brown/Science Photo Library

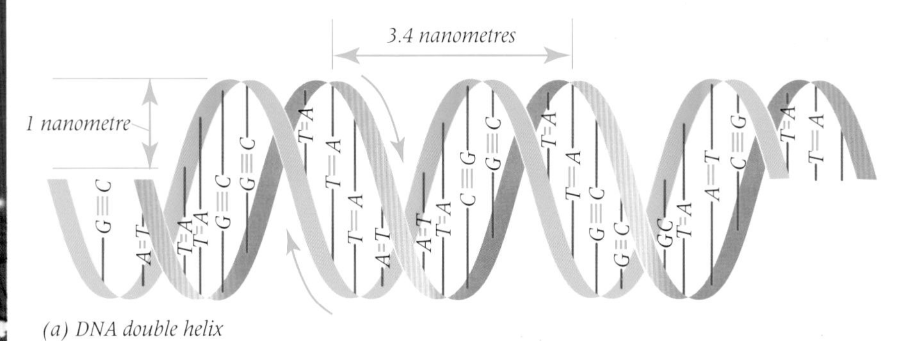

(a) DNA double helix

(b) Base pairing

Figure 17. (a) The DNA double helix and (b) showing how the bases are paired.

RNA differs from DNA in two ways. First the pyrimidine base thymine is replaced by the simpler uracil which retains the ability to bond to adenine through two hydrogen bonds. More important, the nucleotides present in RNA incorporate D-ribose sugars rather than 2′-deoxy ribose

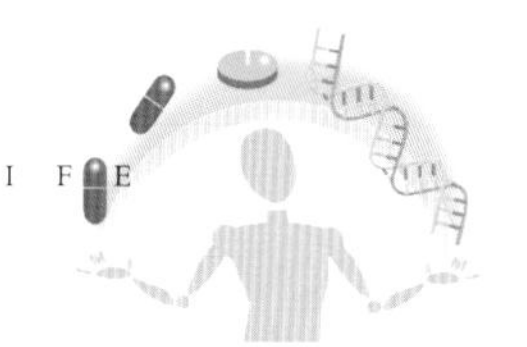

Uracil (U) Ribose

Figure 18. RNA contains uracil rather than thymine and a ribose sugar instead of deoxy ribose.

(Figure 18). This is important chemically in that although DNA is a very robust polymer as required for its function as a repository of genetic information, RNA has several different roles as we shall see shortly. RNA is used in multiple copies to amplify the 'signal' of a single DNA gene sequence and because it is more labile than DNA it can be rapidly recycled within the cell.

The genetic code

Elucidation of the structure of DNA and how it replicated allowed, for the first time, chemists and biologists to answer such crucial questions as how cells divide and still function, how genetic information is stored and transferred from generation to generation. It also led to an understanding of how proteins, the molecules responsible for the day-to-day housekeeping in cells and ultimately for all other life processes, were assembled. In essence, an understanding of the chemistry of life was now within our grasp.

The central dogma, proposed by Crick, for the flow of genetic information, was that DNA provided the blueprint for RNA (actually *messenger-RNA*), which in turn provided the code for protein/enzyme production, and thus established the hierarchy of the cell. It became apparent that not all of the DNA was transcribed into messenger-RNA (m-RNA), but only specific regions called *genes*. Each gene contains the information to make one or more proteins. The process of translating the genes into proteins is called *gene expression*. Flanking the region of DNA that encodes for each gene were regions of DNA known as *operons*, which were responsible for the binding of the RNA polymerase and other proteins necessary for messenger-RNA biosynthesis. Each gene was under the control of an operon and through various methods including positive and negative feedback loops, the amount of a specific m-RNA and thus of one or more proteins produced could be tightly regulated.

Other key developments were also taking place, and the solving of the genetic code by Arthur Kornberg at Washington University, Sydney Brenner and Leslie Barnett at Cambridge University, and Gobind Khorana at the University of Wisconsin at Madison in 1966, warrants special mention. Through a combination of chemical and biochemical studies, these researchers deduced that the genetic code stored in DNA and transcribed in

	Second				
First	U	C	A	G	Third
U	Phenylalanine	Serine	Tyrosine	Cysteine	U
	Phenylalanine	Serine	Tyrosine	Cysteine	C
	Leucine	Serine	CT	CT	A
	Leucine	Serine	CT	Tryptophan	G
C	Leucine	Proline	Histidine	Arginine	U
	Leucine	Proline	Histidine	Arginine	C
	Leucine	Proline	Glutamine	Arginine	A
	Leucine	Proline	Glutamine	Arginine	G
A	Isoleucine	Threonine	Asparagine	Serine	U
	Isoleucine	Threonine	Asparagine	Serine	C
	Isoleucine	Threonine	Lysine	Arginine	A
	Methionine	Threonine	Lysine	Arginine	G
G	Valine	Alanine	Aspartic acid	Glycine	U
	Valine	Alanine	Aspartic acid	Glycine	C
	Valine	Alanine	Glutamic acid	Glycine	A
	Valine	Alanine	Glutamic acid	Glycine	G
RNA bases: U=Uracil, C=Cytosine, A=Adenine, G=Guanine					

Figure 19. The Genetic Code given here indicates the relationship between the three-letter sequence of bases (codon) found in a messenger-RNA linear polymer and its corresponding amino acid which is incorporated into a new peptide. For example, the codon UGC (uracil, guanine, cytosine) is translated into the cysteine amino acid in the new peptide chain. There are three codons that cause peptide biosynthesis to terminate (UUA, UUG, UGA). A single codon exists for the amino acid methionine (AUG). This is always the first amino acid incorporated into a peptide in bacteria (in the form of N-formylmethionine).

messenger-RNA involved triplet sequences of nucleotide bases (A, T, G, C), of which there were 64 possible combinations. Each triplet sequence, or *codon*, is unique to a specific amino acid (Figure 19). This, therefore, was the 'Rosetta Stone' of chemical genetics. Khorana's group actually synthesised all of these combinations, and showed they could then be used in cell cultures to produce polypeptides of defined constitutions, thus demonstrating the specificity of the code.

At the same time, in 1961, Jacques Monod and Francois Jacob of the Pasteur Institute in Paris, proved the existence of messenger RNA, the delivery system for the code. Robert Holley at Cornell and Hans Zachau in Cologne in 1965, isolated and elucidated the structures of several *transfer-RNA* (t-RNA) molecules. These function as 'adapter' or 'interpreter' molecules; they carry both an amino acid at one end of the RNA chain and also the corresponding *three-nucleotide anticodon* (reversed sequence) for that amino acid. This is complementary to the codon on the messenger-RNA molecule. The ribosome, which is a hybrid protein/RNA structure found in the cytoplasm, acts as a scaffold to bring together the m-RNA and appropriate t-RNA units so that the sequence of nucleotide triplets (codons) given by the m-RNA template can be translated into a new peptide through the formation of amide bonds between the amino acids carried on the t-RNA molecules. The latter process is catalysed by the ribosome (Figure 20).

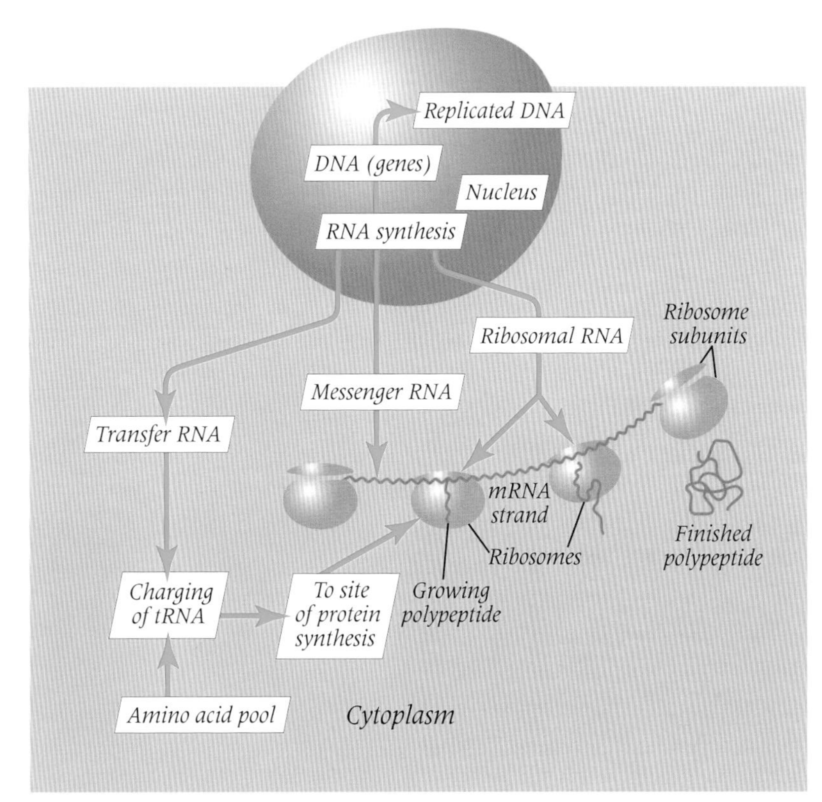

Figure 20. This shows the flow of genetic information from the DNA template in a eukaryotic cell. For cell division, the DNA is directly copied to give daughter DNA strands so that complete copies of the genome are present in both daughter cells. For protein biosynthesis, a messenger-RNA polymer complementary to each gene is transcribed from the DNA and this moves from the nucleus to a ribosome in the cytoplasm. The instructions provided by the m-RNA template are then translated into new polypeptides at the ribosome. This occurs via the codon-anticodon interaction with transfer-RNA units bearing the amino acid building blocks used to from the new polypeptide.

The rise of genetic engineering

It was not long before people considered the possibilities of genetic engineering. Two discoveries ensured that the theory could be turned into reality. First, in 1967 Walter Gilbert at Harvard isolated the first example of an enzyme, DNA ligase, which could join two sections of DNA into a single, longer sequence. Secondly, through the research of Werner Arber from Basel, Switzerland in the early 1960s and later that of Daniel Nathans and Hamilton Smith at Johns Hopkins University in Maryland, the first *restriction enzymes* (enzymes that could cut DNA at specific sites) were identified and used to obtain the sequence of the first genome, that of SV40 virus. Several years later, in 1973, Herbert Boyer and Stanley Cohen at Stanford and the University of California used both restriction enzymes and DNA ligase to cut and join DNA from two different organisms to generate the first *recombinant*

DNA. Leading on from these discoveries, Genetech, the first company to exploit genetic engineering and recombinant DNA technology was founded in 1977.

It was not enough, however, to be able to construct DNA to order, it was also necessary to determine the sequence of natural DNA samples. Novel chemical methods developed by Allan Maxam and Gilbert at Harvard, and the biochemical methods of Fred Sanger at the Medical Research Council laboratories in Cambridge in the 1970s allowed people to determine very long sequences. The current project to determine the entire *human genome* – that is, the complete sequence of genes in a cell's nucleus responsible for its function and growth – would not have been possible without these seminal discoveries.

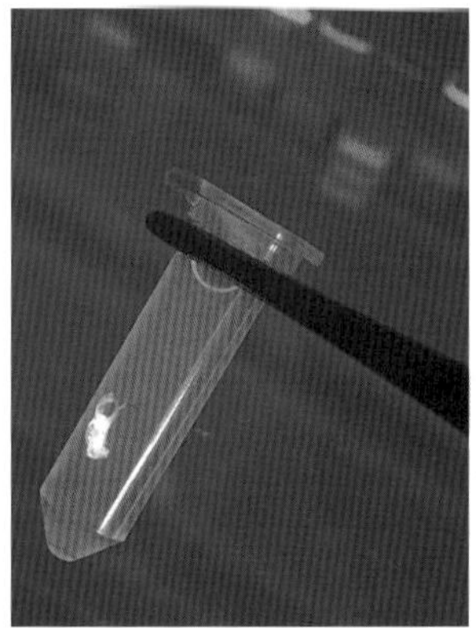

Sample tube containing pellet of human DNA (white).
Philippe Plailly/ Science Photo Library

More recently, it has become possible to produce enormous numbers of copies of a particular DNA, using an enzyme called DNA polymerase. All that is needed is a minute quantity of a master sequence, for instance the DNA from one cell or from a small amount produced synthetically. This *polymerase chain reaction* (PCR) technology, first devised by Kary Mullis, then at the Cetus biotechnology company in California, has significantly accelerated developments in molecular biology, and has revolutionised the field of genetics as well. The well-known technique of *DNA fingerprinting*, discovered by Alec Jeffrey of Leicester University in 1985, also depends upon this technique. This allows a genealogist to trace a family's heritage, or a palaeontologist to map the evolutionary tree of a long-extinct species, or a forensic scientist to identify a criminal using small samples of DNA.

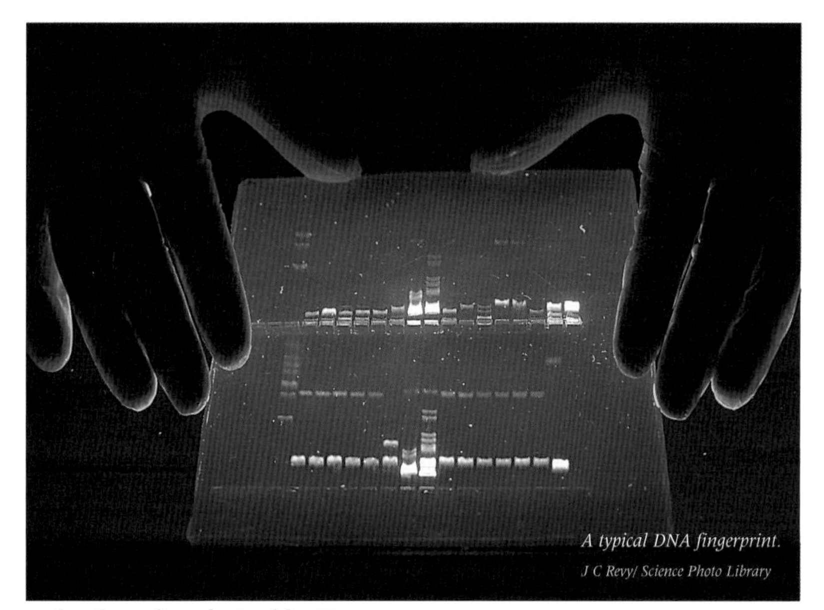

A typical DNA fingerprint.
J C Revy/ Science Photo Library

The role of DNA in drug design

The potential benefits of our growing knowledge of the structure and function of DNA are clearly considerable, and this is perhaps most apparent in the area of drug design. We left the oestrogen-receptor complex about to enter the cell nucleus, so let us now consider the molecular interactions that happen within the nucleus. The endpoint of most of the chemical-signalling cascades, including those involving steroid-receptor complexes, is activation or repression of DNA replication or *transcription* – the process whereby the enzyme RNA polymerase uses the DNA as a blueprint for the production of messenger-RNA. Much recent drug design seeks to identify specific chemical mediators that will modify these processes. For decades, most drugs used for cancer chemotherapy, for example, have interfered with these processes in a nonspecific way, and it is worth comparing the 'old' drugs with some of the new ones that may revolutionise the treatment of this much-feared disease.

Tumours are made up of rogue cells which have lost control over their own growth and replication, so it is very difficult to find agents that target cancer cells whilst leaving healthy tissue

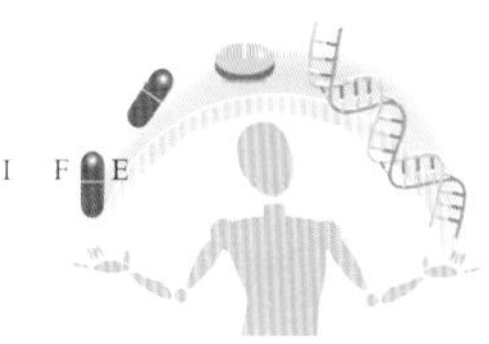

untouched. Many drugs used in cancer chemotherapy show little discrimination. They rely on the fact that cancer cells often grow much more rapidly than normal cells and take up nutrients (and drugs) faster. However, those healthy tissues that undergo rapid replacement, such as the lining of the digestive tract, white blood cell formation and hair follicles, are also seriously effected by such drugs and this leads to nausea, anaemia and hair loss. This seriously limits the dose of drug that can be administered, and has led to efforts to develop drugs and drug-delivery systems that can specifically target tumour cells.

Anticancer drugs

The first anticancer agents were the nitrogen mustards, developed by Alfred Goodman and Fredrick Philips at Edgewood Arsenal in Maryland, from mustard gas used in the First World War (Figure 21) – one of the few positive things to arise from what was otherwise an abuse of chemistry. They were first used clinically in 1942, with some spectacular, if short-lived therapeutic effects. Over the succeeding years, their structures have been modified to optimise their useful activity. Their point of attack is the double helix of DNA and they react chemically with the nucleotide bases (most commonly guanine). Because they mount a two-pronged attack, this results in the formation of cross-links between the DNA strands. Clearly this compromises the unwinding of DNA and thus prevents replication and transcription, with the result that the tumour cell cannot grow and divide.

A similar mode of activity is shared by the platinum anticancer drugs, the most famous of which is cis-platin (Figure 22). This was discovered in 1965, completely by accident, when Barnett Rosenberg at Michigan State University passed an electric current between two platinum electrodes that were immersed in a solution of ammonium chloride containing bacteria. The bacteria produced long, aberrant filamentous structures, then ceased growing – clearly a chemical compound was present that killed cells. Subsequent detective work revealed that small quantities of *cis*-diammonia, platinum dichloride had been produced, and this was developed into the anticancer drug cis-platin. Ten years later this drug (another superb product of what Louis Pasteur called "chance favouring the prepared mind") was revolutionising the treatment of various cancers, most spectacularly testicular teratoma, where a death sentence was translated into a survival rate of more than 90 per cent.

However, like most anticancer drugs, the mustards and cis-platin are not given alone but in combination with other drugs – many of them natural products. One notable family of drugs stop cells dividing by interfering with the functions of an enzyme called tubulin polymerase, which controls the production and destruction of the protein tubulin. This forms the

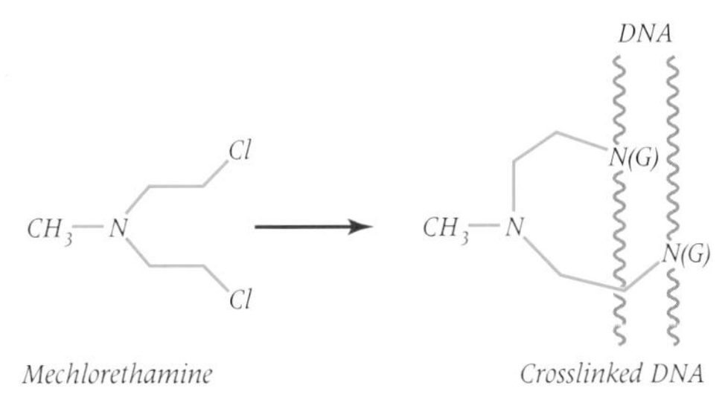

Figure 21. How anticancer agent mechlorethamine, a nitrogen mustard, crosslinks with DNA and prevents it from unwinding.

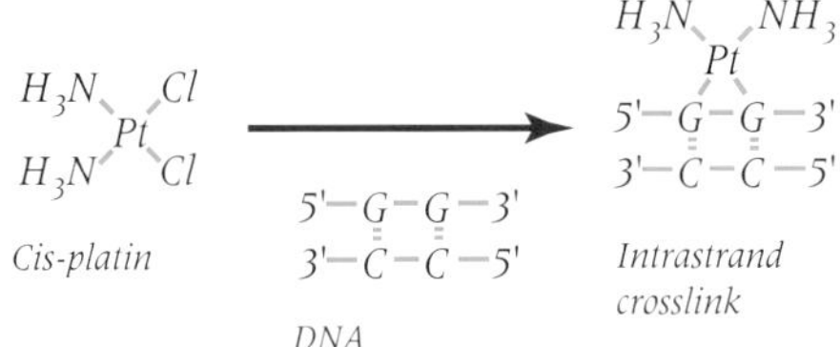

Figure 22. Cis-platin forms crosslinks which prevent replication.

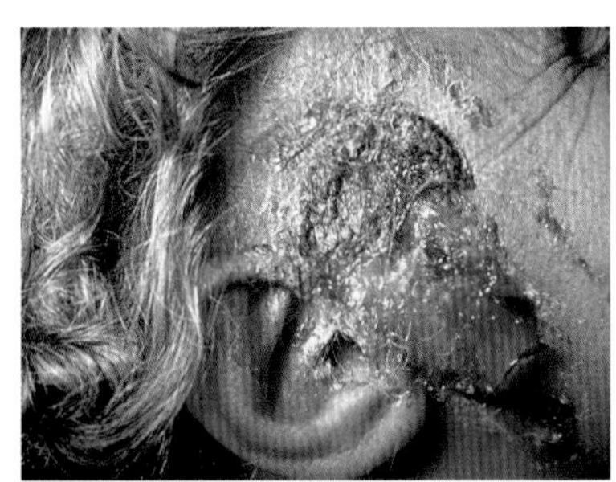

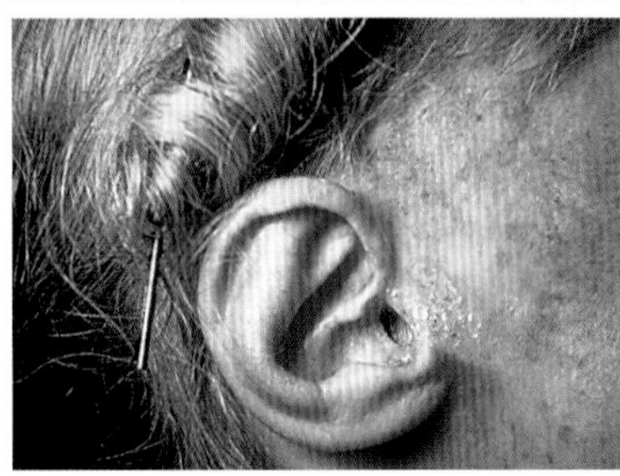

Tumour before and after treatment.
Dr Karol Sikora/ Science Photo Library.

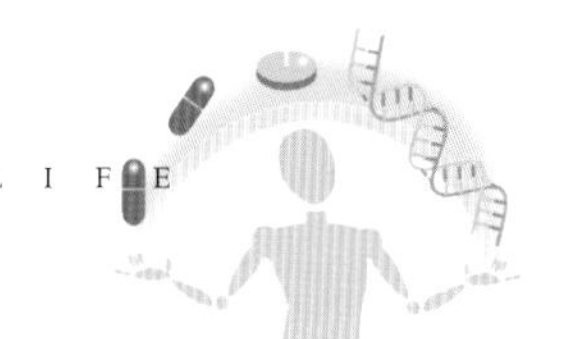

Box 4 Inorganic elements and life

Much of the chemistry of life is inorganic! We now know that the normal growth, development and health of all living systems involves many inorganic elements. Some of them such as iron, calcium, zinc, sodium and potassium are present in most living systems at a reasonably high concentration. However, others exist only in trace amounts but are nevertheless vital for many living processes.

Some transition metals play a part in biology and are incorporated into certain proteins and other large molecular structures. For example, molybdenum and vanadium bound to iron and sulfur are crucial for the catalytic activity of the nitrogenases – the enzymes found in certain bacteria that fix nitrogen from the atmosphere (Figure a); copper is essential for the oxidation of vitamin C (ascorbic acid) – the enzyme ascorbate oxidase involves three different types of copper centre; and zinc is crucial for the activity of a family of proteins (the 'zinc finger' proteins) which control the transcription of the genetic code stored within DNA.

Figure (a). The iron molybdenum centre of the nitrogenase, the enzyme responsible for nitrogen fixation.

Large, flat cyclic carbon-nitrogen structures called porphyrins are ubiquitous in living organisms. They possess four nitrogen atoms in a planar arrangement which strongly bind a metal atom, invariably iron, as the centre of their chemical reactivity. For instance, the haemoglobin molecules of our blood which are responsible for transporting oxygen from our lungs to our muscles, possess four haem centres, each of which binds one molecule of oxygen at the iron centre. Organic molecules which are close relatives of the porphyrins include the corrin and chlorin systems; the former binds cobalt to form the site of catalytic activity of vitamin B_{12} and the latter binds magnesium to form the pigment chlorophyll found in green leaves and which is crucial for photosynthesis. Cytochromes (literally 'cell pigments') also involve an iron atom bound to the four nitrogen atoms of a porphyrin ring plus two other groups from the amino acids of a protein; cytochromes are used widely in living systems as staging posts in electron-transfer relays with the iron centre shuttling reversibly between the two iron oxidation states Fe(III) and Fe(II).

Such metals play a very specific chemical role in the function of these molecules. We have learned that when Nature has a difficult chemical task to perform, a molecule with transition metal centre – or a cluster of metal centres – is invariably employed. Biological processes requiring special control include:

- The selective and reversible binding of oxygen on the iron centres of haemoglobin – and its subsequent reduction to water – on an iron/copper bimetallic centre of cytochrome *c* oxidase. This process releases the maximum free energy and avoids the formation of toxic free radicals.
- The production of oxygen by oxidising two water molecules on the tetranuclear manganese centre on the complex molecular assembly responsible for photosynthesis, Photosystem II, of which chlorophyll is part.
- The fixation of nitrogen, in which the very strong nitrogen-nitrogen bond is cleaved at ambient temperature and pressure to form two ammonium ions.
- The oxidation of methane to methanol on the double iron centre of methane monooxygenase which occurs in some bacteria.

In many cases we are still trying to understand how Nature carries out these reactions under such mild conditions. The important challenge is to make synthetic analogues of these metal-containing biomolecules in order to use them on a laboratory or industrial basis for catalysing similar chemical reactions.

An interesting question is how these elements became incorporated into living organisms in the first place. We are still searching for an intellectually satisfying chemical explanation for the origin of life on Earth. One attractive possibility is that the first biomolecules were formed in the hot conditions produced by volcanoes under oceans, from carbon monoxide and other gases emanating from the magma in reactions catalysed by iron and nickel sulfides. Inorganic elements would then have been incorporated in biological systems from the outset. Certainly all living systems contain molecules with iron-sulfur centres, often in clusters, Fe_2S_2, Fe_3S_4 and Fe_4S_4 (Figure b), which are vital as relays for biological electron transfer processes and as catalytic centres. As the chemistry of the Earth's atmosphere changed from being reducing to oxidising, other elements became available. This is especially notable for molybdenum which is present in the Earth's crust as the insoluble molybdenum sulfide (MoS_2) but, under oxidising conditions, forms the soluble molybdate $[MoO_4]^{2-}$; molybdenum is present at relatively high concentration in seawater and is an essential element for all advanced organisms.

Inorganic compounds also find wide employment as pharmaceutical agents. Some of them have powerful pharmacological effects. The anticancer platinum compound *cis*-platin has already been described. Under careful medical supervision, lithium carbonate

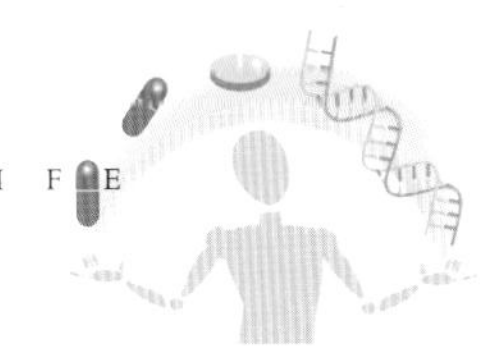

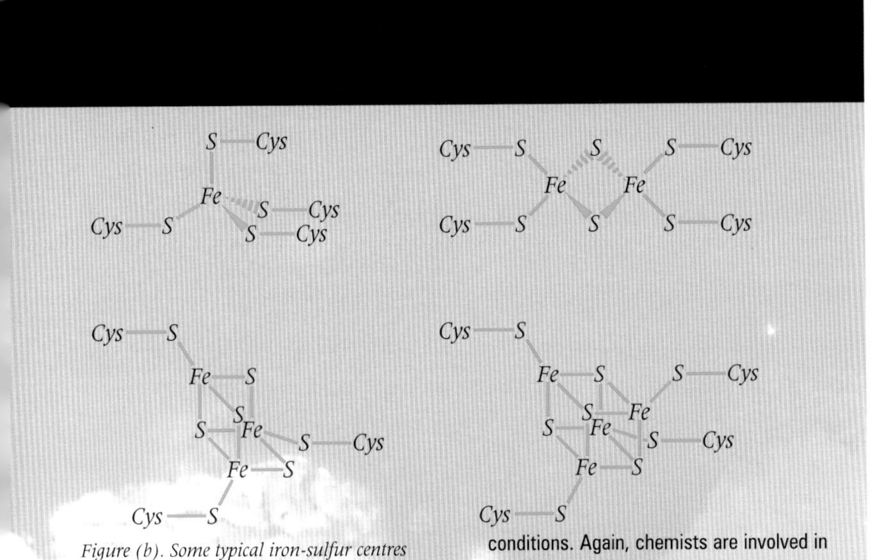

Figure (b). Some typical iron-sulfur centres in metalloproteins.

is prescribed to control the mood swings associated with manic depression and gold compounds can relieve certain rheumatoid conditions. Again, chemists are involved in understanding how these elements interact in biological processes in the human body.

Professor Dave Garner
University of Manchester

The first biomolecules probably formed in the hot conditions like those around 'black smokers' found on the sea bed.
B Murton Southampton Oceanography Centre/Science Photo Library

thin strands that hold dividing cells together until they are ready to lead a discrete existence of their own. Members of this family include the vinca alkaloids (vinblastine and vincristine), the podophyllins, and the more recently famous taxol. Other natural products like doxorubicin and dactinomycin possess flat, multi-ring structures which slip in between the coils of DNA (they are said to intercalate), and act as a kind of 'molecular glue' preventing the separation of the two strands. A final class of drugs, which include the bleomycins and the newly described enediynes like calicheamicin, are metabolised to produce free radicals which interact with oxygen to form dangerously reactive oxygen radicals, and destroy the DNA.

All of these agents are naturally occurring, and as usual their pharmacological activity was discovered before their modes of action were elucidated, so they were definitely not the products of rational design. In the past few years, elegant studies of the interactions of drug molecules with DNA using nuclear magnetic resonance (NMR) and X-ray crystallography, coupled with computer modelling, have allowed medicinal chemists to identify important interactions to DNA in what is known as its minor groove. The phosphate backbone of the DNA duplex divides the exterior surface of the DNA into a narrow minor groove and wider major groove. Whilst many proteins bind in the major groove of DNA using structures with names such as leucine zipper and zinc finger, and these proteins control initiation, replication and transcription, it has been found that a family of compounds called the lexitropsins can bind into the minor groove and so halt these processes and hence have possible use as drugs. These include the polycyclic, but unimaginatively named, Hoechst 43254, and also several analogues of the natural product netropsin. During the next few years, we are likely to hear of very exciting developments in this area of anticancer drug design.

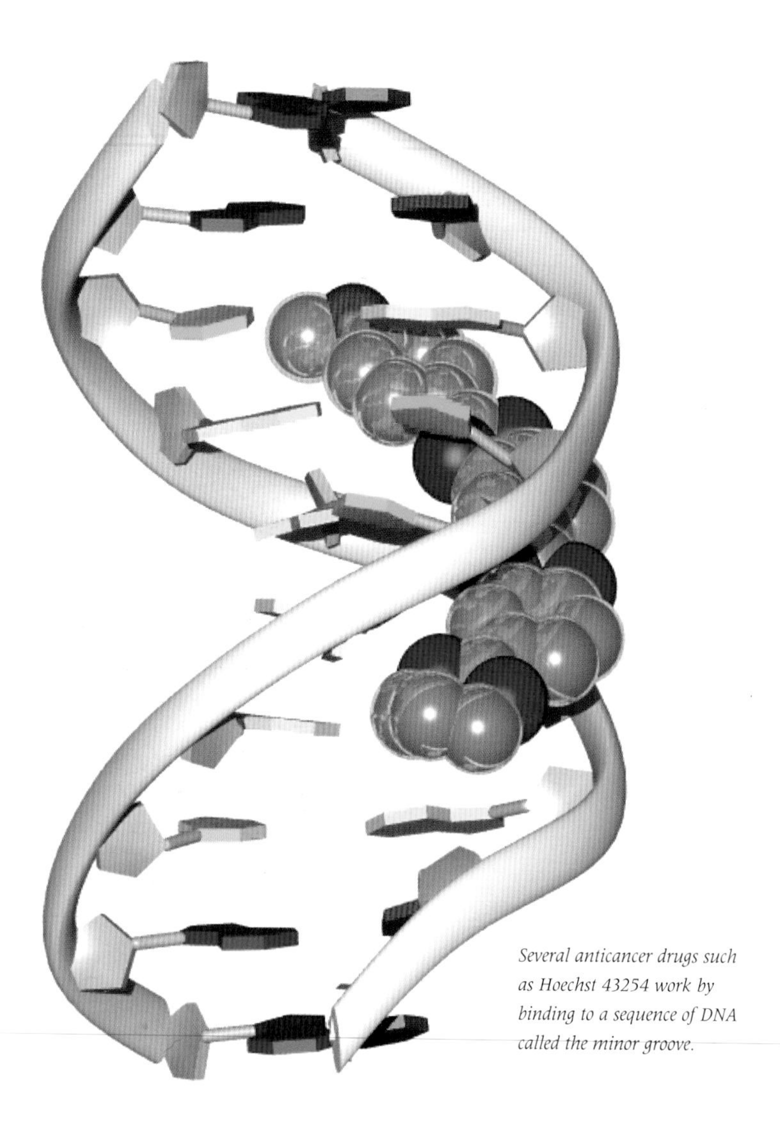

Several anticancer drugs such as Hoechst 43254 work by binding to a sequence of DNA called the minor groove.

Antiviral drugs

A knowledge of the molecular biology of the disease is helping with the design of drugs in antiviral chemotherapy. Viruses can be considered as cellular parasites, since they hijack the DNA replication and protein biosynthesis systems of the host cell, for the production of their own genetic material (DNA or RNA) and their enzymes and coat proteins. Viruses are responsible for a number of life-threatening and fatal diseases including rabies, yellow fever, poliomyelitis, measles, rubella, influenza, AIDS, and certain cancers. For many of these diseases it has been possible to immunise people by injecting them with a modified form of the virus. This stimulates the immune

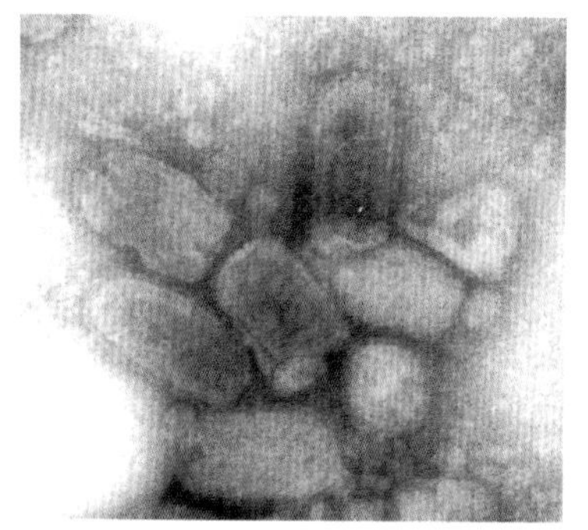

Viruses such as the rabies virus shown, can be treated by immunisation.
Courtesy of New York State Department of Health's Wadsworth Center.

system to produce antibodies which then recruit other members of the white cell family that eventually destroy the invading virus. While this strategy has been immensely successful in some cases – smallpox has been completely eradicated, for example – other viruses have proved to be much more difficult to destroy.

The human immunodeficiency virus (HIV) is estimated to be present in around 30 million people worldwide, and has already caused at least 15 million deaths from acquired immune deficiency syndrome (AIDS). Yet this is a new disease, and the first case of HIV infection probably occurred just after the Second World War. It also provides a superb example of how chemists and biologists can mobilise their forces to tackle a global problem. From the mid-1980s, when the first cases of AIDS were reported, to the late 1990s, when a cure may well be in sight, thousands of scientists around the world have laboured to understand and treat this modern scourge. Infection with the virus depresses the patient's immune system making them more susceptible to opportunistic infections such as pneumonia, tuberculosis, and various cancers. Because the virus undergoes rapid mutation, it avoids recognition by the immune system, and it is also near impossible to produce a vaccine.

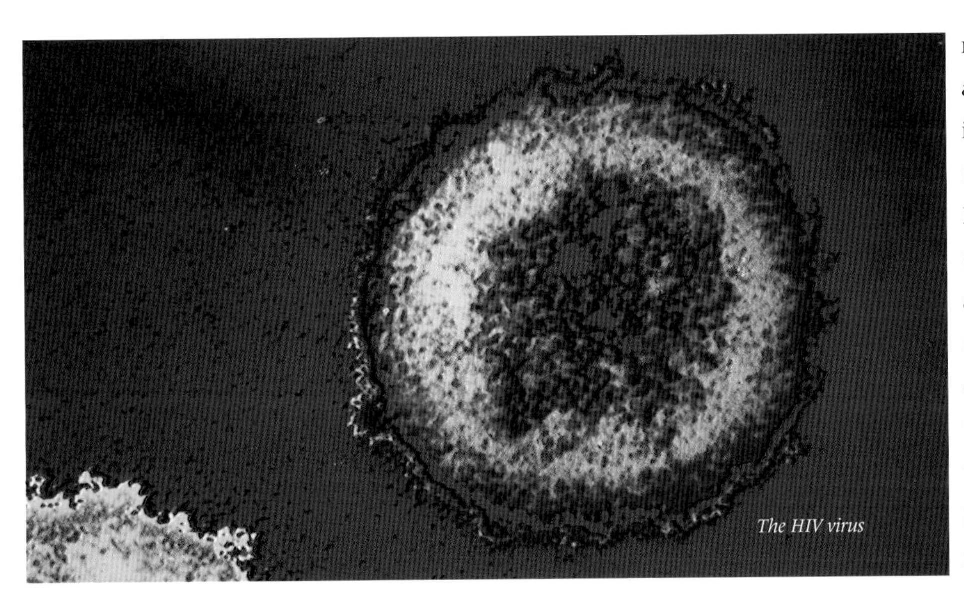

Chemotherapy is the only answer, and this requires an understanding of the biology of the virus. After considerable efforts, scientists identified several viral enzymes that could be targeted by new drugs. These include HIV reverse transcriptase (for HIV is a *retrovirus* which uses RNA rather than DNA as its genetic store, and requires an enzyme to help it produce DNA); HIV protease (which modifies the newly created viral protein to produce the final form of the viral enzymes and coat proteins); and HIV integrase (which catalyses the incorporation of viral DNA into host DNA).

A full understanding of the life-cycle of this virus, and a knowledge of the three-dimensional structures of these key viral enzymes from X-ray crystallographic and nuclear magnetic resonance studies (see *Analysis and structure of molecules*), has allowed the pharmaceutical industry to design drugs to target these enzymes. Several major families of compounds have been identified, and these include the dideoxynucleosides like AZT (Zidovudine developed by Wellcome, 1987) and ddC (Zalcitabine from Hoffmann La Roche, 1991) (Figure 23), which inhibit the functions of viral reverse transcriptase; and the HIV protease inhibitors like Invirase (Saquinavir, Hoffmann La Roche), Novir (Ritinovir, Abbott), and Crixivan (Indinavir, Merck). All of these drugs are totally synthetic, and this provides a cogent reminder of how the major advances made in this area of chemistry, especially with regard to the methodology of preparing stereochemically-defined molecules (see *Make me a molecule*), have enabled the pharmaceutical industry to produce these complex, but vitally important molecules, on the large, multi-tonne scale.

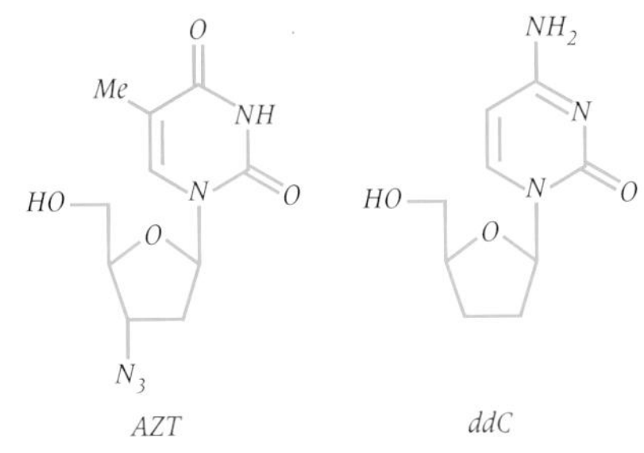

Figure 23. Two antiviral drugs for treating HIV infection.

Chemistry and gene therapy

The extraordinary efficiency with which viruses transfer their genetic material for incorporation into host cell DNA suggests that they could be used as vectors for the transfer of more useful genes. Indeed, clinical trials are underway to test whether retroviruses (like herpes simplex – the virus that causes cold sores) and *adenoviruses* (commonly associated with colds and sore throats) can be used to transfer genes to correct the defects present in Duchenne muscular dystrophy and cystic fibrosis.

One form of cancer chemotherapy, so-called GDEPT (gene directed enzyme activated prodrug therapy) involves injecting a retroviral gene that codes for a particular enzyme into a tumour. This is followed by administration of a *prodrug* that can be activated by the enzyme. For example, injecting a herpes simplex virus-kinase gene and then administering the anti-herpes drug acyclovir (ACV), results in phosphorylation of ACV

Box 5 Designing the perfect pesticide

Human civilisation has been troubled by insect pests since history began. Of the 10 plagues that visited Egypt, as described in the *Book of Exodus*, three were due to insects – locusts, lice and flies. The Bubonic plague, which swept through Europe during the Middle Ages, wiping out one-third of Europe's population, is transmitted from rats to humans by fleas. Malaria is spread by a mosquito-borne parasite and is thought to have been a major cause of the downfall of the Roman Empire. More recently, malaria caused early attempts to build the Panama canal to be abandoned and made large tracts of land uninhabitable by humans.

Today, a major concern is how to protect crops against insect pests, as well as plant diseases and weed infestations. Many of the specially cultivated high-yield varieties of crops are more susceptible to damage than their hardy wild ancestors, so over the past 50 years, there has been a significant effort to understand plant and insect biochemistry so as to help develop new compounds as pesticides and herbicides.

As in the development of pharmaceuticals, the first pesticides were natural products isolated from plants. The most important of these are the pyrethrenoids isolated from the flowers of *Chrysanthemum cinerariaefolium* (Figure a). These compounds have a knockdown effect on insects suggesting they have a neurotoxic action, but are nontoxic to humans. Initially the plants were grown in Persia, and then between the two World Wars Japan became the main producer, with Kenya and Tanzania taking over that role in more recent times. The natural pyrethrenoids are unstable to light but now a wide variety of synthetic pyrethroids such as permithrin and cypermethrin have been produced which overcome this obstacle. Because these pesticides are still not very photostable and are toxic to fish, they are normally used inside rather than outside to control fleas and lice.

(a) Pyrethrin.

Because supplies of pyrethrum from the Far East were restricted during the Second World War, chemists looked for alternatives to pyrethrum. The first compound identified was dichlorodiphenyltrichloroethane (DDT) synthesised by the Swiss chemist Paul Müller at the J R Geigy Dye Factory (Figure b). This was found to be an effective insecticide against many pests and initial tests suggested that it was nontoxic to humans at the concentrations used. DDT enjoyed global use in the 1950s and 1960s and was used to eradicate large numbers of malaria-bearing mosquitoes. Figures produced by the World Health Organisation suggest that 25 million lives were saved through the use of DDT – more than by any other chemical substance. However, even as Müller was receiving his Nobel prize for Physiology and Medicine in 1948 there were growing concerns about the efficacy and safety of DDT. Insects were becoming resistant to DDT, and there was evidence of toxicity to fish, and later birds, as increasing concentrations of the pesticide moved up the food chain. As a result DDT was banned in the US in 1972, and then soon after in other developed countries.

(b) DDT

New pest control agents were desperately needed. Just as certain anticancer agents were developed from chemical warfare agents used in First World War, another major group of insecticides, the organophosphates such as malathion, parathion and diazinon (Figure c), were developed from nerve gases developed in the 1930s and 1940s. These have proved effective against insects resistant to DDT. Because they have reasonably short lifetimes in the environment, are not accumulated as readily as chlorocarbons like DDT. It should be noted that many plants produce their own insecticides such as nicotine and physostigmine, and it is estimated that humans consume around 10,000 times as much natural pesticide as synthetic.

Because of the huge costs involved in manufacturing and delivering pesticides, as well as the potential environmental damage they can cause, researchers have turned to developing

Parathion

Malathion

Diazinon

(c) Some organophosphates.

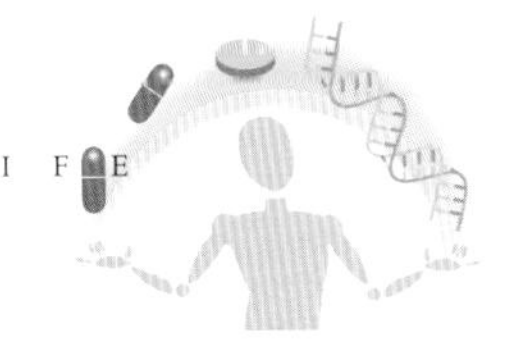

more subtle and sophisticated methods of pest control. Adolf Butenandt, who was partly responsible for the discovery of the sex hormones as described earlier, later turned his attention to the isolation and characterisation of *pheromones*, compounds released by an insect species as a means of communication (from the Greek words *pherin* – to transfer and *hormon* – to excite). In 1957, starting from a million cocoons of the silkworm moth (equivalent to 90 kilograms of silk) Butenandt separated half a million females and from these extracted a total of 6.4 milligrams of a compound bombykol (Figure d) which proved to be a sex attractant for the male moth. He found that a trap containing 0.01 micrograms of this compound would attract every male moth within the radius of a kilometre. Because of the tiny concentrations involved, identifying insect pheromones is a long and laborious task, but does offer a simple and powerful method of pest control with no environmental contamination.

Silkworm moth, Bombyx mori.

(d) The pheromone Bombykol is released by the female silkworm moth to attract male silkworm moths.

to produce its active form ACV-triphosphate, and this then inhibits DNA replication in the dividing tumour cells. The special role for the chemist in these endeavours will be to design highly specific prodrugs and novel delivery systems.

Of more interest to the food and chemical industries are the possibilities of engineering genes in microorganisms and plants so that they will produce large amounts of fine chemicals, plastics or edible proteins. A growing awareness of the environmental problems caused by plastics, now requires that these commodities have both the required physical properties and enhanced biodegradability. Synthetic protein-based polymers modelled on the structures of spider silk and mammalian collagen have already been produced by transferring engineered genes to the bacterium *E. coli*, but the real advances will come when similar genes can be transferred to, and expressed in, plants.

Some idea of the subtleties in molecular design that may be possible through genetic manipulation are provided by some beautiful work carried out with enzymes called polyketide synthases. The *polyketides* comprise a vast array of natural products produced by fungi and filamentous bacteria (*actinomycetes*). They have a plethora of biological activities from antibacterial (erythromycin)(Figure 24),

Figure 24. Erythromycin.

antiparasitic (avermectins), antitumour (adriamycin) to antifungal (strobilurins). The one feature they have in common is that they have long carbon backbones (these are eventually converted into large rings) which are assembled by multienzyme complexes called polyketide

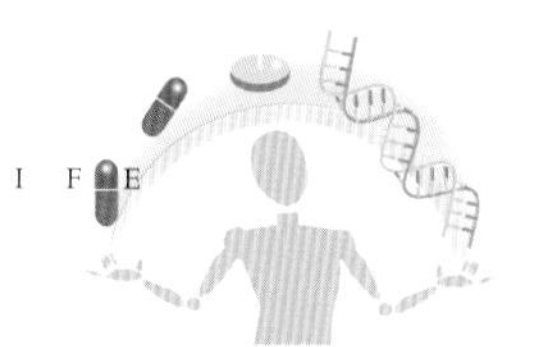

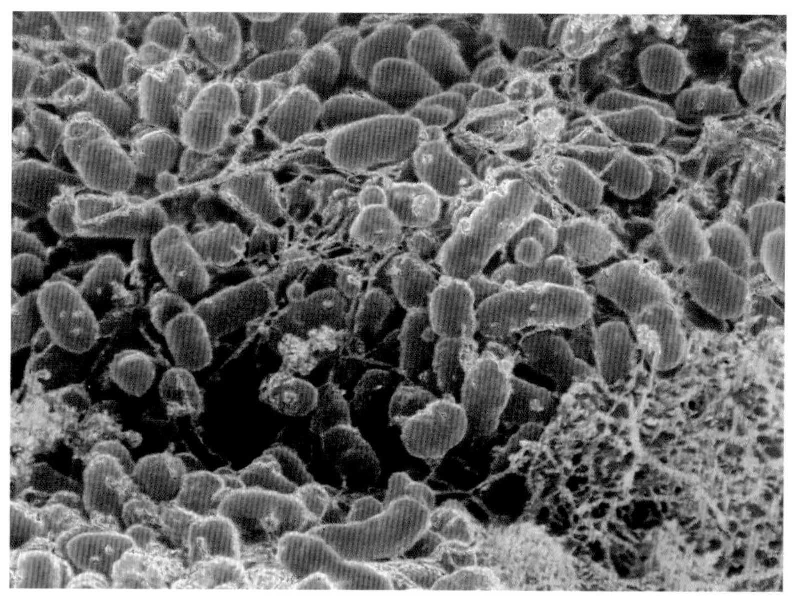

Actinomycetes which produce the biologically-active polyketides.
Dr Kari Lounatmaa/Science Photo Library

synthases. These in turn are coded for by clusters of genes and the complete gene sequence for several of these is now known. It is therefore possible to understand how the complex chemical structures of these polyketides are assembled, since these are determined by the sequence of enzymes that act upon the growing carbon skeleton.

Elegant experiments carried out by Jim Staunton and Peter Leadlay at Cambridge, Chaitan Khosla at Stanford and David Cane at Brown University, Rhode Island have demonstrated that it is possible to 'design' the chemical structures of the final products by manipulating the genes to produce 'mutant' variants, and thus the enzymes for which they are the code. In this way unnatural products have been created, and although to date none of these has had particularly remarkable biological activity, the methodology is now in place that will allow the creation of thousands of novel compounds. Some of these are certain to have interesting and perhaps useful biological activity.

These endeavours emphasise once more, that as the molecular biologists unravel more and more of the secrets of the genomes of organisms from bacteria to humans, it will provide chemists with almost unlimited opportunities to design and synthesise new molecules. If the 20th century has been one in which chemists have designed drugs to treat disease, then the 21st century will surely see the advent of drugs to prevent disease and even eradicate it.

Future challenges for chemistry in biology and medicine

We have outlined some of the important advances brought about through chemistry which have allowed us to probe biological systems, either through the development of drugs to cure diseases, or as tools to help us to understand fundamental life processes. For both of these tasks chemists will be employed in an underpinning role in the 21st century and beyond.

We are in a continuous arms race to develop new antibacterial agents and keep one step ahead of the 'superbugs' which threaten to plunge us back to the 'Victorian 'Dark Age' before we had anti-infective drugs. The advent of new diseases such as AIDS or BSE/CJD which appear to have jumped the species barrier from other animals, or the spread of rare diseases such as the Ebola virus of Central Africa to areas of

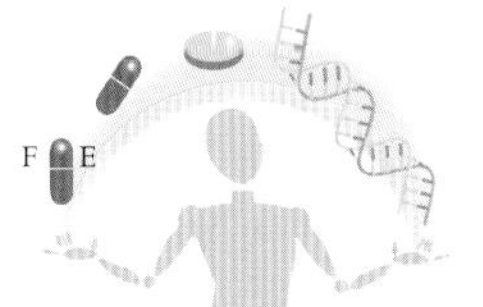

John Greim/ Science Photo Library

greater population will continually throw up new challenges in chemistry, medicine and biology. Recent advances in cancer treatment such as the use of *taxol*, or the epithilones isolated from the myoxobacteria *Sorangium cellulosum* for the treatment of breast or cervical cancer have proved to be effective but still many other cancers prove intractable to chemotherapy.

Combinatorial chemistry, (see *Make me a molecule*) allows us to produce large libraries of new unnatural compounds at the push of a button, and it may be from these new reservoirs of molecules rather than natural sources that we find lead compounds in the coming years.

Nature is still capable of surprising scientists. Fifteen years ago no one would have believed that nitric oxide (NO), a gaseous compound best known as a pollutant in car exhausts, was used in many body tissues as a biological messenger causing smooth muscle to relax. This was first reported by Salvador Moncada and his research group at the Wellcome laboratories in Beckenham, and Lou Ignarro and his colleagues at the University of California in 1987. Once the biological role of nitric oxide had been established, it explained why compounds such as glyceryl trinitrate and amyl nitrite – prescribed for more than a century to treat angina pectoris (narrowing of the arteries to the heart) – worked. These compounds, in fact, decompose in the body to give nitric oxide. The importance of nitric oxide has led to the development of new drugs which interfere with smooth-muscle relaxation including sildenafil (Viagra) (Figure 25) produced by Pfizer which became the top selling drug of 1998 and is used temporarily to overcome penile erection problems.

It is a salutary thought that we still do not know the structures and biological roles of more than 70 per cent of the compounds found in human blood plasma. Our understanding of many biological processes and

Figure 25. Viagra.

their regulation is still rudimentary. The completion of the Human Genome Project within the next five years will provide us with the 'blueprint' for the human body. However, the role played by the majority of genes within this plan and the proteins which they encode still need to be discovered.

The recent speculation about evidence for life on Mars being present in certain meteorite fragments has led to another avenue of active research. This is prebiotic chemistry and explores how life has evolved from simple molecules such as ammonia, water, carbon dioxide and hydrogen sulfide. Why has Nature chosen proteins, sugars and nucleic acids in their current roles primarily as catalysts, sources of energy and repositories of genetic information? Does the discovery of RNA molecules with catalytic function (ribozymes) by Sidney Altman of Yale University and Tom Cech of the University of California in 1986 prove that prior to cellular life as we know it there was an RNA world capable of molecular organisation and replication that only involved nucleic acids? Would life on other planets have to be carbon-based? If not, how would we search for it. These are all challenges that await the chemists of the future.

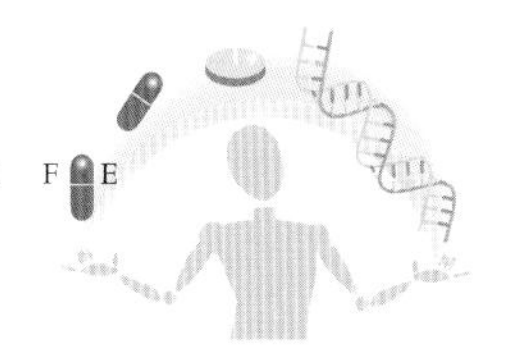

Further reading

1. *Murder, Magic and Medicine*, John Mann, Oxford University Press, Oxford: 1994.
2. *The Elusive Magic Bullet*, John Mann, Oxford University Press, Oxford: 1999.
3. *Molecules at an Exhibition,* John Emsley, Oxford University Press, Oxford: 1998.
4. *Path to the Double Helix; Discovery of DNA,* Robert Olby, Dover Publications, New York: 1994.
5. *Recombinant DNA: a Short Course,* James D. Watson, John Tooze, David T. Kurtz, Scientific American Books, New York: 1992.
6. *Chemical Communication: The Language of Pheromones,* W. C. Agosta, Scientific American Library, New York: 1992.
7. *"Protein folding: a perspective from theory and experiment",* C. M. Dobson, A. Sali and M. Karplus, *Angew. Chem. Int. Eng. Edn.*, 1998, **37**, 868.
8. Up to date reports on HIV/AIDS are available from The World Health Organisation on: http://www.who.int/emc/diseases/hiv/index.html

Glossary

Acquired Immune Deficiency Syndrome (AIDS) A disease in which the immune system of the host is suppressed through the action of the human immunodeficiency virus (HIV) leading to death from opportunistic infections.

Active site A cleft on the surface of an enzyme where the substrate(s) and cofactors bind, and the catalytic reaction takes place.

Agonist A compound that elicits a response when it binds to a receptor.

Alpha helix A type of protein fold in which the linear peptide forms a right-handed corkscrew structure held together by hydrogen bonds.

Antagonist A compound that interacts with a receptor to block the action of an agonist.

Antibiotic A natural product or synthetic derivative from a microbial organism which inhibits the growth of other microbes.

Anticodon A set of three consecutive nucleotides in a transfer RNA molecule which are complementary to a messenger-RNA codon.

Beta pleated sheet A type of protein fold held together by weak intermolecular bonds.

Biosynthesis The synthesis of new compounds by an organism from simpler precursors.

Chaperones Families of proteins which help proteins to fold correctly.

Chemical signalling The transfer of a signal through the movement of a molecule.

Chemotherapy The treatment of pathogenic organisms or control of cellular function by using chemical agents.

Codon A set of three consecutive bases in DNA or RNA, specifying either an amino acid or the termination of translation as defined by the genetic code.

Deoxyribonucleic acid (DNA) A linear polymer made up of the four deoxyribonucleosides; T-deoxyadenosine, T-deoxyguanosine, T-deoxycytosine, and T-deoxythymidine joined together by 3′ to 5′-phosphate diester bonds. DNA is the genome, the primary genetic material of all living organisms and some viruses. It is normally found as a double-stranded double helix.

DNA fingerprinting The unique DNA fragmentation pattern produced by digesting the genome of an organism with a specific cocktail of restriction enzymes.

Enzyme A protein which in combination with certain essential cofactors catalyses one or more chemical reactions.

Gene A sequence of genomic nucleotides in DNA or RNA which carries the code for a single polypeptide which constitutes one or more biologically active proteins.

Genetic engineering The manipulation of the DNA or RNA of an organism to produce modified proteins, or to increase or decrease the amount of a protein produced by the organism.

Gene expression The processing of a specific gene to generate the protein(s) it encodes.

Genome The complete genetic blueprint of an organism containing the genes of all essential proteins. All genomes are composed of DNA except for certain viruses which have RNA genomes.

G-protein Guanine nucleotide – binding protein on the inner face of the cell membrane which forms part of the hormonal signalling system.

Hormone Chemical secreted by a specific gland which elicits a response from one or more other regions or organs of the body. From Greek *hormon* – to excite.

Hydrogen bond A weak bond between a hydrogen atom and a nearby atom with a pair of electrons that attract the proton of the hydrogen atom.

Kinase/phosphatase Enzyme that catalyse the phosphorylation/dephosphorylation of specific hydroxyl groups.

Lock-and-key binding A term coined by Emil Fischer in 1894 to describe the binding of substrates into enzymes, also now applied to ligands and receptors.

Magic Bullet A term coined by Paul Ehrlich in 1908 to describe a drug that interferes only with diseased cells or foreign organisms and leaves healthy tissue untouched.

Messenger RNA (mRNA) The RNA which is transcribed from DNA and is used as the instruction template for the synthesis of proteins.

Metalloprotein A protein that binds metals to perform its catalytic function.

Mutation A change in the DNA sequence of an organism's genome which can lead to a change in the structure or amount of a protein produced.

Operon A genetic unit used to control several genes.

Pathogen Any foreign infectious agent which is harmful to the host animal.

Penicillin An antibiotic isolated from species of *Penicillium* fungi which contains a beta-lactam ring fused to a thiazolidine ring.

Pheromone A compound released by an animal as a way of attracting the opposite sex of the same species. From the Greek word *pherin* – to transfer, and *hormon* – to excite.

Phosphorylation The formation of a phosphate ester bond.

Polymerase An enzyme that catalyses the transcription of nucleic acids from a complementary template.

Polymerase chain reaction (PCR) The generation of multiple copies of DNA (or RNA) using a thermally stable polymerase enzyme *in vitro*.

Polyketide A natural product sequentially built-up from acetate, propionate, or butyrate building blocks that retains oxygen at a variety

of oxidation states within its structure.

Polypeptide A polymer composed of amino acids.

Prodrug A derivative or precursor of an active drug molecule generally of low toxicity or potency which is processed by cellular enzymes or other means into the active drug. A prodrug may be much more easily absorbed or transported by the body than the active drug.

Protein One or more polypeptide chains correctly folded to give a biologically active structure. Proteins may be either structural or catalytic, the latter are known as enzymes.

Protein folding The conversion of one or more linear polypeptides into a biologically active three-dimensional structure held together by disulfide bonds and electrostatic interactions.

Rational drug design The design of new molecules as potential drugs based on an understanding of both the architecture of the biological target (DNA, receptor or enzyme) and its mechanism of action. It is an iterative process utilising X-ray crystallography, computer modelling, structure-reactivity relationships and synthetic organic chemistry.

Receptor A protein in the cell membrane which upon binding of an extracellular messenger substance produces an intracellular biological response.

Recombinant DNA A DNA sequence that has been constructed from fragments of DNA originating from two or more different organisms.

Resistance The change in composition of a pathogen or human so that a drug is no longer effective.

Restriction enzyme Enzymes that can cut DNA at specific sites (restriction sites).

Retrovirus A virus which has a RNA genome and requires this genome to be converted into DNA for it to be replicated.

Ribonucleic acid (RNA) A linear polymer made up of the four ribonucleosides; adenosine, guanosine, cytosine, and uracil joined together by 3′ to 5′-phosphate diester bonds of the ribose sugar. RNA fulfils three roles in organisms as messenger RNA, a transcription of a DNA sequence used as the template for the biosynthesis of one or more proteins; ribosomal RNA, part of the ribosome factory′ which converts mRNA sequences into peptide sequences, by catalysing the formation of peptide (amide) bonds; transfer RNA (tRNA) adapter or interpreter molecules which translate a three-letter nucleotide codon sequence via an anticodon into a specific amino acid in a protein.

Ribosome A hybrid protein/RNA structure found in the cytoplasm of cells. They function as scaffolds to bring together the messenger - RNA and appropriate transfer-RNA units so that a sequence of nucleotide triplets (codons) given by the m-RNA can be translated into a new peptide through the formation of amide bonds between amino acids attached to individual t-RNAs. The bond formation is catalysed by the ribosome.

Ribosomal RNA RNA which together with protein makes up the structure of the ribosome.

Solid-phase synthesis Preparation of a compound by building it up on a bead or solid surface through sequential rounds of reaction and purification. Originally used for peptide synthesis, but now applied to the preparation of a diverse range of materials.

Steroid hormone An animal chemical messenger with a distinctive tetracyclic carbon skeleton.

Transcription The copying of a DNA sequence into an RNA sequence by the enzyme RNA polymerase.

Translation The conversion of a messenger RNA sequence into a new protein sequence.

Transfer RNA The adapter or interpreter molecules which contain both an anticodon sequence for a specific amino acid and that amino acid.

Virus A biological entity on the borderline between living and non-living which is only able to replicate once it has infected a living cell. Many diseases are caused by viral infections.

Biographical details

Professor John Mann is professor of organic chemistry at University of Reading, and is interested in the synthesis of natural products and other biologically interesting molecules, and also in the rational design of anticancer drugs.

Dr Neil Thomas is currently a Royal Society University Research Fellow in the Department of Chemistry at the University of Nottingham. His research interests include the development of new prodrugs and drug targeting systems, the creation of new catalysts based on antibodies and synthetic polymers, and the investigation of how enzymes work.

Epilogue

Chemistry – architecture of the microcosmos

The discovery of new knowledge involves a creative process which harnesses innate human inquisitiveness to carry out experiments that reveal the way the Universe works. It is an intrinsically humanitarian endeavour, encompassing both practical and abstract levels of understanding. Just as architects and builders design and construct edifices from wood, brick, and stone, so chemists create molecules from the elements. Furthermore, chemistry, through materials technology and metallurgy, has added greatly to the inventory of building materials, for example, aluminium, steel, plastics and even concrete.

I will try to uncover some approaches to the way that advances in science – and chemistry in particular – are made. In doing so, I hope to reveal the beauty and elegance of chemistry at a deeper intellectual level. I also hope that it will show how a degree of personal satisfaction can arise from scientific endeavour. My perspective is, of course, based on my own experience and knowledge, and should not be considered as offering an overview of chemistry in any way. For instance, I cannot begin to do justice to the giants of synthetic chemistry such as Emil Fischer and Robert Burns Woodward (mentioned in *Make me a molecule*) or earlier ground-breakers responsible for the underlying theories of chemical bonding and molecular behaviour such as Gilbert Newton Lewis who developed the concept of the electron-pair covalent bond, Linus Pauling who developed valence bond theory, and John Lennard-Jones, one of the pioneers of molecular orbital theory. Nevertheless, I hope that the examples I give – relating to my own knowledge – will add something generally worthwhile to this volume.

Ten revolutionary advances

What a century it has been! Scientific breakthroughs have truly revolutionised all our lives, not only health-wise and socioeconomically but they have also, for the first time, enabled the majority to indulge in cultural and leisure activities. In each and every advance, chemistry in all its aspects – theoretical, organic, inorganic, physical and analytical has been either directly, or indirectly involved.

For me, the greatest intellectual achievement is the development of quantum mechanics – a theory based on probability that describes the behaviour of matter at the atomic level. It mysteriously combines the concept of particles such as atoms – a fundamental notion developed mostly by chemists – with wavelike behaviour. Quantum mechanics has provided an explanation of essentially the whole of chemistry and much of physics. Ernest Rutherford's experiment (for which he won the chemistry Nobel prize) of shooting alpha particles at a thin gold sheet which led him to conclude that an atom consisted of a positively charged nucleus surrounded by negatively charged electrons was, however, one of the crucial prerequisites. It also turns out that analytical chemistry at its most fundamental revealed the fact that the atom could be 'split', in other words, that atomic nuclei could fission and be transmuted into other types of nuclei. Later, a fundamental understanding of the interactions of molecules with light led to the invention of the maser, and later the laser – now found everywhere, from telecommunications, to surgery, to the supermarket.

The greatest new material must be plastic (as in the "Animal, Vegetable, Mineral or Plastic?"), and the greatest invention is probably the transistor. As electronic chips get smaller, science fiction comes ever closer to being realised. The miniaturisation of electronic devices and circuitry used in modern computers was made possible by brilliant advances in ultra-high purity chemistry, ingenious crystal growing, and the chemical manipulation of microstructures at the molecular level. The imminent prospect of a personal computer in every home is a direct result of these advances.

Two triumphs of chemistry epitomise the humanitarian contributions of chemistry. One is the development of synthetic antibiotics, following the discovery of penicillin, which led to a revolution in the treatment of infection; the other is the development of 'the Pill' which has revolutionised social behaviour. The greatest 20th-century discovery, however, is probably the structure of DNA (see *The chemistry of life*) – the molecule that contains all the genetic information necessary to create life. Perhaps it will turn out to be the greatest discovery of any century past, present or future. Certainly its ramifications for the future of the human race are, only now, barely beginning to be appreciated.

Chemistry then and now

The great theoretical physicist Richard Feynman has written that – if in some cataclysmic event, all of scientific knowledge were destroyed, and only one sentence were passed on to the next generations, the statement that would contain the most information in the fewest words is: "All things are made of atoms – little particles that move around in perpetual motion, attracting each other when they are a little distance apart, but repelling upon being squeezed into one another." This statement summarises the atomic hypothesis, imagined by the Ancient Greek philosopher, Democritus and deduced from experiment by John Dalton in 1803, and the potential curve (first suggested by the 18th-century Croatian scientist Rudjer Boskovic). The curve reveals the relationship of distance between interacting particles and their combined potential energy. At very short distances the particles repel each other so the potential energy is high (nuclear energies). At an optimum distance the particles mutually attract each other producing a dip in the curve – a potential energy well (chemical energies). At greater distances the potential energy starts to rise again to a level which represents the situation when the particles break free of the mutual attraction (see *Following chemical reactions*).

These advances were brought into clear historical perspective for me by a couple of prescient sentences, in an old chemistry book published around 1830, which combine the basis of experimental philosophy with a true flavour of the time when Dalton's atomic theory was new. "On the Atomic Theory of Mr Dalton: The brief sketch which has been given of the laws of combination will, I trust, serve to set the importance of this department of chemical science in its true light. It is founded, as will have been seen, on experiment alone, and the laws which have been stated are the pure expression of fact. It is not necessarily connected with any speculation, and may be kept wholly free from it."

Chemistry without a modern understanding of atoms, molecules and potential curves seems impossible, and I am filled with awe for the genius of the fathers of the field. It is today almost as difficult to conceive how chemistry could have been done before the advent of such techniques as spectroscopy, mass spectrometry, chromatography, and in particular X-ray crystallography and NMR (see *Analysis and structure of molecules*). My own favourite technique is microwave spectroscopy which involves the measurement of transitions between rotational energy levels in molecules. Analysis yields molecular moments of inertia, providing parameters such as bond lengths and angles, dipole moments, internal rotation barriers and quadrupole moments. Microwave spectroscopy also led to the discovery, by radioastronomy, of molecules such as ethanol, sulfur dioxide and the long carbon-chain species in the dense dark dust clouds of interstellar space. Indeed, we now know that much of the chemistry in the Universe takes place in the space between the stars and certainly much more than in planetary environments.

Tales of the unexpected

The way in which knowledge may be applied (technology) is usually much easier to describe than the intrinsic nature of knowledge itself or how that knowledge was obtained. As examples, consider non-drip paint and the 'Post-it'® note. The latter is now regarded as a legendary example of how to turn a research failure (in this case, a lousy adhesive!) into an amazing business success. However, the ingenious chemistry needed to create non-drip paint and the imaginative thinking needed to find a use for a semi non-stick substance are much harder to explain. Most people can easily relate to the benefits, especially if they have had to wash their hair after painting a ceiling with ordinary paint or tried to unseal a prematurely licked and sealed envelope. However, some degree of scientific education is necessary in order to appreciate the deeper intellectual pleasures that drive scientists to be creative.

Important advances often occur when conventional wisdom is overturned and problems are solved which require the textbooks to be rewritten. A famous and archetypal example is the discovery of the rare gas compounds first prepared by Neil Bartlett at the University of British Columbia in Vancouver in 1962. The rare gas elements – helium, neon, argon, krypton, xenon, and radon – which have atomic structures with full, stable electronic shells, were thought by nearly everyone to be completely inert. Bartlett, who made the molecule $O_2^+PtF_6^-$, ingeniously reasoned that since the energy needed to remove an electron from the oxygen molecule (O_2) was similar to that required to remove an electron from xenon, then this element might form chemical bonds after all. On mixing xenon and platinum hexafluoride (PtF_6) Bartlett obtained a beautiful red compound, $XePtF_6$, the first true 'inert' gas compound (later it proved to be a little more complicated). This discovery opened up a whole new field leading to the creation of several other molecules, XeF_4, XeO_3 and KrF_2 which are now called noble gas compounds – in other words, no longer inert! This work had great implications for our understanding of chemical bonding.

Another example is the so-called 'double bond rule' found in some books not that long ago, which claimed that second (and third) row elements in the Periodic Table – in particular phosphorus – would not form a double bond because the atoms are too large to engage successfully in "π" bonding. Indeed, this is still being claimed in one recent textbook despite the fact that T. E. Gier at DuPont had made a compound with a carbon-phosphorus triple bond (HC≡P) as early as 1961; other compounds such as $H_2C{=}PH$ and $CH_3C{\equiv}P$ as well as many other related species were made at Sussex University (with John Nixon) in 1974, and a flourishing field of synthetic phosphorus chemistry has evolved ever since.

In my own area of special interest, microwave rotational spectroscopy – according to many textbooks, requires a molecule to have a dipole moment – an asymmetrical distribution of charge. This implies that a highly symmetrical molecule like methane, which is tetrahedral, will not exhibit a spectrum. In 1971, however, James K. G. Watson, then at the University of Reading, reasoned that as a molecule rotates it distorts slightly and this 'centrifugal' distortion effect will in some cases result in a small dipole moment that will allow the rotational transitions to become detectable. Bill Klemperer's research group at Harvard University found evidence for related effects by using molecular beam techniques to show that numerous nonpolar molecules would deflect in an electric field. This happens because the molecules become effectively slightly polar as they vibrate and distort.

Even closer to home is the fact that the commonly given explanation for the lubricating properties of graphite is wrong. The textbook explanation is that the van der Waals forces between the sheets of graphite are so weak that the sheets slide easily over one another. When we discovered C_{60} in 1985, we nonchalantly hypothesised that it would be, among other magical things, a superlubricant – after all, the molecule should roll as well as slide. The moment that I had a sample of C_{60} in my hands in September 1990, the first thing I did was to press a small speck on a glass slide with a spatula. To my intense disappointment I found that it was gritty like dirt and not at all squeezable like graphite, and I immediately began to wonder why graphite itself was a lubricant. I was subsequently informed that graphite cannot be used as a lubricant on the Space Shuttle because in vacuum it loses its lubricating properties. In fact, it has been known for some time that adsorbed gases are needed to reduce the friction between the layers, but this fact has not found its way into many textbooks!

Beauty in science

Beauty is just as fundamental to an appreciation of the sciences as it is of the arts, and recently the intrinsic beauty of molecules has been made much more accessible by computer graphics. One of my favourite structures, the basic metalloporphyrin structure which is a flat symmetrical molecule with a metal atom at its heart, is particularly elegant. The fourfold symmetrical shape is so reminiscent of the petals of a flower that you feel that perhaps there is some deeper intrinsic relationship. The way it is built up in Nature (deduced by Alan Battersby of Liverpool University and then Cambridge) as a key component in several important biological systems is particularly fascinating. It seems remarkable that one structure plays such a key role in oxygen uptake by the amazing haemoglobin molecule as well as the light capture stage in photosynthesis (see *The chemistry of life*).

It seems that one cannot go far in chemistry without symmetry. Indeed, some aspects of symmetry are obvious; the high symmetry seen in flowers, snowflakes, sea urchins and the C_{60} molecule, for example, invariably invokes some sort of deep cathartic experience which seems to be an intrinsic characteristic of the human perception. It is so primal that perhaps the appreciation of symmetry is hard-wired into the receptor networks in our brains at either a very early age or in some more fundamental way. Perhaps symmetry relationships map directly onto analogous simple associative relationships in the neuronal networks. Be that as it may, symmetry has clearly fascinated human beings for millennia as we can see from the artworks and artefacts that have survived.

Symmetry can be a powerful tool in chemical analysis. For instance, sets of magnetic nuclei (such as a proton or the carbon-13 nucleus) in equivalent environments in a molecule result in a single nuclear magnetic resonance signal (see *Analysis and structure of molecules*). Because of its high symmetry, benzene which has a regular hexagonal carbon ring framework has only two types of bond: the carbon-carbon bond and the carbon-hydrogen bond. All the carbon atoms are equivalent so that the NMR (carbon-13) spectrum consists of a single line. It gave us particular delight when we at Sussex were the first to show that C_{60} had a single resonance in the NMR spectrum, thus confirming that all 60 carbon atoms were equivalent, a result consistent with the originally proposed Buckminsterfullerene structure (see *New science from new materials*).

Elegance also plays an important role in mathematics. Symmetry is an intrinsically mathematical idea which is often used to express the relationships between quantities, and indeed underpins many theoretical ideas in physics and chemistry. Symmetry is used to classify crystal structures and as a framework in quantum theory. One aspect of this that fascinated me when I was a young scientist was 'noncommutativity'. This means that two quantities x and y when multiplied in a certain order do not give the same result as when multiplied in the reverse order, in other words xy does not equal yx. The quantities, which are not pure numbers, are called operators and are said 'not to commute'. The difference between xy and yx can be a constant (*ie* $xy - yx = k$) or another operator, say z (*ie* $xy - yx = z$). The most well-known example of noncommuting quantities is the basis of the famous Heisenberg Uncertainty Principle of quantum mechanics which says that you cannot measure the position and momentum of a particle simultaneously with infinite accuracy – position and momentum operators do not commute.

My favourite example of this odd behaviour is the defining relations of quantised angular momentum component operators of the total angular momentum, **J**, one of which is: $J_xJ_y - J_yJ_x = i\hbar J_z$ where J_x, J_y and J_z are quantum operators. The $(2J+1)$ quantised angular momentum components of electronic angular momentum in atoms can be derived from this relation. Thus as $J = 0, 1, 2, 3$ we find that $2J+1 = 1, 3, 5$. These are the degeneracies of the *s, p, d* levels of the hydrogen atom. Thus we rationalise (together with the Pauli Exclusion Principle) the 2(=2x1), 8(=2x{1+3}), 18(=2x{1+3+5})... row-structure of the Periodic Table. I find it amazing that Mendeleef's Periodic Table naturally follows from the odd relation above. It is one of the most beautiful and outstanding of human achievements.

There are also equally subtle beauties in experimental science, for instance there are masterpieces of creativity to be found in synthetic chemistry. Experts in the field would compare the genius displayed in prescient steps in the laboratory syntheses of strychnine, penicillin, vitamin B_{12} and other structurally complex natural products to those in legendary moves made by great chess champions (see *Make me a molecule*).

The conflation of chemistry and physics (and soon biology?)

There are two major discoveries of this century that exemplify (for me) the coming together of physics, chemistry and biology. The impact of the first on our 20th-century world has been little short of revolutionary, and the impact of the second on the 21st century may be still more staggering when it finally arrives.

The first is the discovery by Otto Hahn and Fritz Strassman that their chemical analysis indicated that bombarding the uranium-235 isotope with neutrons produces barium and extra neutrons. This result was rationalised by Lise Meitner (who was forced by Nazi oppression to flee Germany and her experimental study) and Otto Frisch as indicating that uranium-235 had fissioned. The results showed that a nuclear chain reaction, first envisioned by Leo Szilard, was feasible. At the fundamental level this discovery is crucial to an understanding of nuclear chemistry and at the applied level it led to nuclear power and, of course, 'The Bomb'. For some reasons, which are not entirely obvious, only Hahn was awarded the Nobel Prize for this discovery. The prize was awarded for Chemistry, but in retrospect it could have been awarded for Physics or perhaps for Alchemy! (Stanley Kubrick's leading character in his film *Dr Strangelove* would probably have proposed the discovery for the Peace Prize!)

The second major revolution starts with the development of protein crystallography initiated by J. D. Bernal and Michael Polanyi. It continues with the solving of the structures, by X-ray crystallography, of penicillin and vitamin B_{12} by Dorothy Hodgkin, and culminates with the amazing X-ray diffraction pattern of DNA obtained by Rosalind Franklin, Raymond Gosling and Maurice Wilkins and the interpretation of this pattern by Francis Crick and James Watson as a double helix. This is the key that started to unlock Pandora's life-science box. It is an exceptional story of genius at all levels. First, it is a fascinating amalgam of theoretical perception by Wilkins that the structure of genetic material might be determinable by X-ray crystallography; secondly (protein) crystal growing and X-ray analysis expertise by Franklin with Gosling; thirdly, the mathematical insight by Crick as to the X-ray diffraction pattern of a helix; and fourthly, inventive model-building by Watson. The untimely death of Franklin meant that she could not even be a candidate for the Nobel Prize – however, she won a much greater prize – that of our hearts especially through Juliet Stevenson's sensitive portrayal some years ago in the BBC film *Life story*.

This discovery is just now beginning to shake the very roots of our society as ethical and other considerations are beginning to reach previously unheard of scales of complexity. On a different note I wish that more young women knew about Franklin and Meitner so that these two remarkable women could serve as role models. While, one carried out the most important piece of experimental work of the century, the other solved the most important chemical puzzle.

Our origins in stars

Fred Hoyle, together with Geoffrey and Margaret Burbidge and William Fowler (in the celebrated 1957 'B^2FH' paper), showed how, starting from hydrogen, a series of nuclear reactions in stars like our Sun could create all the elements of the Periodic Table up to that with atomic number 56 – iron (elements beyond iron are produced in supernova explosions). This study followed on from Hans Bethe's pre-war study in 1939 of how the Sun has managed to keep us warm for some 4.5 billion years by the gravitationally governed compression and heating, which causes four protons to fuse into a helium nucleus (two protons and two neutrons). The mass of four protons is slightly more than that of a helium nucleus so that the left-over mass is converted into energy, according to Einstein's famous $E = mc^2$ equation, and eventually radiated by the Sun into space so we are kept warm.

This is, of course, one of the main reasons we are here, but there are also other factors at work connected with the properties of atoms. For example, we exist because hydrogen can form weak bonds with other elements (in water, in DNA and proteins, for example), and because carbon can link covalently to itself, forming complex structures with single, double and triple bonds.

However, even more curious is the way carbon atoms are formed in the first place. Helium is readily made in the universe. Hoyle reasoned that when three helium nuclei collide simultaneously inside a star, they must fuse to create a carbon-12 nucleus (note that the fusion of two alpha particles into beryllium-8 is a reversible reaction). Such three-body processes generally happen with very low probability. However, the fact that there is enough carbon around to form complex, self-replicating structures capable of contemplating their own existence and origins indicates that the process must have a higher than generally expected efficiency. Hoyle predicted that the combined energy of three alpha particles must coincide with that of an excited nuclear state of carbon-12 (at 7.65 million electronvolts), enabling a resonant(and thus a favoured) process to take place. Without this 7.65 MeV resonance we simply would not exist.

This crucial coincidence is considered by some to support the so-called 'anthropic' principle. The term relates to an area of philosophy which finds it significant that certain parameters in Nature have exactly the right values for humans to have evolved. Some see in this a sign that a God must have had a hand in it. My own view is that it seems to indicate just the opposite.

For me, the discovery that the carbon atoms in my body were synthesised aeons ago in a star which then exploded to smear my atoms – probably in the form of carbon chains – throughout the space between the stars, ultimately to re-aggregate in a planet near the Sun was a truly wonderful revelation. Why were my atoms, or Picasso's, chosen to end up here on Earth at this or any other time?

The future – single molecule synthesis?

Paul Boyer and John Walker who shared part of the 1997 Chemistry Nobel Prize have made a delightful advance, and truly one of my favourites. It seems, at least to me, to be a portent of fascinating chemistry of the future. The advance has similarities with the breakthrough made by Max Perutz who discovered that deoxyhaemoglobin had developed the knack of readjusting its structure on taking up oxygen in order to facilitate further oxygen uptake (up to four molecules) – a molecular machine as Perutz lovingly describes it. Boyer and Walker found a biological electric motor consisting of a (six-fold) segmented pumpkin-shaped protein with an axial central shaft. As an asymmetrical rotor in the shaft turns, driven by a flow of hydrogen ions through a disc at its base, it distorts cavities in the walls of three of the segments alternately enabling this amazing molecular machine to drive the catalysis of adenosine diphosphate, ADP, to adenosine triphosphate ATP (the compound responsible for delivering energy in our cells for muscle movement and other activities) in the holes. An average person produces (recycles) 35 to 50 kilograms a day! It is a delight to discover that our muscles are powered by myriads of microscopic electric motors. The more we learn about Nature the more we seem to find that we are reinventing the wheel – or in this case the armature.

My colleague David Walton who had devised ways of making a linear 32-atom chain of carbon drew my attention to the fascinating fact that dahlias had devised some clever chemistry to create molecules with chain sections consisting only of carbon atoms. Now I wonder whether, in the future, genetic engineering can modify the process and devise a bacterium that can harness this ability to synthesise a continuous chain of carbon atoms and wind the chain into a helix so that it will undergo controlled cross-linking into a perfect carbon nanotube (DNA is produced with an error rate of only 1 in 10^9). This would be the ultimate nanotube-spinning genetically engineered nanorobot spider! It would generate a single molecule which could be a few nanometres across and infinite in length. Such tubes stacked in bundles would stick together by van der Waals forces to produce the material which would revolutionise civil and electronic engineering. These giant molecular assemblies would be able to eliminate the defect limit of present materials and achieve tensile strengths 50 to 100 times that of steel at one-sixth of the weight and provide a material that can also conduct electricity like copper – there is now some evidence that they can conduct with almost zero resistance at room temperature.

Chemistry as a cultural activity

One way of doing science is to follow your own instincts – even if it seems a bit straightforward and you feel sure that the experiment will only prove you to be correct. It is worth remembering that science has an amazing habit of proving, time and time again, that we are wrong, and even when we are right, some further unexpected feature may be uncovered. Indeed, the times to be most on your guard and self-critical are when you are most certain you are right.

I was once asked what advice I would give a young child who wanted to be where I am now. I was quite taken aback because I had no aspirations to win prizes when I began science. I am sure that it would be unwise

Science Photo Library.

to do something, indeed anything, with the aim of winning a prize. Anything other than trying to solve a problem which one finds personally fascinating is, I think, unwise. In a conversation with a friend I once said that a particular problem was interesting and he pointed out: "What you mean is that *you* think it is interesting". I never forgot that comment and in time learned that that was all that really counted. As far as that young scientist was concerned, I could only advise him/her that seeking reward is, more often than not, the recipe for unhappiness *and* failure and that they should do what they enjoy or fascinates them, and do it as well as they can. If they do it to the best of their ability they will do it better than others – who might be able to do it better but invariably do not because they do not try hard enough. Thus, within reason, they should never give up and try not to let anyone down.

Sir Harold Kroto

Professor Sir Harold Kroto is a Royal Society Research Professor in the School of Chemistry, Physics and Environmental Science at the University of Sussex. He shared the Nobel Prize for Chemistry in 1996 for the discovery of the Fullerenes. His research interests include nanoscale technology and fullerene chemistry. Recently he set up the Vega Science Trust to make serious science programmes for network television.